Molecular Mechanisms and Therapies of Colorectal Cancer

Molecular Mechanisms and Therapies of Colorectal Cancer

Editors

Donatella Delle Cave
Alessandro Ottaiano

MDPI • Basel • Beijing • Wuhan • Barcelona • Belgrade • Manchester • Tokyo • Cluj • Tianjin

Editors

Donatella Delle Cave
Institute of Genetics and Biophysics Adriano Buzzati-Traverso
CNR
Naples
Italy

Alessandro Ottaiano
Department of Abdominal Oncology and CoE ENETS
Istituto Nazionale Tumori
"Fondazione Pascale"
Naples
Italy

Editorial Office
MDPI
St. Alban-Anlage 66
4052 Basel, Switzerland

This is a reprint of articles from the Special Issue published online in the open access journal *International Journal of Molecular Sciences* (ISSN 1422-0067) (available at: www.mdpi.com/journal/ijms/special_issues/Molecular_Colorectal_Cancer).

For citation purposes, cite each article independently as indicated on the article page online and as indicated below:

LastName, A.A.; LastName, B.B.; LastName, C.C. Article Title. *Journal Name* **Year**, *Volume Number*, Page Range.

ISBN 978-3-0365-6547-7 (Hbk)
ISBN 978-3-0365-6546-0 (PDF)

Contents

International Journal of *Molecular Sciences*

MDPI

Review

Relationship between Epithelial-to-Mesenchymal Transition and Tumor-Associated Macrophages in Colorectal Liver Metastases

Aurora Gazzillo [1], Michela Anna Polidoro [1], Cristiana Soldani [1], Barbara Franceschini [1], Ana Lleo [1,2,3] and Matteo Donadon [1,4,5,*]

1 Hepatobiliary Immunopathology Laboratory, IRCCS Humanitas Research Hospital, 20089 Rozzano, MI, Italy
2 Department of Biomedical Sciences, Humanitas University, 20072 Pieve Emanuele, MI, Italy
3 Division of Internal Medicine and Hepatology, Department of Gastroenterology, IRCCS Humanitas Research Hospital, 20089 Rozzano, MI, Italy
4 Department of Health Sciences, Università del Piemonte Orientale, 28100 Novara, NO, Italy
5 Department of General Surgery, University Maggiore Hospital Della Carità, 28100 Novara, NO, Italy
* Correspondence: matteo.donadon@uniupo.it

Abstract: The liver is the most common metastatic site in colorectal cancer (CRC) patients. Indeed, 25–30% of the cases develop colorectal liver metastasis (CLM), showing an extremely poor 5-year survival rate and resistance to conventional anticancer therapies. Tumor-associated macrophages (TAMs) provide a nurturing microenvironment for CRC metastasis, promoting epithelial-to-mesenchymal transition (EMT) through the TGF-β signaling pathway, thus driving tumor cells to acquire mesenchymal properties that allow them to migrate from the primary tumor and invade the new metastatic site. EMT is known to contribute to the disruption of blood vessel integrity and the generation of circulating tumor cells (CTCs), thus being closely related to high metastatic potential in numerous solid cancers. Despite the fact that it is well-recognized that the crosstalk between tumor cells and the inflammatory microenvironment is crucial in the EMT process, the association between the EMT and the role of TAMs is still poorly understood. In this review, we elaborated on the role that TAMs exert in the induction of EMT during CLM development. Since TAMs are the major source of TGF-β in the liver, we also focused on novel insights into their role in TGF-β-induced EMT.

Keywords: colorectal liver metastases; epithelial-to-mesenchymal transition; tumor-associated macrophages; TGF-β signaling

Citation: Gazzillo, A.; Polidoro, M.A.; Soldani, C.; Franceschini, B.; Lleo, A.; Donadon, M. Relationship between Epithelial-to-Mesenchymal Transition and Tumor-Associated Macrophages in Colorectal Liver Metastases. *Int. J. Mol. Sci.* **2022**, *23*, 16197. https://doi.org/10.3390/ijms232416197

Academic Editor: Donatella Delle Cave

Received: 24 October 2022
Accepted: 13 December 2022
Published: 19 December 2022

1. Background

Colorectal cancer (CRC) is the third most commonly diagnosed malignancy, as well as the second leading cause of cancer-related mortality worldwide. It is the second most frequent cancer in women, after breast cancer, and the third most frequent in men, following lung and prostate cancer [1]. Although the incidence and mortality of CRC is showing a steady decrease in population over age 50 due to effective cancer screening measures, the occurrence of this malignancy in people younger than 50 years has been rapidly rising over the past 10 years [2]. Despite significant improvements in its diagnosis and treatments, most deaths are due to the development of distant metastasis characterized by a highly resistance to conventional therapies [3,4]. Indeed, 25–30% of CRC patients already have colorectal liver metastasis (CLM) at the time of diagnosis, or they will develop them after the resection of the primary tumor with an extremely poor 5-year survival rate [5].

In the last years, the crosstalk between cancer cells and the tumor microenvironment (TME) has been shown to play a key role in the clinical outcomes of CRC patients [6,7]. Indeed, tumor-associated macrophages (TAMs), the most abundant cells in TME, have been shown to promote tumor cells invasion and extravasation, to provide a supportive microenvironment for metastases and to be key determinants for the efficacy of anticancer strategies [8,9]. Among the pro-tumor mechanisms promoted by TAMs, TGF-β signaling represents a powerful activator of epithelial-to-mesenchymal transition (EMT), a process that plays a crucial role in CRC metastasis and in the resistance to chemotherapy and immunotherapy drugs [10–13]. Notably, the EMT represents an interesting therapeutic target in the treatment of cancer and could be exploited either to prevent tumor dissemination in patients with high risk to develop metastatic lesions or to eradicate existing metastatic cancer cells in patients with advanced disease [14]. Furthermore, despite it is well-recognized that the crosstalk between tumor cells and the inflammatory microenvironment is crucial in the EMT process, the association between the EMT and the role of TAMs is still poorly understood [15].

In this light, we reviewed the current knowledge about the EMT in CRC invasion, focusing on the role that TAMs play in the development of CLMs by inducing EMT through the TGF-β signaling pathway.

2. The Role of Epithelial-to-Mesenchymal Transition (EMT) in CRC

The liver is the most common CRC metastatic site as it receives and filter the blood from the intestine through the portal vein [16]. CLMs represent the final stage of a multi-step biological process. Firstly, cancer cells begin to migrate to the surrounding tissues near the primary CRC site and then they spread in venules, capillaries and lymphatic vessels until they enter the systemic circulation [17]. Once in the vasculature, circulating tumor cells (CTCs) reach the sinusoidal vessels of the liver [18]. The EMT mechanism provides tumor cells with several dynamic properties that help them to overcome environmental selective limitations of the metastatic translocation [19]. Upon EMT activation, tumor cells undergo to a wide range of biological and molecular changes, which facilitate their dissemination from the primary site, to the formation of metastases in distant organs [20]. Morphologically, EMT leads to the loss of typical polygonal shape of the cancer epithelial cells and the emergence of a more spindle-shaped fibrous mesenchymal-like cells. Epithelial cells are characterized by tight junctions, the expression of intercellular adhesion molecules and apical–basal polarity [21,22]. During EMT, they downregulate the expression of several epithelial proteins, such as Epithelial-cadherin (E-cadherin), claudins, and occludins, while upregulating the mesenchymal proteins, such as neural-cadherin (N-cadherin), fibronectin, and vimentin [23]. The loss of epithelial markers during EMT causes the nuclear translocation of β-catenin, activating the NF-kB pathway and thereby inducing the expression of different matrix metalloproteases (MMPs) [24]. Upon secretion of proteolytic enzymes, MMPs may degrade almost all the components of the extracellular matrix (ECM), such as collagens and laminin, influencing cell proliferation, migration, and adhesion [25]. In CRC, the EMT pathway has been associated with the increased expression of several MMPs, such as MMP2/7/9, by both in vitro studies in CRC cell lines and integrated multi-omics investigations [26,27]. Therefore, EMT may lead to the epithelial cell–cell junction disruption, loss of apical–basal polarity and cytoskeleton remodeling. Meanwhile, mesenchymal characteristics such as enhanced migratory capacity, invasiveness and elevated resistance to apoptosis are acquired [28]. Accordingly, recent studies showed that metastatic CRC patients with CTCs expressing EMT-related genes have worse progression free survival (PFS) and shorter overall survival (OS), suggesting the possibility of exploiting EMT-CTCs as prognostic factors in CRC [29].

CLMs spread can be enhanced by angiogenesis process. It has been shown that, in the presence of the primary tumor, the liver parenchyma adjacent to the liver metastases provides an angiogenic prosperous environment for metastatic tumor growth [30]. However, recent studies demonstrated that CLMs can also metastasize via a non-angiogenic process mediated by EMT, known as vessel co-option [31]. Interestingly, CRC cells can hijack pre-existing vessels of the host liver when conditions are not favorable to form new ones [32]. In CLMs, high levels of EMT markers have been detected in co-opted tumors compared to their angiogenic counterpart [33]. Indeed, Rada et al. recently demonstrated that EMT enhances CRC cells infiltration within liver parenchyma and the motility in hepatocytes, which displaced at the edge of tumor nests to allow the occupation of their space by cancer cells [34]. Another crucial step for metastatic invasion is represented by the disruption of blood vessels integrity [35]. The resulting altered vascular permeability allows tumor cells intravasation in the blood stream and subsequent extravasation in the metastatic site (Figure 1). The adherence of circulating CRC cells to the sinusoidal endothelial cells (SECs) of the liver is a crucial process involved in liver invasion and occurs upon selectins binding [36]. Rigorous blood vessel integrity is normally maintained by adherent junctions, whose basal organization is dictated by vascular endothelial cadherin (VE-cadherin), enforced by p120-catenin (p120) and spans the plasma membrane [37,38]. In such scenario, Dou et al. demonstrated that CRC cells that underwent EMT can interact with endothelial cells to increase vascular permeability. Specifically, EMT-CRC cells produced exosomal miR-27b-3p, which once transferred to endothelial cells, enhanced blood vessel permeability, thus facilitating cancer cell extravasation in the metastatic site by targeting VE-cadherin and p120 [39].

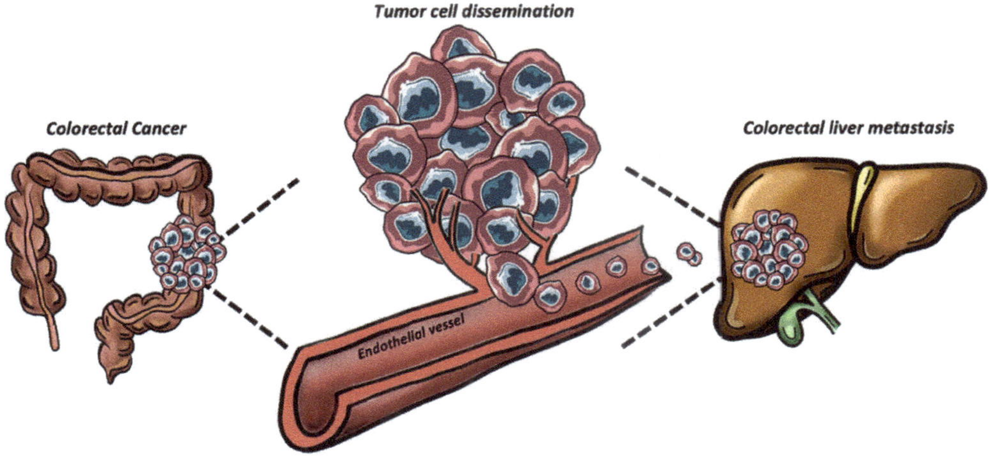

Figure 1. Representative image of the tumor cell dissemination process from the primary site (colon) to the metastatic one (liver).

Once in the circulation, CTCs are subjected to a harsh selective pressure imposed by shear forces within the vessels but also the evasion of immune surveillance by the numerous immune cells in the plasma [40]. Accordingly, recent studies in mouse models showed CTCs survive in blood circulating for a short time, before being rapidly eradicated [41]. In order to survive in the bloodstream and form a niche at secondary sites, CTCs exploit EMT that activates survival pathways (such as activation of Akt, PI3K or EGFR pathways), that enable EMT-shifted cells to better resist apoptosis [19]. Moreover, EMT-mediated overexpression of the cell–cell junction marker plakoglobin may lead to generation of CTCs clusters, comprised of tumor cells with or without other non-malignant cell types, such as mesenchymal cells, epithelial cells, immune cells, platelets, and cancer-associated fibroblasts, that can contribute to the survival and metastatic advantages of CTC clusters [42]. Furthermore, the development of strong intercellular bonds allows invading cancer cells to resist the forces of plasma flow and circulating blood cells, when they adhere to hepatic sinusoids. Tumor cell adhesion and stabilization under the hydrodynamic conditions of blood flow are correlated to the expression of multiple signaling molecules, such as focal adhesion kinase, paxillin, and cytoskeletal proteins, such as actin or microtubules [43].

Subsequently, tumor cells start to interact with the extracellular matrix (ECM) of the invaded tissue, exploiting integrins, essential receptors able to stabilize the tumor cell adhesion [44]. Once disseminated to a distant site, cancer cells are believed to regain epithelial properties in a reverse process, referred to as mesenchymal-epithelial transition (MET), characterized by loss of migratory ability, with cells adopting an apical–basal polarization and expressing junctional complexes [45]. MET leads to the inhibition of cancer cells migration and induces their proliferation, leading to the growth of the new tumor. However, the definition of EMT has now been broadened, based on many new observations of intermediate hybrid epithelial and mesenchymal phenotypes, referred as partial EMT state [46]. These intermediate states that easily induce or reverse EMT process reflect the delicate balance of transcriptional drivers and suppressors of EMT and offer a more dynamic interpretation of the fluidity and plasticity of this phenomenon. Indeed, a recent study performed on CRC tissue samples showed the involvement of partial EMT at the invasive tumor front and the involvement of partial MET in both lymph node and liver metastases, based on the expression patterns of the miR-200 gene family in these critical locations [47]. These results highlighted that CRC cells undergo EMT at the invasive front of primary site along with their ability to recapitulate the phenotype of the primary tumor in the metastatic site.

3. The Molecular Mechanisms of EMT in CRC

One of the most distinguishing features for the establishment of an EMT phenotype is the overexpression of mesenchymal markers and deregulation of structural adhesion proteins. Among the aforementioned mesenchymal proteins, the most relevant ones induced in CRC during EMT are N-cadherin, vimentin, and fibronectin [21]. N-cadherin is a calcium-dependent transmembrane glycoprotein that mediates cell–cell adhesion, whose aberrant expression has been observed in many cancers because of it is closely related to cancer cells transformation and invasiveness [48]. Instead, vimentin is a major constituent of the intermediate filament family of proteins, whose physiological role is to maintain cellular integrity and provide resistance against stress in normal mesenchymal cells [49]. In CRC, the mesenchymal markers, N-cadherin and vimentin, have been shown to drive malignant progression of tumor cells and to correlate with metastasis development and a worse OS in patients [50,51]. Fibronectin is a soluble protein, part of the extracellular matrix (ECM), which in physiological conditions plays a role in wound healing [52]. Downregulation of fibronectin has been shown to inhibit colorectal carcinogenesis by suppressing proliferation, migration, and invasion [53]. Furthermore, fibronectin has been demonstrated to promote tumor cells growth and drugs resistance through a CDC42-YAP-dependent signaling pathway in CRC [54].

On the counterpart, E-cadherin, essential for the maintenance of adherent junctions, is fundamental for the physical integrity and polarization of epithelial cells [55]. Notably, loss of E-cadherin expression has been associated with poor prognosis in stage III CRC patients [56]. E-cadherin expression can be regulated at different levels in response to vary induction signals, including transcriptional repression [57], promoter methylation [58], as well as protein phosphorylation and degradation [59]. Indeed, the discovery that transcriptional repressors of E-cadherin contribute to invasion and metastasis has strengthened the evidence for the importance of the EMT in tumor progression.

EMT is modulated at different levels by epigenetic modifications, transcriptional control, alternative splicing, protein stability, and subcellular localization [21]. The transcriptional control of EMT is mainly driven by three groups of EMT-inducing transcription factors: Snail, Zeb, and Twist families (Table 1). Snail family is characterized by zinc finger transcription factors, all of which bind to a common binding motif known as the E-box and comprise SNAIL1 and SNAIL2 (also known as SLUG) genes [60]. Besides their role to repress E-cadherin expression, Snail transcription factors are involved in promoting the expression of mesenchymal genes, such as vimentin, N-cadherin, and fibronectin [61]. Indeed, Snail1 expression in CRC is associated with tumor progression and metastasis because it leads to the silencing of E-cadherin expression and to the induction of EMT [62]. Franci et al. showed that about 77% of colon cancer samples display Snail1 immunoreactivity both in activated fibroblasts and in carcinoma cells that underwent EMT, suggesting that the presence of Snail1 immunoreactive cells may be exploited as prognostic marker in patients with colon cancer [63,64]. Instead, the Zeb family comprises the 2-handed zinc finger/homeodomain proteins Zeb1 and Zeb2, which are especially overexpressed in CRC [65,66]. These transcription factors regulate EMT by binding E-box on E-cadherin promoter region and by upregulating LAMC2 mesenchymal gene to promote tumor invasion [67]. A recent study showed that the loss of a circadian gene, named Timeless, was able to induce EMT and E-cadherin downregulation in CRC cell lines, via a Zeb1-dependent mechanism [68]. Furthermore, it has been described that transcription factor Zeb1 causes severe alterations in the expression patterns of chromatin-modifying enzymes in CRC [69]. This study reported that the Zeb1-mediated upregulation of histone methyltransferase SETD1B stabilized Zeb1-mediated EMT through a positive feedback loop between Zeb1 and SETD1B, thus each protein can reinforce the activity/expression of the other. Lastly, Twist family, which comprehend Twist1 and Twist2 transcription factors, activate N-cadherin promoter and switch on mesenchymal markers such as N-cadherin and fibronectin, leading to E-cadherin-mediated cell–cell adhesion is lost and thereby promoting EMT [70]. Twist mediates an aggressive phenotype in human CRC cells. Indeed, a recent experiment performed on human CRC cell lines showed that Twist overexpression triggers EMT by E-cadherin downregulation and enhances tumor migration and invasion [71]. In addition, Twist-overexpressing CRC cells were more chemo-resistant to the drug oxaliplatin than control cells (Figure 2) [72].

Table 1. EMT transcription factors and their role in EMT mechanisms to promote CLM development.

EMT-TFs	Members of TF Family	Mechanisms in CLM-Related EMT	References
Snail	*Snail1, Snail2 (SLUG genes)*	■ *E-cadherin downregulation* ■ *Vimentin, N-cadherin, and fibronectin mesenchymal genes upregulation*	[60,61]
Zeb	*Zeb1, Zeb2*	■ *E-cadherin downregulation* ■ *LAMC2 mesenchymal gene upregulation* ■ *Severe alterations in the expression pattern of chromatin-modifying enzymes*	[67–69]
Twist	*Twist1, Twist2*	■ *N-cadherin and fibronectin upregulation* ■ *Loss of E-cadherin-mediated cell–cell adhesion* ■ *Tumor migration and invasion* ■ *Possible induction of chemoresistance to the drug oxaliplatin*	[70–72]

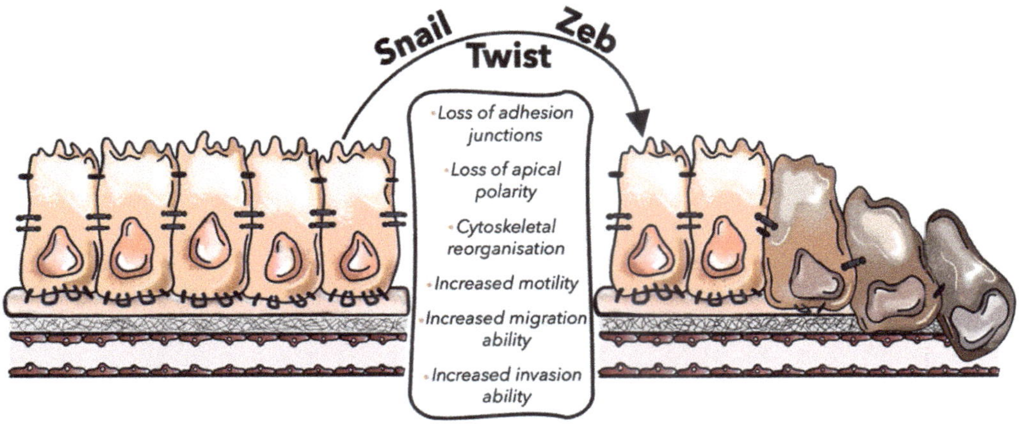

Figure 2. Representative image of the EMT process. Snail, Twist, and Zeb, EMT-promoting transcription factors (EMT-TFs), promote the loss of cellular epithelial characteristics, such as the disassemble of cell–cell junction, the defect of apical–basal polarity and cytoskeletal reorganization, supporting the acquisition of a mesenchymal phenotype of tumor cells.

Multiple miRNAs are thought to even govern EMT. In CRC, it has been found that miR-17-5p overexpression in tumor cell lines significantly decreased vimentin mRNA and protein expression, cell migration, and invasion, whereas downregulation of miR-17-5p in CRC cell lines increased vimentin protein expression, cell migration and invasion in vitro [73]. In addition, the authors' findings suggested that overexpression of miR-17-5p inhibited liver metastasis in an intra-splenic injected mouse model. In another study, it has been found that miR-566 overexpression markedly increased the E-cadherin expression and inhibited the levels of vimentin and N-cadherin in several CRC cell lines (SW480, SW620, LoVo, HT29 and Caco-2) [74]. Therefore, miR-566 overexpression inhibited PSKH1 gene, suppressing cell proliferation, whereas inhibition of its expression enhanced cell survival and proliferation. The correlation between altered expression of specific miRNAs and CRC insurgency have triggered the examination of microRNAs as urinary noninvasive biomarkers, alternatively to invasive colonoscopy to detect early stages of the disease [75].

4. The Relationship between EMT and TGF-β Signaling Pathway in CRC

The transforming growth factor-β (TGF-β) is multifunctional cytokine that plays a role not only in the regulation of EMT, but also in the survival, development and differentiation of almost all cell types and tissues [76]. In cells of epithelial- and endothelial-origins, TGF-β also is a powerful suppressor of cell growth and proliferation [77]. Indeed, in the colon tissue, TGF-β signaling regulates the growth of normal cells in the colonic crypt and villi [78]. However, CRC can evade the tumor suppressing effects of TGF-β pathway, which represents one of the most commonly altered pathways in human cancers [79]. TGF-β signaling exists in three isoforms (TGFβ1, TGFβ2, TGFβ3) and it is mainly divided into two subfamilies: the TGF-β-activin-nodal subfamily and the bone morphogenetic protein (BMP) subfamily. TGF-β pathway is triggered via transmembrane serine/threonine kinase TGF-β type I receptors (TGF-βRI or ALK5) and TGF-β type II receptors (TGF-βRII). Upon TGF-β binding, TGF-βRII recruits and phosphorylates the cytoplasmatic domain of TGF-βRI, leading to the phosphorylation and activation of downstream transcription factors, SMAD2 and SMAD3 in the TGF-β-activin-nodal subfamily and SMAD1/5/8 in BMP subfamily [80]. In both cases, these activations allow them to bind SMAD4, generating SMAD complexes that translocate into the nucleus and bind DNA in a cell-specific manner, thus regulating the transcription of a multitude of TGF-β-responsive genes, including transcription factors belonging to the Snail family (e.g., Snail, Twist, or ZEB1), or of STAT3 [81,82]. The activation of these transcription factors elicits EMT-gene expression and ultimately promotes the prolonged induction of EMT via DNA methylation-mediated silencing of E-cadherin expression, as well as the upregulation of mesenchymal markers [83]. Among the numerous mesenchymal EMT-associated genes upregulated by TGF-β, there are N-cadherin, vimentin, and fibronectin, but also β3 integrin and several matrix metalloproteinases, such as MMP-3 and MMP-9 [84]. Indeed, TGF-β is often used in the cell culture to induce EMT and metabolic reprogramming in various epithelial cells [85,86]. Alternatively, activation of noncanonical TGF-β signaling, such as mitogen-activated protein kinase (MAPK), phosphoinositide 3-kinase (PI3K)/Akt and Rho/Rho-associated protein kinase (ROCK) pathways, also works with TGF-β in its regulation of EMT [83].

Altered functions of TGF-β lead to different gene expression patterns contributing to the development of oncogenic signaling and increasing the invasiveness ability of cancer cells [87]. Indeed, the initiation of oncogenic signaling boosted by TGF-β converts the regulation of physiological EMT in normal epithelial cells to pathological EMT in their malignant counterparts [84]. Loss of SMAD proteins represents one of the leading causes, and almost 25% of patients affected by CRC display a mutation in the SMAD4 protein complex [88]. Notably, the consensus molecular subtype (CMS) 4 of CRC displays the downregulation of SMAD4 [89]. Instead, SMAD4 upregulation suppresses invasion and restores the epithelial phenotype in the SW480 CRC cell line. Knockdown of SMAD4 led to increased levels of endogenous TGF-β cytokines, which upregulated TGF-β signaling and induced EMT [90]. SMAD4 is also a central component of the BMP signaling pathway, implicated in CRC pathogenesis. Hence, it has been shown that loss of SMAD4 alters BMP signaling and promotes CRC metastases via activation of Rho and ROCK pathways, leading BMP signaling to switch from tumor suppressive to metastasis-promoting function [91]. Contrarily, a recent study showed a SMAD4-independent EMT pathway in CRC, in which two epithelial SMAD4^{mut} CRC cell lines were able to acquire mesenchymal characteristics and regulate EMT marker genes in response to Snail1 induction, with phenotype independent from TGF-β and BMP receptor activity. These results suggested that there might be alternative transcription factors taking over the gene regulatory functions of SMAD4 during EMT in CRC [92].

TGF-β signaling has been shown to possess a dual role in the tumor microenvironment (TME), known as the "TGF-β paradox" [93]. In early-stage tumors, the TGF-β pathway acts as a tumor suppressor by inducing apoptosis, triggering cell cycle arrest, and thus inhibiting the proliferation of cancer cells [94]. In contrast, in late-stage tumors, it has pro-tumoral effects by modulating genomic instability, epithelial-mesenchymal transition (EMT), neoangiogenesis, immune evasion, cell motility, and metastasis [95,96]. In this context, in CRC with high microsatellite instability (MSI-H subtype), TGF-βRII mutations interfere with TGF-β-induced EMT and therefore reduce the migratory and invasive capabilities of CRC cells, providing a better prognosis than microsatellite-stable CRCs [97,98]. On the other hand, the CMS4 subtype, with reduced expression of SMAD4, has been associated with poor OS [89,99]. In such a scenario, the bidirectional activity performed by TGF-β signaling in different CRC subtypes reflects the complexity of this signaling pathway in this tumor. Indeed, TGF-β signaling mutations enhance EMT and, subsequently, tumorigenesis and metastases in the CRC-CMS4 subtype, while TGF-βRII mutation impairs EMT and provides a better prognosis [100].

5. The Role of Tumor-Associated Macrophages (TAMs) in CRC Progression

The tumor microenvironment (TME) provides an essential dynamic niche with a key role in cancer initiation and progression [8,101,102]. In CRC, the microenvironment is composed of stromal cells and immune cells such as granulocytes, lymphocytes, and tumor-associated macrophages (TAMs), which affect tumor immune-suppression and inflammation [103,104]. Among the immune cells, TAMs represent the dominant cell type in TME and exhibit different functional polarization, known as M1 or M2 phenotype, in response to various stimuli from both tumor and stromal cells [105,106]. Different methodological approaches have allowed for highlighting the heterogeneity of TAMs in terms of function, polarization, and tissue localization [107]. Within the liver, these technologies include gene expression profiling [108], morphological identification [109], and evaluation of TAMs-specific markers [110]. Moreover, high dimensional analysis exploiting single-cell and spatial genomics technologies, such as single-cell RNA sequencing and mass cytometry by time-of-flight (CyToF), have allowed us to assess TAMs heterogeneity at an unprecedented resolution [111].

The classical M1 polarization is induced by recognition of pathogen-associated moieties, such as lipopolysaccharides (LPS) and Interferon gamma (INF-γ), with a key role in the innate response against pathogenic infection. Indeed, M1 macrophages are mainly involved in proinflammatory responses by producing proinflammatory cytokines (such as IL-12, IL-23) and chemokines (such as CXCL9 and CXCL10) [112,113]. M1 polarization can be identified by overexpression of CD80, CD86 and CD16/32 markers [114]. In contrast, M2 macrophages, which are induced by IL-4 and IL-13, exert a more anti-inflammatory response This population is characterized by elevated expression of arginase-1 (Arg-1), mannose receptor (CD206) and by secretion of anti-inflammatory factors (IL-10), chemokines (CCL17, CCL2) and matrix metalloproteinase 9 (MMP9) [114,115]. The majority of intra-tumoral macrophages has been shown to exhibit an M2 phenotype and to be correlated with poor prognosis in several tumors, including CRC [116,117]. In such scenario, flourishing literature demonstrated that in response to signals released from cancer cells, adaptive immune cells, B cells, fibroblasts, and macrophages themselves, such as IL-10, CCL2/3/4/5/7/8 and CXCL12 (colony stimulus factor 1 (CSF-1)), VEGF and interleukin-6 (IL-6), monocytes are recruited in the tumor niche and differentiate into the TAMs with an M2-like phenotype [118]. This population creates an environment that supports tumor growth and metastases by promoting tissue remodeling, angiogenesis, and secreting immunosuppressive cytokines, thus inhibiting innate and adaptive immune responses [112].

As mentioned above, TAMs play an important role in promoting tumor development and recurrence, as well as in the efficacy of anticancer strategies [8]. Despite the dichotomous M1/M2 classification, recent single-cell-resolution approaches have been useful in recognizing the heterogeneity of TAMs beyond the classic M1-like or M2-like phenotypes [102]. TAMs constitute a diverse macrophage population that shares features of both the M1 and M2 subsets. Therefore, TAMs plasticity could be associated with peculiar roles that macrophages exert in different cancers. With a particular focus on CRC, numerous studies have shown that a high density of macrophages is indicative of favorable outcomes in patients with stage III CRC [119]. The abundance of TAMs, particularly in CRC stage III metastatic lymph-nodes, might modify the efficacy of 5-fluorouracil chemotherapy, increasing CRC cell death and thus leading to better disease-free survival (DFS) [111]. Meanwhile, other data support the opposite findings. *Herrera* et al. reported that infiltration of CD163$^+$ macrophages together with cancer-associated fibroblasts (CAFs) in CRC tissues was related to worse OS and PFS [120]. A recent study showed that a high density of macrophages correlates with worse DFS in patients who underwent chemotherapy for unresectable metastatic CRC after resection of the primary tumor. Indeed, TAMs were observed to induce chemoresistance by promoting malignant angiogenesis [121]. Furthermore, TAMs exert immunosuppressive roles in the CRC microenvironment. They recruit regulatory T cells (Tregs) by secreting the chemokine CCL2 and IL-10. TAMs suppress the antitumor immune response of T cells by metabolic starvation and inappropriately skew dendritic cells toward an immature and tolerogenic state [122]. In addition, immunosuppressive TAMs are characterized by high expression of immune-checkpoint molecules (such as PD-L1), causing T-cell exhaustion [123].

6. Mechanisms Exploited by TAMs to Regulate EMT in Colorectal Liver Metastasis (CLMs): A Focus on TGF-β Signaling Pathway

Despite the huge amounts of studies addressing the role and clinical relevance of TAMs in primary CRC, less is known concerning their role in CLMs, probably due to the different phenotypic profiles expressed by macrophages in the liver [124]. Kupffer cells (KCs) are tissue-resident macrophages localized within the lumen of the liver sinusoids. KCs perform phagocytic and cytokine secretion activities that allow them to eliminate circulating molecules and pathogens [125]. On the other hand, monocyte-derived macrophages are mainly resident in the portal triad, where they contribute to iron and cholesterol metabolism [126]. Therefore, TAMs with different morphologies and molecular fingerprints coexist in CLMs and correlate with clinicopathological variables [112]. Indeed, a recent study demonstrated that the morphology of tumor-associated macrophages (TAMs) correlates with prognosis in CLM patients. Specifically, while TAMs density did not correlate with survival in CLM patients, large (L)-TAMs, characterized by a bigger cell area and perimeter, were associated with statistically significantly worse prognosis and DFS, compared to small (S)-TAM ones [110].

The crosstalk between TAMs and EMT-CRC cells has been experimentally investigated by both in vitro and in vivo approaches. A co-culture assay was recently realized in vitro to evaluate the role of TAMs in CRC EMT, migration, and invasion [127]. This assay allows us to appreciate the changing morphology and gene expression profile of both cellular populations during their crosstalk, as well as to characterize the cytokines involved in cell–cell interactions by the analysis of the co-culture supernatants. Although in vitro metastasis models allow for the manipulation of each metastasis step, they do not provide a comprehensive analysis of the whole metastatic process [128]. In contrast, in vivo metastasis models may be more accurate in the representation of the metastatic process and can be genetically manipulated to mimic human cancer. In CLMs, in vivo experiments have been performed by realizing mouse xenografts, in which HCT116 CRC cell line and in vitro polarized TAMs were subcutaneously injected in 6–8-weeks-old nude mice [127]. Moreover, HCT116 xenografts can be coupled with macrophage depletion, such as by intravenous liposomal clodronate injection [129]. In this context, it is possible to study TAMs–CRC cells interactions in vivo, TAMs migration, and polarization, as well as how this crosstalk can change in the presence of specific microRNAs or interleukins that can be injected into the implanted tumor. However, the disadvantages of this model are the difficulty in assessing the contribution of the immune system in the metastatic process, because the various populations of immune cells may not properly function in the xenograft TME [128]. Moreover, xenografts do not show tumor heterogeneity or histopathologic and genetic characteristics of the original tumor. Therefore, additional research should be performed to develop more precise models to study TAMs and EMT-CRC cell interaction.

In CRC metastatic process, increasing evidence highlighted that TAMs are responsible for the induction of EMT because of the release of multiple factors, such as IL-6 and TGF-β [112]. Particularly, $CD163^+$ TAMs have been observed to secrete IL-6, which, upon binding of IL-6 receptor (IL6R) on the cancer cell surface, phosphorylate STAT3 (pSTAT3) that translocated to the nucleus and regulates the expression of varying microRNAs, including miR-506-3p, which promotes EMT (Figure 3A). Indeed, E-cadherin gene expression was reduced in TAMs-mediated EMT, while the mesenchymal marker, vimentin, was upregulated in CRC cell lines. Therefore, the crosstalk between TAMs and tumor cells has been proposed to play an important role in inducing EMT in CRC and promoting EMT-mediated metastasis [127]. Moreover, IL-6 secretion by TAMs has been observed to induce chemoresistance through the activation of the IL6R/STAT3/miR-204-5p pathway in CRC cells [130]. M1 macrophages exhibited a potential to induce EMT in HCT116 and RKO CRC cell lines by secreting TNF-α and IL-1β [131,132]. Specifically, TNFα expression can improve Snail protein expression and nuclear localization through the AKT pathway, upregulating N-cadherin and fibronectin with a concomitant decrease in E-cadherin [132]. Il-1β secretion, instead, led to EMT of colon cancer cells with loss of E-cadherin, upregulation of Zeb1, and gain of the mesenchymal phenotype in CRC cell lines [131]. Recent evidence highlighted that CRC cells, through the NOTCH2/GATA3 pathway, undergo to EMT process and secrete IL-4, thus polarizing macrophages into an M2-like phenotype [133]. Another study showed that the expression of protein phosphatase of regenerating liver-3 (PRL-3) in CRC cells was able to activate the MAPK pathway in TAMs, thus leading to the release of IL-6 and IL-8 and inducing the EMT in cancer cells [134]. Furthermore, in vivo and in vitro studies showed that mesenchymal CRC cells obtained by Snail-induced EMT were able to secrete CXCL2, which promoted M2 macrophage infiltration and tumor cell metastasis [135]. These results suggest that not only macrophages play a role in EMT, but also tumor cells undergoing EMT could influence TAMs polarization. However, although most of the studies in CRC have focused on TAMs' roles in promoting tumor metastasis and EMT, recent evidence showed that TAMs could inhibit EMT in patients affected by sporadic CRC, thus exerting a protective role against the development of metastases [136]. Strong $CD68^+$ infiltration was reported to inhibit tumor burden, demonstrating that macrophages could have an important role in fighting against CRC cells with EMT traits.

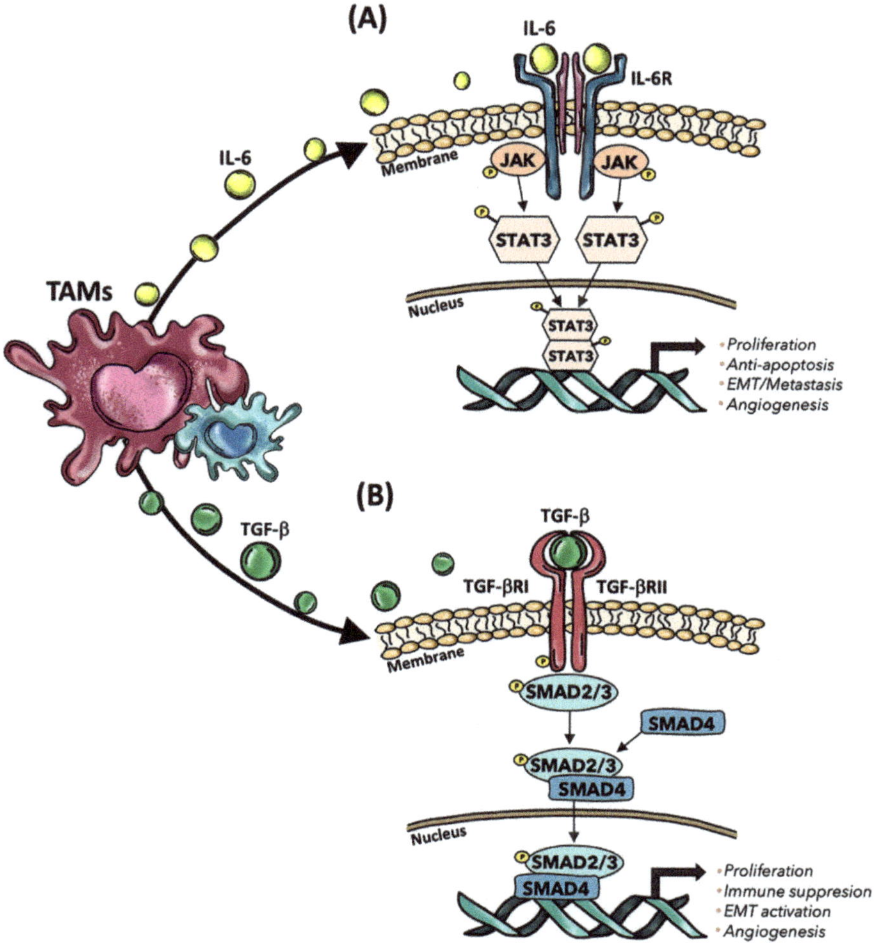

Figure 3. Mechanisms exploited by TAMs to induce cancer progression in CLMs. (**A**) TAMs release IL-6, which then interacts with IL-6R, subsequently activating the JAK-STAT pathway that regulates the expression of genes involved in tumor progression. (**B**) TGF-β secreted by TAMs interacts with TGF-βI and TGF-βII receptors, leading to phosphorylation of SMAD2/3 and subsequent generation of a complex with SMAD4, which finally leads to the expression of cancer-related genes. TAMs: Tumor-associated macrophages; IL-6: Interleukin-6; IL-6R: Interleukin-6 receptor; TGF-b: Transforming growth factor beta; TGF-βR: TGF-β receptor; JAK: Janus kinase; STAT3: signal transducer and activator of transcription 3; SMAD: Suppressor of Mothers against Decapentaplegic.

TGF-β signaling pathway was highlighted to be involved in the crosstalk between TAMs and tumor cells in CLMs tumor microenvironment and to facilitate the induction of EMT in CRC cells [80]. It has been described that TGF-β secreted by TAMs activated the SMAD signaling pathway by binding to the TGF-β receptors, followed by the phosphorylated Smad2/Smad3 to form a complex with Smad4 and regulate transcription of Snail. Once TGF-β triggered EMT, Snail could repress the expression of E-cadherin, resulting in CRC metastasis [137]. Further evidence demonstrated that Collagen Triple Helix Repeat Containing 1 (CTHRC1) produced by CRC cells increased tumor burden and the number of CLMs nodules in mouse models by modulating macrophage polarization to M2 phenotypes through TGF-β signaling. Hence, CTHRC1 bound directly to TGF-βRII and TGF-βRIII in

TAMs, stabilizing the TGF-β receptor complex and activating TGF-β signaling. Moreover, the inhibition of TGF-β signaling in macrophages through CTHRC1 monoclonal antibodies, coupled with PD-1/PD-L1 blockade, effectively led to the reduction of CLMs [138]. Interestingly, different studies showed that CTHRC1 overexpression has been associated with worse OS and DFS in CRC patients, driving the pathogenesis of the EMT process in CRC by activating the TGF-β pathway [138,139]. TAMs and the TGF-β pathways are also involved in the induction of CRC immune evasion [111]. In pediatric patients, it has been observed that CD163$^+$ macrophages promoted the progression of colorectal polyps by inhibiting the local T-cell response through TGF-β production [140]. In another study performed on surgically resected CRC tissues, immunohistochemistry analysis showed that M2 macrophages induced immunosuppressive T-reg cell generation through activation of the TGF-β/SMAD signaling pathway (Figure 3B) [141]. Furthermore, increasing evidence suggests that TGF-β secreted by TAMs could play crucial roles in pathophysiological processes by regulating multiple microRNAs (miRNAs) [142,143]. In this context, TGF-β1 protein levels were found to be highly expressed in CRC tumor tissues and in vitro polarized macrophages. TAMs, obtained by in vitro co-culturing of macrophages with conditioned medium from CRC cells, downregulated the expression of miR-34a in tumor cells by secreting TGF-β1, inducing VEGF upregulation and thereby promoting cell proliferation and invasion of CRC cells [142]. TGF-β1 upregulation has been observed in vessel co-opting CLMs as well. Indeed, recent evidence showed that the TGF-β1 pathway may act as a mediator contributing to hepatocyte displacement during CLM development [34]. Although the role of TAMs remains unclear in this process, recent research showed that M1-macrophages are predicted to interact with pericytes via TGF-β signaling in vessel-co-opted tumors [144]. However, other studies found that the expression of TGF-β1 and SMAD4 in the cytoplasm, as well as the presence of TGF-βRII in the membranes of CRC cells, were associated with lower levels of CD68$^+$ macrophages. Hence, the main components of the TGF-β1 signaling pathway (e.g., TGF-β1, SMAD4, TGF-βRII) could have an immunosuppressive effect via inhibition of macrophage recruitment [143]. Therefore, although TAMs can be one of the sources of TGF-β expression, the effects of TGF-β signaling on TAMs in CRC are not fully understood and requires further investigation.

Among the different mechanisms correlating TAMs and EMT, recent studies showed the potential use of anti-TGF-β strategies to impair CRC development, particularly in its late stages [145]. However, anti-TGF-β therapy alone is insufficient to mediate antitumor immunity in CRC. To overcome this problem, studies on the combination of other biological agents or irradiated tumor vaccine with anti-TGF-β treatment showed a reduction in TGF-β-induced EMT and correlated CRC metastasis [146]. Therefore, the combination therapy of chemotherapy/radiotherapy/targeted therapy with anti-TGF-β might be developed to achieve enhanced antitumor efficacy by regulating the tumor microenvironment. For these reasons, the exploration of the mechanisms of TGF-β signaling to develop TGF-β-based combination therapies might be very crucial for the development of new therapeutic applications in CRC and CLM [147].

7. Concluding Remarks and Future Perspectives

In this review, we discuss the involvement of EMT in the development of colorectal liver metastases (CLMs) and focus on the different signaling pathways exploited by TAMs to promote EMT in CRC, particularly describing the TGF-β signaling pathway.

Curative treatment of CLMs relies on surgical resection and systemic chemotherapy, providing the greatest chance for long-term survival. The 5-year survival after surgical resection is approaching 50%, with a 5-year disease-free survival of 25% [148,149]. Unfortunately, only 10–20% of patients are candidates for curative surgery. Instead, unresectable patients display 3-year survival of only 15% and are left with palliative chemotherapy as their only treatment option [150,151]. Fluoropyrimidine-based combinations (FOLFOX and FOLFIRI) with or without target therapy using anti-vascular endothelial growth factor (VEGF) or anti-epidermal growth factor receptor (EGFR) inhibitors represent the systemic treatments for CLM [152]. This combined approach offers the possibility to undergo hepatic resection even to those patients who would not have been considered for surgery until a few years ago. However, CLM patients display varying degrees of response to therapy, and research over the last decade has aimed to characterize the invasion of immune cells into the TME to stratify patient outcomes [153]. Anti-PD-1 or anti-PD-L1 blocking antibodies have shown significant results in CRC with microsatellite instability (MSI) [154]. Moreover, the neutralization of IL-10 effects in human CLMs has shown therapeutic potential as only treatment and to augment the function of administered CAR-T cells [155]. Nevertheless, the characterization of the immune landscape in CLM patients is complicated by the profound heterogeneity of tumor lesions across patients and the frequent neoadjuvant treatments of CLM patients, which could potentially affect the type of immune infiltrate [153].

Tumor-associated macrophages (TAMs) are essential players in CRC metastatic process, exerting a crucial role in EMT [156]. Upon secretion of various cytokines and other signaling molecules, such as exosomes, macrophages can crosstalk with CRC cells, promoting the EMT of tumor cells. In turn, mesenchymal tumor cells have been observed to enhance the recruitment of TAMs to the tumor site and promote their M2 polarization by secreting IL-10 and IL-4 [157]. In recent years, strategies targeting TAMs, including TAMs depletion, reprogramming, and inhibition of TAMs recruitment, have been investigated [111]. However, due to the significant heterogeneity of TAMs in the liver and in regulating tumor metastasis, results obtained from clinical studies not always were satisfying [158]. Therefore, it is necessary to further investigate the more unknown mechanisms by which TAMs promote CRC metastases, including EMT. Indeed, targeting a single EMT receptor is unlikely to be effective because of the redundant nature of several pathways and the biological features of EMT-transformed cells, such as increased cell mobility, invasiveness, and chemoresistance, further complicating drug development [159]. A wide range of targets associated with EMT is required to be elucidated in the future to overcome therapy resistance. Novel therapies have been proposed in oncology targeting the TGF-β signaling pathway, exploited by TAMs to promote EMT and responsible for resistance to conventional therapies in CRC [160]. Several TGF-β pathway antagonists have advanced to clinical trials and demonstrated acceptable safety profiles and significant therapeutic efficacy in cancer patients. In addition, TGF-β agonists may be exploited in patients resistant to conventional therapies [161]. Chemotherapy drugs ginsenoside Rb2 and tanshinone II A displayed therapeutic effects acting as inhibitors of TGF-β-induced EMT and angiogenesis, respectively [146,162]. Among the numerous pharmacological approaches targeting TGF-β that have undergone preclinical and clinical stages, there are neutralizing antibodies, TGF-β inhibitors, ligand traps, antisense oligonucleotides, and vaccines [145]. In CRC, preclinical trials have shown that the combination with TGF-β inhibitor Galunisertib, a selective inhibitor of TGFβRI, was able to enhance the efficacy of chemotherapy and radiotherapy [163]. Coadministration of TGF-β inhibitors and anti-PD-L1 antibodies displayed effective response in CRC patients by promoting $CD8^+$ T cells penetration into the tumor [164]. Moreover, it has been shown that dual blockade of TAM recruitment and TGFβ signaling significantly augments the therapeutic efficacy of chemotherapy via suppressing PD-L1 expression in metastatic CRC [165]. However, despite the suitable results obtained in the ongoing clinical trials, the mechanisms underlying TGF-β mediated EMT promoted by TAMs remains unclear in CLMs patients. For this reason, further investigations in this field may provide new therapeutic strategies for fighting CLMs, which are still responsible for most of the deaths in CRC patients.

Author Contributions: A.G., M.A.P., C.S. and B.F. researched and compiled the data from the literature and drafted a first draft of the manuscript. A.L. and M.D. revised the manuscript. All authors have read and agreed to the published version of the manuscript.

Funding: This work was supported by "Bando Ricerca Finalizzata 2018 of the Italian Ministry of Health" (ID = RF-2018-12367150). The funding agency had no role in the design of the study or collection and analysis of data.

Conflicts of Interest: The authors declare no conflict of interest.

References

1. Sung, H.; Ferlay, J.; Siegel, R.L.; Laversanne, M.; Soerjomataram, I.; Jemal, A.; Bray, F. Global Cancer Statistics 2020: GLOBOCAN Estimates of Incidence and Mortality Worldwide for 36 Cancers in 185 Countries. *CA Cancer J. Clin.* **2021**, *71*, 209–249. [CrossRef] [PubMed]
2. Siegel, R.L.; Miller, K.D.; Goding Sauer, A.; Fedewa, S.A.; Butterly, L.F.; Anderson, J.C.; Cercek, A.; Smith, R.A.; Jemal, A. Colorectal cancer statistics, 2020. *CA Cancer J. Clin.* **2020**, *70*, 145–164. [CrossRef] [PubMed]
3. Landreau, P.; Drouillard, A.; Launoy, G.; Ortega-Deballon, P.; Jooste, V.; Lepage, C.; Faivre, J.; Facy, O.; Bouvier, A.-M. Incidence and survival in late liver metastases of colorectal cancer: Liver colorectal cancer metastases survival. *J. Gastroenterol. Hepatol.* **2015**, *30*, 82–85. [CrossRef] [PubMed]
4. Mlecnik, B.; Van den Eynde, M.; Bindea, G.; Church, S.E.; Vasaturo, A.; Fredriksen, T.; Lafontaine, L.; Haicheur, N.; Marliot, F.; Debetancourt, D.; et al. Comprehensive Intrametastatic Immune Quantification and Major Impact of Immunoscore on Survival. *J. Natl. Cancer Inst.* **2018**, *110*, 97–108. [CrossRef]
5. Hackl, C.; Neumann, P.; Gerken, M.; Loss, M.; Klinkhammer-Schalke, M.; Schlitt, H.J. Treatment of colorectal liver metastases in Germany: A ten-year population-based analysis of 5772 cases of primary colorectal adenocarcinoma. *BMC Cancer* **2014**, *14*, 810. [CrossRef]
6. Lin, A.; Zhang, J.; Luo, P. Crosstalk Between the MSI Status and Tumor Microenvironment in Colorectal Cancer. *Front. Immunol.* **2020**, *11*, 2039. [CrossRef]
7. Mei, Y.; Xiao, W.; Hu, H.; Lu, G.; Chen, L.; Sun, Z.; Lü, M.; Ma, W.; Jiang, T.; Gao, Y.; et al. Single-cell analyses reveal suppressive tumor microenvironment of human colorectal cancer. *Clin. Transl. Med.* **2021**, *11*, e422. [CrossRef]
8. Mantovani, A.; Marchesi, F.; Malesci, A.; Laghi, L.; Allavena, P. Tumour-associated macrophages as treatment targets in oncology. *Nat. Rev. Clin. Oncol.* **2017**, *14*, 399–416. [CrossRef]
9. Cortese, N.; Donadon, M.; Rigamonti, A.; Marchesi, F. Macrophages at the crossroads of anticancer strategies. *Front. Biosci.* **2019**, *24*, 1271–1283. [CrossRef]
10. Cao, H.; Xu, E.; Liu, H.; Wan, L.; Lai, M. Epithelial–mesenchymal transition in colorectal cancer metastasis: A system review. *Pathol. Res. Pract.* **2015**, *211*, 557–569. [CrossRef]
11. Zheng, X.; Carstens, J.L.; Kim, J.; Scheible, M.; Kaye, J.; Sugimoto, H.; Wu, C.-C.; LeBleu, V.S.; Kalluri, R. Epithelial-to-mesenchymal transition is dispensable for metastasis but induces chemoresistance in pancreatic cancer. *Nature* **2015**, *527*, 525–530. [CrossRef] [PubMed]
12. Fischer, K.R.; Durrans, A.; Lee, S.; Sheng, J.; Li, F.; Wong, S.T.C.; Choi, H.; El Rayes, T.; Ryu, S.; Troeger, J.; et al. Epithelial-to-mesenchymal transition is not required for lung metastasis but contributes to chemoresistance. *Nature* **2015**, *527*, 472–476. [CrossRef] [PubMed]
13. Hugo, W.; Zaretsky, J.M.; Sun, L.; Song, C.; Moreno, B.H.; Hu-Lieskovan, S.; Berent-Maoz, B.; Pang, J.; Chmielowski, B.; Cherry, G.; et al. Genomic and Transcriptomic Features of Response to Anti-PD-1 Therapy in Metastatic Melanoma. *Cell* **2016**, *165*, 35–44. [CrossRef]
14. Singh, M.; Yelle, N.; Venugopal, C.; Singh, S.K. EMT: Mechanisms and therapeutic implications. *Pharmacol. Ther.* **2018**, *182*, 80–94. [CrossRef] [PubMed]
15. Suarez-Carmona, M.; Lesage, J.; Cataldo, D.; Gilles, C. EMT and inflammation: Inseparable actors of cancer progression. *Mol. Oncol.* **2017**, *11*, 805–823. [CrossRef]
16. Tsilimigras, D.I.; Brodt, P.; Clavien, P.-A.; Muschel, R.J.; D'Angelica, M.I.; Endo, I.; Parks, R.W.; Doyle, M.; de Santibañes, E.; Pawlik, T.M. Liver metastases. *Nat. Rev. Dis. Primer* **2021**, *7*, 27. [CrossRef]
17. Nguyen, D.X.; Bos, P.D.; Massagué, J. Metastasis: From dissemination to organ-specific colonization. *Nat. Rev. Cancer* **2009**, *9*, 274–284. [CrossRef]
18. Yang, M.; Zhang, C. The role of liver sinusoidal endothelial cells in cancer liver metastasis. *Am. J. Cancer Res.* **2021**, *11*, 1845–1860. [CrossRef]
19. Genna, A.; Vanwynsberghe, A.M.; Villard, A.V.; Pottier, C.; Ancel, J.; Polette, M.; Gilles, C. EMT-Associated Heterogeneity in Circulating Tumor Cells: Sticky Friends on the Road to Metastasis. *Cancers* **2020**, *12*, 1632. [CrossRef]
20. Terry, S.; Savagner, P.; Ortiz-Cuaran, S.; Mahjoubi, L.; Saintigny, P.; Thiery, J.-P.; Chouaib, S. New insights into the role of EMT in tumor immune escape. *Mol. Oncol.* **2017**, *11*, 824–846. [CrossRef]
21. Nieto, M.A.; Huang, R.Y.-J.; Jackson, R.A.; Thiery, J.P. EMT: 2016. *Cell* **2016**, *166*, 21–45. [CrossRef] [PubMed]
22. Polyak, K.; Weinberg, R.A. Transitions between epithelial and mesenchymal states: Acquisition of malignant and stem cell traits. *Nat. Rev. Cancer* **2009**, *9*, 265–273. [CrossRef] [PubMed]
23. Jie, X.-X.; Zhang, X.-Y.; Xu, C.-J. Epithelial-to-mesenchymal transition, circulating tumor cells and cancer metastasis: Mechanisms and clinical applications. *Oncotarget* **2017**, *8*, 81558–81571. [CrossRef] [PubMed]
24. Barillari, G. The Impact of Matrix Metalloproteinase-9 on the Sequential Steps of the Metastatic Process. *Int. J. Mol. Sci.* **2020**, *21*, 4526. [CrossRef]
25. Alba, J.; Barcia, R.; Gutiérrez-Berzal, J.; Ramos-Martínez, J.I. Could inhibition of metalloproteinases be used to block the process of metastasis? *Cell Biochem. Funct.* **2022**, *40*, 600–607. [CrossRef]

26. Ding, C.; Luo, J.; Li, L.; Li, S.; Yang, L.; Pan, H.; Liu, Q.; Qin, H.; Chen, C.; Feng, J. Gab2 facilitates epithelial-to-mesenchymal transition via the MEK/ERK/MMP signaling in colorectal cancer. *J. Exp. Clin. Cancer Res.* **2016**, *35*, 5. [CrossRef]
27. Buttacavoli, M.; Di Cara, G.; Roz, E.; Pucci-Minafra, I.; Feo, S.; Cancemi, P. Integrated Multi-Omics Investigations of Metalloproteinases in Colon Cancer: Focus on MMP2 and MMP9. *Int. J. Mol. Sci.* **2021**, *22*, 12389. [CrossRef]
28. Lamouille, S.; Xu, J.; Derynck, R. Molecular mechanisms of epithelial–mesenchymal transition. *Nat. Rev. Mol. Cell Biol.* **2014**, *15*, 178–196. [CrossRef]
29. Kozuka, M.; Battaglin, F.; Jayachandran, P.; Wang, J.; Arai, H.; Soni, S.; Zhang, W.; Hirai, M.; Matsusaka, S.; Lenz, H.-J. Clinical Significance of Circulating Tumor Cell Induced Epithelial-Mesenchymal Transition in Patients with Metastatic Colorectal Cancer by Single-Cell RNA-Sequencing. *Cancers* **2021**, *13*, 4862. [CrossRef]
30. Wal, G.; Gouw, A.; Kamps, J.; Moorlag, H.; Bulthuis, M.; Molema, G.; de Jong, K. Angiogenesis in Synchronous and Metachronous Colorectal Liver Metastases the Liver as a Permissive Soil. *Ann. Surg.* **2011**, *255*, 86–94. [CrossRef]
31. Kuczynski, E.A.; Vermeulen, P.B.; Pezzella, F.; Kerbel, R.S.; Reynolds, A.R. Vessel co-option in cancer. *Nat. Rev. Clin. Oncol.* **2019**, *16*, 469–493. [CrossRef] [PubMed]
32. Rada, M.; Lazaris, A.; Kapelanski-Lamoureux, A.; Mayer, T.Z.; Metrakos, P. Tumor microenvironment conditions that favor vessel co-option in colorectal cancer liver metastases: A theoretical model. *Semin. Cancer Biol.* **2021**, *71*, 52–64. [CrossRef] [PubMed]
33. Rada, M.; Kapelanski-Lamoureux, A.; Petrillo, S.; Tabariès, S.; Siegel, P.; Reynolds, A.R.; Lazaris, A.; Metrakos, P. Runt related transcription factor-1 plays a central role in vessel co-option of colorectal cancer liver metastases. *Commun. Biol.* **2021**, *4*, 950. [CrossRef] [PubMed]
34. Rada, M.; Tsamchoe, M.; Kapelanski-Lamoureux, A.; Hassan, N.; Bloom, J.; Petrillo, S.; Kim, D.H.; Lazaris, A.; Metrakos, P. Cancer Cells Promote Phenotypic Alterations in Hepatocytes at the Edge of Cancer Cell Nests to Facilitate Vessel Co-Option Establishment in Colorectal Cancer Liver Metastases. *Cancers* **2022**, *14*, 1318. [CrossRef] [PubMed]
35. Reymond, N.; d'Água, B.B.; Ridley, A.J. Crossing the endothelial barrier during metastasis. *Nat. Rev. Cancer* **2013**, *13*, 858–870. [CrossRef]
36. Paschos, K.A. Natural history of hepatic metastases from colorectal cancer—Pathobiological pathways with clinical significance. *World J. Gastroenterol.* **2014**, *20*, 3719–3737. [CrossRef]
37. Rahimi, N. Defenders and Challengers of Endothelial Barrier Function. *Front. Immunol.* **2017**, *8*, 1847. [CrossRef]
38. Giannotta, M.; Trani, M.; Dejana, E. VE-Cadherin and Endothelial Adherens Junctions: Active Guardians of Vascular Integrity. *Dev. Cell* **2013**, *26*, 441–454. [CrossRef]
39. Dou, R.; Liu, K.; Yang, C.; Zheng, J.; Shi, D.; Lin, X.; Wei, C.; Zhang, C.; Fang, Y.; Huang, S.; et al. EMT-cancer cells-derived exosomal miR-27b-3p promotes circulating tumour cells-mediated metastasis by modulating vascular permeability in colorectal cancer. *Clin. Transl. Med.* **2021**, *11*, e595. [CrossRef]
40. Heeke, S.; Mograbi, B.; Alix-Panabières, C.; Hofman, P. Never Travel Alone: The Crosstalk of Circulating Tumor Cells and the Blood Microenvironment. *Cells* **2019**, *8*, 714. [CrossRef]
41. Chambers, A.F.; Groom, A.C.; MacDonald, I.C. Dissemination and growth of cancer cells in metastatic sites. *Nat. Rev. Cancer* **2002**, *2*, 563–572. [CrossRef] [PubMed]
42. Aceto, N.; Toner, M.; Maheswaran, S.; Haber, D.A. En Route to Metastasis: Circulating Tumor Cell Clusters and Epithelial-to-Mesenchymal Transition. *Trends Cancer* **2015**, *1*, 44–52. [CrossRef] [PubMed]
43. Haier, J.; Nicolson, G.L. Tumor cell adhesion under hydrodynamic conditions of fluid flow. *APMIS* **2001**, *109*, 241–262. [CrossRef] [PubMed]
44. Hamidi, H.; Ivaska, J. Every step of the way: Integrins in cancer progression and metastasis. *Nat. Rev. Cancer* **2018**, *18*, 533–548. [CrossRef]
45. Bakir, B.; Chiarella, A.M.; Pitarresi, J.R.; Rustgi, A.K. EMT, MET, Plasticity, and Tumor Metastasis. *Trends Cell Biol.* **2020**, *30*, 764–776. [CrossRef]
46. Haerinck, J.; Berx, G. Partial EMT takes the lead in cancer metastasis. *Dev. Cell* **2021**, *56*, 3174–3176. [CrossRef]
47. Pavlič, A.; Urh, K.; Štajer, K.; Boštjančič, E.; Zidar, N. Epithelial-Mesenchymal Transition in Colorectal Carcinoma: Comparison Between Primary Tumor, Lymph Node and Liver Metastases. *Front. Oncol.* **2021**, *11*, 662806. [CrossRef]
48. Cao, Z.-Q.; Wang, Z.; Leng, P. Aberrant N-cadherin expression in cancer. *Biomed. Pharmacother.* **2019**, *118*, 109320. [CrossRef]
49. Paulin, D.; Lilienbaum, A.; Kardjian, S.; Agbulut, O.; Li, Z. Vimentin: Regulation and pathogenesis. *Biochimie* **2022**, *197*, 96–112. [CrossRef]
50. Wang, Q.; Zhu, G.; Lin, C.; Lin, P.; Chen, H.; He, R.; Huang, Y.; Yang, S.; Ye, J. Vimentin affects colorectal cancer proliferation, invasion, and migration via regulated by activator protein 1. *J. Cell Physiol.* **2021**, *236*, 7591–7604. [CrossRef]
51. Niknami, Z.; Muhammadnejad, A.; Ebrahimi, A.; Harsani, Z.; Shirkoohi, R. Significance of E-cadherin and Vimentin as epithelial-mesenchymal transition markers in colorectal carcinoma prognosis. *EXCLI J.* **2020**, *19*, 917–926. [CrossRef] [PubMed]
52. Patten, J.; Wang, K. Fibronectin in development and wound healing. *Adv. Drug Deliv. Rev.* **2021**, *170*, 353–368. [CrossRef] [PubMed]
53. Cai, X.; Liu, C.; Zhang, T.; Zhu, Y.; Dong, X.; Xue, P. Down-regulation of FN1 inhibits colorectal carcinogenesis by suppressing proliferation, migration, and invasion. *J. Cell Biochem.* **2018**, *119*, 4717–4728. [CrossRef] [PubMed]
54. Ye, Y.; Zhang, R.; Feng, H. Fibronectin promotes tumor cells growth and drugs resistance through a CDC42-YAP-dependent signaling pathway in colorectal cancer. *Cell Biol. Int.* **2020**, *44*, 1840–1849. [CrossRef] [PubMed]

55. Daulagala, A.C.; Bridges, M.C.; Kourtidis, A. E-cadherin Beyond Structure: A Signaling Hub in Colon Homeostasis and Disease. *Int. J. Mol. Sci.* **2019**, *20*, 2756. [CrossRef]
56. Yun, J.-A.; Kim, S.-H.; Hong, H.K.; Yun, S.H.; Kim, H.C.; Chun, H.-K.; Cho, Y.B.; Lee, W.Y. Loss of E-Cadherin Expression Is Associated with a Poor Prognosis in Stage III Colorectal Cancer. *Oncology* **2014**, *86*, 318–328. [CrossRef]
57. Cho, H.-J.; Oh, N.; Park, J.-H.; Kim, K.-S.; Kim, H.-K.; Lee, E.; Hwang, S.; Kim, S.-J.; Park, K.-S. ZEB1 Collaborates with ELK3 to Repress E-Cadherin Expression in Triple-Negative Breast Cancer Cells. *Mol. Cancer Res.* **2019**, *17*, 2257–2266. [CrossRef]
58. Cui, H.; Hu, Y.; Guo, D.; Zhang, A.; Gu, Y.; Zhang, S.; Zhao, C.; Gong, P.; Shen, X.; Li, Y.; et al. DNA methyltransferase 3A isoform b contributes to repressing E-cadherin through cooperation of DNA methylation and H3K27/H3K9 methylation in EMT-related metastasis of gastric cancer. *Oncogene* **2018**, *37*, 4358–4371. [CrossRef]
59. Venhuizen, J.-H.; Jacobs, F.J.C.; Span, P.N.; Zegers, M.M. P120 and E-cadherin: Double-edged swords in tumor metastasis. *Semin. Cancer Biol.* **2020**, *60*, 107–120. [CrossRef]
60. Huang, Y.; Hong, W.; Wei, X. The molecular mechanisms and therapeutic strategies of EMT in tumor progression and metastasis. *J. Hematol Oncol* **2022**, *15*, 129. [CrossRef]
61. Villarejo, A.; Cortés-Cabrera, Á.; Molina-Ortíz, P.; Portillo, F.; Cano, A. Differential Role of Snail1 and Snail2 Zinc Fingers in E-cadherin Repression and Epithelial to Mesenchymal Transition. *J. Biol. Chem.* **2014**, *289*, 930–941. [CrossRef] [PubMed]
62. Brzozowa, M.; Michalski, M.; Wyrobiec, G.; Piecuch, A.; Dittfeld, A.; Harabin-Słowińska, M.; Boroń, D.; Wojnicz, R. The role of Snail1 transcription factor in colorectal cancer progression and metastasis. *Contemp. Oncol.* **2015**, *19*, 265–270. [CrossRef] [PubMed]
63. Francí, C.; Gallén, M.; Alameda, F.; Baró, T.; Iglesias, M.; Virtanen, I.; García de Herreros, A. Snail1 Protein in the Stroma as a New Putative Prognosis Marker for Colon Tumours. *PLoS ONE* **2009**, *4*, e5595. [CrossRef] [PubMed]
64. Francı, C.; Takkunen, M.; Dave, N.; Alameda, F. Expression of Snail protein in tumor–stroma interface. *Oncogene* **2006**, *25*, 5134–5144. [CrossRef] [PubMed]
65. Larsen, J.E.; Nathan, V.; Osborne, J.K.; Farrow, R.K.; Deb, D.; Sullivan, J.P.; Dospoy, P.D.; Augustyn, A.; Hight, S.K.; Sato, M.; et al. ZEB1 drives epithelial-to-mesenchymal transition in lung cancer. *J. Clin. Investig.* **2016**, *126*, 3219–3235. [CrossRef]
66. Zhang, N.; Ng, A.S.; Cai, S.; Li, Q.; Yang, L.; Kerr, D. Novel therapeutic strategies: Targeting epithelial–mesenchymal transition in colorectal cancer. *Lancet Oncol.* **2021**, *22*, e358–e368. [CrossRef] [PubMed]
67. Moon, Y.W.; Rao, G.; Kim, J.J.; Shim, H.-S.; Park, K.-S.; An, S.S.; Kim, B.; Steeg, P.S.; Sarfaraz, S.; Changwoo Lee, L.; et al. LAMC2 enhances the metastatic potential of lung adenocarcinoma. *Cell Death Differ.* **2015**, *22*, 1341–1352. [CrossRef]
68. Colangelo, T.; Carbone, A.; Mazzarelli, F.; Cuttano, R.; Dama, E.; Nittoli, T.; Albanesi, J.; Barisciano, G.; Forte, N.; Palumbo, O.; et al. Loss of circadian gene Timeless induces EMT and tumor progression in colorectal cancer via Zeb1-dependent mechanism. *Cell Death Differ.* **2022**, *29*, 1552–1568. [CrossRef]
69. Lindner, P.; Paul, S.; Eckstein, M.; Hampel, C.; Muenzner, J.K.; Erlenbach-Wuensch, K.; Ahmed, H.P.; Mahadevan, V.; Brabletz, T.; Hartmann, A.; et al. EMT transcription factor ZEB1 alters the epigenetic landscape of colorectal cancer cells. *Cell Death Dis.* **2020**, *11*, 147. [CrossRef]
70. Franco, H.L.; Casasnovas, J.; Rodriguez-Medina, J.R.; Cadilla, C.L. Redundant or separate entities?–roles of Twist1 and Twist2 as molecular switches during gene transcription. *Nucleic Acids Res.* **2011**, *39*, 1177–1186. [CrossRef]
71. Abdelmaksoud-Dammak, R.; Chamtouri, N.; Triki, M.; Saadallah-Kallel, A.; Ayadi, W.; Charfi, S.; Khabir, A.; Ayadi, L.; Sallemi-Boudawara, T.; Mokdad-Gargouri, R. Overexpression of miR-10b in colorectal cancer patients: Correlation with *TWIST-1* and E-cadherin expression. *Tumor Biol.* **2017**, *39*, 1010428317695916. [CrossRef] [PubMed]
72. Deng, J.-J.; Zhang, W.; Xu, X.-M.; Zhang, F.; Tao, W.-P.; Ye, J.-J.; Ge, W. Twist mediates an aggressive phenotype in human colorectal cancer cells. *Int. J. Oncol.* **2016**, *48*, 1117–1124. [CrossRef] [PubMed]
73. Kim, T.W.; Lee, Y.S.; Yun, N.H.; Shin, C.H.; Hong, H.K.; Kim, H.H.; Cho, Y.B. MicroRNA-17-5p regulates EMT by targeting vimentin in colorectal cancer. *Br. J. Cancer* **2020**, *123*, 1123–1130. [CrossRef] [PubMed]
74. Zhang, Y.; Zhang, S.; Yin, J.; Xu, R. MiR-566 mediates cell migration and invasion in colon cancer cells by direct targeting of PSKH1. *Cancer Cell Int.* **2019**, *19*, 333. [CrossRef] [PubMed]
75. Iwasaki, H.; Shimura, T.; Kitagawa, M.; Yamada, T.; Nishigaki, R.; Fukusada, S.; Okuda, Y.; Katano, T.; Horike, S.; Kataoka, H. A Novel Urinary miRNA Biomarker for Early Detection of Colorectal Cancer. *Cancers* **2022**, *14*, 461. [CrossRef]
76. Zhao, H.; Wei, J.; Sun, J. Roles of TGF-β signaling pathway in tumor microenvirionment and cancer therapy. *Int. Immunopharmacol.* **2020**, *89*, 107101. [CrossRef]
77. Siegel, P.M.; Massagué, J. Cytostatic and apoptotic actions of TGF-β in homeostasis and cancer. *Nat. Rev. Cancer* **2003**, *3*, 807–820. [CrossRef]
78. Bach, S.P.; Renehan, A.G.; Potten, C.S. Stem cells: The intestinal stem cell as a paradigm. *Carcinogenesis* **2000**, *21*, 469–476. [CrossRef]
79. Pellatt, A.J.; Mullany, L.E.; Herrick, J.S.; Sakoda, L.C.; Wolff, R.K.; Samowitz, W.S.; Slattery, M.L. The TGFβ-signaling pathway and colorectal cancer: Associations between dysregulated genes and miRNAs. *J. Transl. Med.* **2018**, *16*, 191. [CrossRef]
80. Itatani, Y.; Kawada, K.; Sakai, Y. Transforming Growth Factor-β Signaling Pathway in Colorectal Cancer and Its Tumor Microenvironment. *Int. J. Mol. Sci.* **2019**, *20*, 5822. [CrossRef]
81. Peinado, H.; Olmeda, D.; Cano, A. Snail, Zeb and bHLH factors in tumour progression: An alliance against the epithelial phenotype? *Nat. Rev. Cancer* **2007**, *7*, 415–428. [CrossRef] [PubMed]

82. Moustakas, A.; Heldin, C.-H. Induction of epithelial–mesenchymal transition by transforming growth factor β. *Semin. Cancer Biol.* **2012**, *22*, 446–454. [CrossRef] [PubMed]
83. Hao, Y.; Baker, D.; ten Dijke, P. TGF-β-Mediated Epithelial-Mesenchymal Transition and Cancer Metastasis. *Int. J. Mol. Sci.* **2019**, *20*, 2767. [CrossRef]
84. Wendt, M.K.; Allington, T.M.; Schiemann, W.P. Mechanisms of the epithelial–mesenchymal transition by TGF-β. *Future Oncol.* **2009**, *5*, 1145–1168. [CrossRef]
85. Hua, W.; Kostidis, S.; Mayboroda, O.; Giera, M.; Hornsveld, M.; ten Dijke, P. Metabolic Reprogramming of Mammary Epithelial Cells during TGF-β-Induced Epithelial-to-Mesenchymal Transition. *Metabolites* **2021**, *11*, 626. [CrossRef] [PubMed]
86. Zhang, J.; Thorikay, M.; van der Zon, G.; van Dinther, M.; ten Dijke, P. Studying TGF-β; Signaling and TGF-β-induced Epithelial-to-mesenchymal Transition in Breast Cancer and Normal Cells. *J. Vis. Exp.* **2020**, *164*, e61830. [CrossRef]
87. Bierie, B.; Moses, H. TGF-β and cancer. *Cytokine Growth Factor Rev.* **2006**, *17*, 29–40. [CrossRef]
88. Chruścik, A.; Gopalan, V.; Lam, A.K. The clinical and biological roles of transforming growth factor beta in colon cancer stem cells: A systematic review. *Eur. J. Cell Biol.* **2018**, *97*, 15–22. [CrossRef]
89. Mooi, J.K.; Wirapati, P.; Asher, R.; Lee, C.K.; Savas, P.; Price, T.J.; Townsend, A.; Hardingham, J.; Buchanan, D.; Williams, D.; et al. The prognostic impact of consensus molecular subtypes (CMS) and its predictive effects for bevacizumab benefit in metastatic colorectal cancer: Molecular analysis of the AGITG MAX clinical trial. *Ann. Oncol.* **2018**, *29*, 2240–2246. [CrossRef]
90. Pohl, M.; Radacz, Y.; Pawlik, N.; Schoeneck, A.; Baldus, S.E.; Munding, J.; Schmiegel, W.; Schwarte-Waldhoff, I.; Reinacher-Schick, A. SMAD4 Mediates Mesenchymal–Epithelial Reversion in SW480 Colon Carcinoma Cells. *Anticancer Res.* **2010**, *18*, 601–608. [CrossRef]
91. Voorneveld, P.W.; Kodach, L.L.; Jacobs, R.J.; Liv, N.; Zonnevylle, A.C.; Hoogenboom, J.P.; Biemond, I.; Verspaget, H.W.; Hommes, D.W.; de Rooij, K.; et al. Loss of SMAD4 Alters BMP Signaling to Promote Colorectal Cancer Cell Metastasis via Activation of Rho and ROCK. *Gastroenterology* **2014**, *147*, 196–208.e13. [CrossRef] [PubMed]
92. Frey, P.; Devisme, A.; Rose, K.; Schrempp, M.; Freihen, V.; Andrieux, G.; Boerries, M.; Hecht, A. SMAD4 mutations do not preclude epithelial–mesenchymal transition in colorectal cancer. *Oncogene* **2022**, *41*, 824–837. [CrossRef] [PubMed]
93. Wu, F.; Weigel, K.J.; Zhou, H.; Wang, X.-J. Paradoxical roles of TGF-β signaling in suppressing and promoting squamous cell carcinoma. *Acta Biochim. Biophys. Sin.* **2018**, *50*, 98–105. [CrossRef] [PubMed]
94. Seoane, J.; Gomis, R.R. TGF-β Family Signaling in Tumor Suppression and Cancer Progression. *Cold Spring Harb. Perspect. Biol.* **2017**, *9*, a022277. [CrossRef] [PubMed]
95. Lin, Y.-T.; Wu, K.-J. Epigenetic regulation of epithelial-mesenchymal transition: Focusing on hypoxia and TGF-β signaling. *J. Biomed. Sci.* **2020**, *27*, 39. [CrossRef] [PubMed]
96. Bagati, A.; Kumar, S.; Jiang, P.; Pyrdol, J.; Zou, A.E.; Godicelj, A.; Mathewson, N.D.; Cartwright, A.N.R.; Cejas, P.; Brown, M.; et al. Integrin αvβ6–TGFβ–SOX4 Pathway Drives Immune Evasion in Triple-Negative Breast Cancer. *Cancer Cell* **2021**, *39*, 54–67.e9. [CrossRef] [PubMed]
97. Taieb, J.; Svrcek, M.; Cohen, R.; Basile, D.; Tougeron, D.; Phelip, J.-M. Deficient mismatch repair/microsatellite unstable colorectal cancer: Diagnosis, prognosis and treatment. *Eur. J. Cancer* **2022**, *175*, 136–157. [CrossRef]
98. Pino, M.S.; Kikuchi, H.; Zeng, M.; Herraiz, M.; Sperduti, I.; Berger, D.; Park, D.; Iafrate, A.J.; Zukerberg, L.R.; Chung, D.C. Epithelial to Mesenchymal Transition Is Impaired in Colon Cancer Cells with Microsatellite Instability. *Gastroenterology* **2010**, *138*, 1406–1417. [CrossRef]
99. Mizuno, T.; Cloyd, J.M.; Vicente, D.; Omichi, K.; Chun, Y.S.; Kopetz, S.E.; Maru, D.; Conrad, C.; Tzeng, C.-W.D.; Wei, S.H.; et al. SMAD4 gene mutation predicts poor prognosis in patients undergoing resection for colorectal liver metastases. *Eur. J. Surg. Oncol.* **2018**, *44*, 684–692. [CrossRef]
100. Okita, A.; Takahashi, S.; Ouchi, K.; Inoue, M.; Watanabe, M.; Endo, M.; Honda, H.; Yamada, Y.; Ishioka, C. Consensus molecular subtypes classification of colorectal cancer as a predictive factor for chemotherapeutic efficacy against metastatic colorectal cancer. *Oncotarget* **2018**, *9*, 18698–18711. [CrossRef]
101. Jahchan, N.S.; Mujal, A.M.; Pollack, J.L.; Binnewies, M.; Sriram, V.; Reyno, L.; Krummel, M.F. Tuning the Tumor Myeloid Microenvironment to Fight Cancer. *Front. Immunol.* **2019**, *10*, 1611. [CrossRef] [PubMed]
102. Locati, M.; Curtale, G.; Mantovani, A. Diversity, Mechanisms, and Significance of Macrophage Plasticity. *Annu. Rev. Pathol. Mech. Dis.* **2020**, *15*, 123–147. [CrossRef] [PubMed]
103. Coussens, L.M.; Zitvogel, L.; Palucka, A.K. Neutralizing Tumor-Promoting Chronic Inflammation: A Magic Bullet? *Science* **2013**, *339*, 286–291. [CrossRef] [PubMed]
104. Morrison, C. Immuno-oncologists eye up macrophage targets. *Nat. Rev. Drug Discov.* **2016**, *15*, 373–374. [CrossRef]
105. Qian, B.-Z.; Pollard, J.W. Macrophage Diversity Enhances Tumor Progression and Metastasis. *Cell* **2010**, *141*, 39–51. [CrossRef]
106. Tamura, R.; Tanaka, T.; Yamamoto, Y.; Akasaki, Y.; Sasaki, H. Dual role of macrophage in tumor immunity. *Immunotherapy* **2018**, *10*, 899–909. [CrossRef]
107. Laoui, D.; Van Overmeire, E.; Di Conza, G.; Aldeni, C.; Keirsse, J.; Morias, Y.; Movahedi, K.; Houbracken, I.; Schouppe, E.; Elkrim, Y.; et al. Tumor Hypoxia Does Not Drive Differentiation of Tumor-Associated Macrophages but Rather Fine-Tunes the M2-like Macrophage Population. *Cancer Res.* **2014**, *74*, 24–30. [CrossRef]
108. Galon, J.; Angell, H.K.; Bedognetti, D.; Marincola, F.M. The Continuum of Cancer Immunosurveillance: Prognostic, Predictive, and Mechanistic Signatures. *Immunity* **2013**, *39*, 11–26. [CrossRef]

109. Donadon, M.; Torzilli, G.; Cortese, N.; Soldani, C.; Di Tommaso, L.; Franceschini, B.; Carriero, R.; Barbagallo, M.; Rigamonti, A.; Anselmo, A.; et al. Macrophage morphology correlates with single-cell diversity and prognosis in colorectal liver metastasis. *J. Exp. Med.* **2020**, *217*, e20191847. [CrossRef]
110. Malesci, A.; Bianchi, P.; Celesti, G.; Basso, G.; Marchesi, F.; Grizzi, F.; Di Caro, G.; Cavalleri, T.; Rimassa, L.; Palmqvist, R.; et al. Tumor-associated macrophages and response to 5-fluorouracil adjuvant therapy in stage III colorectal cancer. *OncoImmunology* **2017**, *6*, e1342918. [CrossRef]
111. Mantovani, A.; Allavena, P.; Marchesi, F.; Garlanda, C. Macrophages as tools and targets in cancer therapy. *Nat. Rev. Drug Discov.* **2022**, *21*, 799–820. [CrossRef] [PubMed]
112. Boutilier, A.J.; Elsawa, S.F. Macrophage Polarization States in the Tumor Microenvironment. *Int. J. Mol. Sci.* **2021**, *22*, 6995. [CrossRef] [PubMed]
113. Sica, A.; Mantovani, A. Macrophage plasticity and polarization: In vivo veritas. *J. Clin. Investig.* **2012**, *122*, 787–795. [CrossRef] [PubMed]
114. Mantovani, A.; Sozzani, S.; Locati, M.; Allavena, P.; Sica, A. Macrophage polarization: Tumor-associated macrophages as a paradigm for polarized M2 mononuclear phagocytes. *Trends Immunol.* **2002**, *23*, 549–555. [CrossRef] [PubMed]
115. Atri, C.; Guerfali, F.; Laouini, D. Role of Human Macrophage Polarization in Inflammation during Infectious Diseases. *Int. J. Mol. Sci.* **2018**, *19*, 1801. [CrossRef]
116. Väyrynen, J.P.; Haruki, K.; Lau, M.C.; Väyrynen, S.A.; Zhong, R.; Dias Costa, A.; Borowsky, J.; Zhao, M.; Fujiyoshi, K.; Arima, K.; et al. The Prognostic Role of Macrophage Polarization in the Colorectal Cancer Microenvironment. *Cancer Immunol. Res.* **2021**, *9*, 8–19. [CrossRef]
117. Macciò, A.; Gramignano, G.; Cherchi, M.C.; Tanca, L.; Melis, L.; Madeddu, C. Role of M1-polarized tumor-associated macrophages in the prognosis of advanced ovarian cancer patients. *Sci. Rep.* **2020**, *10*, 6096. [CrossRef]
118. Yunna, C.; Mengru, H.; Lei, W.; Weidong, C. Macrophage M1/M2 polarization. *Eur. J. Pharmacol.* **2020**, *877*, 173090. [CrossRef]
119. Cavalleri, T.; Greco, L.; Rubbino, F.; Hamada, T.; Quaranta, M.; Grizzi, F.; Sauta, E.; Craviotto, V.; Bossi, P.; Vetrano, S.; et al. Tumor-associated macrophages and risk of recurrence in stage III colorectal cancer. *J. Pathol. Clin. Res.* **2022**, *8*, 307–312. [CrossRef]
120. Herrera, M.; Herrera, A.; Domínguez, G.; Silva, J.; García, V.; García, J.M.; Gómez, I.; Soldevilla, B.; Muñoz, C.; Provencio, M. Cancer-associated fibroblast and M 2 macrophage markers together predict outcome in colorectal cancer patients. *Cancer Sci.* **2013**, *104*, 437–444. [CrossRef]
121. Shibutani, M.; Nakao, S.; Maeda, K.; Nagahara, H.; Kashiwagi, S.; Hirakawa, K.; Ohira, M. The Impact of Tumor-associated Macrophages on Chemoresistance via Angiogenesis in Colorectal Cancer. *Anticancer Res.* **2021**, *41*, 4447–4453. [CrossRef] [PubMed]
122. DeNardo, D.G.; Ruffell, B. Macrophages as regulators of tumour immunity and immunotherapy. *Nat. Rev. Immunol.* **2019**, *19*, 369–382. [CrossRef] [PubMed]
123. Sugimura-Nagata, A.; Koshino, A.; Inoue, S.; Matsuo-Nagano, A.; Komura, M.; Riku, M.; Ito, H.; Inoko, A.; Murakami, H.; Ebi, M.; et al. Expression and Prognostic Significance of CD47–SIRPA Macrophage Checkpoint Molecules in Colorectal Cancer. *Int. J. Mol. Sci.* **2021**, *22*, 2690. [CrossRef] [PubMed]
124. Cortese, N.; Soldani, C.; Franceschini, B.; Barbagallo, M.; Marchesi, F.; Torzilli, G.; Donadon, M. Macrophages in Colorectal Cancer Liver Metastases. *Cancers* **2019**, *11*, 633. [CrossRef]
125. Guilliams, M.; Scott, C.L. Liver macrophages in health and disease. *Immunity* **2022**, *55*, 1515–1529. [CrossRef]
126. Krenkel, O.; Tacke, F. Liver macrophages in tissue homeostasis and disease. *Nat. Rev. Immunol.* **2017**, *17*, 306–321. [CrossRef]
127. Wei, C.; Yang, C.; Wang, S.; Shi, D.; Zhang, C.; Lin, X.; Liu, Q.; Dou, R.; Xiong, B. Crosstalk between cancer cells and tumor associated macrophages is required for mesenchymal circulating tumor cell-mediated colorectal cancer metastasis. *Mol. Cancer* **2019**, *18*, 64. [CrossRef]
128. Heijstek, M.W.; Kranenburg, O.; Borel Rinkes, I.H.M. Mouse Models of Colorectal Cancer and Liver Metastases. *Dig. Surg.* **2005**, *22*, 16–25. [CrossRef]
129. Yang, C.; Dou, R.; Wei, C.; Liu, K.; Shi, D.; Zhang, C.; Liu, Q.; Wang, S.; Xiong, B. Tumor-derived exosomal microRNA-106b-5p activates EMT-cancer cell and M2-subtype TAM interaction to facilitate CRC metastasis. *Mol. Ther.* **2021**, *29*, 2088–2107. [CrossRef]
130. Yin, Y.; Yao, S.; Hu, Y.; Feng, Y.; Li, M.; Bian, Z.; Zhang, J.; Qin, Y.; Qi, X.; Zhou, L.; et al. The Immune-microenvironment Confers Chemoresistance of Colorectal Cancer through Macrophage-Derived IL6. *Clin. Cancer Res.* **2017**, *23*, 7375–7387. [CrossRef]
131. Li, Y.; Wang, L.; Pappan, L.; Galliher-Beckley, A.; Shi, J. IL-1β promotes stemness and invasiveness of colon cancer cells through Zeb1 activation. *Mol. Cancer* **2012**, *11*, 87. [CrossRef] [PubMed]
132. Wang, H.; Wang, H.-S.; Zhou, B.-H.; Li, C.-L.; Zhang, F.; Wang, X.-F.; Zhang, G.; Bu, X.-Z.; Cai, S.-H.; Du, J. Epithelial–Mesenchymal Transition (EMT) Induced by TNF-α Requires AKT/GSK-3β-Mediated Stabilization of Snail in Colorectal Cancer. *PLoS ONE* **2013**, *8*, e56664. [CrossRef] [PubMed]
133. Lin, X.; Wang, S.; Sun, M.; Zhang, C.; Wei, C.; Yang, C.; Dou, R.; Liu, Q.; Xiong, B. miR-195-5p/NOTCH2-mediated EMT modulates IL-4 secretion in colorectal cancer to affect M2-like TAM polarization. *J. Hematol. Oncol.* **2019**, *12*, 20. [CrossRef] [PubMed]
134. Zhang, T.; Liu, L.; Lai, W.; Zeng, Y.; Xu, H.; Lan, Q.; Su, P.; Chu, Z. Interaction with tumor-associated macrophages promotes PRL-3-induced invasion of colorectal cancer cells via MAPK pathway-induced EMT and NF-κB signaling-induced angiogenesis. *Oncol. Rep.* **2019**, *41*, 2790–2802. [CrossRef] [PubMed]

135. Bao, Z.; Zeng, W.; Zhang, D.; Wang, L.; Deng, X.; Lai, J.; Li, J.; Gong, J.; Xiang, G. SNAIL Induces EMT and Lung Metastasis of Tumours Secreting CXCL2 to Promote the Invasion of M2-Type Immunosuppressed Macrophages in Colorectal Cancer. *Int. J. Biol. Sci.* **2022**, *18*, 2867–2881. [CrossRef]
136. Li, S.; Xu, F.; Zhang, J.; Wang, L.; Zheng, Y.; Wu, X.; Wang, J.; Huang, Q.; Lai, M. Tumor-associated macrophages remodeling EMT and predicting survival in colorectal carcinoma. *OncoImmunology* **2018**, *7*, e1380765. [CrossRef]
137. Cai, J.; Xia, L.; Li, J.; Ni, S.; Song, H.; Wu, X. Tumor-Associated Macrophages Derived TGF-β–Induced Epithelial to Mesenchymal Transition in Colorectal Cancer Cells through Smad2,3-4/Snail Signaling Pathway. *Cancer Res. Treat.* **2019**, *51*, 252–266. [CrossRef]
138. Zhang, X.-L.; Hu, L.-P.; Yang, Q.; Qin, W.-T.; Wang, X.; Xu, C.-J.; Tian, G.-A.; Yang, X.-M.; Yao, L.-L.; Zhu, L.; et al. CTHRC1 promotes liver metastasis by reshaping infiltrated macrophages through physical interactions with TGF-β receptors in colorectal cancer. *Oncogene* **2021**, *40*, 3959–3973. [CrossRef]
139. Ni, S.; Ren, F.; Xu, M.; Tan, C.; Weng, W.; Huang, Z.; Sheng, W.; Huang, D. CTHRC1 overexpression predicts poor survival and enhances epithelial-mesenchymal transition in colorectal cancer. *Cancer Med.* **2018**, *7*, 5643–5654. [CrossRef]
140. Zhang, S.; Li, X.; Zhu, L.; Ming, S.; Wang, H.; Xie, J.; Ren, L.; Huang, J.; Liang, D.; Xiong, L.; et al. CD163+ macrophages suppress T cell response by producing TGF-β in pediatric colorectal polyps. *Int. Immunopharmacol.* **2021**, *96*, 107644. [CrossRef]
141. Ma, X.; Gao, Y.; Chen, Y.; Liu, J.; Yang, C.; Bao, C.; Wang, Y.; Feng, Y.; Song, X.; Qiao, S. M2-Type Macrophages Induce Tregs Generation by Activating the TGF-β/Smad Signalling Pathway to Promote Colorectal Cancer Development. *OncoTargets Ther.* **2021**, *14*, 5391–5402. [CrossRef] [PubMed]
142. Zhang, D.; Qiu, X.; Li, J.; Zheng, S.; Li, L.; Zhao, H. TGF-β secreted by tumor-associated macrophages promotes proliferation and invasion of colorectal cancer via miR-34a-VEGF axis. *Cell Cycle* **2018**, *17*, 2766–2778. [CrossRef] [PubMed]
143. Gulubova, M.; Ananiev, J.; Yovchev, Y.; Julianov, A.; Karashmalakov, A.; Vlaykova, T. The density of macrophages in colorectal cancer is inversely correlated to TGF-β1 expression and patients' survival. *J. Mol. Histol.* **2013**, *44*, 679–692. [CrossRef] [PubMed]
144. Teuwen, L.-A.; De Rooij, L.P.M.H.; Cuypers, A.; Rohlenova, K.; Dumas, S.J.; García-Caballero, M.; Meta, E.; Amersfoort, J.; Taverna, F.; Becker, L.M.; et al. Tumor vessel co-option probed by single-cell analysis. *Cell Rep.* **2021**, *35*, 109253. [CrossRef]
145. Peng, D.; Fu, M.; Wang, M.; Wei, Y.; Wei, X. Targeting TGF-β signal transduction for fibrosis and cancer therapy. *Mol. Cancer* **2022**, *21*, 104. [CrossRef]
146. Dai, G.; Sun, B.; Gong, T.; Pan, Z.; Meng, Q.; Ju, W. Ginsenoside Rb2 inhibits epithelial-mesenchymal transition of colorectal cancer cells by suppressing TGF-β/Smad signaling. *Phytomedicine* **2019**, *56*, 126–135. [CrossRef]
147. Villalba, M.; Evans, S.R.; Vidal-Vanaclocha, F.; Calvo, A. Role of TGF-β in metastatic colon cancer: It is finally time for targeted therapy. *Cell Tissue Res.* **2017**, *370*, 29–39. [CrossRef]
148. House, M.G.; Ito, H.; Gönen, M.; Fong, Y.; Allen, P.J.; DeMatteo, R.P.; Brennan, M.F.; Blumgart, L.H.; Jarnagin, W.R.; D'Angelica, M.I. Survival after Hepatic Resection for Metastatic Colorectal Cancer: Trends in Outcomes for 1600 Patients during Two Decades at a Single Institution. *J. Am. Coll. Surg.* **2010**, *210*, 744–752. [CrossRef]
149. Hughes, K.; Simon, R.; Songhorabodi, S.; Adson, M.; Ilstrup, D. Resection of the liver for colorectal carcinoma metastases: A multi-institutional study of patterns of recurrence. *Surgery* **1986**, *31*, 278–284. [CrossRef]
150. Hołówko, W.; Grąt, M.; Hinderer, B.; Orlińska, I.; Krawczyk, M. Prediction of Survival in Patients with Unresectable Colorectal Liver Metastases. *Pol. J. Surg.* **2014**, *86*, 319–324. [CrossRef]
151. Wagner, J.; Adson, M.; Van Heerden, J.; Adson, M.; Ilstrup, D. The natural history of hepatic metastases from colorectal cancer. A comparison with resective treatment. *Ann. Surg.* **1984**, *199*, 502–508. [CrossRef] [PubMed]
152. Kopetz, S.; Chang, G.J.; Overman, M.J.; Eng, C.; Sargent, D.J.; Larson, D.W.; Grothey, A.; Vauthey, J.-N.; Nagorney, D.M.; McWilliams, R.R. Improved Survival in Metastatic Colorectal Cancer Is Associated with Adoption of Hepatic Resection and Improved Chemotherapy. *J. Clin. Oncol.* **2009**, *27*, 3677–3683. [CrossRef] [PubMed]
153. Donadon, M.; Cortese, N.; Marchesi, F.; Cimino, M.; Mantovani, A.; Torzilli, G. Hepatobiliary surgeons meet immunologists: The case of colorectal liver metastases patients. *HepatoBiliary Surg. Nutr.* **2019**, *8*, 370–377. [CrossRef]
154. Li, Q.; Cheng, X.; Zhou, C.; Tang, Y.; Li, F.; Zhang, B.; Huang, T.; Wang, J.; Tu, S. Fruquintinib Enhances the Antitumor Immune Responses of Anti-Programmed Death Receptor-1 in Colorectal Cancer. *Front. Oncol.* **2022**, *12*, 841977. [CrossRef] [PubMed]
155. Sullivan, K.M.; Jiang, X.; Guha, P.; Lausted, C.; Carter, J.A.; Hsu, C.; Labadie, K.P.; Kohli, K.; Kenerson, H.L.; Daniel, S.K.; et al. Blockade of interleukin 10 potentiates antitumour immune function in human colorectal cancer liver metastases. *Gut* **2022**, 1–13. [CrossRef] [PubMed]
156. Li, X.; Chen, L.; Peng, X.; Zhan, X. Progress of tumor-associated macrophages in the epithelial-mesenchymal transition of tumor. *Front. Oncol.* **2022**, *12*, 911410. [CrossRef]
157. Zhang, J.; Zhou, X.; Hao, H. Macrophage phenotype-switching in cancer. *Eur. J. Pharmacol.* **2022**, *931*, 175229. [CrossRef]
158. Xu, W.; Cheng, Y.; Guo, Y.; Yao, W.; Qian, H. Targeting tumor associated macrophages in hepatocellular carcinoma. *Biochem. Pharmacol.* **2022**, *199*, 114990. [CrossRef]
159. Xu, Z.; Zhang, Y.; Dai, H.; Han, B. Epithelial–Mesenchymal Transition-Mediated Tumor Therapeutic Resistance. *Molecules* **2022**, *27*, 4750. [CrossRef]
160. Maslankova, J.; Vecurkovska, I.; Rabajdova, M.; Katuchova, J.; Kicka, M.; Gayova, M.; Katuch, V. Regulation of transforming growth factor-β signaling as a therapeutic approach to treating colorectal cancer. *World J. Gastroenterol.* **2022**, *28*, 4744–4761. [CrossRef]

161. Zhang, M.; Zhang, Y.Y.; Chen, Y.; Wang, J.; Wang, Q.; Lu, H. TGF-β Signaling and Resistance to Cancer Therapy. *Front. Cell Dev. Biol.* **2021**, *9*, 786728. [CrossRef] [PubMed]
162. Sui, H.; Zhao, J.; Zhou, L.; Wen, H.; Deng, W.; Li, C.; Ji, Q.; Liu, X.; Feng, Y.; Chai, N.; et al. Tanshinone IIA inhibits β-catenin/VEGF-mediated angiogenesis by targeting TGF-β1 in normoxic and HIF-1α in hypoxic microenvironments in human colorectal cancer. *Cancer Lett.* **2017**, *403*, 86–97. [CrossRef] [PubMed]
163. Yamazaki, T.; Gunderson, A.J.; Gilchrist, M.; Whiteford, M.; Kiely, M.X.; Hayman, A.; O'Brien, D.; Ahmad, R.; Manchio, J.V.; Fox, N.; et al. Galunisertib plus neoadjuvant chemoradiotherapy in patients with locally advanced rectal cancer: A single-arm, phase 2 trial. *Lancet Oncol.* **2022**, *23*, 1189–1200. [CrossRef] [PubMed]
164. Wang, R.; Liu, H.; He, P.; An, D.; Guo, X.; Zhang, X.; Feng, M. Inhibition of PCSK9 enhances the antitumor effect of PD-1 inhibitor in colorectal cancer by promoting the infiltration of CD8+ T cells and the exclusion of Treg cells. *Front. Immunol.* **2022**, *13*, 947756. [CrossRef]
165. Chen, T.-W.; Hung, W.-Z.; Chiang, S.-F.; Chen, W.T.-L.; Ke, T.-W.; Liang, J.-A.; Huang, C.-Y.; Yang, P.-C.; Huang, K.C.-Y.; Chao, K.S.C. Dual inhibition of TGFβ signaling and CSF1/CSF1R reprograms tumor-infiltrating macrophages and improves response to chemotherapy via suppressing PD-L1. *Cancer Lett.* **2022**, *543*, 215795. [CrossRef] [PubMed]

International Journal of *Molecular Sciences*

MDPI

Article

Selective Role of TNFα and IL10 in Regulation of Barrier Properties of the Colon in DMH-Induced Tumor and Healthy Rats

Viktoria Bekusova [1,*], Tatiana Zudova [1], Ilyas Fatyykhov [1], Arina Fedorova [1], Salah Amasheh [2] and Alexander G. Markov [1]

[1] Department of General Physiology, Faculty of Biology, Saint Petersburg State University, 199034 Saint Petersburg, Russia

[2] Department of Veterinary Medicine, Institute of Veterinary Physiology, Freie Universität Berlin, 14163 Berlin, Germany

* Correspondence: v.bekusova@spbu.ru; Tel.: +7-911-722-7837

Abstract: Recently it has been reported that the tumor adjacent colon tissues of 1,2-dymethylhydrazine induced (DMH)-rats revealed a high paracellular permeability. We hypothesized that the changes might be induced by cytokines. Colorectal cancer is accompanied by an increase in tumor necrosis factor alpha (TNFα) and interleukin 10 (IL10) that exert opposite regulatory effects on barrier properties of the colon, which is characterized by morphological and functional segmental heterogeneity. The aim of this study was to analyze the level of TNFα and IL10 in the colon segments of DMH-rats and to investigate their effects on barrier properties of the proximal and distal parts of the colon in healthy rats. Enzyme immunoassay analysis showed decreased TNFα in tumors in the distal part of the colon and increased IL10 in proximal tumors and in non-tumor tissues. Four-hour intraluminal exposure of the colon of healthy rats with cytokines showed reduced colon barrier function dependent on the cytokine: TNFα decreased it mainly in the distal part of the colon, whereas IL10 decreased it only in the proximal part. Western blot analysis revealed a more pronounced influence of IL10 on tight junction (TJ) proteins expression by down-regulation of the TJ proteins claudin-1, -2 and -4, and up-regulation of occludin only in the proximal part of the colon. These data may indicate a selective role of the cytokines in regulation of the barrier properties of the colon and a prominent role of IL10 in carcinogenesis in its proximal part.

Keywords: barrier properties; colon; heterogeneity; cytokines; tumor necrosis factor alpha; interleukin 10; tight junction proteins; 1,2-dimethylhydrazine; colorectal cancer; rat

Citation: Bekusova, V.; Zudova, T.; Fatyykhov, I.; Fedorova, A.; Amasheh, S.; Markov, A.G. Selective Role of TNFα and IL10 in Regulation of Barrier Properties of the Colon in DMH-Induced Tumor and Healthy Rats. *Int. J. Mol. Sci.* **2022**, *23*, 15610. https://doi.org/10.3390/ijms232415610

Academic Editors: Alessandro Ottaiano and Donatella Delle Cave

Received: 29 October 2022

Accepted: 6 December 2022

Published: 9 December 2022

1. Introduction

The large intestine is characterized by morphological and functional segmental heterogeneity [1–3]. The specificity of its two main different parts—the proximal and the distal colon, is manifested in many aspects, particularly, in development of right- or left-side colorectal carcinogenesis (CRC) [4–13]. Tumors more often appears in the proximal colon in adults [14], whereas in adolescents and young adults these are predominantly located in the distal colon [15]. A precancerous lesion of CRC as serrated adenoma was predominantly found in the distal part of the colon [16].

The most adequate to human CRC model of experimental carcinogenesis is 1,2-dymethylhydrazine (DMH)-induced CRC. DMH is a pro-carcinogenic agent that is activated in the liver and transported to the intestine by bile and blood. The mechanism of action of DMH is associated primarily with DNA methylation of the stem colonocytes which are located at the base of the intestinal crypts, with the subsequent development of colon adenocarcinomas [17–19]. The majority of colon tumors in the DMH-rats are located in the distal colon, while are less frequent in the proximal colon [17,18,20]. The cause of this preference remains unclear.

One of the reasons for the development of CRC is considered to be an imbalance between pro- and anti-inflammatory cytokines [21]. Pre-neoplastic lesions under DMH carcinogenesis may be connected with an increase in pro-inflammatory cytokines, especially tumor necrosis factor alpha (TNFα) [22]. It can be assumed that the different incidences of the colon may be associated with heterogeneity of pro- and anti-inflammatory cytokine production. The first group includes TNFα, while the second group includes interleukin 10 (IL10). There is no data available regarding the concentration of the cytokines in intestinal tissue in DMH-induced carcinogenesis.

CRC is accompanied by destroying the interactions between neighboring epithelial cells, cell dedifferentiation, loss of polarity, and metastasis. Impairment of the barrier properties of the colon and changes in the molecular composition of tight junctions (TJs) may lead to disturbance of the epithelial integrity [23].

TJs are composed of various transmembrane proteins, which belong to the MARVEL protein family, such as occludin and tricellulin, and the claudin family, which is divided in two groups according to their contribution to paracellular permeability [24]. Whereas some claudins as claudin-1, -3, -4 reduce epithelial permeability [25], others, particularly claudin-2, form pores and increase paracellular permeability for ions, water, and macromolecules [26]. One of the functions of TJs is to combine epithelial cells into unified sheets and maintain the integrity of the epithelial layer [27]. Under conditions of possible disintegration of TJs, the expression of claudins changes [28]. Therefore, one of the features of carcinogenesis, in particular in the colon, is a change in the expression of TJ proteins [29–32]. The physiological parameter for analysis of the destroying of the epithelial barrier function, consider the change in transepithelial resistance (TEER), short-circuit current (Isc) and paracellular permeability to some macromolecules such as sodium fluorescein [33].

Previously we have shown that DMH changed the intestinal permeability and induced alteration in expression of TJ proteins in the rat colon and IPEC-J2 cells [34,35]. We hypothesized that different incidence of the colon may be due to the different effects of the cytokines in carcinogenesis. TNFα decreased barrier properties of the colon ex vivo in Ussing chambers [36] and monolayers of epithelial cells in vitro [37–40]. IL10, on the contrary, had a protective effect, preventing disruption of the intestinal barrier function [41,42] and, perhaps, IL10 prevents the development of tumors in the proximal colon. Yet, the effects of cytokines on the barrier properties of the proximal and the distal parts of the colon have not been studied before. The aim of this study was to analyze the levels of TNFα and IL10 in the different segments of rat colon in DMH-induced carcinogenesis and to investigate the barrier properties of the main parts of the colon after intraluminal incubation with the cytokines in healthy rats.

2. Results

2.1. Study of the Level of TNFα and IL10 in Segments of Different Parts of the Colon during DMH-Induced Carcinogenesis

In the colon tissues, TNFα was different significantly only in the distal part. It was lower in the tumors compared to the control and tumor-adjacent tissues (Figure 1A). In comparison, IL10 was different only in the proximal part of the colon, where it was higher in tumors and non-tumor tissues compared to the control (Figure 1B).

2.2. Study of TEER, Isc and Paracellular Permeability to Sodium Fluorescein in the Proximal and Distal Parts of the Colon

The incubation with TNFα revealed an increase in paracellular permeability with a more pronounced effect in the distal part compared to controls (from 0.6 ± 0.1 to $1.4 \pm 0.3 \cdot 10^{-4}$ cm/s and 1.5 ± 0.3 to $4.5 \pm 0.4 \cdot 10^{-4}$ cm/s accordingly) (Figure 2A). TEER did not differ significantly compared to the control. Isc increased (from 15 ± 2 to 25 ± 3 $\mu A/cm^2$) only in the distal part of the colon (Figure 2C).

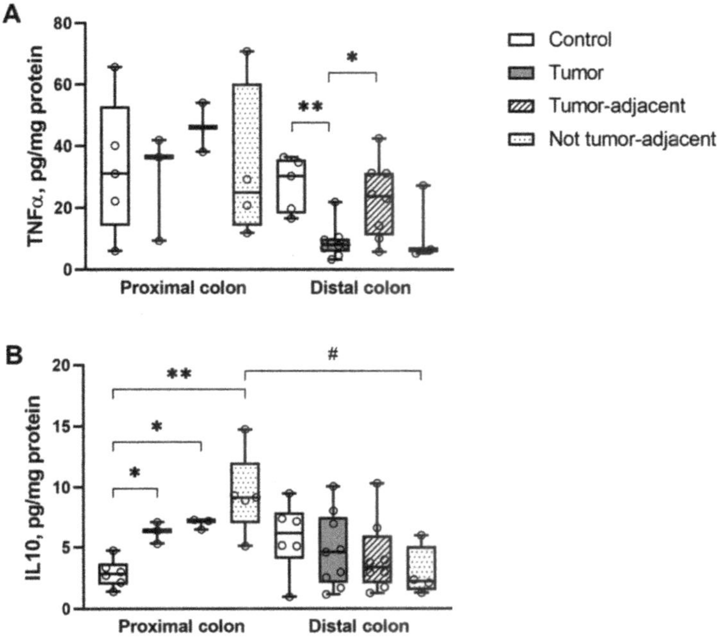

Figure 1. Elisa kit analysis of (**A**) TNFα and (**B**) IL10 in tumor, tumor-adjacent and non- tumor segments in the proximal and distal parts of the colon in DMH-rats. * $p < 0.05$, ** $p < 0.005$, # $p < 0.05$, Mann–Whitney U-test. The number of symbols corresponds to the number of samples.

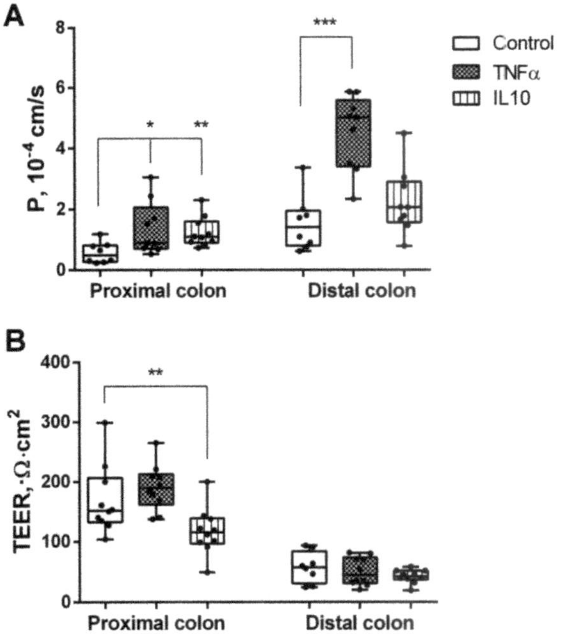

Figure 2. *Cont.*

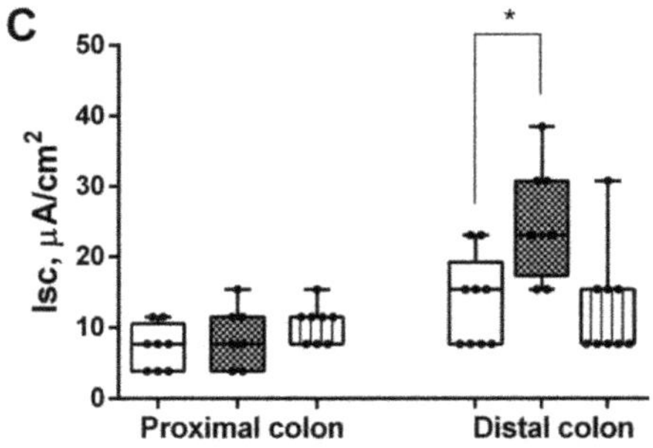

Figure 2. (**A**) Paracellular permeability for sodium fluorescein (P), (**B**) TEER, (**C**) Isc of the proximal and distal segments of the rat colon under the action of TNFα and IL10, * $p < 0.05$, ** $p < 0.01$, *** $p < 0.001$, Mann–Whitney U–test. The number of symbols corresponds to the number of samples.

In the proximal colon, incubation with IL10 revealed an increase in paracellular permeability (from 0.6 ± 0.1·to 1.2 ± 0.2·10^{-4} cm/s) (Figure 2A) and a decrease in TEER (from 170 ± 18 to 118 ± 12 $\Omega \cdot cm^2$) (Figure 2B), whereas Isc did not show a significant change compared to the control.

2.3. Western Blotting in the Colon Tissues

The expression of TJ proteins, namely claudin-1, -2, -3, -4, occludin and tricellulin, after incubation with the cytokines was determined by immunoblotting, revealing that the level of TJ proteins significantly changed only in the proximal part of the colon (Figure 3). There were not significant differences in the distal part (Figure 4). TNFα reduced the level of claudin-1 (55 ± 22% vs. control, n = 4), while IL10 reduced the level of claudin-1 (32 ± 16% vs. control, n = 4), claudin-2 (36 ± 13% vs. control, n = 4) and claudin-4 (47 ± 17% vs. control, n = 4). At the same time IL10 increased the level of occludin (367 ± 209% vs. control, n = 4) (Figure 3). Other TJ proteins, namely, claudin-3 and tricellulin, did not show significant differences in all studied groups.

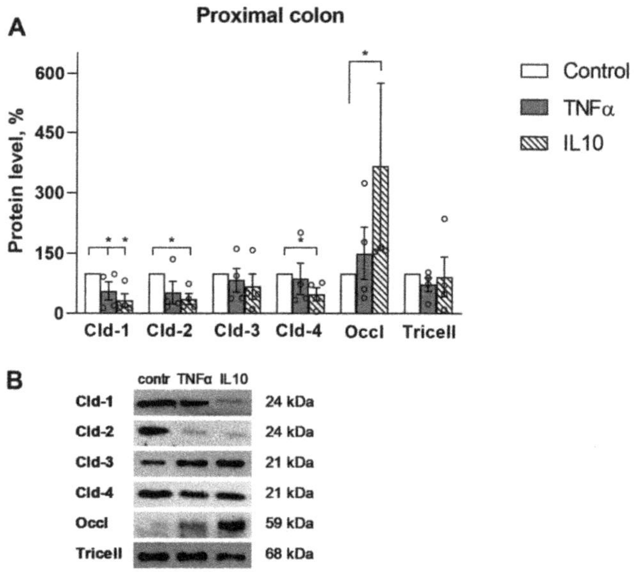

Figure 3. Western blot analysis of TJ proteins in the proximal colon. (**A**) Densitometric analysis revealed decreased Cld-1 under the action of TNFα, decreased claudin (Cld)-1, -2 and -4 and increased

occludin (Occl) after IL10 action, Tricellulin (Tricell), * $p < 0.05$, Mann–Whitney U-test, n for each protein = 4. The values were normalized to total protein amount. The number of symbols corresponds to the number of samples. (**B**) Representative Western blot bands.

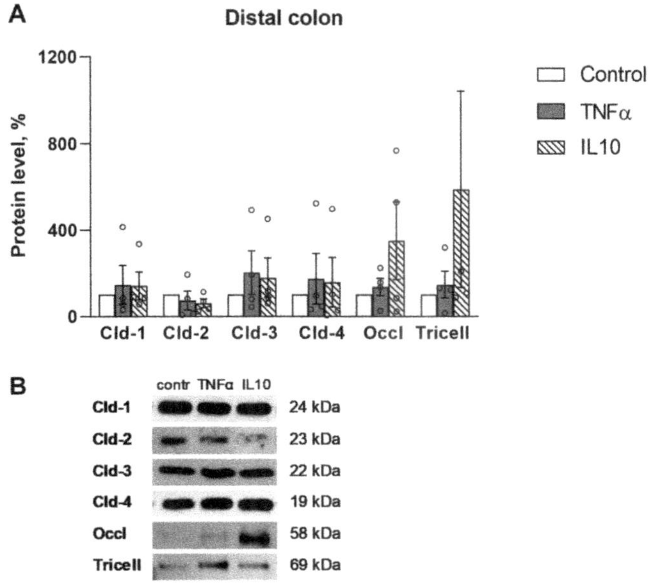

Figure 4. Western blot analysis of TJ proteins in the distal colon. (**A**) Densitometric analysis did not reveal any significant differences, $p < 0.05$, Mann–Whitney U-test, n for each protein = 4. The values were normalized to total protein amount. The number of symbols corresponds to the number of samples. (**B**) Representative Western blot bands.

Thus, the cytokines induced a change in TJ proteins level only in the proximal part of the colon, and IL10 showed the most pronounced effect.

3. Discussion

Previously, it was found that the tumor-adjacent tissues in DMH-rats were characterized by a high paracellular permeability [34]. We hypothesized that these changes might be induced by cytokines. Therefore, we chose pro-inflammatory TNFα and anti-inflammatory IL10, which play opposite roles in the development of CRC, as well as regarding participation in the regulation of the barrier properties of the colon [36–42].

The role of cytokines in the development of CRC is ambiguous. An amount of TNFα in human blood serum was negatively correlated with proliferation of tumor cells [43]; however, increased circulating TNFα was associated with poor overall and cancer-specific patient survival [44]. In contrast, the level of IL10 has been shown to be positive correlated with the intensity of tumor proliferation and apoptosis [43]. However, it decreased in colon tissues with increased tumor invasion and lesion of lymph nodes at the late stages of CRC, corresponding to poor patient survival [45].

Most authors indicate that the concentration of TNFα in human blood serum in CRC is increased compared to the control [46–50]. The concentration of circulating TNFα was significantly higher compared to the control at all stages of CRC and showed highest values at the last, IV stage [46]. Expression of TNFα mRNA in tumor was significantly higher than in neighboring tissues [51]. Some authors reported that the level of IL10 was increased in human blood serum in CRC [49,50,52], while others showed that its concentration remained practically unchanged [53].

We investigated the cytokines level in the segments of the colon in DMH-rats in accordance with the mapping scheme we used earlier [34]. The study of the level of the cytokines into three types of segments—tumors, adjacent to tumors, and not adjacent to tumors, allowed us to analyze their concentration in the entire organ in detail, making a note of the possible influence in the tumor and tumor microenvironment on the colon tissues.

The level of TNFα was changed only in the distal part of the colon, it was lower in tumors compared to the control and tumor-adjacent tissues. We suppose that these results are explained by the features of our experimental model. Increased production of TNFα is usually associated with inflammation, and TNFα is a key cytokine that links inflammation and carcinogenesis [54]. DMH-induced carcinogenesis is not the direct result of an inflammatory process. It is believed that in the distal colon, histogenesis follows aberrant crypt foci-adenoma-carcinoma sequences, while in the proximal colon, carcinomas arise de novo without an intermediate stage of colon carcinogenesis [17]. Taking into account the inhibitory effect of TNFα on tumor cell proliferation [43] and that its pro-tumorigenic properties may rather be invoked by low chronic TNFα production than by an intensive outburst that activates reactive oxygen species and kills malignant cells [55], reduced concentration of TNFα in the tumors should have promoted tumor growth in the distal part of the colon. In fact, our hypothesis about the potential role of TNFα in the impairment of barrier properties of the colon and the possible negative effect of the tumor on neighboring tissues through the production of TNFα in DMH-induced carcinogenesis was not confirmed. That does not exclude its possible role in the disturbance of the barrier function with other types of CRC.

In our study IL10 was higher compared to the control in tumors, tumor-adjacent and not tumor-adjacent segments in the proximal part of the colon. It was most pronounced in the proximal not tumor-adjacent segments that significantly differed from the distal not tumor-adjacent segments. Thus, IL10 was significantly increased in all the segments of the proximal part of the colon in DMH-rats with right-side carcinogenesis. The studies revealed that IL10 can be a protective factor in animal CRC models [56,57]. Based on the fact that a decrease in IL10 in cancerous tissue is an independent risk factor for poor survival [45], elevated IL10 could have a protective effect in CRC. However, given that cytokines can play a dual role in tumor development—they can either participate in the suppression of carcinogenesis, or contribute to its progress [58], increased IL10 associated with tumors in the proximal part of the colon where tumors develop only in 20–25% of cases of CRC, could stimulate their development. Some studies suggest that IL10 serum levels are lower in the control group than in the CRC patients [59]. Patients in the fourth clinical stage of CRC have a higher level of serum IL10 when compared to lower stages, while a high serum concentration of IL10 correlates with poor survival of patients with CRC [60,61]. Additionally, IL10 overexpression was positively correlated with metastasis occurrence [62].

Previously, we conducted a segmental analysis of the barrier properties of the colon [63]. It was shown that the barrier properties were more pronounced in the proximal part of the colon compared to the distal part. We assume that the heterogeneity of the barrier properties of the colon and different effects of cytokines which are elevated in pathologies may determine how pathological processes develop and contribute to different incidences of CRC in two main parts of the colon.

Based on the literature, most studies of the barrier properties of the intestinal epithelium under cytokines action were carried out in vitro on cell cultures, with stimulation mainly from the basolateral side of the epithelium [37,64,65]. We stimulated the colon in situ from the apical side of the epithelium according to the approach we successfully used in our previous study [63]. We supposed that the addition of the cytokines to the mucosal side ensured their prolonged action and prevented their rapid enzymatic degradation by tissue cells.

We have shown that the four-hour prolonged action of cytokines on the colon of healthy rats led to the changes in the parameters of its barrier properties—a decrease in TEER, an increase in Isc, and an increase in paracellular permeability of the colon. Thus, under the action of TNFα and IL10 on the apical side of the epithelium, the barrier properties of the colon were reduced.

Regarding the main hypothesis that we checked, whether cytokines have the same regulatory effect on different parts of the colon, we analyzed barrier properties of the segments taken from its proximal and distal part. It was shown that TNFα changed the barrier properties mainly of the distal part of the colon, and IL10—only of the proximal one.

TJ proteins only showed changes in the proximal colon: TNFα reduced claudin-1, while IL10 decreased claudin-1, -2, -4, and increased occludin, demonstrating the most pronounced effect. The decrease of claudin-1 and -4, which increases the barrier properties of the colon, is consistent with the data obtained in Ussing chambers—an increase in paracellular permeability under the action of TNFα and IL10 and a decrease in TEER under the action of IL10 for the proximal colon segments. At the same time, increased occludin and decreased claudin-2 indicated an increase in the barrier properties of the proximal part of the colon and indicated a different effect of IL10 on the expression of TJ proteins in this region. Changes in TJ proteins expression only in the proximal part of the colon could indicate a greater sensitivity of this region to the action of cytokines and its ability to more pronounced molecular rearrangements.

Data on low TNFα in tumors in the distal part of the colon indicated that TNFα was probably not involved in the impairment of the barrier properties of the colon in DMH-induced carcinogenesis. Increased IL10 in the proximal segments of the colon of DMH-rats with proximal tumor location indicated that IL10 could play an important role in regulating the barrier properties in this colon segment. IL10 decreased the barrier properties and contributed to the restructuring of the TJ proteins only in the proximal part of the colon. Thus, IL10 altered intercellular interactions and could influence the development of CRC in the proximal part of the colon.

Our results have provided following novel findings: (1) cytokines have selective effects on the regulation of barrier properties of the colon: TNFα decreases barrier properties mainly in the distal part of the colon, while IL10 decreases the barrier properties only in the proximal one, (2) IL10 plays a prominent role in regulation of carcinogenesis and barrier function of the proximal part of the colon and may contribute to different incidences of CRC in the proximal and distal parts of the colon.

The heterogeneity of the cytokines production during cancer development and the selective effects of the cytokines on barrier properties of the proximal and the distal parts of the colon may contribute to different involvement of the colon in CRC.

4. Materials and Methods

4.1. Animals

Male Wistar rats for Experiment 1 were obtained from the Animal Laboratory of the I.P. Pavlov Institute of Physiology (Experiment 1, below: 5.1.) and from the vivarium of St. Petersburg University (Experiment 2, below: 5.2.). They were kept under a standard light/dark cycle (12 h light:12 h dark) at 22 ± 2 °C with ad libitum access to tap water and complete pelleted feed (Delta Feeds, BioPro, Novosibirsk, Russia). The studies were carried out in accordance with the guidelines of the FELASA [66] and approved by the Ethics Committee for Animal Research of St. Petersburg State University (Conclusion No. 131-03-1 dated 2 February 2021).

4.2. Chemicals

The cytokines were obtained from Sigma-Aldrich (Taufkirchen, Germany): TNFα from rat, recombinant, expressed in *E. coli*, IL10 human, recombinant, expressed in HEK 293 cells, HumanKine®.

5. Experimental Design

5.1. Experiment 1

Animals weighing 120–150 g (n = 20) were randomly subdivided into 2 groups—a control group (n = 6) and an experimental group (n = 14).

The rats in the control group were not exposed to the carcinogen, whereas the rats in the experimental group were administered 5 subcutaneous injections of DMH weekly at 21 mg/kg of body weight (each dose). DMH was obtained from Sigma-Aldrich (Tokyo, Japan). Six months after the first carcinogen injection, the rats were decapitated with a guillotine (Open Science, Moscow, Russia).

Mapping of the colon segments in the experimental group of DMH-induced rats depended on tumor location and was carried out according to the previously described method [34] (Figure 5).

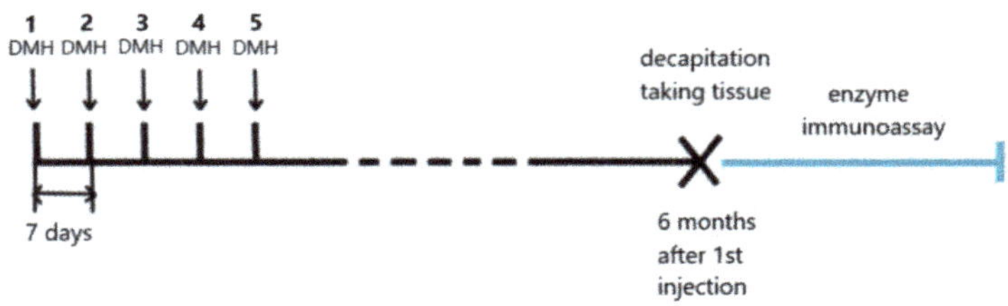

Figure 5. Design of experiment 1 (scheme).

Study of TNFα and IL10 in Colon Tissues in DMH-Rats by Enzyme Immunoassay

The colon segments were immediately frozen at −80 °C and stored at this temperature until analysis. A total of 50–60 mg colon tissue samples were homogenized in 1 mL RIPA buffer with protease inhibitors (150 mM NaCl; 10 mM Tris-HCl, pH 7.4; 0.5% Triton X-100; 0.1% SDS), the suspension was sonicated and centrifuged (10,000× *g*, 5 min at 4 °C). In the supernatants, the concentration of total protein was determined. The cytokines in the samples were determined following the manufacturers' instructions by sandwich ELISA and commercial reagent kits (High Sensitive ELISA Kit for IL10 and TNFα, Cloud-Clone Corp., Wuhan, China) using spectrophotometer SPECTROstar Nano (BMG LABTECH, Ortenberg, Germany). TNFα and IL10 concentration values were recalculated per mg of total protein.

5.2. Experiment 2

Animals weighing 280–380 g (n = 20) were randomly subdivided into four groups—one control group and three experimental groups with five animals in each group. The rats were anesthetized intraperitoneally with Zoletil 100 (Virbac, Carros, France, 100 mg/kg of the body weight). Narcosis was verified by the disappearance of the reaction to a painful stimulus (tail prick). An incision was made along the midline of the abdomen from the processus xiphoideus of the sternum in the distal direction. Two ligatures isolated the colon loop between the cecum and the anus, thus limiting the colon. Two tubes were inserted at the beginning and at the end of the loop, and the system was filled hermetically with the test solution. In the control group, the loop was luminally filled with Krebs-Ringer's solution, while in the experimental groups it was filled with one of the cytokine solutions with 200 ng/mL TNFα and 100 ng/mL IL10. After 4 h of incubation the rats were decapitated, the large intestine was divided into proximal and distal segments, as described in our previous studies [34] and then investigated in Ussing chambers for 1 h (Figure 6).

5.2.1. Electrophysiological Assay of the Colon Segments

Isc and TEER of the large intestine wall were studied according to the previously described protocol [67]. Briefly, the segments of the large intestine were mounted in Ussing chambers filled with Krebs-Ringer solution at 37 °C, which was maintained throughout the experiment, and permanently oxygenated with a mixture of 95% oxygen and 5% carbon dioxide. The Krebs-Ringer solution was composed as follows (mM): NaCl (119), KCl (5), $MgCl_2 \cdot 6H_2O$ (1.2), $NaHCO_3$ (25), $NaH_2PO_4 \cdot H_2O$ (0.4), $Na_2HPO_4 \cdot 7H_2O$ (1.6), $CaCl_2$ (1.2), and D-glucose (10). The short circuit current was recorded when the voltage was set at 0 mV. To evaluate TEER, we recorded voltage fluctuations when the current was 10 μA and calculated it using Ohm's law: R = U/I (Ω). The size of the examined tissue was calculated by using the diameter of the slotted opening between the two chamber halves (4 mm) and was equal to 0.13 cm^2. With regard to the size of the examined tissue, we adjusted the obtained transepithelial voltage values for 1 cm^2 of tissue ($\Omega \cdot cm^2$).

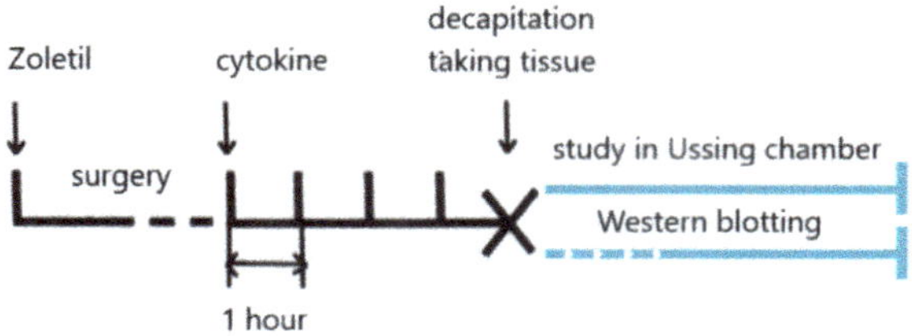

Figure 6. Design of experiment 2 (scheme).

5.2.2. Assessment of the Paracellular Permeability of the Colon Segments

To study the paracellular permeability of the intestine in the Ussing chamber, we added sodium fluorescein to the mucosal bathing solution at a final concentration of 100 μM. This concentration was determined from earlier published reports [68,69]. Thirty minutes after the experiment started, the serosal bathing solution was removed to analyze the concentration of the diffused sodium fluorescein. To assess the optical density of this solution, we used Cary Eclipse Fluorescence Spectrophotometer (Agilent Technologies, CA, USA). The excitation and emission wavelengths were 460 and 515 nm, respectively. The permeability coefficient (P_{app}, cm/s) was calculated using the following equitation: $P_{app} = (dQ/dt)/(A \cdot C_0)$, with: dQ/dt as the concentration of sodium fluorescein in the serosal bathing solution (mol/s), A as the size of the examined tissue (cm^2), and C_0 as the concentration of sodium fluorescein in the mucosal bathing solution.

5.2.3. Western Blotting of Colon Tissues

TJ protein levels were analyzed in the colon segments, as described in detail earlier [70], and stain-free immunoblotting was performed, as described previously [71–73].

Briefly, tissues were homogenized in RIPA buffer (150 mM NaCl; 10 mM Tris-HCl, pH 7.4; 0.5% Triton X-100; 0.1% SDS) with protease inhibitor (Complete ULTRA Tablets, Mini; Roche, Mannheim, Germany), then centrifuged (15 min, 15,000× *g*, 4 °C), and a quantitative protein analysis using Thermo BCA assay kit (Thermo Fisher Scientific, Waltham, MA, USA) was performed by a spectrophotometer SPECTROstar Nano (BMG Labtech, Ortenberg, Germany). SDS buffer (Laemmli) was added to extracted proteins, and samples were loaded on 10% Stain-Free gels and electrophoresis was performed. Proteins from the gels were transferred to PVDF membranes with a 0.2 μm pore size (Bio-Rad, Hercules, CA, USA), which were first incubated with primary antibodies raised against claudin-1, -2, -3, -4, occludin or tricellulin and then visualized using secondary goat anti-rabbit and anti-mouse IgG antibodies and chemiluminescence reaction (Bio-Rad). The following antibodies were used: claudin-1 (#51-9000, Invitrogen, Carlsbad, CA, USA), claudin-2 (#32-5600, Invitrogen), claudin-3 (#34-1700, Invitrogen), claudin-4 (#36-4800, Invitrogen), occludin (#DF7504, Affinity Biosciences, Cincinnati, OH, USA), tricellulin (#48-8400, Invitrogen). The protein bands were detected and identified using Clarity Western ECL Substrate and the ChemiDoc XRS+ imager (Bio-Rad). Normalization of detected proteins was performed using the Image Lab 6.1 Software (Bio-Rad) to the total protein load measured in the membrane. The signal density in the control group was set as 100% (Supplementary Materials).

5.3. Statistical Analysis

Statistical analysis was performed using the Anova group analysis in GraphPad Prism 8.4.3 (Graphpad Software Inc., San Diego, CA, USA). The data were analyzed using Mann–Whitney U-test. The results of the analyses are presented as mean ± standard error (M ± SEM). Statistically reliable differences were reported with a probability value of 95% ($p < 0.05$).

Supplementary Materials: The following supporting information can be downloaded at: https://www.mdpi.com/article/10.3390/ijms232415610/s1, Figure S1: Original images.

Author Contributions: Conceptualization, A.G.M. and V.B.; methodology, A.G.M. and V.B.; investigation, V.B., T.Z., I.F. and A.F.; writing—original draft preparation, V.B. and T.Z.; writing—review and editing, S.A. and A.G.M.; supervision, A.G.M.; funding acquisition, S.A. and A.G.M. All authors have read and agreed to the published version of the manuscript.

Funding: This work was supported by a grant from the Russian Foundation for Basic Research (Grant No. 20-04-01050) and the German Research Association (Grant No. AM141/11-2).

Institutional Review Board Statement: The studies were approved by the Ethics Committee for Animal Research of St. Petersburg State University (Conclusion No. 131-03-1 dated 2 February 2021).

Informed Consent Statement: Not applicable.

Data Availability Statement: Date is contained in the article. The datasets analyzed in the study are available from the corresponding author upon reasonable request.

Acknowledgments: Scientific research was performed using the Research Park of St. Petersburg State University "Center for Molecular and Cell Technologies". We are grateful to Irina A. Lekomtseva, Department of English Philology and Translation, St Petersburg University, for help during translation of this article.

Conflicts of Interest: The authors declare no conflict of interest.

References

1. Mirakhur, M.; Diener, M. Proteinase-activated receptors regulate intestinal functions in a segment-dependent manner in rats. *Eur. J. Pharmacol.* **2022**, *933*, 175264. [CrossRef]
2. Garg, S.; Zheng, J.; Wang, J.; Authier, S.; Pouliot, M.; Hauer-Jensen, M. Segmental differences in radiation-induced alterations of tight junction-related proteins in non-human primate jejunum, ileum and colon. *Radiat. Res.* **2016**, *185*, 50–59. [CrossRef] [PubMed]
3. Markov, A.; Veshnyakova, A.; Fromm, M.; Amasheh, M.; Amasheh, S. Segmental expression of claudin proteins correlates with tight junction barrier properties in rat intestine. *J. Comp. Physiol. B-Bioch. Syst. Envir. Physiol.* **2010**, *180*, 591–598. [CrossRef] [PubMed]
4. Lee, M.; Menter, D.; Kopetz, S. Right Versus left colon cancer biology: Integrating the consensus molecular subtypes. *J. Natl. Compr. Canc. Netw.* **2017**, *15*, 411–419. [CrossRef] [PubMed]
5. Mendis, S.; Beck, S.; Lee, B.; Lee, M.; Wong, R.; Kosmider, S.; Shapiro, J.; Yip, D.; Steel, S.; Nott, L.; et al. Right versus left sided metastatic colorectal cancer: Teasing out clinicopathologic drivers of disparity in survival. *Asia Pac. J. Clin. Oncol.* **2019**, *15*, 136–143. [CrossRef]
6. Testa, U.; Pelosi, E.; Castelli, G. Colorectal cancer: Genetic abnormalities, tumor progression, tumor heterogeneity, clonal evolution and tumor-initiating cells. *Med. Sci.* **2018**, *6*, 31. [CrossRef]
7. Miyake, T.; Mori, H.; Yasukawa, D.; Hexun, Z.; Maehira, H.; Ueki, T.; Kojima, M.; Kaida, S.; Iida, H.; Shimizu, T.; et al. The comparison of fecal microbiota in left-side and right-side human colorectal cancer. *Eur. Surg. Res.* **2021**, *62*, 248–254. [CrossRef]
8. Sugai, T.; Habano, W.; Jiao, Y.; Tsukahara, M.; Takeda, Y.; Otsuka, K.; Nakamura, S. Analysis of molecular alterations in left- and right-sided colorectal carcinomas reveals distinct pathways of carcinogenesis: Proposal for new molecular profile of colorectal carcinomas. *J. Mol. Diagn.* **2006**, *8*, 193–201. [CrossRef]
9. Patel, M.; McSorley, S.; Park, J.; Roxburgh, C.; Edwards, J.; Horgan, P.; McMillan, D. The relationship between right-sided tumour location, tumour microenvironment, systemic inflammation, adjuvant therapy and survival in patients undergoing surgery for colon and rectal cancer. *Br. J. Cancer* **2018**, *118*, 705–712. [CrossRef]

10. Natsume, S.; Yamaguchi, T.; Takao, M.; Iijima, T.; Wakaume, R.; Takahashi, K.; Matsumoto, H.; Nakano, D.; Horiguchi, S.; Koizumi, K.; et al. Clinicopathological and molecular differences between right-sided and left-sided colorectal cancer in Japanese patients. *Jpn. J. Clin. Oncol.* **2018**, *48*, 609–618. [CrossRef]
11. Mesa, H.; Manivel, J.; Larson, W.; Dachel, S.; Reinink, A.; Jessurun, J. Immunophenotypic comparison of neoplasms of the appendix, right colon, and left colon in search of a site-specific phenotypic signature. *Int. J. Surg. Pathol.* **2020**, *28*, 20–30. [CrossRef]
12. Kanno, H.; Miyoshi, H.; Yoshida, N.; Sudo, T.; Nakashima, K.; Takeuchi, M.; Nomura, Y.; Seto, M.; Hisaka, T.; Tanaka, H.; et al. Differences in the immunosurveillance pattern associated with DNA mismatch repair status between right-sided and left-sided colorectal cancer. *Cancer Sci.* **2020**, *111*, 3032–3044. [CrossRef]
13. Guo, D.; Li, X.; Xie, A.; Cao, Q.; Zhang, J.; Zhang, F.; Li, W.; Chen, J. Differences in oncological outcomes and inflammatory biomarkers between right-sided and left-sided stage I-III colorectal adenocarcinoma. *J. Clin. Lab. Anal.* **2020**, *34*, e23132. [CrossRef]
14. Siegel, R.; Miller, K.; Goding Sauer, A.; Fedewa, S.; Butterly, L.; Anderson, J.; Cercek, A.; Smith, R. Colorectal cancer statistics, 2020. *CA Cancer J. Clin.* **2020**, *70*, 145–164. [CrossRef]
15. Salem, M.; Battaglin, F.; Goldberg, R.; Puccini, A.; Shields, A.; Arguello, D.; Korn, W.; Marshall, J.; Grothey, A.; Lenz, H. Molecular analyses of left- and right-sided tumors in adolescents and young adults with colorectal cancer. *Oncologist* **2020**, *25*, 404–413. [CrossRef]
16. Zhao, X.; Dou, L.; Zhang, Y.; Liu, Y.; He, S.; Ke, Y.; Liu, X.; Liu, Y.; Wang, G. Clinicopathological features of the colorectal serrated adenoma and analysis on influencing factors of malignancy. *Zhonghua Wei Chang Wai Ke Za Zhi* **2021**, *24*, 75–80. [CrossRef]
17. Perse, M.; Cerar, A. Morphological and molecular alterations in 1,2 dimethylhydrazine and azoxymethane induced colon carcinogenesis in rats. *J. Biomed. Biotechnol.* **2011**, *2011*, 473964. [CrossRef]
18. Venkatachalam, K.; Vinayagam, R.; Anand, M.; Isa, N.; Ponnaiyan, R. Biochemical and molecular aspects of 1,2-dimethylhydrazine (DMH)-induced colon carcinogenesis: A review. *Toxicol. Res.* **2020**, *9*, 2–18. [CrossRef]
19. De-Souza, A.; Costa-Casagrande, T. Animal models for colorectal cancer. *Abcd-Arq. Bras. Cir. Dig. Braz. Arch. Dig. Surg.* **2018**, *31*, e1369. [CrossRef]
20. Bekusova, V.; Patsanovskii, V.; Nozdrachev, A.; Trashkov, A.; Artemenko, M.; Anisimov, V. Metformin prevents hormonal and metabolic disturbances and 1,2-dimethylhydrazine-induced colon carcinogenesis in non-diabetic rats. *Cancer Biol. Med.* **2017**, *14*, 100–107. [CrossRef]
21. Borowczak, J.; Szczerbowski, K.; Maniewski, M.; Kowalewski, A.; Janiczek-Polewska, M.; Szylberg, A.; Marszałek, A.; Szylberg, Ł. The Role of inflammatory cytokines in the pathogenesis of colorectal carcinoma-recent findings and review. *Biomedicines* **2022**, *10*, 1670. [CrossRef]
22. Umesalma, S.; Sudhandiran, G. Differential inhibitory effects of the polyphenol ellagic acid on inflammatory mediators NF-kappa B, iNOS, COX-2, TNF-alpha, and IL-6 in 1,2-dimethylhydrazine-induced rat colon carcinogenesis. *Bas. Clinic. Pharm. Toxicol.* **2010**, *107*, 650–655. [CrossRef]
23. Tsukita, S.; Tanaka, H.; Tamura, A. The claudins: From tight junctions to biological systems. *Trends Biochem. Sci.* **2019**, *44*, 141–152. [CrossRef]
24. Gunzel, D.; Fromm, M. Claudins and other tight junction proteins. *Compr. Physiol.* **2012**, *2*, 1819–1852. [CrossRef]
25. Gunzel, D.; Yu, A. Claudins and the modulation of tight junction permeability. *Physiol. Rev.* **2013**, *93*, 525–569. [CrossRef] [PubMed]
26. Amasheh, S.; Meiri, N.; Gitter, A.; Schöneberg, T.; Mankertz, J.; Schulzke, J.; Fromm, M. Claudin-2 expression induces cation-selective channels in tight junctions of epithelial cells. *J. Cell Sci.* **2002**, *115*, 4969–4976. [CrossRef] [PubMed]
27. Markov, A.; Aschenbach, J.; Amasheh, S. The epithelial barrier and beyond: Claudins as amplifiers of physiological organ functions. *Iubmb. Life* **2017**, *69*, 290–296. [CrossRef] [PubMed]
28. Kruglova, N.; Razgovorova, I.; Amasheh, S.; Markov, A. Accumulation of milk increases the width of tight junctions in the epithelium of mouse mammary alveoli. *Biol. Comm.* **2020**, *65*, 277–280. [CrossRef]
29. Landy, J.; Ronde, E.; English, N.; Clark, S.; Hart, A.; Knight, S.; Ciclitira, P.; Al-Hassi, H. Tight junctions in inflammatory bowel diseases and inflammatory bowel disease associated colorectal cancer. *World J. Gastroenterol.* **2016**, *22*, 3117–3126. [CrossRef]
30. Barmeyer, C.; Fromm, M.; Schulzke, J. Active and passive involvement of claudins in the pathophysiology of intestinal inflammatory diseases. *Pflug. Arch. Europ. J. Physiol.* **2017**, *469*, 15–26. [CrossRef]
31. Wang, X.; Tully, O.; Ngo, B.; Zitin, M.; Mullin, J. Epithelial tight junctional changes in colorectal cancer tissues. *Sci. World J.* **2011**, *11*, 826–841. [CrossRef]
32. Turksen, K.; Troy, T. Junctions gone bad: Claudins and loss of the barrier in cancer. *Biochim. Et Biophys. Acta-Rev. Cancer* **2011**, *1816*, 73–79. [CrossRef]
33. Thomson, A.; Smart, K.; Somerville, M.; Lauder, S.; Appanna, G.; Horwood, J.; Raj, L.; Srivastava, B.; Durai, D.; Scurr, M.; et al. The Ussing chamber system for measuring intestinal permeability in health and disease. *BM. Gastroenterol.* **2019**, *19*, 1–14. [CrossRef]
34. Bekusova, V.; Falchuk, E.; Okorokova, L.; Kruglova, N.; Nozdrachev, A.; Markov, A. Increased paracellular permeability of tumor-adjacent areas in 1,2-dimethylhydrazine-induced colon carcinogenesis in rats. *Cancer Biol. Med.* **2018**, *15*, 251–259. [CrossRef]

35. Bekusova, V.; Droessler, L.; Amasheh, S.; Markov, A. Effects of 1,2-dimethylhydrazine on barrier properties of rat large intestine and IPEC-J2 cells. *Int. J. Mol. Sci.* **2021**, *22*, 10278. [CrossRef]
36. Amasheh, M.; Grotjohann, I.; Amasheh, S.; Fromm, A.; Soderholm, J.; Zeitz, M.; Fromm, M.; Schulzke, J. Regulation of mucosal structure and barrier function in rat colon exposed to tumor necrosis factor alpha and interferon gamma in vitro: A novel model for studying the pathomechanisms of inflammatory bowel disease cytokines. *Scand. J. Gastroenterol.* **2009**, *44*, 1226–1235. [CrossRef]
37. Al-Sadi, R.; Guo, S.; Ye, D.; Ma, T. TNF-alpha modulation of intestinal epithelial tight junction barrier is regulated by ERK1/2 activation of Elk-1. *Am. J. Pathol.* **2013**, *183*, 1871–1884. [CrossRef]
38. He, F.; Peng, J.; Deng, X.; Yang, L.; Camara, A.; Omran, A.; Wang, G.; Wu, L.; Zhang, C.; Yin, F. Mechanisms of tumor necrosis factor-alpha-induced leaks in intestine epithelial barrier. *Cytokine* **2012**, *59*, 264–272. [CrossRef]
39. Khan, M.; Uwada, J.; Yazawa, T.; Islam, M.; Krug, S.; Fromm, M.; Karaki, S.; Suzuki, Y.; Kuwahara, A.; Yoshiki, H.; et al. Activation of muscarinic cholinoceptor ameliorates tumor necrosis factor-α-induced barrier dysfunction in intestinal epithelial cells. *FEBS Lett.* **2015**, *589*, 3640–3647. [CrossRef]
40. Droessler, L.; Cornelius, V.; Markov, A.; Amasheh, S. Tumor necrosis factor alpha effects on the porcine intestinal epithelial barrier include enhanced expression of TNF receptor 1. *Int. J. Mol. Sci.* **2021**, *22*, 8746. [CrossRef]
41. Sun, X.; Yang, H.; Nose, K.; Nose, S.; Haxhija, E.; Koga, H.; Feng, Y.; Teitelbaum, D. Decline in intestinal mucosal IL-10 expression and decreased intestinal barrier function in a mouse model of total parenteral nutrition. *Am. J. Physiol. Gastrointest Liver Physiol.* **2008**, *294*, G139–G147. [CrossRef] [PubMed]
42. Al-Sadi, R.; Boivin, M.; Ma, T. Mechanism of cytokine modulation of epithelial tight junction barrier. *Front. Biosci.-Landmark* **2009**, *14*, 2765–2778. [CrossRef] [PubMed]
43. Evans, C.; Morrison, I.; Heriot, A.; Bartlett, J.; Finlayson, C.; Dalgleish, A.; Kumar, D. The correlation between colorectal cancer rates of proliferation and apoptosis and systemic cytokine levels; plus their influence upon survival. *Br. J. Cancer* **2006**, *94*, 1412–1419. [CrossRef] [PubMed]
44. Yu, Y.; Fan, C.; Tseng, W.; Chang, P.; Kuo, H.; Pan, Y.; Yeh, K. Correlation between the Glasgow prognostic score and the serum cytokine profile in Taiwanese patients with colorectal cancer. *Int. J. Biol. Markers* **2021**, *36*, 40–49. [CrossRef] [PubMed]
45. Toiyama, Y.; Miki, C.; Inoue, Y.; Minobe, S.; Urano, H.; Kusunoki, M. Loss of tissue expression of interleukin-10 promotes the disease progression of colorectal carcinoma. *Surg. Today* **2010**, *40*, 46–53. [CrossRef]
46. Stanilov, N.; Miteva, L.; Dobreva, Z.; Stanilova, S. Colorectal cancer severity and survival in correlation with tumour necrosis factor-alpha. *Biotech. Equip.* **2014**, *28*, 911–917. [CrossRef]
47. Nikiteas, N.; Tzanakis, N.; Gazouli, M.; Rallis, G.; Daniilidis, K.; Theodoropoulos, G.; Kostakis, A.; Peros, G. Serum IL-6, TNFalpha and CRP levels in Greek colorectal cancer patients: Prognostic implications. *World J. Gastroenterol.* **2005**, *11*, 1639–1643. [CrossRef]
48. Coşkun, Ö.; Öztopuz, Ö.; Özkan, Ö. Determination of IL-6, TNF-α and VEGF levels in the serums of patients with colorectal cancer. *Cell Mol. Biol.* **2017**, *63*, 97–101. [CrossRef]
49. Pengjun, Z.; Xinyu, W.; Feng, G.; Xinxin, D.; Yulan, L.; Juan, L.; Xingwang, J.; Zhennan, D.; Yaping, T. Multiplexed cytokine profiling of serum for detection of colorectal cancer. *Future Oncol.* **2013**, *9*, 1017–1027. [CrossRef]
50. Galizia, G.; Orditura, M.; Romano, C.; Lieto, E.; Castellano, P.; Pelosio, L.; Imperatore, V.; Catalano, G.; Pignatelli, C.; De Vita, F. Prognostic significance of circulating IL-10 and IL-6 serum levels in colon cancer patients undergoing surgery. *Clin. Immunol.* **2002**, *102*, 169–178. [CrossRef]
51. Al Obeed, O.; Alkhayal, K.; Al Sheikh, A.; Zubaidi, A.; Vaali-Mohammed, M.; Boushey, R.; Mckerrow, J.; Abdulla, M. Increased expression of tumor necrosis factor-α is associated with advanced colorectal cancer stages. *World J. Gastroenterol.* **2014**, *20*, 18390–18396. [CrossRef]
52. Czajka-Francuz, P.; Francuz, T.; Cisoń-Jurek, S.; Czajka, A.; Fajkis, M.; Szymczak, B.; Kozaczka, M.; Malinowski, K.; Zasada, W.; Wojnar, J.; et al. Serum cytokine profile as a potential prognostic tool in colorectal cancer patients–one center study. *Rep. Pract. Oncol. Radiother.* **2020**, *25*, 867–875. [CrossRef]
53. Szkaradkiewicz, A.; Marciniak, R.; Chudzicka-Strugała, I.; Wasilewska, A.; Drews, M.; Majewski, P.; Karpiński, T.; Zwoździak, B. Proinflammatory cytokines and IL-10 in inflammatory bowel disease and colorectal cancer patients. *Arch. Immunol. Ther. Exp.* **2009**, *57*, 291–294. [CrossRef]
54. Mager, L.; Wasmer, M.; Rau, T.; Krebs, P. Cytokine-induced modulation of colorectal cancer. *Front. Oncol.* **2016**, *6*, 96. [CrossRef]
55. Monteleone, G.; Pallone, F.; Stolfi, C. The dual role of inflammation in colon carcinogenesis. *Int. J. Mol. Sci.* **2012**, *13*, 11071–11084. [CrossRef]
56. Gounaris, E.; Blatner, N.; Dennis, K.; Magnusson, F.; Gurish, M.; Strom, T.; Beckhove, P.; Gounari, F.; Khazaie, K. T-regulatory cells shift from a protective anti-inflammatory to a cancer-promoting proinflammatory phenotype in polyposis. *Cancer Res.* **2009**, *69*, 5490–5497. [CrossRef]
57. Krause, P.; Morris, V.; Greenbaum, J.; Park, Y.; Bjoerheden, U.; Mikulski, Z.; Muffley, T.; Shui, J.; Kim, G.; Cheroutre, H.; et al. IL-10-producing intestinal macrophages prevent excessive antibacterial innate immunity by limiting IL-23 synthesis. *Nat. Commun.* **2015**, *6*, 7055. [CrossRef]
58. Shrihari, T. Dual role of inflammatory mediators in cancer. *Ecancermedicalscience* **2017**, *11*, 721. [CrossRef]
59. Abtahi, S.; Davani, F.; Mojtahedi, Z.; Hosseini, S.; Bananzadeh, A.; Ghaderi, A. Dual association of serum interleukin-10 levels with colorectal cancer. *J. Cancer Res. Ther.* **2017**, *13*, 252–256. [CrossRef]

60. Stanilov, N.; Miteva, L.; Stankova, N.; Jovchev, J.; Deliyski, T.; Stanilova, S. Role of IL-12P40 and IL-10 in progression of colorectal cancer. *Khirurgiia* **2010**, *4*, 26–29.
61. Miteva, L.; Stanilov, N.; Deliysky, T.; Stanilova, S. Significance of -1082A/G polymorphism of IL10 gene for progression of colorectal cancer and IL-10 expression. *Tumour Biol.* **2014**, *35*, 12655–12664. [CrossRef]
62. Townsend, M.; Felsted, A.; Piccolo, S.; Robison, R.; O'Neill, K. Metastatic colon adenocarcinoma has a significantly elevated expression of IL-10 compared with primary colon adenocarcinoma tumors. *Cancer Biol. Ther.* **2018**, *19*, 913–920. [CrossRef] [PubMed]
63. Bekusova, V.; Fatyykhov, I.; Amasheh, S.; Markov, A. Heterogeneity of the barrier properties of the colon in rat. *Biol. Comm.* **2021**, *66*, 160–170. [CrossRef]
64. Ma, T.; Boivin, M.; Ye, D.; Pedram, A.; Said, H. Mechanism of TNF-alpha modulation of Caco-2 intestinal epithelial tight junction barrier: Role of myosin light-chain kinase protein expression. *Am. J. Physiol. Gastrointest. Liver Physiol.* **2005**, *288*, G422–G430. [CrossRef] [PubMed]
65. Freour, T.; Jarry, A.; Bach-Ngohou, K.; Dejoie, T.; Bou-Hanna, C.; Denis, M.; Mosnier, J.; Laboisse, C.; Masson, D. TACE inhibition amplifies TNF-alpha-mediated colonic epithelial barrier disruption. *Int. J. Mol. Med.* **2009**, *23*, 41–48. [CrossRef] [PubMed]
66. Mahler, M.; Berard, M.; Feinstein, R.; Gallagher, A.; Illgen-Wilcke, B.; Pritchett-Corning, K.; Raspa, M. FELASA recommendations for the health monitoring of mouse, rat, hamster, guinea pig and rabbit colonies in breeding and experimental units. *Lab. Anim.* **2015**, *49*, 88. [CrossRef]
67. Markov, A.; Falchuk, E.; Kruglova, N.; Radloff, J.; Amasheh, S. Claudin expression in follicle-associated epithelium of rat Peyer's patches defines a major restriction of the paracellular pathway. *Acta Physiol.* **2016**, *216*, 112–119. [CrossRef]
68. Zakelj, S.; Legen, I.; Veber, M.; Kristl, A. The influence of buffer composition on tissue integrity during permeability experiments "in vitro". *Int. J. Pharm.* **2004**, *272*, 173–180. [CrossRef]
69. Molenda, N.; Urbanova, K.; Weiser, N.; Kusche-Vihrog, K.; Gunzel, D.; Schillers, H. Paracellular transport through healthy and cystic fibrosis bronchial epithelial cell lines–Do we have a proper model? *PLoS ONE* **2014**, *9*, e100621. [CrossRef]
70. Amasheh, S.; Milatz, S.; Krug, S.M.; Bergs, M.; Arnasheh, M.; Schulzke, J.; Fromm, M. "Na+ absorption defends from paracellular back-leakage by claudin-8 upregulation. *Biochem. Biophys. Res. Comm.* **2009**, *378*, 45–50. [CrossRef]
71. Stein, L.; Brunner, N.; Amasheh, S. Functional analysis of gastric tight junction proteins in Xenopus laevis oocytes. *Membranes* **2022**, *12*, 731. [CrossRef]
72. Droessler, L.; Cornelius, V.; Boehm, E.; Stein, L.; Brunner, N.; Amasheh, S. Barrier perturbation in porcine Peyer's patches by tumor necrosis factor is associated with a dysregulation of claudins. *Front. Physiol.* **2022**, *13*, 889552. [CrossRef]
73. Markov, A.; Fedorova, A.; Kravtsova, V.; Bikmurzina, A.; Okorokova, L.; Matchkov, V.; Cornelius, V.; Amasheh, S.; Krivoi, I. Circulating ouabain modulates expression of claudins in rat intestine and cerebral blood vessels. *Int. J. Mol. Sci.* **2020**, *17*, 5067. [CrossRef]

International Journal of
Molecular Sciences

MDPI

Review

TGF-β Signaling in Metastatic Colorectal Cancer (mCRC): From Underlying Mechanism to Potential Applications in Clinical Development

Xiaoshuang Li †, Yanmin Wu † and Tian Tian *

College of Life Science and Bioengineering, Beijing Jiaotong University, Beijing 100044, China
* Correspondence: ttian@bjtu.edu.cn
† These authors contributed equally to this work.

Abstract: Colorectal cancer (CRC) is a serious public health issue, and it has the leading incidence and mortality among malignant tumors worldwide. CRC patients with metastasis in the liver, lung or other distant sites always have poor prognosis. Thus, there is an urgent need to discover the underlying mechanisms of metastatic colorectal cancer (mCRC) and to develop optimal therapy for mCRC. Transforming growth factor-β (TGF-β) signaling plays a significant role in various physiologic and pathologic processes, and aberrant TGF-β signal transduction contributes to mCRC progression. In this review, we summarize the alterations of the TGF-β signaling pathway in mCRC patients, the functional mechanisms of TGF-β signaling, its promotion of epithelial–mesenchymal transition, its facilitation of angiogenesis, its suppression of anti-tumor activity of immune cells in the microenvironment and its contribution to stemness of CRC cells. We also discuss the possible applications of TGF-β signaling in mCRC diagnosis, prognosis and targeted therapies in clinical trials. Hopefully, these research advances in TGF-β signaling in mCRC will improve the development of new strategies that can be combined with molecular targeted therapy, immunotherapy and traditional therapies to achieve better efficacy and benefit mCRC patients in the near future.

Keywords: colorectal cancer; metastasis; TGF-β signaling; targeting therapy; immune-suppressive

Citation: Li, X.; Wu, Y.; Tian, T. TGF-β Signaling in Metastatic Colorectal Cancer (mCRC): From Underlying Mechanism to Potential Applications in Clinical Development. *Int. J. Mol. Sci.* **2022**, 23, 14436. https://doi.org/10.3390/ijms232214436

Academic Editors: Alessandro Ottaiano and Donatella Delle Cave

Received: 3 October 2022
Accepted: 17 November 2022
Published: 20 November 2022

1. Introduction

Colorectal cancer (CRC) is a type of cancer in which abnormal cells grow out of control in the large intestine. According to global cancer statistics, more than 1.9 million new cases and 935,000 deaths from CRC occurred in 2020. CRC is the third most commonly diagnosed malignancy worldwide and ranks second in cancer-related mortality. In other words, CRC accounts for 10% of all cancer cases and deaths [1]. In the United States, CRC ranked fourth in estimated new cases and second in estimated cancer-related deaths (thus far) in 2022 [2,3]. Due to the prevalence of obesity and lack of exercise in recent decades, the incidence of CRC is on the rise among the entire population in China, and it is currently the fifth leading cause of cancer death there [4]. The accumulation of genetic mutations and environmental risk factors are the main causes of CRC [5].

CRC metastasis is always a thorny problem in clinical situations. At the time of diagnosis, about 20% of CRC patients already have metastasis and 35–45% succumb to recurrence within five years after surgery [6]. The five-year survival rate of stage I-III CRC patients can be as high as 80%, whereas it drops to roughly 13% for patients with stage IV CRC [7]. It is reported that up to 60% of patients with stage IV CRC develop liver metastasis, demonstrating that the liver is the most common site for CRC metastatic spread [8,9]. The lung is the second most common metastatic target organ for CRC. Consequently, although modern surgical techniques and multidisciplinary systematic care have led to significant improvements in survival, long-term remission can only be achieved in 20% of patients with metastasis, and relapse occurred in 60–70% of patients [10,11]. Therefore, there is an urgent need to identify the underlying mechanisms of metastatic colorectal cancer (mCRC) and for new optimal therapeutic strategies for mCRC to be developed.

The transforming growth factor-β (TGF-β) signaling pathway plays a multifaceted role in various biological processes, such as cell growth and differentiation, apoptosis, cell motility, epithelial–mesenchymal transition (EMT), extracellular matrix (ECM) remodeling, angiogenesis and cellular immune responses [12,13]. Therefore, malfunction of the TGF-β signal pathway, either via genetic mutation or misexpression, is associated with many diseases, including cancer, fibrosis, inflammation, cardiovascular diseases, myelodysplastic syndrome, Marfan syndrome, scleroderma, endometriosis and more [14–16]. Currently, 33 members in the TGF-β superfamily have been identified in human beings, including three TGF-β isoforms, three activins, nodal, growth and differentiation factor (GDF) and the bone morphogenetic protein (BMP) subfamily, which are involved in various physiologic and pathologic mechanisms [17]. There are three isoforms of TGF-β: TGF-β1, TGF-β2, and TGF-β3. They manifest different expression patterns, bioavailability and physiological functions in organisms, respectively [18]. In addition to these ligands, downstream intracellular effectors, termed SMAD, are a group of proteins that include eight different members in mammalian cells and can transduce extracellular signals to the nucleus [19]. Aberrant signal transduction of TGF-β signaling may lead to a variety of tumors, including esophageal cancer, hepatocellular carcinoma, pancreatic cancer, gastric cancer, CRC, etc. [20]. TGF-β signaling can suppress tumor development by inhibiting cell proliferation and stimulating cell differentiation in the early stages of cancer. However, it induces tumor progression and metastasis in late stages of cancer, which is known as the "TGF-β paradox" [21]. Variation in the TGF-β pathway is also a common event in CRC tumorigenesis and metastasis. When TGF-β or SMAD are mutated, an abnormal TGF-β signaling pathway would contribute to CRC metastasis [22].

The present review mainly focuses on the role of altered TGF-β signaling in mCRC, the mechanisms through which TGF-β affects CRC metastasis and the clinical application of the key components in TGF-β signaling as potential therapeutic targets for mCRC. These research advances will surely shed new light on TGF-β targeting therapy and benefit the mCRC patients in the near future.

2. Alterations in TGF-β Signaling Pathway in mCRC

2.1. *TGF-β Signaling Pathway*

Research has shown that TGF-β signaling is transduced from cell membrane surface receptors to the nucleus. TGF-β ligands secreted by cells disperses in the matrix in an inactive form, and it can be activated in an integrin-dependent manner [23]. TGF-β1 and TGF-β3 can be activated by avβ6 or avβ8 integrins while TGF-β2 cannot, which implies a different mechanism for TGF-β2 [24]. There are three receptors in the TGF-β signaling pathway: TGFBR1, TGFBR2 and TGFBR3. The TGF-β ligand first binds to the TGFBR2 and induces the formation of a hetero-tetrameric complex of TGFBR2 and TGFBR1 [25,26]. Subsequently, this complex causes the TGFBR2 kinase domain to phosphorylate TGFBR1 in a region of the juxtamembrane domain that is rich in glycine and serine residues. This then activates TGFBR1 and subsequently phosphorylates SMAD2/3 [27,28]. Following this, phosphorylated SMAD2 and SMAD3 can be assembled into complexes with SMAD4 and then translocated to the nucleus where they can regulate the expression of target genes [28]. SMAD7, a negative regulator of the TGF-β pathway, competes with SMAD2/3 for the catalytic site of TGFBR1 phosphorylation and thereby inhibits the phosphorylation of SMAD2/3 [29]. SMAD proteins can be divided into three categories, including the common-mediator SMAD (Co-SMAD), the receptor-regulated SMAD (R-SMAD) and the inhibitory SMAD (I-SMAD). The Co-SMAD (SMAD4) is the central mediator of the TGF-β signaling pathway. The R-SMAD (SMAD1, -2, -3, -5 and -8/9) can be phosphorylated by activated type I receptor kinases. The I-SMAD (SMAD6/7) can competitively inhibit R-SMAD phosphorylation and thereby antagonize TGF-β signaling [19].

In addition to the canonical SMAD-dependent signaling pathway, there are also several non-canonical pathways within the TGF-β superfamily, such as the Rho-associated kinase (ROCK) pathway, the phosphoinositide 3-kinase (PI3K)/protein kinase B (AKT) and the mitogen-activated protein kinase (MAPK) pathway [30]. These activated non-canonical SMAD pathways also crosstalk with the canonical SMAD pathway.

2.2. Aberrant TGF-β Pathway Signals in mCRC

It is generally accepted that the occurrence of cancer is accompanied by the accumulation of gene mutations [31]. Mutations in TGF-β receptors and SMAD proteins occur more frequently in CRC resulting in malignant phenotypes, whereas mutations in TGF-β ligands are relatively rare.

With advances in next-generation sequencing technology, variations in TGF-β signaling in mCRC on different levels have become more accessible for clinical investigators. According to the Ingenuity Pathway Analysis, Wnt, PI3K/AKT and TGF-β/SMAD signaling are the most commonly mutated pathways in colorectal cancer metastasis [32]. Carcinoembryonic antigen (CEA) is widely used as a prognostic clinical marker of metastasis, and the TGF-β signaling pathway is significantly enriched in CEA-induced colorectal liver metastases (CRLM) according to the Kyoto Encyclopedia of Genes and Genomes (KEGG) pathway analysis [33]. In a study utilizing targeted next-generation sequencing (NGS) to assess 128 patients with mCRC, alterations of TGF-β pathways were identified in 17% of the mCRC tissues [34]. In another study involving 579 patients undergoing CRLM resection, 11.2% of patients were found to have TGF-β mutations [35]. Furthermore, aberrant DNA-methylation-regulated genes showed enrichment in TGF-β signaling pathway based on data of DNA methylation (GSE90709, GSE77955) downloaded from the Gene Expression Omnibus database [36].

Although most studies utilize genetic and pharmacological strategies to investigate TGF-β signaling of all three isoforms, these three isoforms actually function through distinct mechanisms. The knockout mice of the three isoforms demonstrated non-overlapping defects: TGF-β1-null mice showed inflammatory disease, TGF-β2-null mice exhibit multiple developmental defects in a wide range of organs, while TGF-β3 knockout led to defective palatogenesis [37–39]. TGF-β1 is expressed more abundantly in the tumor microenvironment (TME) in various human tumors than the other two isoforms and contributes to resistance to checkpoint blockade therapy [40]. TGF-β2 was shown to be involved in neutrophil recruitment in an organoid model of mCRC [41]. In addition, TGF-β1 and TGF-β3 are reported to both be activated in stroma cells and to contribute to the prometastatic process in CRC [42].

The receptor of the TGF-β signaling pathway is indispensable and its change can lead to abnormalities of the pathway. Reports indicated that TGFBR1*6A can switch TGF-β1 growth-inhibitory functions into growth-stimulatory functions, which significantly increased the invasion of SW48 and DLD-1 cells compared with transfected TGFBR1*9A cell lines [43]. The germline allele-specific expression (ASE) of TGFBR1 increases CRC risk for the Caucasian-dominated population in the United States [44]. In the majority of microsatellite instability (MSI) CRC tumors, the gene encoding TGFBR2 has a very high frequency of uniquely inactivating mutations. According to public databases, tumors harboring TGFBR2 mutations showed a greater degree of vascular invasion than tumors without such mutations, which contributes to tumor progression in MSI-positive CRC [45]. In addition, frameshift mutations of TGFBR2 were present in three quarters of late-stage MSI CRC, and this mutation might mediate CRC progression from the early to late stage [46]. In an MSI CRC model cell line, inactivating frameshift mutations of TGFBR2 can reprogram the protein content and regulate the cytokine secretion profile. These changes are related to tumor angiogenesis, migration, metastasis and immune escape of recipient cells [47]. In the HCT116-TGFBR2 MSI CRC cell line model system, which reflects the inverse situation of the TGFBR2-deficient MSI CRC, sialylated β1-integrin is significantly decreased, and variant sialylation could affect metastasis and migration of CRC cells [48]. Additionally, in a cohort of 184 CRC patients and 307 healthy volunteers, male CRC patients with TGFBR2-875A genotypes had a lower risk of CRC progression and metastasis compared with CRC patients with TGFBR2-875G [49].

Because SMAD proteins are key factors in the transduction of the classical TGF-β signaling pathway, alterations in SMAD proteins play a crucial role in late stages of CRC by contributing to migration and metastasis. As reported, CRC patients who lose SMAD activity are more likely to have lymph node metastasis resulting in a poor prognosis [50]. As reported in an analysis of exome capture DNA sequencing from 224 participants, both those with tumors and without, SMAD4 and TGFBR2 are two commonly mutated genes. The mutation frequency of SMAD4 and SMAD2 in non-hypermutated tumors is 10%, while that of TGFBR2 in hypermutated tumors (including MSI-high) is 51% [51]. According to the sequencing analysis of SMAD4, SMAD2 and SMAD3 in a group of 744 primary CRC patients and 36 CRC cell lines, the prevalence of SMAD4, SMAD2 and SMAD3 mutations was found to be 8.6%, 3.4% and 4.3% in sporadic CRC, respectively. In addition, the mutation spectra of SMAD2/3 were highly similar to that of SMAD4, and joint biallelic hits in SMAD2/3 were highly frequent and mutually exclusive to SMAD4 mutation, indicating the crucial roles of these three SMAD proteins in the TGF-β signaling pathway [52].

According to a series of high-throughput analyses, SMAD4 was one of the most commonly mutated genes in mCRC, which will now be further discussed. The results of one study that used targeted NGS sequencing involving 123 non-MSI-high mCRC patients showed a 22.8% mutation frequency of SMAD4 [34]. Similarly, in another study of 32 mCRC patients, the SMAD4 mutation frequency rate was approximately 6% [53]. Similar results were achieved for SMAD4 mutation rates (15% vs. 14%) in primary and metastatic CRCs by comparing genetic profiles [54]. In CRC patients, SMAD4 mutation and deletion detected with NGS were significantly associated with invasive-front pathological markers [55]. In 330 early onset (EO) mCRC patients, SMAD4 was recurrently mutated, resulting in aberrance of the TGF-β pathway in 30% of patients [56]. Using samples obtained from 32 patients, Lopez-Gomez et al. found that SMAD4 expression was at similar levels and was positively associated between the formalin-fixed paraffin-embedded (FFPE) mCRC tumor and their matched liver metastases [57]. Moreover, there is an increased frequency of SMAD4 alterations in ovarian metastases from CRC, suggesting that the oncogenic properties conferred by aberrant TGF-β signaling may contribute to CRC metastasis to the ovaries [58].

SMAD7 is a crucial negative regulator of the TGF-β signaling pathway [59]. Reports indicated that the expression of SMAD7 was remarkably lower in mCRC tissues than in non-tumor tissues [60]. Compared with control mice, mice injected with SMAD7-expressing clones had elevated levels of TGFBR2 expression and TGF-β secretion in liver metastases, which could then lead to phosphorylation and nuclear accumulation of SMAD2. In the nude mouse CRC model, ectopic expression of SMAD7 promoted CRC metastasis to the liver in the splenic injection model [61].

Furthermore, TGF-β mutations always occur concurrently with variations in other signaling pathways, demonstrating that the accumulation of these mutations in mCRC has a synergistic effect on CRC metastasis. For example, $KRAS^{G12D}$ mutation can induce an EMT-like morphology of tumors when combined with mutations in $Tgfbr2^{-/-}$. Moreover, KRAS activation promotes liver metastasis when combined with adenomatous polyposis coli $(APC)^{\Delta 716}$ and TGFBR2 mutations [62]. In the colon epithelium of a CRC mouse model, combined inactivation of APC and TGFBR2 promoted development of adenocarcinoma in the proximal colon, and gasdermin C expression was upregulated by TGFBR2 mutation, resulting in increased CRC cells proliferation [63]. Based on the mCRC mouse model that harbored a $KRAS^{mut}$ allele, conditional null alleles of APC and transformation-related protein 53 (Trp53), the TGF-β pathway was a critical mediator of $KRAS^{mut}$-driven invasiveness, as proven by system-level and functional analysis [64]. Fumagalli et al. found that the accumulation of genetic mutations in the Wnt, epidermal growth factor receptor (EGFR), P53 and TGF-β signaling pathways can drive CRC cells to migrate and grow at distant sites in an orthotopic organoid transplantation model and in engineered human colon tumor organoids [65]. In a Chinese CRLM cohort, CRLM patients with differing primary tumor sites had differences in survival rates, which could be driven by combined variations in the TGF-β, PI3K and RAS signaling pathways [66]. Reports showed patients with mutated SMAD4 had shorter progression-free survival (PFS) than patients with wild-type SMAD4 after receiving anti-EGFR therapy, which may imply a synergistic effect of SMAD4 loss and EGFR in mCRC [67].

In addition, some key proteins or factors can act on the TGF-β signaling pathway and affect CRC progression. Compared with a control group, CRC cells overexpressing Tripartite Motif Containing 25 (TRIM25) exhibit a two-fold higher migration rate. TRIM25 also promotes CRC tumor progression in a nude mice xenograft model by positively regulating the TGF-β signaling pathway [68]. The overexpression of Helicase-like Transcription Factor (HLTF) and activation of Slit2/Robo1 signaling can suppress both CRC cell migration and invasion through the TGF-β/SMAD pathway [69,70]. Prolyl 4-Hydroxylase Subunit Alpha 3 (P4HA3) promotes subcutaneous tumorigenesis in nude mice by upregulating the TGF-β/SMAD signaling pathway, and the knockdown of P4HA3 strongly inhibits the proliferation and invasion abilities of CRC cells [71]. ETS homologous factor could activate the canonical TGF-β pathway through directly upregulating TGF-β1 expression at the transcriptional level and could promote CRC cell proliferation and migration in vitro and in vivo [72]. Glypicans 1 (GPC1) knockdown significantly suppressed levels of TGF-β1 and p-SMAD2, resulting in inhibition of the migration of CRC cells [73]. In vivo experiments showed ZIC2, a protein involved in the advancement of many types of tumors, induced TGF-β1 expression and SMAD3 phosphorylation, resulting in CRLM progression [74].

3. Mechanism of TGF-β Functions in mCRC

CRC metastasis is a dynamic, multistep and multifactorial process, which includes the following successive steps: detachment from the primary CRC site, infiltration into adjacent tissues, invasion into blood/lymphatic circulation, transportation through the circulatory system, intravasation from vasculature and formation of CRC colonies in distant sites. Three critical factors contribute to CRC cells migration (pivotal for early metastasis): regulating the EMT process, stemness and the microenvironment of CRC cells. Additionally, angiogenesis facilitates CRC cell transportation to distal locations. TGF-β signaling contributes to mCRC mainly through the following four mechanisms: promoting EMT, facilitating angiogenesis, creating an immunosuppressive microenvironment and regulating the stemness of mCRC (as shown in Figure 1) [21,28].

3.1. TGF-β Signaling in EMT in mCRC

Epithelial cells undergoing EMT will lose their apicobasal polarity and adhesion, acquire motile mesenchymal characteristics and become more invasive, which contributes to the onset of CRC metastasis [75]. Epithelial and mesenchymal cells can be distinguished by specific molecular markers expressed in cells. For instance, epithelial cells express E-cadherin and cytokeratins, while N-cadherin, Snail, Slug and Vimentin are markers of mesenchymal cells [76]. Cells undergoing the EMT process are distinguished by the loss of E-cadherin expression, a decrease of epithelial cell junctions and cytoskeleton and display a mesenchymal pattern with enhanced cell motility and invasiveness [77].

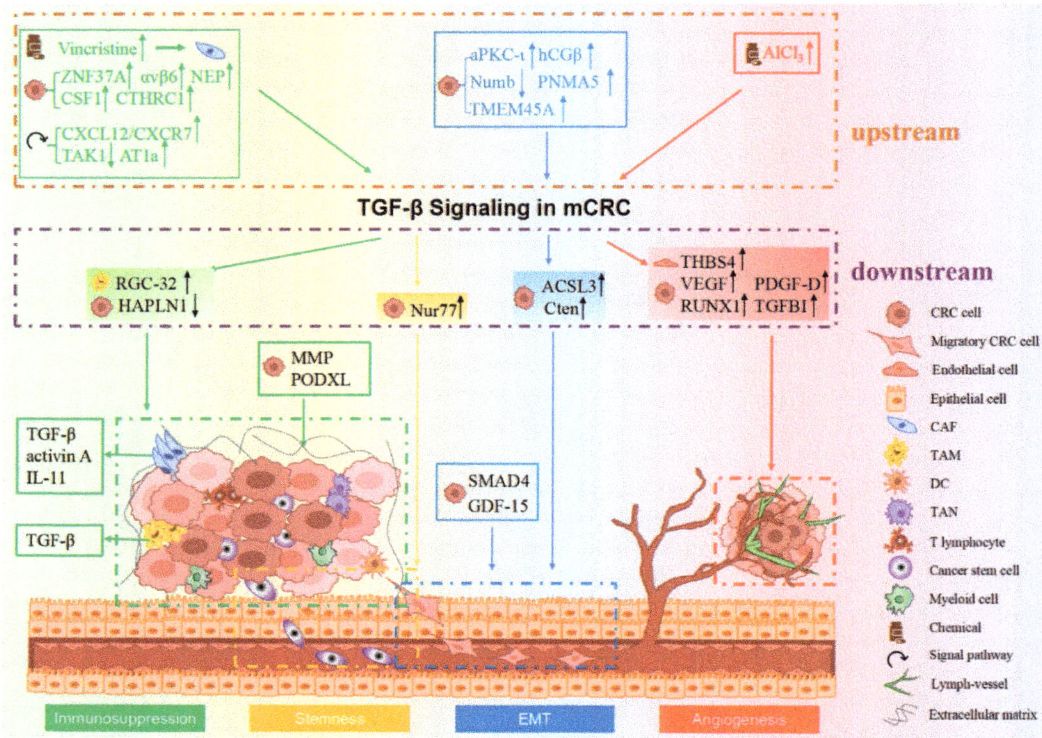

Figure 1. The mechanism of TGF-β signaling in CRC metastasis. TGF-β mainly affects CRC metastasis in four different ways: EMT, angiogenesis, immunosuppression and stemness. Together, they work to facilitate the metastasis of CRC. Tumor cells undergo the EMT process, acquire a mesenchymal-like phenotype in response to TGF-β signaling and then becoming more invasive and spread to distant sites. TGF-β signaling can mediate the formation of new blood vessels, which can promote intravasation of tumor cells from primary lesions into the blood vessels, resulting in tumor metastasis. In the tumor microenvironment (TME), immune cells such as CAFs and TAMs contribute to the immunosuppressive microenvironment and induce dissemination of tumor cells to distant places through TGF-β signaling. Moreover, TGF-β signaling can regulate CSCs in CRC, further promoting tumor metastasis. The molecules upstream or downstream of TGF-β signaling have been enclosed by dotted lines of different colors, and they function through EMT, angiogenesis, immunosuppression and stemness, respectively (distinguished by four different background colors). The symbols in front of the molecules represent whether it is a chemical, a signal factor or a molecule secreted by CRC cells, epithelial cells or immune cells in the TME. EMT, epithelial-to-mesenchymal transition. The legend for different cell types is shown in the lower right.

TGF-β signaling is an essential regulator of the process of EMT. As reported, TGF-β can induce the EMT process by downregulating the expression of tight junction proteins, resulting in weakened tight junctions, which is the key point for EMT induction of TGF-β signaling [78]. SMAD4 is demonstrated to downregulate the expression of Claudin 1, which contributes to CRC metastasis [79]. Most of these observations were made in vitro, although the results of in vivo experiments are more convincing and important. In light of research in human SW480 CRC cells, TGF-β1 can induce Alu RNA expression, the accumulation of which promotes the EMT process, and Alu expression significantly correlates with CRC progression [80]. TGF-β1 upregulates the expression of C-terminal tensin-like (Cten) and EMT markers, and it promotes the cell motility of the CRC cell lines SW620 and HCT116 [81]. TGF-β1 can also induce the upregulation of acyl-CoA synthetases 3 (ACSL3) which produces ATP and reduces NADPH, thus sustaining redox homeostasis and mediating the EMT and metastasis of CRC cells [82]. A functional study indicated that TGF-β can induce SMAD4-dependent EMT followed by apoptosis in HCT-116 and DLD1 CRC cell lines [83]. As reported in CRC cell assays and murine models, acidosis-induced TGF-β2 activation promotes the formation of lipid droplets, which provides energy for cancer cell metastasis and partially promotes EMT [84]. SMAD4 in TGF-β signaling is frequently inactivated in human CRC, and SMAD4 codes for a transcription factor central to canonical TGF-β signaling. Therefore, it is generally understood that EMT will not occur in SMAD4-mutant tumors. However, in SMAD4-mutant CRC cell lines and analyses of human CRC transcriptomes, EMT is not categorically precluded. Possible explanations for this may be that SMAD4-mutant tumors escape the tumor-suppressive function of TGF-β or undergo SMAD4-independent EMT [85]. Moreover, CRC patient tissues exhibited higher GDF-15 expression compared with non-cancerous controls, and in the human CRC cell line LoVo, the overexpression of GDF-15 could upregulate the marker genes of mesenchymal cells. Thus, GDF-15 could lead to EMT and promote CRC cell invasion and migration [86]. Based on the systematic analysis of samples from seven CRC patients, it was found that some potential EMT biomarkers were enriched in TGF-β/Snail and TNF-α/nuclear factor-κB (NF-κB) pathways, and the integrated pathway may be the main axis connecting cancer cells with their TME during EMT [87]. In an immunohistochemical study of 48 resected CRC specimens, SMAD4 was positively linked with the expression of Snail-1, Slug and Twist-1, while it was negatively correlated with E-cadherin expression, implying that SMAD4 promotes the process of EMT [88].

There are also other factors that affect EMT by regulating the TGF-β pathway. For example, atypical protein kinase C-ι (aPKC-ι) knockdown inhibits TGF-β1-induced EMT and cell migration in CRC cells [89]. Furthermore, in 5-fluorouracil (5-FU)-resistant CRC cell lines, knockdown of transmembrane protein 45A (TMEM45A) attenuated multidrug-resistance-enhanced EMT by suppressing the TGF-β/SMAD signaling pathway [90]. In studies utilizing cell line experiments and nude mouse models, Numb expression was negatively correlated with TNM stage and lymph node metastasis, and inhibiting Numb expression promoted the EMT process and the invasion of CRC cells induced by TGF-β [91]. It was found that Paraneoplastic antigen Ma family number 5 (PNMA5) accelerated CRC cell proliferation, invasion and migration in nude mice lung metastasis models, and the knockdown of PNMA5 attenuated TGF-β-induced EMT in CRC cells [92]. As reported in cell assays and mouse xenograft tumors, beta human chorionic gonadotropin (hCGβ) changed expression of EMT-associated genes, and these changes could be reversed by TGFBR1 and TGFBR2 inhibitors, indicating that hCGβ induces EMT in a manner that depends on the TGF-β pathway [93].

3.2. TGF-β Signaling in Angiogenesis in mCRC

Angiogenesis in the TME is a pivotal process that promotes tumor development and metastasis [94]. Newly formed blood vessels can provide oxygen and nutrients to tumor cells as well as allow them to enter into blood circulation and metastasize to distant sites [95].

First, the TGF-β pathway can regulate tumor metastasis by affecting vascular endothelial growth factor (VEGF). In the CRC HCT116 cell line, the upregulation of VEGF expression caused by the absence of SMAD4 enhanced vascular density and promoted the development of metastasis [96]. Additionally, SMAD4 overexpression can inhibit CRC growth by inhibiting VEGF-A and VEGF-C expression in the HCT116 cell line and an promote tumor cell apoptosis in HCT116 cells and nude mouse models [97]. There are primarily two histopathological patterns of vascular changes in CRLM: angiogenic desmoplastic and non-angiogenic replacement [98]. Overexpression of Runt-Related Transcription Factor-1 (RUNX1) in cancer cells of the replacement lesions, which is mediated by TGF-β1 and thrombospondin 1 (TSP1), enhances cell motility to achieve vessel co-option [98]. TGF-β expression is increased in the $AlCl_3$-exposed human CRC cell line HT-29, and this particularly promoted endothelial cell angiogenesis via the induction of VEGF secretion [99]. In an orthotopic mouse model of liver metastasis, the inhibition of TGF-β-induced protein ig-h3 (TGFBI) suppressed angiogenesis of CRC cells and inhibited the progression of CRLM [100]. Second, the synergistic effects of TGF-β and other signal cascades can stimulate angiogenesis by accelerating endothelial cell migration and proliferation [77]. TGF-β can interact with other proteins or pathways to foster angiogenesis in mCRC. The downregulation of platelet-derived growth factor-D (PDGF-D), a downstream signal of TGF-β, inhibited the growth, migration and angiogenesis of CRC cells in vitro and in vivo [101]. Thrombospondin-4 (THBS4), an ECM protein, plays an essential role in the TME and augments the effects of TGF-β1 on angiogenesis [102,103].

3.3. TGF-β Signaling in Immunosuppressive Microenvironment in mCRC

Growing evidence has shown that the TME performs a significant role in tumor initiation, progression and metastasis. The TME comprises non-cancerous cells in the tumor, including cancer-associated fibroblasts (CAFs), endothelial cells, pericytes and different types of immune cells (dendritic cells (DCs), tumor-associated macrophages (TAMs), tumor-associated neutrophils (TANs), natural killer (NK) cells, myeloid cells, T cells, B cells, monocytes etc.), as well as non-cellular components, including ECM and soluble products such as collagen, various cytokines, chemokines and other factors that contribute to CRC metastasis [104–106]. Direct cell-to-cell contact between cancer cells and secretion of cytokines in the TME caused crosstalk, resulting in CRC progression and ultimately metastasis. It was reported that the activity of TGF-β signaling in TME cells such as T cells, macrophages, endothelial cells and fibroblasts improved the organ colonization efficiency of CRC cells, while treating the mice with the TGFBR1-spesific inhibitor LY2157299 inhibited CRC metastasis formation [42]. Elevated TGF-β expression levels is an important feature in the TME of CRC, and TGF-β signaling can regulate the development of CRC, form the system structure of tumors and inhibit the activity of anti-tumor immune cells, which results in an immunosuppressive microenvironment [28,107,108].

Here we summarize recent research and find that most of studies focused on CAFs and immune cells such as TAMs, TANs, DCs, T cells, myeloid cells and monocytes. Only a few studies on TGF-β signaling-mediated CRC progression and metastasis were related to collagen (discussed in the CAF section). TGF-β-signaling-related CRC metastasis involving CAFs and immune cells will be further discussed in detail in the following sections.

3.3.1. CAFs

CAFs are the most numerous cells in the TME, and they affect CRC metastasis by regulating TGF-β signaling directly or indirectly [108,109]. TGF-β is mainly produced by CAFs in CRC, and increased TGF-β promotes T cell exclusion and inhibits the effector phenotype acquisition of type 1 T helper cells (TH1). It has been reported that inhibition of TGF-β enhances the cytotoxic T cell response to tumor cells, thus suppressing liver metastasis [110]. TGF-β activates CAFs to secrete activin A, a TGF-β family member, which induces colon epithelial cell migration and EMT, resulting in a more metastatic phenotype of CRC [111]. Wang et al. have recently reported that the activation of C-X-C motif chemokine ligand 12 (CXCL12)/CXCR7 axis drove CRC cells to secrete exosomal miR-146a-5p and miR-155-5p, which could be taken up by CAFs, thus enhancing CAF activation via JAK2-STAT3/NF-κB signaling. CAFs could secrete more inflammatory cytokines, including TGF-β, further promoting EMT and CRC metastasis to the lung in vivo [112]. ZNF37A, which is upregulated in CRC, is reported to facilitate tumor cell metastasis to the lung and liver via the activation of Thrombospondin Type-1 Domain-Containing protein 4 (THSD4)/TGF-β signaling, and increased TGF-β secretion contributes to transforming fibroblasts to CAFs in the TME, further promoting CRC metastasis [113]. Integrin αvβ6 secreted by CRC cells induced the expression of TGF-β, thereby converting fibroblasts into CAFs and promoting CRC metastasis through the stromal cell derived factor-1 (SDF-1)/C-

X-C motif chemokine receptor type 4 (CXCR4) axis [114]. Treatment of co-cultured CRC and CAF-like cells with vincristine, which is a chemotherapy drug used widely in mCRC clinical treatment, increased the secretion of TGF-βs, induced EMT and promoted the formation of CAFs, thereby enhancing the invasion and metastasis of CRC [115]. Interleukin-11 (IL-11) secreted by TGF-β-stimulated CAFs is a TGF-β target gene, and it activated GP130/signal transducer and activator of transcription 3 (STAT3) signaling in CRC cells and promoted the initiation of CRC cells to metastasis [42]. Endoglin, a TGF-β family coreceptor produced by CAFs, enhanced CRC cell metastasis to the liver in both zebrafish and mouse models [116]. In addition, it has been demonstrated that tumor necrosis factor-related apoptosis-inducing ligand (TRAIL) secreted by SMAD4-deficient CRC cells promotes fibroblasts to produce BMP2, resulting in CRC cell invasion and metastasis [117].

Moreover, TGF-β1 can be secreted by tumor cells in metastasis. Neutral endopeptidase (NEP) co-culturing human colon cancer cell line SW620 (derived from metastatic tumors) with normal colon fibroblasts induced a significant increase in expression of TGF-β1 in SW620 cells, and this effect could be reversed by deletion of NEP [118]. As reported, TGF-β1 promoted the co-migration of colon cancer cells and CAFs, resulting in enhanced liver metastasis and tumor burden [119]. CAF-derived exosomal microRNA (miR)-17-5p caused CRC cells to secrete TGF-β1 into the TME through RUNX3/MYC/TGF-β1 signaling, which triggered CAFs to release more exosomal miR-17-5p to CRC cells, thus establishing a positive feedback loop for CRC metastasis [120]. In CRC, fibroblasts could be converted to CAFs via IL-1β/TGF-β1 signaling, and both TGF-β-activated kinase 1 (TAK1) and TGFBR1 inhibitors suppressed CRC metastasis and CAF accumulation [121]. Two additional studies revealed that CXCR4/TGF-β1 signaling plays an important role in the transformation of mesenchymal stem cells or hepatic stellate cells into CAFs, further promoting CRLM [122,123].

However, CAFs can also suppress CRC progression in some situations. In a genetically modified metastatic CRC mouse model, depletion of alpha smooth muscle actin $(\alpha SMA)^+$ CAFs resulted in an increase of forkhead box protein 3 $(Foxp3)^+$ regulatory T cells (Tregs) and suppression of $CD8^+$ T cells via BMP4/TGF-β1 paracrine signaling, ultimately promoting CRC invasiveness and lymph node metastasis [124]. A recent study showed that gremlin 1 (GREM1) and the immunoglobulin superfamily contain leucine-rich repeat (ISLR), representing two different types of fibroblast subpopulations that exert opposing roles in the signal transduction of BMP. Neutralization of GREM1 or overexpression of ISLR in fibroblasts could reduce CRC hepatic metastasis [125].

Furthermore, decreased expression of hyaluronan and proteoglycan link protein-1 (HAPLN1) regulated collagen deposition in CRC via the TGF-β signaling pathway, and increased collagen resulted in TME changes and CRC cell proliferation, migration and invasion [126].

3.3.2. Immune Cells

TAMs, one of the most common immune cells in the TME, have been reported as key contributors to promote tumor metastasis [127,128]. Liu et al. found severe TAM infiltration in tumor tissues of mCRC patients, and TAM-derived TGF-β could activate HIF1α/TRIB3/β-catenin/Wnt signaling to enhance CRC progression [129]. GDF-15, secreted by macrophages, is a divergent member of the human TGF-β superfamily, and it can increase expression of EMT genes, thereby promoting the invasion and metastasis of CRC via the ERK1/2/c-Fos signaling pathway [130]. Shimizu et al. found that Kupffer cells, known to be resident hepatic macrophages, released TGF-β1 and promoted liver metastasis of CRC through angiotensin II subtype receptor 1a (AT1a) signaling. Moreover, depletion of Kupffer cells reduced metastatic areas [131]. It has been proven that TGF-β1 secretion of CRC cells upregulated macrophage expression of Response Gene to Complement 32 (RGC-32) and thus enhanced macrophage migration and promoted tumor progression [132]. Recently, Chen et al. found that oxaliplatin-based chemotherapy induced TAM recruitment to release TGF-β, which was mediated by CRC-cell-derived CSF1, resulting in programmed cell death-Ligand 1 (PD-L1) upregulation and an immunosuppressive TME. Inhibition of PD-L1 expression in CRC could make cancer cells sensitive to chemotherapy, reduce CRC lung metastasis and increase infiltration of $CD8^+$ T cells. Both $CSF1R^+$ TAM depletion and TGF-β receptor blockade combined with chemotherapy could inhibit tumor growth significantly [133]. Through specific differentiation, macrophages can be polarized into two different phenotypes: activated M1-type and alternatively activated M2-type. M1-type macrophages inhibit tumor growth and progression, whereas M2-type macrophages induce the progression and metastasis of tumors in CRC [127,134]. Ma et al. found that M2-type macrophages were positively correlated with infiltrating $Foxp3^+$ Tregs in CRC, which may promote the development of CRC via the TGF-β/SMAD signaling pathway [135]. Cai et al. reported that M2-type macrophages that secreted TGF-β promoted EMT by activating the SMAD2,3-4/Snail/E-cadherin signaling pathway, resulting in CRC lung metastasis [136]. Zhang et al. revealed that Collagen Triple Helix Repeat Containing 1 (CTHRC1) secreted by CRC cells induced macrophages to the M2-type through activation of TGF-β signaling, further enhancing CRC liver metastasis [137]. Recently, Li et al. developed a thermosensitive hydrogel called Gel/(regorafenib + NG/LY3200882 (LY)), which could sequentially release regorafenib and LY (a selective TGF-β inhibitor) in tumor cells. Using colorectal tumor-bearing mouse models, they found that Gel/(regorafenib + NG/LY) can effectively inhibit tumor growth and liver metastasis, which was achieved by increasing levels of $CD8^+$ T cells, reducing infiltration of TAMs and myeloid-derived suppressor cells and shifting macrophage polarization from M2-type to M1-type in TME [138].

Apart from CAFs and TAMs, other cellular components, such as TANs, myeloid cells, monocytes, DCs and T cells, in the TME can also affect CRC metastasis through TGF-β signaling. TAN infiltration was demonstrated to be positively correlated with the clinical stage of CRC patients [139]. Anti-TGF-β treatment attenuated tumor growth, which was mediated by inhibition of PI3K/AKT signaling pathways in TANs and TGF-β/SMAD signaling pathways in CRC cells [140]. Activation of epithelial NOTCH1 enhanced epithelial TGF-β2 expression and facilitated liver metastasis of CRC through TAN infiltration, which was mediated by TGF-β signaling. Neutrophil depletion led to increased $CD8^+$ T cells in both primary tumors and livers and decreased metastasis. In addition, blocking TGF-β signaling in neutrophils can effectively reduce CRC metastasis [41]. Using mouse xenograft models, Itatani et al. found that a deficiency of SMAD4 in human CRC cells upregulated CCL15 expression, thus recruiting $CCR1^+$ myeloid cells and promoting liver metastasis of CRC [141]. Furthermore, inflammation is an important driver for CRC development and metastasis. CRC cells treated with lipopolysaccharide-stimulated monocyte conditioned medium showed reduced expression of Growth Factor Independence 1 and enhanced EMT and CRC cell metastatic formation, which might have been mediated by TGF-β signaling [142]. Wang et al. suggested that silencing poly (ADP-ribose) glycohydrolase (PARG) in CT26 cells could suppress liver metastasis of colon carcinoma by suppression of poly (ADP-ribose) polymerase (PARP) and NF-κB and that it could reduce secretion of IL-10 and TGF-β, thus promoting the proliferation and differentiation of DCs and T cells, resulting in inhibition of metastasis by changes in immune function [143]. Treg and T helper 17 (Th17)-related genes seem to contribute greatly to CRC development and progression. Miteva et al. investigated the expression of Treg and Th17-related genes in CRC tissues and found that Foxp3, IL-10 and TGF-β1 expression was increased in CRC metastases in contrast to IL17A and NOS2. Treg and Th17-related gene expression in both primary tumor and regional lymph nodes might provide a suitable microenvironment for accelerating CRC metastasis [144]. The mechanism by which other cells in the TME influence T cells via TGF-β signaling directly or indirectly was covered in the previous section [41,110,124,133,135,138,143].

3.4. TGF-β Signaling in Stemness in mCRC

Most tumors, including CRC, contain a small population of cancer stem cells (CSCs) which are regarded as key contributors to tumor generation, progression, recurrence, metastasis and chemotherapy drug resistance [145,146]. According to recent studies, the TGF-β signaling pathway can affect metastasis of CRC by affecting CSCs in CRC or the stemness of CRC cells.

Mesenchymal stem cells co-cultured with CRC cells showed enhanced invasive ability, which was mediated by increased expression of TGF-β1 and decreased expression of p53, resulting in effective inhibition of CRC metastasis [147]. Reports suggested that TGF-β could convert Nur77's role from cancer inhibition to cancer promotion, which is associated with CRC stemness, metastasis and oxaliplatin resistance [148]. CSCs have specific markers on their surface. CD51, a novel functional marker for colorectal CSCs, could increase the sphere-forming abilities, tumorigenic capacities and migratory potentials of CRC cells, and it may regulate EMT and chemoresistance through TGF-β/SMAD signaling [149]. In a novel mouse model of CRLM, proteomic analysis revealed that the expression of CRC stem cell markers in CRC cells was elevated compared with the non-metastatic model, and the expression of these markers was regulated negatively by the TGF-β/SMAD4 pathways [150].

3.5. Other Mechanisms of TGF-β in mCRC

In addition to the four mechanisms mentioned above, TGF-β can also affect the metastasis of CRC through some additional mechanisms. TGF-β regulates matrix metalloproteinase (MMP) expression in cancer cells, while MMPs produced by either cancer cells or stroma cells activate latent TGF-β, together facilitating progression of CRC [151]. The expression of TGF-β and the podocalyxin-like (PODXL) protein in CRC cells could increase under radiation and then promote ECM deposition, resulting in cell migration and invasiveness [152]. Bioinformatic analysis and functional characterization indicated that TGF-β and Snail promoted CRC migration by preventing degradation of the non-coding RNA LOC113230-related argininosuccinate synthase 1 (ASS1) [153]. Reports indicate that cancer epithelial cells show a robust outward apical pole throughout the process of dissemination, which is referred to as tumor spheres with inverted polarity (TSIPs). TSIPs form and propagate via the collective apical budding of hypermethylated CRCs downstream of TGF-β signaling, which could drive the formation of peritoneal metastases [154]. Moreover, TrkC, which was overexpressed in CRC, could also increase the ability to form tumor spheroids, thus enhancing the metastatic potential of CRC by activation of AKT and suppression of TGF-β signaling [155]. It has been demonstrated that TGF-β inhibits lymph angiogenesis by inhibiting collagen and calcium-binding EGF domain-1 (CCBE1) expression, and CCBE1 has a pro-tumorigenic role in lymphatic metastasis of CRC [156]. TGF-β2 could enhance the metastatic potential of human CRC cell lines via upregulating the expression of catalase and controlling H_2O_2 output [157].

4. Potential Application of TGF-β Signaling in Diagnosis and Prognosis of mCRC

Specific molecular biomarkers are significant tools for early diagnosis and prognosis of mCRC, and early prognosis is the most successful and effective method to improve the survival rate of CRC patients [158]. Through integrative clustering of the expression profiles of miRNA-correlated genes and methylation-correlated genes, four molecular subtypes (S-I, S-II, S-III and S-IV) were confirmed in CRC patients from The Cancer Genome Atlas (TCGA) [159].

Due to the poor prognosis of patients with mCRC, it is crucial to make a rapid and accurate diagnosis of mCRC based on specific biomarkers as early as possible. Many reports on applications of TGF-β signaling in the diagnosis of mCRC have been published. According to relative mRNA quantification, the expression of TGF-β1 in CRC distant metastases is significantly increased compared with primary tumor tissues [160]. The increased expression of TGF-β2 is a dependable predictor of lymph node metastasis in CRC patients [161]. Compared with healthy controls, serum levels of GDF-15, a member of the TGF-β superfamily, were remarkably upregulated in mCRC patients and had the same sensitivity as the standard tumor marker CEA, indicating that GDF-15 could be an effective biomarker in mCRC patients [162]. The morphogen nodal is dramatically overexpressed in malignant transformation in CRC, and it could be a potential marker for the consensus molecular subtype 4 (CMS4) subtype of CRC [163].

More reports of TGF-β signaling on prognosis have been shown in mCRC patients. The combined activin and TGF-β ligand expression score was utilized to predict shorter OS in a group of 40 CRC tumors, 10 metastasis and 10 control samples [164]. In EO mCRC patients, mutated TGF-β pathways were found to be associated with unfavorable OS by capture-based targeted sequencing [56]. For the TGFB1-509C/T single nucleotide polymorphism (SNP), CRC patients with the TT allele have the shortest median survival, which is due to malignant progression in advanced stages [165]. High levels of TGF-β in blood samples are negatively correlated with PFS in mCRC patients before treatment with regorafenib, and these results suggest that the cytokine signature can distinguish whether patients respond to regorafenib treatment or not [166]. The survival analysis demonstrated that MYC and TGF-β pathway alterations were related to a shorter OS in mCRC patients, and this negative prognostic impact was maintained after receiving an anti-EGFR antibody [34]. Using tumor tissues from 230 mCRC patients treated with oxaliplatin combined with 5-FU chemotherapy, Baraniskin et al. reported that SMAD4 expression was decreased in 34% of mCRC samples, and these patients had a shorter PFS and OS compared with patients in which SMAD4 is stably expressed [167]. In an analysis of multiple gene mutation assessments in 123 regorafenib-treated mCRC patients, researchers detected a SMAD4 mutation in one patient who had long response to regorafenib [168]. In addition, SMAD4-mutated patients performed significantly worse in terms of PFS than those without SMAD4 mutations in a study with 76 regorafenib-treated mCRC patients [169]. Whole exome sequencing analysis of 77 mCRC patients revealed that SMAD4 mutations were significantly correlated with poor prognosis [170]. Patients with SMAD4 mutations developed CRLM and had worse OS after hepatic resection [171]. In general, changes in TGF-β signaling pathways in CRC cause cancer cells to become more aggressive and more likely to metastasize; thus, patients harboring mutations in TGF-β signaling components often have a poor prognosis.

Other members associated with the TGF-β family can also be utilized as prognostic biomarkers in mCRC. Studies imply that high tumor expression of activin A (a homodimer of inhibin beta A) is associated with poor prognosis in patients with CRC, and activin A receptor type 2A (ACVR2A) (a membrane receptor in the TGF-β signaling pathway) depletion plays an important role in CRC distant metastasis and may be recommended as a prognostic biomarker in CRC patients [172,173]. As mentioned in the previous section, the entry level of GDF-15 might be a prognostic factor that is strongly relevant to OS in mCRC patients [162]. Overexpression of inhibin subunit beta B (INHBB) (a protein-coding gene that participates in the synthesis of TGF-β family members) was positively associated with CRC invasion and distant metastasis, suggesting that it could be a potential prognostic biomarker for mCRC [174]. Furthermore, knockdown of high inhibin, beta A (INHBA) in vitro can inhibit CRC cell migration and invasion by inhibiting the TGF-β pathway, and INHBA expression is closely related to poor prognosis in CRC patients [175]. Overexpression of lncRNA-activated by TGF-β (lncRNA-ATB) was significantly associated with CRC metastasis, and lncRNA-ATB expression could be a prognosis biomarker of OS in CRC patients [176]. Additionally, TGFBR2 deficiency is positively correlated with upregulation of miR-31-3p [177], which is a predictive biomarker for the efficacy of anti-EGFR treatment that mCRC patients received [178].

5. Targeting TGF-β Signaling Pathway in mCRC

As we have discussed in previous sections, TGF-β signaling plays a significant role in CRC metastasis by promoting EMT, facilitating angiogenesis, contributing to an immunosuppressive TME, regulating stemness of mCRC cells and other mechanisms. These research achievements led us to explore more strategies targeting TGF-β signaling which may have promising application prospects in mCRC therapy. So far, long non-coding RNAs (lncRNAs), miRNAs, kinase inhibitors and natural compounds are common strategies that have been utilized for TGF-β targeting in clinical trials, and new sequencing techniques could facilitate the development of personalized medicine for mCRC patients. These various factors targeting TGF-β signaling in CRC metastasis are summarized in Table 1.

5.1. LncRNAs

LncRNAs are a class of multifunctional noncoding RNAs whose sizes are greater than 200 nucleotides. Many recent studies have shown that lncRNAs play a critical role in regulating progression and metastasis in CRC [179,180]. In CRC patients, outlier expression of the lncRNA MIR31HG was observed and was characterized by elevated EMT, TGF-β and IFN-α/γ gene expression signatures in pre-clinical models [181]. The lncRNA CTBP1-AS2 increased CRC cell invasion and decreased apoptosis by activating the TGF-β/SMAD2/3 pathway and was closely associated with worse survival rate in CRC patients [182]. Additionally, the lncRNAs TP73-AS1 and MIR503HG inhibited the migration and invasion of CRC cells by inactivating TGF-β1 and downregulating TGF-β2, respectively [183,184]. Transwell assays showed that TGF-β2 overexpression increased cell invasion, while overexpression of the lncRNA HOXC-AS3 could reverse the effect of overexpression of TGF-β2 [185]. According to an experiment in CRC cell lines, silencing of the lncRNA ezrin antisense RNA 1 (lncRNA EZR-AS1) accelerated CRC cell apoptosis and inhibited the migration and EMT of CRC cells by blocking TGF-β signaling [186]. Silencing of the lncRNA MIR22HG promoted CRC cell proliferation and tumor metastasis in vitro and in vivo by competitively interacting with SMAD2 [187]. Moreover, LINC00941 enhanced invasive capacity and accelerated lung metastasis by activating EMT by directly binding SMAD4 and preventing SMAD4 protein degradation in mCRC [188].

5.2. MiRNAs

Increasing evidence has shown that miRNAs that regulate TGF-β signals have significant roles in the progression and metastasis of CRC; they act as oncogenes or tumor suppressors to regulate expression of specific targets [189]. Compared with primary CRC, a series of studies on related miRNAs reported epigenetic alternations in CRLM [36]. MiR-425 and miR-576 were significantly upregulated in CRLM based on GSE81581 and GSE44121 datasets, and the two miRNAs were associated with CRC metastasis by co-participating in inhibition of the TGF-β signaling pathway [190]. It has been proven that upregulation of miR-329 suppresses CRC cell invasion by inhibiting TGF-β1, and low expression of miR-329 is correlated with lymph node metastasis in CRC patients [191]. Upregulated expression of plasma miR-211 and 25, which are relevant to the high expression of TGF-β1 in CRC patients, was positively correlated with lymph node metastasis [192]. In HCT116 colon cancer cells in which kallikrein 6 was knocked down, miR-203 was demonstrated to inhibit migration and invasion of CRC cells by inhibiting the EMT through suppression of TGF-β2 [193].

Some miRNAs function through targeting the TGF-β receptors in mCRC. For instance, downregulation of miR-301a was shown to inhibit CRC migration and invasion both in vitro and in vivo by repressing TGFBR2 protein expression in an analysis containing 48 cases of CRC tissues, adjacent non-tumor tissues and five CRC cell lines [194]. TGFBR2 repression by overexpression of the entire miR-371~373 cluster decreased tumor-initiating potential in tumor-initiating cells [195]. Reports indicated that miR-3191 promoted CRC cells migration and invasion by downregulating TGFBR2 [196]. Artificial overexpression of miR-490-3p inhibited cell migration and invasion in CRC cell lines through the suppression of TGFBR1 and MMP2/9 [197]. CircFAM120B overexpression blocked CRC cell migration and reduced the expression of miR-645. In addition, TGFBR2 was a target of miR-645, whose inhibition suppressed CRC cell migration and can be restored by TGFBR2 knockdown [198]. As reported in the LoVo cell experiment and subcutaneous tumor model, the inhibition of miR-424 suppressed migration and invasion of CRC cells as well as arrested CRC cells at the G0/G1 phase by repressing TGFBR3 [199].

There are also many reports on miRNAs that affect CRC metastasis through SMAD proteins in the TGF-β signaling pathway. Functional studies showed that miR-27a inhibited SMAD2 expression at transcriptional and translational levels and that it promoted colon cancer cell apoptosis and attenuated cell migration [200]. Ectopic expression of miR-140 inhibited EMT partially through downregulating SMAD3, and it enhanced invasive capacities of CRC cells in vitro, while overexpression of miR-140 inhibited the metastasis of CRC in vivo [201]. Studies confirm that miR-20a-5p promoted the invasion and metastasis ability of CRC cells and liver metastasis, as well as accelerated the EMT process by reducing SMAD4 expression, which is slightly controversial compared with most other reports [202]. Furthermore, bioinformatic predictions and experimental validation demonstrated that SMAD7 is a direct target of miR-25 in mCRC, and miR-25 inhibition could promote the migratory ability of CRC cells via the suppression of SMAD7 [203]. High miR-4775 expression promoted CRC cell metastasis and EMT via downregulating SMAD7 and thereby activated the TGF-β pathway both in vitro and in vivo [204]. Wang et al. demonstrated that miR-21-mediated inhibition of SMAD7 accelerated TGF-β-dependent EMT in CRC, indicating that loss or inhibition of SMAD7 could promote CRC metastasis [205]. This implies that the overexpression of circTBL1XR1 enhances the proliferation and migration of CRC cells by binding to miR-424, which inhibits SMAD7 [206]. All these examples show that miRNAs can inhibit SMAD7, promote TGF-β-dependent EMT and contribute to CRC metastasis.

5.3. Kinase Inhibitors

Some kinase inhibitors for the TGF-β signaling pathway have been evaluated in various models for mCRC combination treatment to improve the efficacy of therapy. TGFBR1, TGFBR2 and TGFBR3 mutations were found in mCRC patients who responded to regorafenib, suggesting that the TGF-β signaling pathway may play a leading role in the regorafenib response [207]. While using regorafenib, a novel oral multikinase inhibitor, mCRC patients with SMAD4 mutations or activation of the TGF-β pathway showed a worse PFS, which was demonstrated by NGS-based cancer panel tests [169]. Based on a TGF-β-inducible reporter system, Zhang et, al. showed that the TGF-β receptor kinase inhibitor LY2109761 inhibited CRLM by blocking the tumor promoting function of TGF-β in vivo [208]. In a colon cancer liver metastases murine model, mice were treated with adoptive natural killer cells combined with the TGF-β receptor kinase inhibitor LY2157299, and a significant eradication of liver metastases occurred [209]. A study using human CRC cell lines demonstrated that sitagliptin can inhibit CRC cell metastasis by partially blocking TGF-β1-driven EMT [210]. According to the targeted NGS analysis of tumor samples with pre- and post-cetuximab treatment, the copy number of the SMAD4 gene changed, while the TGF-β signaling pathway had various recurrent mutations [211]. The therapeutic potential of these variants requires further clarification.

Correspondingly, dual treatments with the TGF-β galunisertib (LY2157299) inhibitor and AXL inhibitor prominently reduced migration capabilities of human CRC cell lines [212]. Moreover, melatonin, hyperbaric oxygen and combined treatments inhibited CRC metastasis through a variety of mechanisms, including restraining cancer stemness [213]. However, the application of inhibitors should be taken under careful consideration, as it has been reported that epithelial truncation of TGFBR2 leads to fatal inflammatory diseases and invasive CRC in APC mice (a model of intestinal neoplastic disease). Moreover, APC mice with global suppression of TGF-β signaling present with an overall increase in inflammation and tumor formation, suggesting that CRC patients treated with TGF-β inhibitors may have a worse outcome by enhancing inflammatory responses [214].

5.4. Natural Compounds and Chinese Herbal Formulas

Some natural compounds and Chinese herbal formulas can also be utilized as indirect approaches for targeting TGF-β. In vitro results from transwell and scratch wound assays demonstrated that solasodine inhibited CRC cell invasion and migration, which was strengthened by TGF-β1. Solasodine also attenuated TGF-β1-induced EMT in vivo [215]. A traditional Chinese herbal medicine, Hedyotis diffusa Willd, may develop its anti-metastatic activity by restraining TGF-β/SMAD4 pathway-mediated EMT in 5-FU-resistant CRC cells [216]. In addition, baicalin caused cell cycle arrest in the G1 phase and EMT inhibition through inhibiting the TGF-β/SMAD pathway in CRC RKO and HCT116 cell lines [217]. Celastrol significantly inhibited human CRC cells growth, adhesion and metastasis by repressing the TGF-β1/SMAD signaling pathway [218]. The ethanol extract of *Scutellaria barbata* D. Don (EESB) significantly reduced the migration ability of HCT-8 cells in a dose-dependent manner. Furthermore, EESB decreased expression of MMPs and proteins involved in PI3K/AKT and TGF-β/SMAD signaling [219]. Mechanistically, metformin can block the activation of TGF-β signaling by INHBA, which is an important ligand of TGF-β signaling. It can then downregulate the activity of the PI3K/AKT pathway, leading to cell cycle arrest and inhibition of the proliferation of CRC [220]. In vitro, ursolic acid inhibited the migration and invasion of human CRC HCT116 and HCT-8 cells by interfering with the TGF-β1/ZEB1/miR-200c signaling network [221]. Compared with control mice, Modified Shenlingbaizhu Decoction (MSD) treatment significantly reduced the size of CRC tumors and the serum content of TGF-β1. Similarly, MSD inhibited CRC cell migration and invasion by limiting TGF-β/SMAD signaling [222]. Qingjie Fuzheng granule (QFG), a traditional Chinese medicine, suppressed the growth, wound-healing abilities and migration of HCT-8 and HCT116 cells. Moreover, QFG decreased the expression of lncRNA ANRIL, TGF-β1, p-SMAD2/3, SMAD4 and N-cadherin in CRC cells, suggesting that QFG inhibits the metastasis of CRC through the TGF-β1/SMAD axis [223]. Combined with the TCGA database results and previous network pharmacology, reports indicated that Fuzheng Xiaojijinzhan might play an anti-CRC metastasis role by inhibiting the TGF-β-Snail1 pathway [224].

Since strong genetic heterogeneity exists in CRC patients, development of personalized medicine for CRC patients is of extraordinary significance and value in clinical trials [225]. Based on large-scale data sharing and analytics, CRC is divided into four CMSs with distinguishing features: CMS1 (microsatellite instability immune, 14%), CMS2 (canonical, 37%), CMS3 (metabolic, 13%) and CMS4 (mesenchymal, 23%). Among them, CMS4 has prominent TGF-β activation, stromal invasion, angiogenesis and an immunosuppressive phenotype [226]. In the past decade, more and more efforts have been made to select the appropriate patient subsets for specific treatment of mCRC. Although the development of novel biological agents for therapies such as VEGF and EGFR has further changed the prospects for the treatments of mCRC, not all patients respond similarly to these therapies, so individualized medical treatments are in great need [227]. Designing personalized medicine targeting TGF-β signaling is definitely a good choice for CMS4 subtypes patients of CRC.

Table 1. Summary of factors targeting TGF-β signaling and acting on CRC metastasis.

Types		Targets	Involvements in Metastasis	Clinical Application	References
LncRNAs	MIR22HG	TGF-β pathway	Interact with SMAD2 and inhibit EMT	Facilitating immunotherapy in CRC	[187]
	MIR31HG	TGF-β pathway	Promote CRC cell migration and immunosuppression	Biomarker of cellular state	[181]
	EZR-AS1	TGF-β pathway	Promote CRC cell migration, proliferation and EMT	—	[186]
	TP73-AS1	TGF-β1	Promote CRC cell migration	Prognosis marker in CRC	[183]
	MIR503HG	TGF-β2	Inhibit CRC cell migration	Prognosis marker in CRC	[184]
	HOXC-AS3	TGF-β2	Reverse the effect of overexpression of TGF-β2	—	[185]
	CTBP1-AS2	TGF-β/SMAD2/3 pathway	Promote CRC cell migration and inhibit apoptosis	Prognosis marker in CRC	[182]
	LINC00941	TGF-β/SMAD2/3 pathway	Prevent SMAD4 protein degradation and activate EMT	Prognosis marker in CRC	[188]

Table 1. *Cont.*

Types		Targets	Involvements in Metastasis	Clinical Application	References
miRNAs	miR-425	PTEN-P53/TGF-β	Inhibit cellular immune function	Shortened overall survival	[190]
	miR-576	PTEN-P53/TGF-β	Inhibit cellular immune function	Shortened overall survival	[190]
	miR-329	TGF-β1	Inhibit CRC cell migration	—	[191]
	miR-203	TGF-β2	Inhibit EMT	—	[193]
	miR-490-3p	TGFBR1	Inhibit CRC cell migration	Associated with poor prognosis of survival	[197]
	miR-301a	TGFBR2	Promote CRC cell migration	—	[194]
	miR-371~373	TGFBR2	Decrease tumor-initiating potential of CRC cells	—	[195]
	miR-3191	TGFBR2	Promote CRC cell migration	—	[196]
	miR-645	TGFBR2	Promote CRC cell migration and glycolysis	—	[198]
	miR-424	TGFBR3/SMAD7	Promote CRC cell migration and arrest cell cycle/promote proliferation	—	[199,206]
	miR-27a	SMAD2 and SGPP1	Inhibit CRC cell migration and promote apoptosis	Biomarker for monitoring CRC development and progression	[200]
	miR-140	SMAD3	Inhibit CRC cell migration	—	[201]
	miR-20a-5p	SMAD4	Promote CRC cell migration and EMT	Predicts poor prognosis in CRC patients	[202]
	miR-25	SMAD7	Inhibit CRC cell migration	—	[203]
	miR-4775	SMAD7	Promote CRC cell migration and EMT	Predicts poor survival	[204]
	miR-21	SMAD7	Accelerate TGF-β dependent EMT	—	[205]
circRNAs	circFAM120B	miR-645	Inhibit CRC cell migration and glycolysis	—	[198]
	circTBL1XR1	miR-424	Promote CRC cell migration	—	[206]
Kinase Inhibitors	Sitagliptin	TGF-β1	Inhibit EMT and impair cell cycle	Prevents colon cancer and lung metastasis in animal models and humans	[210]
	LY2157299	TGF-β receptor	Mitigate TGF-β driven impairment of NK cell cytotoxicity	Currently in clinical trials for various malignancies	[209]
	LY2109761	TGF-β receptor	Downregulated the phosphorylation of SMAD2	Applied mostly in preclinical animal experiments	[208]
	regorafenib	TGF-β/SMAD4 pathway	—	While using regorafenib, patients with SMAD4 mutation or activation of TGF-β pathway showed a worse PFS	[169]
Natural com-pounds and Chinese herbal formulas	solasodine	TGF-β1	Inhibit CRC cell stemness and EMT	—	[215]
	MSD	TGF-β1	Inhibit CRC cell migration	Reduced the size of CRC tumors in mouse model	[222]
	Celastrol	TGF-β1/SMAD pathway	Inhibit CRC cell migration	—	[218]
	baicalin	TGF-β/SMAD pathway	Inhibit EMT, stemness and cell cycle	—	[217]
	QFG	TGF-β/SMAD pathway	Inhibit CRC cells growth and migration	—	[223]
	Hedyotis diffusa Willd	TGF-β/SMAD4 pathway	Inhibit CRC cell migration and EMT	—	[216]
	ursolic acid	TGF-β1/ZEB1/ miR-200c	Inhibit CRC cell migration	—	[221]
	metformin	TGF-β ligand and PI3K/AKT pathway	Arrest cell cycle and inhibit cell proliferation	—	[220]
	EESB	PI3K/AKT and TGF-β/SMAD signaling	Inhibit CRC cell migration and decrease the expression of MMPs	—	[219]
	Fuzheng Xiaojijinzhan	TGF-β-Snail1	Anti-CRC metastasis role	—	[224]

6. Conclusions and Future Perspective

Metastatic CRC is an intractable disease due to its poor prognosis, high mortality and limited optimal therapies in clinical situations worldwide, even in developed countries. Several key areas in mCRC research include: early identification of metastasis, recognition of specific prognostic and predictive biomarkers, discovery of new molecular targets, development of new drugs and clinical operations. TGF-β represents a conserved signaling pathway that is widely involved in various physiological and pathological processes. In this review, we summarized the changes in the TGF-β signaling pathway in mCRC patients, its functional mechanisms and its possible applications in mCRC diagnosis, prognosis and potential targeted therapies in clinical trials. We explained in detail that TGF-β signaling functions to promote EMT, facilitate angiogenesis, suppress anti-tumor activity of the immune cells in the microenvironment and contribute to stemness of CRC cells in mCRC (as shown in Figure 1). Following these working mechanisms of TGF-β signaling in mCRC, molecular targeting therapies aimed at the different key factors upstream and downstream of TGF-β signaling could be accordingly developed to improve the efficacy and safety of treatments, especially in MCS4 subtypes of mCRC.

Various remaining problems regarding TGF-β signaling in CRC metastasis still need to be clarified. Initially, TGF-β was not considered as a good target for tumor treatments because of its dual roles in tumor early development and late-stage metastasis. However, recent years witnessed an increasing number of new therapies that target TGF-β signaling in CRC metastasis. These include suppressing TGF-β or downstream components of the signaling, blocking crosstalk between TGF-β signaling pathways and other signal pathways and redirecting TGF-β signaling from pro-tumor to anti-tumor functions in CRC metastasis [228]. Additionally, because mCRC is a complicated disease, we should attach importance to not only TGF-β signals from inside but also outside of CRC cells, namely from the stroma cells and microenvironment of mCRC. Hopefully, in-depth and systematic studies in this research field will help us understand more in the future.

Due to the heterogeneity of CRC cells and individual differences among mCRC patients, it is impossible to find an optimal treatment strategy that fits everyone. Thanks to improved knowledge of the molecular mechanisms underlying CRC metastasis, promising advances help us modify traditional treatments. Recent reports demonstrated that targeting TGF-β could be combined with other signal inhibitors such as combinatorial synergy, reverse therapy resistance or sensitize radiotherapy to achieve a sustained therapy response in CRC patients [28,229]. Moreover, since TGF-β signaling is an immunosuppressive regulator in the TME of mCRC, there is great potentials in combining TGF-β targeting with immunotherapy agents to enhance the efficacy and benefit for patients. In order to move forward, applications of NGS and genetic profiling in clinical trials will help us characterize the molecular subtypes of TGF-β signaling in mCRC patients, and appropriate personalized medicine specifically targeting TGF-β can be definitely and smoothly translated into mCRC treatments.

Author Contributions: Writing—original draft preparation, X.L., Y.W. and T.T.; writing—review and editing, X.L., Y.W. and T.T.; supervision and funding acquisition, T.T. All authors have read and agreed to the published version of the manuscript.

Funding: This research was funded by the Natural Science Foundation of Beijing (grant number 5222019).

Acknowledgments: We thank Juan Liu and Tongtong Cui for reviewing and editing the manuscript. We also thank Angela Papierski for proofreading the manuscript.

Conflicts of Interest: The authors declare no conflict of interest.

Abbreviations

5-FU	5-fluorouracil
αSMA	alpha smooth muscle actin
ACSL3	acyl-CoA synthetases 3
ACVR2A	activin A receptor type 2A
AKT	protein kinase B
APC	adenomatous polyposis coli
aPKC-ι	atypical protein kinase C-ι
ASE	allele-specific expression
ASS1	argininosuccinate synthase 1
AT1a	angiotensin II subtype receptor 1a
BMP	bone morphogenetic protein
CAFs	cancer-associated fibroblasts
CCBE1	collagen and calcium-binding EGF domain-1
CEA	Carcinoembryonic antigen
CMS	consensus molecular subtype
Co-SMAD	common-mediator SMAD
CRC	colorectal cancer
CRLM	colorectal liver metastases
CSCs	cancer stem cells
Cten	C-terminal tensin-like
CTHRC1	collagen triple helix repeat containing 1
CXCL12	C-X-C motif chemokine ligand 12
CXCR4	C-X-C motif chemokine receptor type 4
DCs	dendritic cells
ECM	extracellular matrix
EESB	ethanol extract of Scutellaria barbata D. Don
EGFR	epidermal growth factor receptor
EMT	epithelial-mesenchymal transition
EO	early onset
FFPE	formalin-fixed paraffin-embedded
Foxp3	forkhead box protein 3
GDF	growth and differentiation factor
GPC1	Glypicans 1
GREM1	gremlin 1
HAPLN1	hyaluronan and proteoglycan link protein-1
hCGβ	beta human chorionic gonadotropin
HLTF	helicase-like transcription factor
IL-11	interleukin-11
INHBA	inhibin, beta A
INHBB	inhibin subunit beta B
ISLR	immunoglobulin superfamily containing leucine-rich repeat
I-SMAD	inhibitory SMAD
KEGG	Kyoto Encyclopedia of Genes and Genomes
lncRNA-ATB	lncRNA-activated by TGF-β
lncRNA EZR-AS1	lncRNA ezrin antisense RNA 1
lncRNAs	long non-coding RNAs
MAPK	mitogen-activated protein kinase
mCRC	metastatic colorectal cancer
MiR	microRMA
MMPs	matrix metalloproteinases
MSD	Modified Shenlingbaizhu Decoction
MSI	microsatellite instability
NEP	neutral endopeptidase
NF-κB	nuclear factor-κB
NGS	next-generation sequencing
OS	overall survival

PARG	poly (ADP-ribose) glycohydrolase
PDGF-D	platelet-derived growth factor-D
PD-L1	programmed cell death-Ligand 1
PFS	progression-free survival
P4HA3	prolyl 4-hydroxylase subunit alpha 3
PI3K	phosphoinositide 3-kinase
PNMA5	Paraneoplastic antigen Ma family number 5
PODXL	podocalyxin-like
QFG	Qingjie Fuzheng granule
RGC-32	response gene to complement 32
ROCK	Rho-associated kinase
R-SMAD	receptor-regulated SMAD
RUNX	runt related transcription factor
SDF-1	stromal cell derived factor-1
SNP	single nucleotide polymorphism
STAT3	signal transducer and activator of transcription 3
TAK1	TGF-β-activated kinase 1
TAMs	tumor-associated macrophages
TANs	tumor-associated neutrophils
TCGA	The Cancer Genome Atlas
TGF-β	transforming growth factor-β
TGFBI	TGF-β-induced protein ig-h3
TH1	type 1 T-helper cell
Th17	T helper 17
THBS4	thrombospondin-4
THSD4	thrombospondin type-1 domain-containing protein 4
TME	tumor microenvironment
TMEM45A	transmembrane protein 45A
TRAIL	tumor necrosis factor-related apoptosis-inducing ligand
Tregs	regulatory T cells
TRIM25	tripartite motif containing 25
Trp53	transformation related protein 53
TSIPs	tumor spheres with inverted polarity
TSP1	thrombospondin 1
VEGF	vascular endothelial growth factor

References

1. Sung, H.; Ferlay, J.; Siegel, R.L.; Laversanne, M.; Soerjomataram, I.; Jemal, A.; Bray, F. Global Cancer Statistics 2020: GLOBOCAN Estimates of Incidence and Mortality Worldwide for 36 Cancers in 185 Countries. *CA Cancer J. Clin.* **2021**, *71*, 209–249. [CrossRef] [PubMed]
2. Siegel, R.L.; Miller, K.D.; Fuchs, H.E.; Jemal, A. Cancer statistics, 2022. *CA Cancer J. Clin.* **2022**, *72*, 7–33. [CrossRef] [PubMed]
3. Xia, C.; Dong, X.; Li, H.; Cao, M.; Sun, D.; He, S.; Yang, F.; Yan, X.; Zhang, S.; Li, N.; et al. Cancer statistics in China and United States, 2022: Profiles, trends, and determinants. *Chin. Med. J.* **2022**, *135*, 584–590. [CrossRef] [PubMed]
4. Chen, W.; Zheng, R.; Baade, P.D.; Zhang, S.; Zeng, H.; Bray, F.; Jemal, A.; Yu, X.Q.; He, J. Cancer statistics in China, 2015. *CA Cancer J. Clin.* **2016**, *66*, 115–132. [CrossRef] [PubMed]
5. Hutter, C.M.; Chang-Claude, J.; Slattery, M.L.; Pflugeisen, B.M.; Lin, Y.; Duggan, D.; Nan, H.; Lemire, M.; Rangrej, J.; Figueiredo, J.C.; et al. Characterization of gene-environment interactions for colorectal cancer susceptibility loci. *Cancer Res.* **2012**, *72*, 2036–2044. [CrossRef] [PubMed]
6. Tauriello, D.V.; Calon, A.; Lonardo, E.; Batlle, E. Determinants of metastatic competency in colorectal cancer. *Mol. Oncol.* **2017**, *11*, 97–119. [CrossRef]
7. Siegel, R.L.; Miller, K.D.; Goding Sauer, A.; Fedewa, S.A.; Butterly, L.F.; Anderson, J.C.; Cercek, A.; Smith, R.A.; Jemal, A. Colorectal cancer statistics, 2020. *CA Cancer J. Clin.* **2020**, *70*, 145–164. [CrossRef]
8. Oki, E.; Ando, K.; Nakanishi, R.; Sugiyama, M.; Nakashima, Y.; Kubo, N.; Kudou, K.; Saeki, H.; Nozoe, T.; Emi, Y.; et al. Recent advances in treatment for colorectal liver metastasis. *Ann. Gastroenterol. Surg.* **2018**, *2*, 167–175. [CrossRef]
9. Zarour, L.R.; Anand, S.; Billingsley, K.G.; Bisson, W.H.; Cercek, A.; Clarke, M.F.; Coussens, L.M.; Gast, C.E.; Geltzeiler, C.B.; Hansen, L.; et al. Colorectal Cancer Liver Metastasis: Evolving Paradigms and Future Directions. *Cell Mol. Gastroenterol. Hepatol.* **2017**, *3*, 163–173. [CrossRef]

10. Modest, D.P.; Pant, S.; Sartore-Bianchi, A. Treatment sequencing in metastatic colorectal cancer. *Eur. J. Cancer* **2019**, *109*, 70–83. [CrossRef]
11. Jones, R.P.; Jackson, R.; Dunne, D.F.; Malik, H.Z.; Fenwick, S.W.; Poston, G.J.; Ghaneh, P. Systematic review and meta-analysis of follow-up after hepatectomy for colorectal liver metastases. *Br. J. Surg.* **2012**, *99*, 477–486. [CrossRef] [PubMed]
12. Derynck, R.; Akhurst, R.J.; Balmain, A. TGF-beta signaling in tumor suppression and cancer progression. *Nat. Genet.* **2001**, *29*, 117–129. [CrossRef] [PubMed]
13. Marinelli Busilacchi, E.; Costantini, A.; Mancini, G.; Tossetta, G.; Olivieri, J.; Poloni, A.; Viola, N.; Butini, L.; Campanati, A.; Goteri, G.; et al. Nilotinib Treatment of Patients Affected by Chronic Graft-versus-Host Disease Reduces Collagen Production and Skin Fibrosis by Downmodulating the TGF-beta and p-SMAD Pathway. *Biol. Blood Marrow Transpl.* **2020**, *26*, 823–834. [CrossRef] [PubMed]
14. Akhurst, R.J.; Hata, A. Targeting the TGFbeta signalling pathway in disease. *Nat. Rev. Drug Discov.* **2012**, *11*, 790–811. [CrossRef]
15. Goteri, G.; Altobelli, E.; Tossetta, G.; Zizzi, A.; Avellini, C.; Licini, C.; Lorenzi, T.; Castellucci, M.; Ciavattini, A.; Marzioni, D. High temperature requirement A1, transforming growth factor beta1, phosphoSmad2 and Ki67 in eutopic and ectopic endometrium of women with endometriosis. *Eur. J. Histochem.* **2015**, *59*, 2570. [CrossRef]
16. Pardali, E.; Ten Dijke, P. TGFbeta signaling and cardiovascular diseases. *Int. J. Biol. Sci.* **2012**, *8*, 195–213. [CrossRef]
17. Morikawa, M.; Derynck, R.; Miyazono, K. TGF-beta and the TGF-beta Family: Context-Dependent Roles in Cell and Tissue Physiology. *Cold Spring Harb. Perspect. Biol.* **2016**, *8*, a021873. [CrossRef]
18. Tauriello, D.V.; Sancho, E.; Batlle, E. Overcoming TGFβ-mediated immune evasion in cancer. *Nat. Rev. Cancer* **2022**, *22*, 25–44. [CrossRef]
19. Attisano, L.; Wrana, J.L. Signal transduction by the TGF-beta superfamily. *Science* **2002**, *296*, 1646–1647. [CrossRef]
20. Gough, N.R.; Xiang, X.; Mishra, L. TGF-beta Signaling in Liver, Pancreas, and Gastrointestinal Diseases and Cancer. *Gastroenterology* **2021**, *161*, 434–452.e15. [CrossRef]
21. Haque, S.; Morris, J.C. Transforming growth factor-beta: A therapeutic target for cancer. *Hum. Vacc. Immunother.* **2017**, *13*, 1741–1750. [CrossRef] [PubMed]
22. Jung, B.; Staudacher, J.J.; Beauchamp, D. Transforming Growth Factor beta Superfamily Signaling in Development of Colorectal Cancer. *Gastroenterology* **2017**, *152*, 36–52. [CrossRef] [PubMed]
23. Robertson, I.B.; Rifkin, D.B. Regulation of the Bioavailability of TGF-beta and TGF-beta-Related Proteins. *Cold Spring Harb. Perspect. Biol.* **2016**, *8*, a021907. [CrossRef] [PubMed]
24. Justin, P.A.; Daniel, B.R.; John, S.M. The integrin alphaVbeta6 binds and activates latent TGFbeta3. *FEBS Lett.* **2002**, *511*, 65–68.
25. Nagaraj, N.S.; Datta, P.K. Targeting the transforming growth factor-beta signaling pathway in human cancer. *Expert Opin. Investig. Drugs* **2010**, *19*, 77–91. [CrossRef]
26. Gu, S.; Feng, X.H. TGF-beta signaling in cancer. *Acta Biochim. Biophys. Sin.* **2018**, *50*, 941–949. [CrossRef]
27. Heldin, C.H.; Moustakas, A. Signaling Receptors for TGF-beta Family Members. *Cold Spring Harb. Perspect. Biol.* **2016**, *8*, a022053. [CrossRef]
28. Colak, S.; Ten Dijke, P. Targeting TGF-beta Signaling in Cancer. *Trends Cancer* **2017**, *3*, 56–71. [CrossRef]
29. Nakao, A.; Afrakhte, M.; Morén, A.; Nakayama, T.; Christian, J.L.; Heuchel, R.; Itoh, S.; Kawabata, M.; Heldin, N.E.; Heldin, C.H.; et al. Identification of Smad7, a TGFbeta-inducible antagonist of TGF-beta signalling. *Nature* **1997**, *389*, 631–635. [CrossRef]
30. Derynck, R.; Muthusamy, B.P.; Saeteurn, K.Y. Signaling pathway cooperation in TGF-beta-induced epithelial-mesenchymal transition. *Curr. Opin. Cell Biol.* **2014**, *31*, 56–66. [CrossRef]
31. Croce, C.M. Oncogenes and cancer. *N. Engl. J. Med.* **2008**, *358*, 502–511. [CrossRef]
32. Fang, W.; Radovich, M.; Zheng, Y.; Fu, C.Y.; Zhao, P.; Mao, C.; Zheng, Y.; Zheng, S. Druggable alterations detected by Ion Torrent in metastatic colorectal cancer patients. *Oncol. Lett.* **2014**, *7*, 1761–1766. [CrossRef] [PubMed]
33. Bajenova, O.; Gorbunova, A.; Evsyukov, I.; Rayko, M.; Gapon, S.; Bozhokina, E.; Shishkin, A.; O'Brien, S.J. The Genome-Wide Analysis of Carcinoembryonic Antigen Signaling by Colorectal Cancer Cells Using RNA Sequencing. *PLoS ONE* **2016**, *11*, e0161256. [CrossRef] [PubMed]
34. Huang, Y.H.; Lin, P.C.; Su, W.C.; Chan, R.H.; Chen, P.C.; Lin, B.W.; Shen, M.R.; Chen, S.H.; Yeh, Y.M. Association between Altered Oncogenic Signaling Pathways and Overall Survival of Patients with Metastatic Colorectal Cancer. *Diagnostics* **2021**, *11*, 2308. [CrossRef] [PubMed]
35. Kawaguchi, Y.; Kopetz, S.; Kwong, L.; Xiao, L.; Morris, J.S.; Tran Cao, H.S.; Tzeng, C.D.; Chun, Y.S.; Lee, J.E.; Vauthey, J.N. Genomic Sequencing and Insight into Clinical Heterogeneity and Prognostic Pathway Genes in Patients with Metastatic Colorectal Cancer. *J. Am. Coll. Surg.* **2021**, *233*, 272–284.e13. [CrossRef] [PubMed]
36. Liu, J.; Li, H.; Sun, L.; Shen, S.; Zhou, Q.; Yuan, Y.; Xing, C. Epigenetic Alternations of MicroRNAs and DNA Methylation Contribute to Liver Metastasis of Colorectal Cancer. *Dig. Dis. Sci.* **2019**, *64*, 1523–1534. [CrossRef]
37. Kulkarni, A.B.; Huh, C.G.; Becker, D.; Geiser, A.; Lyght, M.; Flanders, K.C.; Roberts, A.B.; Sporn, M.B.; Ward, J.M.; Karlsson, S. Transforming growth factor beta 1 null mutation in mice causes excessive inflammatory response and early death. *Proc. Natl. Acad. Sci. USA* **1993**, *90*, 770–774. [CrossRef]
38. Sanford, L.P.; Ormsby, I.; Gittenberger-de Groot, A.C.; Sariola, H.; Friedman, R.; Boivin, G.P.; Cardell, E.L.; Doetschman, T. TGFbeta2 knockout mice have multiple developmental defects that are non-overlapping with other TGFbeta knockout phenotypes. *Development* **1997**, *124*, 2659–2670. [CrossRef]

39. Kaartinen, V.; Voncken, J.W.; Shuler, C.; Warburton, D.; Bu, D.; Heisterkamp, N.; Groffen, J. Abnormal lung development and cleft palate in mice lacking TGF-beta 3 indicates defects of epithelial-mesenchymal interaction. *Nat. Genet.* **1995**, *11*, 415–421. [CrossRef]
40. Martin, C.J.; Datta, A.; Littlefield, C.; Kalra, A.; Chapron, C.; Wawersik, S.; Dagbay, K.B.; Brueckner, C.T.; Nikiforov, A.; Danehy, F.T.J.; et al. Selective inhibition of TGFβ1 activation overcomes primary resistance to checkpoint blockade therapy by altering tumor immune landscape. *Sci. Transl. Med.* **2020**, *12*, eaay8456. [CrossRef]
41. Jackstadt, R.; van Hooff, S.R.; Leach, J.D.; Cortes-Lavaud, X.; Lohuis, J.O.; Ridgway, R.A.; Wouters, V.M.; Roper, J.; Kendall, T.J.; Roxburgh, C.S.; et al. Epithelial NOTCH Signaling Rewires the Tumor Microenvironment of Colorectal Cancer to Drive Poor-Prognosis Subtypes and Metastasis. *Cancer Cell* **2019**, *36*, 319–336.e7. [CrossRef] [PubMed]
42. Calon, A.; Espinet, E.; Palomo-Ponce, S.; Tauriello, D.V.; Iglesias, M.; Cespedes, M.V.; Sevillano, M.; Nadal, C.; Jung, P.; Zhang, X.H.; et al. Dependency of colorectal cancer on a TGF-β-driven program in stromal cells for metastasis initiation. *Cancer Cell* **2012**, *22*, 571–584. [CrossRef]
43. Zhou, R.; Huang, Y.; Cheng, B.; Wang, Y.; Xiong, B. TGFBR1*6A is a potential modifier of migration and invasion in colorectal cancer cells. *Oncol. Lett.* **2018**, *15*, 3971–3976. [CrossRef] [PubMed]
44. Valle, L.; Serena-Acedo, T.; Liyanarachchi, S.; Hampel, H.; Comeras, I.; Li, Z.; Zeng, Q.; Zhang, H.T.; Pennison, M.J.; Sadim, M.; et al. Germline allele-specific expression of TGFBR1 confers an increased risk of colorectal cancer. *Science* **2008**, *321*, 1361–1365. [CrossRef] [PubMed]
45. Fernandez-Peralta, A.M.; Nejda, N.; Oliart, S.; Medina, V.; Azcoita, M.M.; Gonzalez-Aguilera, J.J. Significance of mutations in TGFBR2 and BAX in neoplastic progression and patient outcome in sporadic colorectal tumors with high-frequency microsatellite instability. *Cancer Genet. Cytogenet.* **2005**, *157*, 18–24. [CrossRef]
46. Yashiro, M.; Hirakawa, K.; Boland, C.R. Mutations in TGFbeta-RII and BAX mediate tumor progression in the later stages of colorectal cancer with microsatellite instability. *BMC Cancer* **2010**, *10*, 303. [CrossRef]
47. Fricke, F.; Lee, J.; Michalak, M.; Warnken, U.; Hausser, I.; Suarez-Carmona, M.; Halama, N.; Schnolzer, M.; Kopitz, J.; Gebert, J. TGFBR2-dependent alterations of exosomal cargo and functions in DNA mismatch repair-deficient HCT116 colorectal cancer cells. *Cell Commun. Signal.* **2017**, *15*, 14. [CrossRef]
48. Lee, J.; Ballikaya, S.; Schonig, K.; Ball, C.R.; Glimm, H.; Kopitz, J.; Gebert, J. Transforming growth factor beta receptor 2 (TGFBR2) changes sialylation in the microsatellite unstable (MSI) Colorectal cancer cell line HCT116. *PLoS ONE* **2013**, *8*, e57074. [CrossRef]
49. Stanilov, N.; Grigorova, A.; Velikova, T.; Stanilova, S.A. Genetic variation of TGF-BetaR2 as a protective genotype for the development of colorectal cancer in men. *World J. Gastrointest. Oncol.* **2021**, *13*, 1766–1780. [CrossRef]
50. Xie, W.; Rimm, D.L.; Lin, Y.; Shih, W.J.; Reiss, M. Loss of Smad signaling in human colorectal cancer is associated with advanced disease and poor prognosis. *Cancer J.* **2003**, *9*, 302–312. [CrossRef]
51. Cancer Genome Atlas, N. Comprehensive molecular characterization of human colon and rectal cancer. *Nature* **2012**, *487*, 330–337. [CrossRef] [PubMed]
52. Fleming, N.I.; Jorissen, R.N.; Mouradov, D.; Christie, M.; Sakthianandeswaren, A.; Palmieri, M.; Day, F.; Li, S.; Tsui, C.; Lipton, L.; et al. SMAD2, SMAD3 and SMAD4 mutations in colorectal cancer. *Cancer Res.* **2013**, *73*, 725–735. [CrossRef] [PubMed]
53. Johnson, B.; Cooke, L.; Mahadevan, D. Next generation sequencing identifies 'interactome' signatures in relapsed and refractory metastatic colorectal cancer. *J. Gastrointest. Oncol.* **2017**, *8*, 20–31. [CrossRef] [PubMed]
54. Lee, S.E.; Park, H.Y.; Hwang, D.Y.; Han, H.S. High Concordance of Genomic Profiles between Primary and Metastatic Colorectal Cancer. *Int. J. Mol. Sci.* **2021**, *22*, 5561. [CrossRef] [PubMed]
55. Oyanagi, H.; Shimada, Y.; Nagahashi, M.; Ichikawa, H.; Tajima, Y.; Abe, K.; Nakano, M.; Kameyama, H.; Takii, Y.; Kawasaki, T.; et al. SMAD4 alteration associates with invasive-front pathological markers and poor prognosis in colorectal cancer. *Histopathology* **2019**, *74*, 873–882. [CrossRef]
56. Xu, T.; Zhang, Y.; Zhang, J.; Qi, C.; Liu, D.; Wang, Z.; Li, Y.; Ji, C.; Li, J.; Lin, X.; et al. Germline Profiling and Molecular Characterization of Early Onset Metastatic Colorectal Cancer. *Front. Oncol.* **2020**, *10*, 568911. [CrossRef]
57. Lopez-Gomez, M.; Moreno-Rubio, J.; Suarez-Garcia, I.; Cejas, P.; Madero, R.; Casado, E.; Jimenez, A.M.; Sereno, M.; Gomez-Raposo, C.; Zambrana, F.; et al. Gene expression differences in primary colorectal tumors and matched liver metastases: Chemotherapy related or tumoral heterogeneity? *Clin. Transl. Oncol.* **2015**, *17*, 322–329. [CrossRef]
58. Ganesh, K.; Shah, R.H.; Vakiani, E.; Nash, G.M.; Skottowe, H.P.; Yaeger, R.; Cercek, A.; Lincoln, A.; Tran, C.; Segal, N.H.; et al. Clinical and genetic determinants of ovarian metastases from colorectal cancer. *Cancer* **2017**, *123*, 1134–1143. [CrossRef]
59. Abd El-Fattah, A.A.; Sadik, N.A.H.; Shaker, O.G.; Mohamed Kamal, A. Single Nucleotide Polymorphism in SMAD7 and CHI3L1 and Colorectal Cancer Risk. *Mediat. Inflamm.* **2018**, *2018*, 9853192. [CrossRef]
60. Rosic, J.; Dragicevic, S.; Miladinov, M.; Despotovic, J.; Bogdanovic, A.; Krivokapic, Z.; Nikolic, A. SMAD7 and SMAD4 expression in colorectal cancer progression and therapy response. *Exp. Mol. Pathol.* **2021**, *123*, 104714. [CrossRef]
61. Halder, S.K.; Rachakonda, G.; Deane, N.G.; Datta, P.K. Smad7 induces hepatic metastasis in colorectal cancer. *Br. J. Cancer* **2008**, *99*, 957–965. [CrossRef] [PubMed]
62. Sakai, E.; Nakayama, M.; Oshima, H.; Kouyama, Y.; Niida, A.; Fujii, S.; Ochiai, A.; Nakayama, K.I.; Mimori, K.; Suzuki, Y.; et al. Combined Mutation of Apc, Kras, and Tgfbr2 Effectively Drives Metastasis of Intestinal Cancer. *Cancer Res.* **2018**, *78*, 1334–1346. [CrossRef] [PubMed]

63. Miguchi, M.; Hinoi, T.; Shimomura, M.; Adachi, T.; Saito, Y.; Niitsu, H.; Kochi, M.; Sada, H.; Sotomaru, Y.; Ikenoue, T.; et al. Gasdermin C Is Upregulated by Inactivation of Transforming Growth Factor beta Receptor Type II in the Presence of Mutated Apc, Promoting Colorectal Cancer Proliferation. *PLoS ONE* **2016**, *11*, e0166422. [CrossRef] [PubMed]
64. Boutin, A.T.; Liao, W.T.; Wang, M.; Hwang, S.S.; Karpinets, T.V.; Cheung, H.; Chu, G.C.; Jiang, S.; Hu, J.; Chang, K.; et al. Oncogenic Kras drives invasion and maintains metastases in colorectal cancer. *Genes Dev.* **2017**, *31*, 370–382. [CrossRef]
65. Fumagalli, A.; Drost, J.; Suijkerbuijk, S.J.; van Boxtel, R.; de Ligt, J.; Offerhaus, G.J.; Begthel, H.; Beerling, E.; Tan, E.H.; Sansom, O.J.; et al. Genetic dissection of colorectal cancer progression by orthotopic transplantation of engineered cancer organoids. *Proc. Natl. Acad. Sci. USA* **2017**, *114*, E2357–E2364. [CrossRef]
66. Wang, H.W.; Yan, X.L.; Wang, L.J.; Zhang, M.H.; Yang, C.H.; Wei, L.; Jin, K.M.; Bao, Q.; Li, J.; Wang, K.; et al. Characterization of genomic alterations in Chinese colorectal cancer patients with liver metastases. *J. Transl. Med.* **2021**, *19*, 313. [CrossRef]
67. Mehrvarz Sarshekeh, A.; Advani, S.; Overman, M.J.; Manyam, G.; Kee, B.K.; Fogelman, D.R.; Dasari, A.; Raghav, K.; Vilar, E.; Manuel, S.; et al. Association of SMAD4 mutation with patient demographics, tumor characteristics, and clinical outcomes in colorectal cancer. *PLoS ONE* **2017**, *12*, e0173345.
68. Sun, N.; Xue, Y.; Dai, T.; Li, X.; Zheng, N. Tripartite motif containing 25 promotes proliferation and invasion of colorectal cancer cells through TGF-beta signaling. *Biosci. Rep.* **2017**, *37*, BSR20170805. [CrossRef]
69. Yao, Y.; Zhou, Z.; Li, L.; Li, J.; Huang, L.; Li, J.; Qi, C.; Zheng, L.; Wang, L.; Zhang, Q.Q. Activation of Slit2/Robo1 Signaling Promotes Tumor Metastasis in Colorectal Carcinoma through Activation of the TGF-beta/Smads Pathway. *Cells* **2019**, *8*, 635. [CrossRef]
70. Liu, L.; Liu, H.; Zhou, Y.; He, J.; Liu, Q.; Wang, J.; Zeng, M.; Yuan, D.; Tan, F.; Zhou, Y.; et al. HLTF suppresses the migration and invasion of colorectal cancer cells via TGFbeta/SMAD signaling in vitro. *Int. J. Oncol.* **2018**, *53*, 2780–2788.
71. Zhou, H.; Zou, J.; Shao, C.; Zhou, A.; Yu, J.; Chen, S.; Xu, C. Prolyl 4-hydroxylase subunit alpha 3 facilitates human colon cancer growth and metastasis through the TGF-beta/Smad signaling pathway. *Pathol. Res. Pract.* **2022**, *230*, 153749. [CrossRef] [PubMed]
72. Wang, L.; Ai, M.; Nie, M.; Zhao, L.; Deng, G.; Hu, S.; Han, Y.; Zeng, W.; Wang, Y.; Yang, M.; et al. EHF promotes colorectal carcinoma progression by activating TGF-beta1 transcription and canonical TGF-beta signaling. *Cancer Sci.* **2020**, *111*, 2310–2324. [CrossRef] [PubMed]
73. Lu, F.; Chen, S.; Shi, W.; Su, X.; Wu, H.; Liu, M. GPC1 promotes the growth and migration of colorectal cancer cells through regulating the TGF-beta1/SMAD2 signaling pathway. *PLoS ONE* **2022**, *17*, e0269094.
74. Liu, F.; Shi, Z.; Bao, W.; Zheng, J.; Chen, K.; Lin, Z.; Song, H.N.; Luo, X.; Dong, Q.; Jiang, L.; et al. ZIC2 promotes colorectal cancer growth and metastasis through the TGF-beta signaling pathway. *Exp. Cell Res.* **2022**, *415*, 113118. [CrossRef] [PubMed]
75. Vu, T.; Datta, P.K. Regulation of EMT in Colorectal Cancer: A Culprit in Metastasis. *Cancers* **2017**, *9*, 171. [CrossRef] [PubMed]
76. Lee, J.M.; Dedhar, S.; Kalluri, R.; Thompson, E.W. The epithelial-mesenchymal transition: New insights in signaling, development, and disease. *J. Cell Biol.* **2006**, *172*, 973–981. [CrossRef]
77. Seoane, J.; Gomis, R.R. TGF-beta Family Signaling in Tumor Suppression and Cancer Progression. *Cold Spring Harb. Perspect. Biol.* **2017**, *9*, a022277. [CrossRef]
78. Tossetta, G.; Paolinelli, F.; Avellini, C.; Salvolini, E.; Ciarmela, P.; Lorenzi, T.; Emanuelli, M.; Toti, P.; Giuliante, R.; Gesuita, R.; et al. IL-1beta and TGF-beta weaken the placental barrier through destruction of tight junctions: An in vivo and in vitro study. *Placenta* **2014**, *35*, 509–516. [CrossRef]
79. Shiou, S.R.; Singh, A.B.; Moorthy, K.; Datta, P.K.; Washington, M.K.; Beauchamp, R.D.; Dhawan, P. Smad4 regulates claudin-1 expression in a transforming growth factor-beta-independent manner in colon cancer cells. *Cancer Res.* **2007**, *67*, 1571–1579. [CrossRef]
80. Di Ruocco, F.; Basso, V.; Rivoire, M.; Mehlen, P.; Ambati, J.; De Falco, S.; Tarallo, V. Alu RNA accumulation induces epithelial-to-mesenchymal transition by modulating miR-566 and is associated with cancer progression. *Oncogene* **2018**, *37*, 627–637. [CrossRef]
81. Asiri, A.; Raposo, T.P.; Alfahed, A.; Ilyas, M. TGFbeta1-induced cell motility but not cell proliferation is mediated through Cten in colorectal cancer. *Int. J. Exp. Pathol.* **2018**, *99*, 323–330. [CrossRef] [PubMed]
82. Quan, J.; Cheng, C.; Tan, Y.; Jiang, N.; Liao, C.; Liao, W.; Cao, Y.; Luo, X. Acyl-CoA synthetase long-chain 3-mediated fatty acid oxidation is required for TGFbeta1-induced epithelial-mesenchymal transition and metastasis of colorectal carcinoma. *Int. J. Biol. Sci.* **2022**, *18*, 2484–2496. [CrossRef] [PubMed]
83. Siraj, A.K.; Pratheeshkumar, P.; Divya, S.P.; Parvathareddy, S.K.; Bu, R.; Masoodi, T.; Kong, Y.; Thangavel, S.; Al-Sanea, N.; Ashari, L.H.; et al. TGFbeta-induced SMAD4-dependent Apoptosis Proceeded by EMT in CRC. *Mol. Cancer Ther.* **2019**, *18*, 1312–1322. [CrossRef] [PubMed]
84. Corbet, C.; Bastien, E.; Santiago de Jesus, J.P.; Dierge, E.; Martherus, R.; Vander Linden, C.; Doix, B.; Degavre, C.; Guilbaud, C.; Petit, L.; et al. TGFbeta2-induced formation of lipid droplets supports acidosis-driven EMT and the metastatic spreading of cancer cells. *Nat. Commun.* **2020**, *11*, 454. [CrossRef] [PubMed]
85. Frey, P.; Devisme, A.; Rose, K.; Schrempp, M.; Freihen, V.; Andrieux, G.; Boerries, M.; Hecht, A. SMAD4 mutations do not preclude epithelial-mesenchymal transition in colorectal cancer. *Oncogene* **2022**, *41*, 824–837. [CrossRef]
86. Zhang, Y.; Wang, X.; Zhang, M.; Zhang, Z.; Jiang, L.; Li, L. GDF15 promotes epithelial-to-mesenchymal transition in colorectal [corrected]. *Artif. Cell. Nanomed. Biotechnol.* **2018**, *46* (Suppl. 2), 652–658. [CrossRef]

87. Li, H.; Zhong, A.; Li, S.; Meng, X.; Wang, X.; Xu, F.; Lai, M. The integrated pathway of TGFbeta/Snail with TNFalpha/NFkappaB may facilitate the tumor-stroma interaction in the EMT process and colorectal cancer prognosis. *Sci. Rep.* **2017**, *7*, 4915. [CrossRef]
88. Ioannou, M.; Kouvaras, E.; Papamichali, R.; Samara, M.; Chiotoglou, I.; Koukoulis, G. Smad4 and epithelial-mesenchymal transition proteins in colorectal carcinoma: An immunohistochemical study. *J. Mol. Histol.* **2018**, *49*, 235–244. [CrossRef]
89. Du, G.S.; Qiu, Y.; Wang, W.S.; Peng, K.; Zhang, Z.C.; Li, X.S.; Xiao, W.D.; Yang, H. Knockdown on aPKC-iota inhibits epithelial-mesenchymal transition, migration and invasion of colorectal cancer cells through Rac1-JNK pathway. *Exp. Mol. Pathol.* **2019**, *107*, 57–67. [CrossRef]
90. Zhu, M.; Jiang, B.; Yan, D.; Wang, X.; Ge, H.; Sun, Y. Knockdown of TMEM45A overcomes multidrug resistance and epithelial-mesenchymal transition in human colorectal cancer cells through inhibition of TGF-beta signalling pathway. *Clin. Exp. Pharmacol. Physiol.* **2020**, *47*, 503–516. [CrossRef]
91. Yang, Y.; Li, L.; He, H.; Shi, M.; He, L.; Liang, S.; Qi, J.; Chen, W. Numb inhibits migration and promotes proliferation of colon cancer cells via RhoA/ROCK signaling pathway repression. *Exp. Cell Res.* **2022**, *411*, 113004. [CrossRef] [PubMed]
92. Lin, J.; Zhang, X.; Meng, F.; Zeng, F.; Liu, W.; He, X. PNMA5 accelerated cellular proliferation, invasion and migration in colorectal cancer. *Am. J. Transl. Res.* **2022**, *14*, 2231–2243. [PubMed]
93. Kawamata, F.; Nishihara, H.; Homma, S.; Kato, Y.; Tsuda, M.; Konishi, Y.; Wang, L.; Kohsaka, S.; Liu, C.; Yoshida, T.; et al. Chorionic Gonadotropin-beta Modulates Epithelial-Mesenchymal Transition in Colorectal Carcinoma Metastasis. *Am. J. Pathol.* **2018**, *188*, 204–215. [CrossRef] [PubMed]
94. Lee, H.-J. Recent Advances in the Development of TGF-β Signaling Inhibitors for Anticancer Therapy. *J. Cancer Prev.* **2020**, *25*, 213–222. [CrossRef]
95. Akhurst, R.J.; Derynck, R. TGF-beta signaling in cancer–a double-edged sword. *Trends Cell Biol.* **2001**, *11*, S44–S51.
96. Papageorgis, P.; Cheng, K.; Ozturk, S.; Gong, Y.; Lambert, A.W.; Abdolmaleky, H.M.; Zhou, J.R.; Thiagalingam, S. Smad4 inactivation promotes malignancy and drug resistance of colon cancer. *Cancer Res.* **2011**, *71*, 998–1008. [CrossRef]
97. Li, X.; Li, X.; Lv, X.; Xiao, J.; Liu, B.; Zhang, Y. Smad4 Inhibits VEGF-A and VEGF-C Expressions via Enhancing Smad3 Phosphorylation in Colon Cancer. *Anat. Rec.* **2017**, *300*, 1560–1569. [CrossRef]
98. Rada, M.; Kapelanski-Lamoureux, A.; Petrillo, S.; Tabaries, S.; Siegel, P.; Reynolds, A.R.; Lazaris, A.; Metrakos, P. Runt related transcription factor-1 plays a central role in vessel co-option of colorectal cancer liver metastases. *Commun. Biol.* **2021**, *4*, 950. [CrossRef]
99. Jeong, C.H.; Kwon, H.C.; Cheng, W.N.; Kim, D.H.; Choi, Y.; Han, S.G. Aluminum exposure promotes the metastatic proclivity of human colorectal cancer cells through matrix metalloproteinases and the TGF-beta/Smad signaling pathway. *Food Chem. Toxicol.* **2020**, *141*, 111402. [CrossRef]
100. Chiavarina, B.; Costanza, B.; Ronca, R.; Blomme, A.; Rezzola, S.; Chiodelli, P.; Giguelay, A.; Belthier, G.; Doumont, G.; Van Simaeys, G.; et al. Metastatic colorectal cancer cells maintain the TGFbeta program and use TGFBI to fuel angiogenesis. *Theranostics* **2021**, *11*, 1626–1640. [CrossRef]
101. Chen, J.; Yuan, W.; Wu, L.; Tang, Q.; Xia, Q.; Ji, J.; Liu, Z.; Ma, Z.; Zhou, Z.; Cheng, Y.; et al. PDGF-D promotes cell growth, aggressiveness, angiogenesis and EMT transformation of colorectal cancer by activation of Notch1/Twist1 pathway. *Oncotarget* **2017**, *8*, 9961–9973. [CrossRef] [PubMed]
102. Muppala, S.; Frolova, E.; Xiao, R.; Krukovets, I.; Yoon, S.; Hoppe, G.; Vasanji, A.; Plow, E.; Stenina-Adognravi, O. Proangiogenic Properties of Thrombospondin-4. *Arterioscler. Thromb. Vasc. Biol.* **2015**, *35*, 1975–1986. [CrossRef] [PubMed]
103. Muppala, S.; Xiao, R.; Krukovets, I.; Verbovetsky, D.; Yendamuri, R.; Habib, N.; Raman, P.; Plow, E.; Stenina-Adognravi, O. Thrombospondin-4 mediates TGF-beta-induced angiogenesis. *Oncogene* **2017**, *36*, 5189–5198. [CrossRef] [PubMed]
104. Derynck, R.; Turley, S.J.; Akhurst, R.J. TGFβ biology in cancer progression and immunotherapy. *Nat. Rev. Clin. Oncol.* **2021**, *18*, 9–34. [CrossRef]
105. Tommelein, J.; Verset, L.; Boterberg, T.; Demetter, P.; Bracke, M.; De Wever, O. Cancer-associated fibroblasts connect metastasis-promoting communication in colorectal cancer. *Front. Oncol.* **2015**, *5*, 63. [CrossRef]
106. Hinshaw, D.C.; Shevde, L.A. The Tumor Microenvironment Innately Modulates Cancer Progression. *Cancer Res.* **2019**, *79*, 4557–4566. [CrossRef]
107. Calon, A.; Lonardo, E.; Berenguer-Llergo, A.; Espinet, E.; Hernando-Momblona, X.; Iglesias, M.; Sevillano, M.; Palomo-Ponce, S.; Tauriello, D.V.; Byrom, D.; et al. Stromal gene expression defines poor-prognosis subtypes in colorectal cancer. *Nat. Genet.* **2015**, *47*, 320–329. [CrossRef]
108. Itatani, Y.; Kawada, K.; Sakai, Y. Transforming Growth Factor-beta Signaling Pathway in Colorectal Cancer and Its Tumor Microenvironment. *Int. J. Mol. Sci.* **2019**, *20*, 5822. [CrossRef]
109. Koliaraki, V.; Pallangyo, C.K.; Greten, F.R.; Kollias, G. Mesenchymal Cells in Colon Cancer. *Gastroenterology* **2017**, *152*, 964–979. [CrossRef]
110. Tauriello, D.V.F.; Palomo-Ponce, S.; Stork, D.; Berenguer-Llergo, A.; Badia-Ramentol, J.; Iglesias, M.; Sevillano, M.; Ibiza, S.; Canellas, A.; Hernando-Momblona, X.; et al. TGFbeta drives immune evasion in genetically reconstituted colon cancer metastasis. *Nature* **2018**, *554*, 538–543. [CrossRef]
111. Bauer, J.; Emon, M.A.B.; Staudacher, J.J.; Thomas, A.L.; Zessner-Spitzenberg, J.; Mancinelli, G.; Krett, N.; Saif, M.T.; Jung, B. Increased stiffness of the tumor microenvironment in colon cancer stimulates cancer associated fibroblast-mediated prometastatic activin A signaling. *Sci. Rep.* **2020**, *10*, 50. [CrossRef] [PubMed]

112. Wang, D.; Wang, X.; Song, Y.; Si, M.; Sun, Y.; Liu, X.; Cui, S.; Qu, X.; Yu, X. Exosomal miR-146a-5p and miR-155-5p promote CXCL12/CXCR7-induced metastasis of colorectal cancer by crosstalk with cancer-associated fibroblasts. *Cell Death Dis.* **2022**, *13*, 380. [CrossRef] [PubMed]
113. Liu, J.; Huang, Z.; Chen, H.-N.; Qin, S.; Chen, Y.; Jiang, J.; Zhang, Z.; Luo, M.; Ye, Q.; Xie, N.; et al. ZNF37A promotes tumor metastasis through transcriptional control of THSD4/TGF-β axis in colorectal cancer. *Oncogene* **2021**, *40*, 3394–3407. [CrossRef] [PubMed]
114. Peng, C.; Zou, X.; Xia, W.; Gao, H.; Li, Z.; Liu, N.; Xu, Z.; Gao, C.; He, Z.; Niu, W.; et al. Integrin αvβ6 plays a bi-directional regulation role between colon cancer cells and cancer-associated fibroblasts. *Biosci. Rep.* **2018**, *38*, BSR20180243. [CrossRef]
115. Wawro, M.E.; Sobierajska, K.; Ciszewski, W.M.; Niewiarowska, J. Nonsteroidal Anti-Inflammatory Drugs Prevent Vincristine-Dependent Cancer-Associated Fibroblasts Formation. *Int. J. Mol. Sci.* **2019**, *20*, 1941. [CrossRef]
116. Paauwe, M.; Schoonderwoerd, M.J.A.; Helderman, R.F.C.P.; Harryvan, T.J.; Groenewoud, A.; van Pelt, G.W.; Bor, R.; Hemmer, D.M.; Versteeg, H.H.; Snaar-Jagalska, B.E.; et al. Endoglin Expression on Cancer-Associated Fibroblasts Regulates Invasion and Stimulates Colorectal Cancer Metastasis. *Clin. Cancer Res.* **2018**, *24*, 6331–6344. [CrossRef]
117. Ouahoud, S.; Voorneveld, P.W.; van der Burg, L.R.A.; de Jonge-Muller, E.S.M.; Schoonderwoerd, M.J.A.; Paauwe, M.; de Vos, T.; de Wit, S.; van Pelt, G.W.; Mesker, W.E.; et al. Bidirectional tumor/stroma crosstalk promotes metastasis in mesenchymal colorectal cancer. *Oncogene* **2020**, *39*, 2453–2466. [CrossRef]
118. Mizerska-Kowalska, M.; Sawa-Wejksza, K.; Slawinska-Brych, A.; Kandefer-Szerszen, M.; Zdzisinska, B. Neutral endopeptidase depletion decreases colon cancer cell proliferation and TGF-beta1 synthesis in indirect co-cultures with normal colon fibroblasts. *Clin. Transl. Oncol.* **2021**, *23*, 1405–1414. [CrossRef]
119. Gonzalez-Zubeldia, I.; Dotor, J.; Redrado, M.; Bleau, A.-M.; Manrique, I.; de Aberasturi, A.L.; Villalba, M.; Calvo, A. Co-migration of colon cancer cells and CAFs induced by TGFβ1 enhances liver metastasis. *Cell Tissue Res.* **2015**, *359*, 829–839. [CrossRef]
120. Zhang, Y.; Wang, S.; Lai, Q.; Fang, Y.; Wu, C.; Liu, Y.; Li, Q.; Wang, X.; Gu, C.; Chen, J.; et al. Cancer-associated fibroblasts-derived exosomal miR-17-5p promotes colorectal cancer aggressive phenotype by initiating a RUNX3/MYC/TGF-β1 positive feedback loop. *Cancer Lett.* **2020**, *491*, 22–35. [CrossRef]
121. Guillén Díaz-Maroto, N.; Sanz-Pamplona, R.; Berdiel-Acer, M.; Cimas, F.J.; García, E.; Gonçalves-Ribeiro, S.; Albert, N.; Garcia-Vicién, G.; Capella, G.; Moreno, V.; et al. Noncanonical TGFβ Pathway Relieves the Blockade of IL1β/TGFβ-Mediated Crosstalk between Tumor and Stroma: TGFBR1 and TAK1 Inhibition in Colorectal Cancer. *Clin. Cancer Res.* **2019**, *25*, 4466–4479. [CrossRef] [PubMed]
122. Tan, H.-X.; Gong, W.-Z.; Zhou, K.; Xiao, Z.-G.; Hou, F.-T.; Huang, T.; Zhang, L.; Dong, H.-Y.; Zhang, W.-L.; Liu, Y.; et al. CXCR4/TGF-β1 mediated hepatic stellate cells differentiation into carcinoma-associated fibroblasts and promoted liver metastasis of colon cancer. *Cancer Biol. Ther.* **2020**, *21*, 258–268. [CrossRef] [PubMed]
123. Tan, H.-X.; Xiao, Z.-G.; Huang, T.; Fang, Z.-X.; Liu, Y.; Huang, Z.-C. CXCR4/TGF-β1 mediated self-differentiation of human mesenchymal stem cells to carcinoma-associated fibroblasts and promoted colorectal carcinoma development. *Cancer Biol. Ther.* **2020**, *21*, 248–257. [CrossRef] [PubMed]
124. McAndrews, K.M.; Vazquez-Arreguin, K.; Kwak, C.; Sugimoto, H.; Zheng, X.; Li, B.; Kirtley, M.L.; LeBleu, V.S.; Kalluri, R. alphaSMA(+) fibroblasts suppress Lgr5(+) cancer stem cells and restrain colorectal cancer progression. *Oncogene* **2021**, *40*, 4440–4452. [CrossRef] [PubMed]
125. Kobayashi, H.; Gieniec, K.A.; Wright, J.A.; Wang, T.; Asai, N.; Mizutani, Y.; Lida, T.; Ando, R.; Suzuki, N.; Lannagan, T.R.M.; et al. The Balance of Stromal BMP Signaling Mediated by GREM1 and ISLR Drives Colorectal Carcinogenesis. *Gastroenterology* **2021**, *160*, 1224–1239.e30. [CrossRef] [PubMed]
126. Wang, Y.; Xu, X.; Marshall, J.E.; Gong, M.; Zhao, Y.; Dua, K.; Hansbro, P.M.; Xu, J.; Liu, G. Loss of Hyaluronan and Proteoglycan Link Protein-1 Induces Tumorigenesis in Colorectal Cancer. *Front. Oncol.* **2021**, *11*, 754240. [CrossRef] [PubMed]
127. Wang, H.; Tian, T.; Zhang, J. Tumor-Associated Macrophages (TAMs) in Colorectal Cancer (CRC): From Mechanism to Therapy and Prognosis. *Int. J. Mol. Sci.* **2021**, *22*, 8470. [CrossRef]
128. Cassetta, L.; Pollard, J.W. Targeting macrophages: Therapeutic approaches in cancer. *Nat Rev. Drug Discov.* **2018**, *17*, 887–904. [CrossRef]
129. Liu, C.; Zhang, W.; Wang, J.; Si, T.; Xing, W. Tumor-associated macrophage-derived transforming growth factor-β promotes colorectal cancer progression through HIF1-TRIB3 signaling. *Cancer Sci.* **2021**, *112*, 4198–4207. [CrossRef]
130. Ding, Y.; Hao, K.; Li, Z.; Ma, R.; Zhou, Y.; Zhou, Z.; Wei, M.; Liao, Y.; Dai, Y.; Yang, Y.; et al. c-Fos separation from Lamin A/C by GDF15 promotes colon cancer invasion and metastasis in inflammatory microenvironment. *J. Cell. Physiol.* **2020**, *235*, 4407–4421. [CrossRef]
131. Shimizu, Y.; Amano, H.; Ito, Y.; Betto, T.; Yamane, S.; Inoue, T.; Nishizawa, N.; Matsui, Y.; Kamata, M.; Nakamura, M.; et al. Angiotensin II subtype 1a receptor signaling in resident hepatic macrophages induces liver metastasis formation. *Cancer Sci.* **2017**, *108*, 1757–1768. [CrossRef] [PubMed]
132. Zhao, P.; Wang, B.; Zhang, Z.; Zhang, W.; Liu, Y. Response gene to complement 32 expression in macrophages augments paracrine stimulation-mediated colon cancer progression. *Cell Death Dis.* **2019**, *10*, 776. [CrossRef] [PubMed]
133. Chen, T.W.; Hung, W.Z.; Chiang, S.F.; Chen, W.T.; Ke, T.W.; Liang, J.A.; Huang, C.Y.; Yang, P.C.; Huang, K.C.; Chao, K.S.C. Dual inhibition of TGFbeta signaling and CSF1 CSF1R reprograms tumor-infiltrating macrophages and improves response to chemotherapy via suppressing PD-L1. *Cancer Lett.* **2022**, *543*, 215795. [CrossRef]

134. Shapouri-Moghaddam, A.; Mohammadian, S.; Vazini, H.; Taghadosi, M.; Esmaeili, S.-A.; Mardani, F.; Seifi, B.; Mohammadi, A.; Afshari, J.T.; Sahebkar, A. Macrophage plasticity, polarization, and function in health and disease. *J. Cell. Physiol.* **2018**, *233*, 6425–6440. [CrossRef] [PubMed]
135. Ma, X.; Gao, Y.; Chen, Y.; Liu, J.; Yang, C.; Bao, C.; Wang, Y.; Feng, Y.; Song, X.; Qiao, S. M2-Type Macrophages Induce Tregs Generation by Activating the TGF-beta/Smad Signalling Pathway to Promote Colorectal Cancer Development. *OncoTargets Ther.* **2021**, *14*, 5391–5402. [CrossRef]
136. Cai, J.; Xia, L.; Li, J.; Ni, S.; Song, H.; Wu, X. Tumor-Associated Macrophages Derived TGF-β–Induced Epithelial to Mesenchymal Transition in Colorectal Cancer Cells through Smad2,3-4/Snail Signaling Pathway. *Cancer Res. Treat.* **2019**, *51*, 252–266. [CrossRef] [PubMed]
137. Zhang, X.-L.; Hu, L.-P.; Yang, Q.; Qin, W.-T.; Wang, X.; Xu, C.-J.; Tian, G.-A.; Yang, X.-M.; Yao, L.-L.; Zhu, L.; et al. CTHRC1 promotes liver metastasis by reshaping infiltrated macrophages through physical interactions with TGF-β receptors in colorectal cancer. *Oncogene* **2021**, *40*, 3959–3973. [CrossRef]
138. Li, Z.; Xu, W.; Yang, J.; Wang, J.; Wang, J.; Zhu, G.; Li, D.; Ding, J.; Sun, T. A Tumor Microenvironments-Adapted Polypeptide Hydrogel/Nanogel Composite Boosts Antitumor Molecularly Targeted Inhibition and Immunoactivation. *Adv. Mater.* **2022**, *34*, e2200449. [CrossRef]
139. Rao, H.-L.; Chen, J.-W.; Li, M.; Xiao, Y.-B.; Fu, J.; Zeng, Y.-X.; Cai, M.-Y.; Xie, D. Increased intratumoral neutrophil in colorectal carcinomas correlates closely with malignant phenotype and predicts patients' adverse prognosis. *PLoS ONE* **2012**, *7*, e30806. [CrossRef]
140. Qin, F.; Liu, X.; Chen, J.; Huang, S.; Wei, W.; Zou, Y.; Liu, X.; Deng, K.; Mo, S.; Chen, J.; et al. Anti-TGF-β attenuates tumor growth via polarization of tumor associated neutrophils towards an anti-tumor phenotype in colorectal cancer. *J. Cancer* **2020**, *11*, 2580–2592. [CrossRef]
141. Itatani, Y.; Kawada, K.; Fujishita, T.; Kakizaki, F.; Hirai, H.; Matsumoto, T.; Iwamoto, M.; Inamoto, S.; Hatano, E.; Hasegawa, S.; et al. Loss of SMAD4 from colorectal cancer cells promotes CCL15 expression to recruit CCR1+ myeloid cells and facilitate liver metastasis. *Gastroenterology* **2013**, *145*, 1064–1075.e11. [CrossRef] [PubMed]
142. Xing, W.; Xiao, Y.; Lu, X.; Zhu, H.; He, X.; Huang, W.; Lopez, E.S.; Wong, J.; Ju, H.; Tian, L.; et al. GFI1 downregulation promotes inflammation-linked metastasis of colorectal cancer. *Cell Death Differ.* **2017**, *24*, 929–943. [CrossRef] [PubMed]
143. Wang, J.Q.; Tang, Y.; Li, Q.S.; Xiao, M.; Li, M.; Sheng, Y.T.; Yang, Y.; Wang, Y.L. PARG regulates the proliferation and differentiation of DCs and T cells via PARP/NFkappaB in tumour metastases of colon carcinoma. *Oncol. Rep.* **2019**, *41*, 2657–2666.
144. Miteva, L.D.; Stanilov, N.S.; Cirovski Gcapital Em, C.; Stanilova, S.A. Upregulation of Treg-Related Genes in Addition with IL6 Showed the Significant Role for the Distant Metastasis in Colorectal Cancer. *Cancer Microenviron.* **2017**, *10*, 69–76. [CrossRef] [PubMed]
145. Batlle, E.; Clevers, H. Cancer stem cells revisited. *Nat. Med.* **2017**, *23*, 1124–1134. [CrossRef]
146. Zeuner, A.; Todaro, M.; Stassi, G.; De Maria, R. Colorectal cancer stem cells: From the crypt to the clinic. *Cell Stem Cell* **2014**, *15*, 692–705. [CrossRef]
147. Oh, I.R.; Raymundo, B.; Kim, M.; Kim, C.W. Mesenchymal stem cells co-cultured with colorectal cancer cells showed increased invasive and proliferative abilities due to its altered p53/TGF-beta1 levels. *Biosci. Biotechnol. Biochem.* **2020**, *84*, 256–267. [CrossRef]
148. Niu, B.; Liu, J.; Lv, B.; Lin, J.; Li, X.; Wu, C.; Jiang, X.; Zeng, Z.; Zhang, X.K.; Zhou, H. Interplay between transforming growth factor-beta and Nur77 in dual regulations of inhibitor of differentiation 1 for colonic tumorigenesis. *Nat. Commun.* **2021**, *12*, 2809. [CrossRef]
149. Wang, J.; Zhang, B.; Wu, H.; Cai, J.; Sui, X.; Wang, Y.; Li, H.; Qiu, Y.; Wang, T.; Chen, Z.; et al. CD51 correlates with the TGF-beta pathway and is a functional marker for colorectal cancer stem cells. *Oncogene* **2017**, *36*, 1351–1363. [CrossRef]
150. Fujishita, T.; Kojima, Y.; Kajino-Sakamoto, R.; Mishiro-Sato, E.; Shimizu, Y.; Hosoda, W.; Yamaguchi, R.; Taketo, M.M.; Aoki, M. The cAMP/PKA/CREB and TGF-β/SMAD4 pathways regulate stemness and metastatic potential in colorectal cancer cells. *Cancer Res.* **2022**, *82*, 4179–4190. [CrossRef]
151. Krstic, J.; Santibanez, J.F. Transforming growth factor-beta and matrix metalloproteinases: Functional interactions in tumor stroma-infiltrating myeloid cells. *Sci. World J.* **2014**, *2014*, 521754. [CrossRef] [PubMed]
152. Lee, H.; Kong, J.S.; Lee, S.S.; Kim, A. Radiation-Induced Overexpression of TGFbeta and PODXL Contributes to Colorectal Cancer Cell Radioresistance through Enhanced Motility. *Cells* **2021**, *10*, 2087. [CrossRef] [PubMed]
153. Jia, H.; Yang, Y.; Li, M.; Chu, Y.; Song, H.; Zhang, J.; Zhang, D.; Zhang, Q.; Xu, Y.; Wang, J.; et al. Snail enhances arginine synthesis by inhibiting ubiquitination-mediated degradation of ASS1. *EMBO Rep.* **2021**, *22*, e51780. [CrossRef] [PubMed]
154. Zajac, O.; Raingeaud, J.; Libanje, F.; Lefebvre, C.; Sabino, D.; Martins, I.; Roy, P.; Benatar, C.; Canet-Jourdan, C.; Azorin, P.; et al. Tumour spheres with inverted polarity drive the formation of peritoneal metastases in patients with hypermethylated colorectal carcinomas. *Nat. Cell Biol.* **2018**, *20*, 296–306. [CrossRef] [PubMed]
155. Kim, M.S.; Suh, K.W.; Hong, S.; Jin, W. TrkC promotes colorectal cancer growth and metastasis. *Oncotarget* **2017**, *8*, 41319–41333. [CrossRef] [PubMed]
156. Song, J.; Chen, W.; Cui, X.; Huang, Z.; Wen, D.; Yang, Y.; Yu, W.; Cui, L.; Liu, C.Y. CCBE1 promotes tumor lymphangiogenesis and is negatively regulated by TGFbeta signaling in colorectal cancer. *Theranostics* **2020**, *10*, 2327–2341. [CrossRef] [PubMed]

157. Haidar, M.; Metheni, M.; Batteux, F.; Langsley, G. TGF-beta2, catalase activity, H2O2 output and metastatic potential of diverse types of tumour. *Free Radic. Biol. Med.* **2019**, *134*, 282–287. [CrossRef]
158. Gutierrez, A.; Demond, H.; Brebi, P.; Ili, C.G. Novel Methylation Biomarkers for Colorectal Cancer Prognosis. *Biomolecules* **2021**, *11*, 1722. [CrossRef]
159. Wang, X.; Liu, J.; Wang, D.; Feng, M.; Wu, X. Epigenetically regulated gene expression profiles reveal four molecular subtypes with prognostic and therapeutic implications in colorectal cancer. *Brief. Bioinform.* **2021**, *22*, bbaa309. [CrossRef]
160. Stanilova, S.; Stanilov, N.; Julianov, A.; Manolova, I.; Miteva, L. Transforming growth factor-beta1 gene promoter -509C/T polymorphism in association with expression affects colorectal cancer development and depends on gender. *PLoS ONE* **2018**, *13*, e0201775. [CrossRef]
161. Tu, Y.; Han, J.; Dong, Q.; Chai, R.; Li, N.; Lu, Q.; Xiao, Z.; Guo, Y.; Wan, Z.; Xu, Q. TGF-beta2 is a Prognostic Biomarker Correlated with Immune Cell Infiltration in Colorectal Cancer: A STROBE-compliant article. *Medicine* **2020**, *99*, e23024. [CrossRef] [PubMed]
162. Vocka, M.; Langer, D.; Fryba, V.; Petrtyl, J.; Hanus, T.; Kalousova, M.; Zima, T.; Petruzelka, L. Growth/differentiation factor 15 (GDF-15) as new potential serum marker in patients with metastatic colorectal cancer. *Cancer Biomark.* **2018**, *21*, 869–874. [CrossRef] [PubMed]
163. Wang, X.; Liu, S.; Cao, H.; Li, X.; Rong, Y.; Liu, G.; Du, H.; Shen, H. Increasing Embryonic Morphogen Nodal Expression Suggests Malignant Transformation in Colorectal Lesions and as a Potential Marker for CMS4 Subtype of Colorectal Cancer. *Pathol. Oncol. Res.* **2021**, *27*, 587029. [CrossRef]
164. Staudacher, J.J.; Bauer, J.; Jana, A.; Tian, J.; Carroll, T.; Mancinelli, G.; Ozden, O.; Krett, N.; Guzman, G.; Kerr, D.; et al. Activin signaling is an essential component of the TGF-beta induced pro-metastatic phenotype in colorectal cancer. *Sci. Rep.* **2017**, *7*, 5569. [CrossRef] [PubMed]
165. Gulubova, M.; Aleksandrova, E.; Vlaykova, T. Promoter polymorphisms in TGFB1 and IL10 genes influence tumor dendritic cells infiltration, development and prognosis of colorectal cancer. *J. Genet. Med.* **2018**, *20*, e3005. [CrossRef] [PubMed]
166. Ricci, V.; Granetto, C.; Falletta, A.; Paccagnella, M.; Abbona, A.; Fea, E.; Fabozzi, T.; Lo Nigro, C.; Merlano, M.C. Circulating cytokines and outcome in metastatic colorectal cancer patients treated with regorafenib. *World J. Gastrointest. Oncol.* **2020**, *12*, 301–310. [CrossRef] [PubMed]
167. Baraniskin, A.; Munding, J.; Schulmann, K.; Meier, D.; Porschen, R.; Arkenau, H.T.; Graeven, U.; Schmiegel, W.; Tannapfel, A.; Reinacher-Schick, A. Prognostic value of reduced SMAD4 expression in patients with metastatic colorectal cancer under oxaliplatin-containing chemotherapy: A translational study of the AIO colorectal study group. *Clin. Colorectal Cancer* **2011**, *10*, 24–29. [CrossRef]
168. Martinelli, E.; Sforza, V.; Cardone, C.; Capasso, A.; Nappi, A.; Martini, G.; Napolitano, S.; Rachiglio, A.M.; Normanno, N.; Cappabianca, S.; et al. Clinical outcome and molecular characterisation of chemorefractory metastatic colorectal cancer patients with long-term efficacy of regorafenib treatment. *ESMO Open* **2017**, *2*, e000177. [CrossRef]
169. Lee, M.S.; Cho, H.J.; Hong, J.Y.; Lee, J.; Park, S.H.; Park, J.O.; Park, Y.S.; Lim, H.Y.; Kang, W.K.; Cho, Y.B.; et al. Clinical and molecular distinctions in patients with refractory colon cancer who benefit from regorafenib treatment. *Ther. Adv. Med. Oncol.* **2020**, *12*, 1758835920965842. [CrossRef]
170. D'Agay, M.G.; Galland, L.; Tharin, Z.; Truntzer, C.; Ghiringhelli, F. Utility of exome sequencing in routine care for metastatic colorectal cancer. *Mol. Clin. Oncol.* **2021**, *15*, 229. [CrossRef]
171. Mizuno, T.; Cloyd, J.M.; Vicente, D.; Omichi, K.; Chun, Y.S.; Kopetz, S.E.; Maru, D.; Conrad, C.; Tzeng, C.D.; Wei, S.H.; et al. SMAD4 gene mutation predicts poor prognosis in patients undergoing resection for colorectal liver metastases. *Eur. J. Surg. Oncol.* **2018**, *44*, 684–692. [CrossRef] [PubMed]
172. Zhuo, C.; Hu, D.; Li, J.; Yu, H.; Lin, X.; Chen, Y.; Zhuang, Y.; Li, Q.; Zheng, X.; Yang, C. Downregulation of Activin A Receptor Type 2A Is Associated with Metastatic Potential and Poor Prognosis of Colon Cancer. *J. Cancer* **2018**, *9*, 3626–3633. [CrossRef] [PubMed]
173. Daitoku, N.; Miyamoto, Y.; Hiyoshi, Y.; Tokunaga, R.; Sakamoto, Y.; Sawayama, H.; Ishimoto, T.; Baba, Y.; Yoshida, N.; Baba, H. Activin A promotes cell proliferation, invasion and migration and predicts poor prognosis in patients with colorectal cancer. *Oncol. Rep.* **2022**, *47*, 107. [CrossRef] [PubMed]
174. Yuan, J.; Xie, A.; Cao, Q.; Li, X.; Chen, J. INHBB Is a Novel Prognostic Biomarker Associated with Cancer-Promoting Pathways in Colorectal Cancer. *Biomed. Res. Int.* **2020**, *2020*, 6909672. [CrossRef]
175. He, Z.; Liang, J.; Wang, B. Inhibin, beta A regulates the transforming growth factor-beta pathway to promote malignant biological behaviour in colorectal cancer. *Cell Biochem. Funct.* **2020**, *39*, 258–266. [CrossRef]
176. Liu, X.; Wang, C. Long non-coding RNA ATB is associated with metastases and promotes cell invasion in colorectal cancer via sponging miR-141-3p. *Exp. Ther. Med.* **2020**, *20*, 261. [CrossRef]
177. Fricke, F.; Mussack, V.; Buschmann, D.; Hausser, I.; Pfaffl, M.W.; Kopitz, J.; Gebert, J. TGFBR2dependent alterations of microRNA profiles in extracellular vesicles and parental colorectal cancer cells. *Int. J. Oncol.* **2019**, *55*, 925–937.
178. Nosho, K.; Igarashi, H.; Nojima, M.; Ito, M.; Maruyama, R.; Yoshii, S.; Naito, T.; Sukawa, Y.; Mikami, M.; Sumioka, W.; et al. Association of microRNA-31 with BRAF mutation, colorectal cancer survival and serrated pathway. *Carcinogenesis* **2014**, *35*, 776–783. [CrossRef]
179. Schwarzmueller, L.; Bril, O.; Vermeulen, L.; Leveille, N. Emerging Role and Therapeutic Potential of lncRNAs in Colorectal Cancer. *Cancers* **2020**, *12*, 3843. [CrossRef]

180. Wang, D.Z.; Chen, G.Y.; Li, Y.F.; Zhang, N.W. Comprehensive analysis of long non-coding RNA and mRNA expression profile in rectal cancer. *Chin. Med. J.* **2020**, *133*, 1312–1321. [CrossRef]
181. Eide, P.W.; Eilertsen, I.A.; Sveen, A.; Lothe, R.A. Long noncoding RNA MIR31HG is a bona fide prognostic marker with colorectal cancer cell-intrinsic properties. *Int. J. Cancer* **2019**, *144*, 2843–2853. [CrossRef] [PubMed]
182. Li, Q.; Yue, W.; Li, M.; Jiang, Z.; Hou, Z.; Liu, W.; Ma, N.; Gan, W.; Li, Y.; Zhou, T.; et al. Downregulating Long Non-coding RNAs CTBP1-AS2 Inhibits Colorectal Cancer Development by Modulating the miR-93-5p/TGF-beta/SMAD2/3 Pathway. *Front. Oncol.* **2021**, *11*, 626620. [CrossRef] [PubMed]
183. Li, M.; Jin, Y.; Li, Y. LncRNA TP73-AS1 Activates TGF-beta1 to Promote the Migration and Invasion of Colorectal Cancer Cell. *Cancer Manag. Res.* **2019**, *11*, 10523–10529. [CrossRef] [PubMed]
184. Chuo, D.; Liu, F.; Chen, Y.; Yin, M. LncRNA MIR503HG is downregulated in Han Chinese with colorectal cancer and inhibits cell migration and invasion mediated by TGF-beta2. *Gene* **2019**, *713*, 143960. [CrossRef] [PubMed]
185. Zhang, T.T.; Chen, H.P.; Yu, S.Y.; Zhao, S.P. LncRNA HOXC-AS3 overexpression inhibits TGF-beta2-induced colorectal cancer cell migration and invasion by sponging miR-1269. *Hum. Exp. Toxicol.* **2022**, *41*, 9603271221093630. [CrossRef]
186. Liu, Z.; Wang, N.; Wang, F.; Zhang, S.; Ding, J. Silencing of lncRNA EZR-AS1 inhibits proliferation, invasion, and migration of colorectal cancer cells through blocking transforming growth factor beta signaling. *Biosci. Rep.* **2019**, *39*, BSR20191199. [CrossRef]
187. Xu, J.; Shao, T.; Song, M.; Xie, Y.; Zhou, J.; Yin, J.; Ding, N.; Zou, H.; Li, Y.; Zhang, J. MIR22HG acts as a tumor suppressor via TGFbeta/SMAD signaling and facilitates immunotherapy in colorectal cancer. *Mol. Cancer* **2020**, *19*, 51. [CrossRef]
188. Wu, N.; Jiang, M.; Liu, H.; Chu, Y.; Wang, D.; Cao, J.; Wang, Z.; Xie, X.; Han, Y.; Xu, B. LINC00941 promotes CRC metastasis through preventing SMAD4 protein degradation and activating the TGF-beta/SMAD2/3 signaling pathway. *Cell Death Differ.* **2021**, *28*, 219–232. [CrossRef]
189. Balacescu, O.; Sur, D.; Cainap, C.; Visan, S.; Cruceriu, D.; Manzat-Saplacan, R.; Muresan, M.S.; Balacescu, L.; Lisencu, C.; Irimie, A. The Impact of miRNA in Colorectal Cancer Progression and Its Liver Metastases. *Int. J. Mol. Sci.* **2018**, *19*, 3711. [CrossRef]
190. Hu, X.; Chen, Q.; Guo, H.; Li, K.; Fu, B.; Chen, Y.; Zhao, H.; Wei, M.; Li, Y.; Wu, H. Identification of Target PTEN-Based miR-425 and miR-576 as Potential Diagnostic and Immunotherapeutic Biomarkers of Colorectal Cancer With Liver Metastasis. *Front. Oncol.* **2021**, *11*, 657984. [CrossRef]
191. Li, B.; Huang, M.; Liu, M.; Wen, S.; Sun, F. MicroRNA329 serves a tumor suppressive role in colorectal cancer by directly targeting transforming growth factor beta1. *Mol. Med. Rep.* **2017**, *16*, 3825–3832. [CrossRef] [PubMed]
192. Radwan, E.; Shaltout, A.S.; Mansor, S.G.; Shafik, E.A.; Abbas, W.A.; Shehata, M.R.; Ali, M. Evaluation of circulating microRNAs-211 and 25 as diagnostic biomarkers of colorectal cancer. *Mol. Biol. Rep.* **2021**, *48*, 4601–4610. [CrossRef]
193. Sells, E.; Pandey, R.; Chen, H.; Skovan, B.A.; Cui, H.; Ignatenko, N.A. Specific microRNA-mRNA Regulatory Network of Colon Cancer Invasion Mediated by Tissue Kallikrein-Related Peptidase 6. *Neoplasia* **2017**, *19*, 396–411. [CrossRef] [PubMed]
194. Zhang, W.; Zhang, T.; Jin, R.; Zhao, H.; Hu, J.; Feng, B.; Zang, L.; Zheng, M.; Wang, M. MicroRNA-301a promotes migration and invasion by targeting TGFBR2 in human colorectal cancer. *J. Exp. Clin. Cancer Res.* **2014**, *33*, 113. [CrossRef] [PubMed]
195. Ullmann, P.; Rodriguez, F.; Schmitz, M.; Meurer, S.K.; Qureshi-Baig, K.; Felten, P.; Ginolhac, A.; Antunes, L.; Frasquilho, S.; Zugel, N.; et al. The miR-371 approximately 373 Cluster Represses Colon Cancer Initiation and Metastatic Colonization by Inhibiting the TGFBR2/ID1 Signaling Axis. *Cancer Res.* **2018**, *78*, 3793–3808. [CrossRef]
196. He, H.; Zhao, X.; Zhu, Z.; Du, L.; Chen, E.; Liu, S.; Li, Q.; Dong, J.; Yang, J.; Lei, L. MicroRNA-3191 promotes migration and invasion by downregulating TGFBR2 in colorectal cancer. *J. Biochem. Mol. Toxicol.* **2019**, *33*, e22308. [CrossRef]
197. Xu, X.; Chen, R.; Li, Z.; Huang, N.; Wu, X.; Li, S.; Li, Y.; Wu, S. MicroRNA-490-3p inhibits colorectal cancer metastasis by targeting TGFbetaR1. *BMC Cancer* **2015**, *15*, 1023. [CrossRef]
198. Yu, Y.; Lei, X. CircFAM120B Blocks the Development of Colorectal Cancer by Activating TGF-Beta Receptor II Expression via Targeting miR-645. *Front. Cell Dev. Biol.* **2021**, *9*, 682543. [CrossRef]
199. Zhang, N.; Li, L.; Luo, J.; Tan, J.; Hu, W.; Li, Z.; Wang, X.; Ye, T. Inhibiting microRNA-424 in bone marrow mesenchymal stem cells-derived exosomes suppresses tumor growth in colorectal cancer by upregulating TGFBR3. *Arch. Biochem. Biophys.* **2021**, *709*, 108965. [CrossRef]
200. Bao, Y.; Chen, Z.; Guo, Y.; Feng, Y.; Li, Z.; Han, W.; Wang, J.; Zhao, W.; Jiao, Y.; Li, K.; et al. Tumor suppressor microRNA-27a in colorectal carcinogenesis and progression by targeting SGPP1 and Smad2. *PLoS ONE* **2014**, *9*, e105991. [CrossRef]
201. Li, J.; Zou, K.; Yu, L.; Zhao, W.; Lu, Y.; Mao, J.; Wang, B.; Wang, L.; Fan, S.; Song, B.; et al. MicroRNA-140 Inhibits the Epithelial-Mesenchymal Transition and Metastasis in Colorectal Cancer. *Mol. Ther. Nucl. Acids* **2018**, *10*, 426–437. [CrossRef]
202. Cheng, D.; Zhao, S.; Tang, H.; Zhang, D.; Sun, H.; Yu, F.; Jiang, W.; Yue, B.; Wang, J.; Zhang, M.; et al. MicroRNA-20a-5p promotes colorectal cancer invasion and metastasis by downregulating Smad4. *Oncotarget* **2016**, *7*, 45199–45213. [CrossRef] [PubMed]
203. Li, Q.; Zou, C.; Zou, C.; Han, Z.; Xiao, H.; Wei, H.; Wang, W.; Zhang, L.; Zhang, X.; Tang, Q.; et al. MicroRNA-25 functions as a potential tumor suppressor in colon cancer by targeting Smad7. *Cancer Lett.* **2013**, *335*, 168–174. [CrossRef] [PubMed]
204. Zhao, S.; Sun, H.; Jiang, W.; Mi, Y.; Zhang, D.; Wen, Y.; Cheng, D.; Tang, H.; Wu, S.; Yu, Y.; et al. miR-4775 promotes colorectal cancer invasion and metastasis via the Smad7/TGFbeta-mediated epithelial to mesenchymal transition. *Mol. Cancer* **2017**, *16*, 12. [CrossRef]
205. Wang, H.; Nie, L.; Wu, L.; Liu, Q.; Guo, X. NR2F2 inhibits Smad7 expression and promotes TGF-beta-dependent epithelial-mesenchymal transition of CRC via transactivation of miR-21. *Biochem. Biophys. Res. Commun.* **2017**, *485*, 181–188. [CrossRef] [PubMed]

206. Li, N. CircTBL1XR1/miR-424 axis regulates Smad7 to promote the proliferation and metastasis of colorectal cancer. *J. Gastrointest. Oncol.* **2020**, *11*, 918–931. [CrossRef]
207. De Summa, S.; Danza, K.; Pilato, B.; Matera, G.; Fasano, R.; Calabrese, A.; Lacalamita, R.; Silvestris, N.; Tommasi, S.; Argentiero, A.; et al. A Promising Role of TGF-beta Pathway in Response to Regorafenib in Metastatic Colorectal Cancer: A Case Report. *Medicina* **2021**, *57*, 1241. [CrossRef]
208. Zhang, B.H.; Wang, C.; Dong, W.; Chen, X.; Leng, C.; Luo, X.; Dong, S.L.; Yin, P.; Zhang, B.X.; Datta, P.K.; et al. A novel approach for monitoring TGF-beta signaling in vivo in colon cancer. *Carcinogenesis* **2021**, *42*, 631–639. [CrossRef]
209. Otegbeye, F.; Ojo, E.; Moreton, S.; Mackowski, N.; Lee, D.A.; de Lima, M.; Wald, D.N. Inhibiting TGF-beta signaling preserves the function of highly activated, in vitro expanded natural killer cells in AML and colon cancer models. *PLoS ONE* **2018**, *13*, e0191358.
210. Varela-Calvino, R.; Rodriguez-Quiroga, M.; Dias Carvalho, P.; Martins, F.; Serra-Roma, A.; Vazquez-Iglesias, L.; Paez de la Cadena, M.; Velho, S.; Cordero, O.J. The mechanism of sitagliptin inhibition of colorectal cancer cell lines' metastatic functionalities. *IUBMB Life* **2021**, *73*, 761–773. [CrossRef]
211. Kim, S.Y.; Kim, K.; Cho, S.H.; Chun, S.M.; Tak, E.; Hong, Y.S.; Kim, J.E.; Kim, T.W. Longitudinal change of genetic variations in cetuximab-treated metastatic colorectal cancer. *Cancer Genet.* **2021**, *258–259*, 27–36. [CrossRef] [PubMed]
212. Ciardiello, D.; Blauensteiner, B.; Matrone, N.; Belli, V.; Mohr, T.; Vitiello, P.P.; Martini, G.; Poliero, L.; Cardone, C.; Napolitano, S.; et al. Dual inhibition of TGFbeta and AXL as a novel therapy for human colorectal adenocarcinoma with mesenchymal phenotype. *Med. Oncol.* **2021**, *38*, 24. [CrossRef]
213. Li, Y.C.; Chen, C.H.; Chang, C.L.; Chiang, J.Y.; Chu, C.H.; Chen, H.H.; Yip, H.K. Melatonin and hyperbaric oxygen therapies suppress colorectal carcinogenesis through pleiotropic effects and multifaceted mechanisms. *Int. J. Biol. Sci.* **2021**, *17*, 3728–3744. [CrossRef] [PubMed]
214. Principe, D.R.; DeCant, B.; Staudacher, J.; Vitello, D.; Mangan, R.J.; Wayne, E.A.; Mascariñas, E.; Diaz, A.M.; Bauer, J.; McKinney, R.D.; et al. Loss of TGFβ signaling promotes colon cancer progression and tumor-associated inflammation. *Oncotarget* **2017**, *8*, 3826–3839. [CrossRef] [PubMed]
215. Zhuang, Y.W.; Wu, C.E.; Zhou, J.Y.; Zhao, Z.M.; Liu, C.L.; Shen, J.Y.; Cai, H.; Liu, S.L. Solasodine reverses stemness and epithelial-mesenchymal transition in human colorectal cancer. *Biochem. Biophys. Res. Commun.* **2018**, *505*, 485–491. [CrossRef] [PubMed]
216. Lai, Z.; Yan, Z.; Chen, W.; Peng, J.; Feng, J.; Li, Q.; Jin, Y.; Lin, J. Hedyotis diffusa Willd suppresses metastasis in 5fluorouracilresistant colorectal cancer cells by regulating the TGFbeta signaling pathway. *Mol. Med. Rep.* **2017**, *16*, 7752–7758. [CrossRef]
217. Yang, B.; Bai, H.; Sa, Y.; Zhu, P.; Liu, P. Inhibiting EMT, stemness and cell cycle involved in baicalin-induced growth inhibition and apoptosis in colorectal cancer cells. *J. Cancer* **2020**, *11*, 2303–2317. [CrossRef]
218. Jiang, Z.; Cao, Q.; Dai, G.; Wang, J.; Liu, C.; Lv, L.; Pan, J. Celastrol inhibits colorectal cancer through TGF-beta1/Smad signaling. *OncoTargets Ther.* **2019**, *12*, 509–518. [CrossRef]
219. Jin, Y.; Chen, W.; Yang, H.; Yan, Z.; Lai, Z.; Feng, J.; Peng, J.; Lin, J. Scutellaria barbata D. Don inhibits migration and invasion of colorectal cancer cells via suppression of PI3K/AKT and TGF-beta/Smad signaling pathways. *Exp. Ther. Med.* **2017**, *14*, 5527–5534.
220. Xiao, Q.; Xiao, J.; Liu, J.; Liu, J.; Shu, G.; Yin, G. Metformin suppresses the growth of colorectal cancer by targeting INHBA to inhibit TGF-beta/PI3K/AKT signaling transduction. *Cell Death Dis.* **2022**, *13*, 202. [CrossRef]
221. Zhang, L.; Cai, Q.Y.; Liu, J.; Peng, J.; Chen, Y.Q.; Sferra, T.J.; Lin, J.M. Ursolic acid suppresses the invasive potential of colorectal cancer cells by regulating the TGF-beta1/ZEB1/miR-200c signaling pathway. *Oncol. Lett.* **2019**, *18*, 3274–3282. [PubMed]
222. Dai, Y.; Wang, H.; Sun, R.; Diao, J.; Ma, Y.; Shao, M.; Xu, Y.; Zhang, Q.; Gao, Z.; Zeng, Z.; et al. Modified Shenlingbaizhu Decoction represses the pluripotency of colorectal cancer stem cells by inhibiting TGF-beta mediated EMT program. *Phytomedicine* **2022**, *103*, 154234. [CrossRef] [PubMed]
223. Zhang, L.; Liu, J.; Lin, S.; Tan, J.; Huang, B.; Lin, J. Qingjie Fuzheng Granule Inhibited the Migration and Invasion of Colorectal Cancer Cells by Regulating the lncRNA ANRIL/let-7a/TGF-beta1/Smad Axis. *Evid. Based Compl. Altern. Med.* **2020**, *2020*, 5264651. [CrossRef] [PubMed]
224. Li, Q.; Chen, J.X.; Wu, Y.; Lv, L.L.; Ying, H.F.; Zhu, W.H.; Xu, J.Y.; Ruan, M.; Guo, Y.; Zhu, W.R.; et al. The mechanism of FZXJJZ decoction suppresses colorectal liver metastasis via the VDR/TGF-beta/Snail1 signaling pathways based on network pharmacology-TCGA data-transcriptomics analysis. *J. Ethnopharmacol.* **2022**, *287*, 114904. [CrossRef]
225. Roerink, S.F.; Sasaki, N.; Lee-Six, H.; Young, M.D.; Alexandrov, L.B.; Behjati, S.; Mitchell, T.J.; Grossmann, S.; Lightfoot, H.; Egan, D.A.; et al. Intra-tumour diversification in colorectal cancer at the single-cell level. *Nature* **2018**, *556*, 457–462. [CrossRef] [PubMed]
226. Guinney, J.; Dienstmann, R.; Wang, X.; de Reynies, A.; Schlicker, A.; Soneson, C.; Marisa, L.; Roepman, P.; Nyamundanda, G.; Angelino, P.; et al. The consensus molecular subtypes of colorectal cancer. *Nat. Med.* **2015**, *21*, 1350–1356. [CrossRef]
227. Yu, I.S.; Cheung, W.Y. Metastatic Colorectal Cancer in the Era of Personalized Medicine: A More Tailored Approach to Systemic Therapy. *Can. J. Gastroenterol. Hepatol.* **2018**, *2018*, 9450754. [CrossRef]
228. Xie, F.; Ling, L.; van Dam, H.; Zhou, F.; Zhang, L. TGF-beta signaling in cancer metastasis. *Acta Biochim. Biophys. Sin.* **2018**, *50*, 121–132. [CrossRef]
229. Binefa, G.; Rodriguez-Moranta, F.; Teule, A.; Medina-Hayas, M. Colorectal cancer: From prevention to personalized medicine. *World J. Gastroenterol.* **2014**, *20*, 6786–6808. [CrossRef]

International Journal of
Molecular Sciences

MDPI

Article

A Transcriptome and Methylome Study Comparing Tissues of Early and Late Onset Colorectal Carcinoma

Muhammad G Kibriya [1,*], Maruf Raza [2], Anthony Quinn [1], Mohammed Kamal [3], Habibul Ahsan [1] and Farzana Jasmine [1]

1 Institute for Population and Precision Health (IPPH), Biological Sciences Division, The University of Chicago, Chicago, IL 60637, USA
2 Department of Pathology, Jahurul Islam Medical College, Kishoregonj 2336, Bangladesh
3 Department of Pathology, The Laboratory Dhaka, Dhaka 1205, Bangladesh
* Correspondence: mkibriya@bsd.uchicago.edu

Abstract: There is an increase in the incidence of early onset colorectal carcinoma (EOCRC). To better understand if there is any difference in molecular pathogenesis of EOCRC and late onset colorectal carcinoma (LOCRC), we compared the clinical, histological, transcriptome, and methylome profile of paired CRC and healthy colonic tissue from 67 EOCRC and 98 LOCRC patients. The frequency of stage 3 CRC, lymph node involvement, lymphovascular invasion, and perineural invasion was higher in the EOCRC group. Many of the cancer related pathways were differentially expressed in CRC tissue in both EOCRC and LOCRC patients. However, the magnitude of differential expression for some groups of genes, such as DNA damage repair genes and replication stress genes, were significantly less pronounced in the EOCRC group, suggesting less efficient DNA damage repair to be associated with EOCRC. A more marked methylation of "growth factor receptor" genes in LOCRC correlated with a more pronounced down-regulation of those genes in that group. From a therapeutic point of view, more over-expression of fatty acid synthase (*FASN*) among the LOCRC patients may suggest a better response of *FASN* targeted therapy in that group. The age of onset of CRC did not appear to modify the response of cis-platin or certain immune checkpoint inhibitors. We found some differences in the molecular pathogenesis in EOCRC and LOCRC that may have some biological and therapeutic significance.

Keywords: colorectal carcinoma; early onset CRC; multi-omics; lymphovascular invasion; perineural invasion; DNA damage repair; DNA replication repair; MMR; base excision repair; FASN; *CTLA4*; *HAVCR2*

Citation: Kibriya, M.G.; Raza, M.; Quinn, A.; Kamal, M.; Ahsan, H.; Jasmine, F. A Transcriptome and Methylome Study Comparing Tissues of Early and Late Onset Colorectal Carcinoma. *Int. J. Mol. Sci.* **2022**, *23*, 14261. https://doi.org/10.3390/ijms232214261

Academic Editors: Alessandro Ottaiano and Donatella Delle Cave

Received: 28 October 2022
Accepted: 15 November 2022
Published: 17 November 2022

1. Introduction

Incidence of colorectal cancer (CRC) among young adults has increased over the last 25 years [1]. In a study involving a large cohort of 143.7 million people aged 20–49 years from 20 European countries, a total of 187,918 individuals (0.13%) were diagnosed with CRC. From 2004 to 2016, on average, CRC incidence incrteased by 7.9% per year among subjects aged 20–29 years, 4.9% per year among subjects aged 30–39 years, and 1.6% per year in the age group of 40–49 years [1]. Considering the current practice of a preventive colonoscopy recommendation at 50 years of age, some considered CRC before the age of 50 years as early onset CRC (EOCRC) [1–5]. Other studies have used 40 years as a cut-off for EOCRC and late onset CRC (LOCRC) [6–8]. There are some interesting findings, such as the association of EOCRC and obesity [9,10], the cause of which is undetermined. In a previous study, we also reported a high proportion of CRC patients in Bangladesh presenting at or below the age of 40 years [11]. There is a growing interest in exploring the underlying causes of EOCRC. Like some other studies [6–8], we divided the cases as EOCRC and LOCRC as =<40 years and >40 years, respectively.

The clinical features of EOCRC may be different from those of LOCRC. Some studies suggested that EOCRC is usually detected in the distal colon and rectum, whereas in the older age group it is in the proximal colon [3,12]. The anatomical location of EOCRC may provide important insights into the underlying disease processes and treatment responses, since embryologically the proximal and distal colon are different. EOCRC display different histopathological features when compared to late onset cases. Poor differentiation, perineural invasion (PNI), venous invasion, and mucinous and/or signet cell morphology, all of which are suggestive of an unfavorable tumor biology and associated with worse oncological outcomes, are more common among patients with EOCRC [6,8]. It was also found that EOCRC patients present with higher rates of metastasis and recurrence compared to their older counterparts. Regarding the outcome of EOCRC and LOCRC, some studies reported a worse prognosis for EOCRC, while others found equivalent or superior outcomes among the younger patients [8,13–16].

A study showed that, among EOCRC, approximately 30% of patients were affected by tumors harboring mutations causing hereditary cancer predisposing syndromes, and 20% have familial CRC [17]. Others also found this incidence among 19.7% to 29.4% of EOCRC, depending on the age of onset [7,18]. Chang et al. found that, among EOCRC, 17% demonstrated abnormalities in DNA mismatch repair and 5% had known germline genetic disorders [6]. Therefore, the remaining 50% of cases were of sporadic EOCRC. The rate of EOCRC is increasing, and it shows some distinct characteristics regarding the location, anatomical features, and pathological features. There is a growing interest to explore the findings at the molecular level, which can potentially lead to improved treatment and survival.

It is largely unclear if there are any differences in the transcriptome or methylome profile among the EOCRC and LOCRC patients that can (a) shed light on the difference in the molecular pathogenesis in these two groups and (b) provide a molecular basis of potential use of different treatment options. To address these research questions, in this study, we compared the clinical, histological, transcriptome, and methylome profiles of paired CRC and surrounding healthy colonic tissue from the same patients who presented with CRC at =<40 years of age and those who presented after 40 years of age.

2. Results

In this collection of 165 CRC patients from Bangladesh, we included all consecutive patients during the multiple time periods spanning between December 2009 and May 2016, irrespective of the age of onset of CRC. Of the 165 patients, 67 of them (40% of the patients) were of age =< 40 years. The comparison of clinical and histological features among the patients with LOCRC (age > 40 years) and EOCRC (age =< 40 years) is shown in Table 1. We did not find any statistically significant difference for sex, tumor location, tumor grading, presence of tumor infiltrating leukocyte (TIL) status, presence of signet ring appearance, CEA level, microsatellite instability (MSI), KRAS mutation, BRAF mutation, EGFR mutation, and relative telomere length (RTL) shortening. It may be noted that the frequency of BRAF mutation was low in this population, and we did not find an EGFR mutation in any patient. The analysis showed that the EOCRC patients more frequently had advanced stage CRC and lymph node involvement than LOCRC patients (see Table 1). In fact, we also found that the EOCRC patients more frequently had a lympho-vascular invasion (LVI) and PNI than the LOCRC patients did (see Table 1).

Histological staging has a strong association with LVI (13.5%, 23.8%, and 50% in stage 1, stage 2, and stage 3, respectively, $p = 0.0001$, Chi square test) and PNI (2.7%, 7.1%, and 20.9% in stage 1, stage 2, and stage 3, respectively, $p = 0.01$, Chi-square test). Therefore, we looked at the association between the age of onset and LVI and PNI by stratifying the data by stage (see Table 2). Compared to LOCRC, EOCRC more frequently had LVI only in the presence of the stage 2 disease and more frequently had PNI only in the presence of stage 3 disease. In other words, these associations of LVI and EOCRC or PNI and EOCRC were not independent of staging.

Table 1. Comparison of histological and molecular changes in LOCRC and EOCRC.

Characteristic	Category	Late-Onset (>40 Years)	Early-Onset (<40 Years)	*Chi-Square Test* *p*-Value
Sex	Male	57 (58.2%)	39 (58.2%)	0.995
	Female	41 (41.8%)	28 (41.8%)	
Location	Right Colon	22 (22.4%)	11 (16.4%)	0.327
	Left Colon	20 (20.4%)	10 (14.9%)	
	Rectum	56 (57.1%)	46 (68.7%)	
Stage	Stage-1	30 (30.6%)	7 (10.4%)	0.004 *
	Stage-2	26 (26.5%)	16 (23.9%)	
	Stage-3	42 (42.9%)	44 (65.7%)	
Grade	Low	51 (52%)	27 (40.3%)	0.138
	High	47 (48%)	40 (59.7%)	
Lymph Node	Present	56 (57.1%)	23 (34.3%)	0.004 *
	Absent	42 (42.9%)	44 (65.7%)	
TIL	0	58 (59.2%)	41 (61.2%)	0.796
	1	40 (40.8%)	26 (38.8%)	
Signet Ring	Absent	70 (71.4%)	43 (64.2%)	0.325
	Present	28 (28.6%)	24 (35.8%)	
LV Invasion	Absent	71 (72.4%)	36 (53.7%)	0.013 *
	Present	27 (27.6%)	31 (46.3%)	
PN Invasion	Absent	90 (91.8%)	53 (79.1%)	0.018 *
	Present	8 (8.2%)	14 (20.9%)	
CEA (ng/mL)	Mean	50.221	30.133	0.180 #
	(SD)	100.584	65.732	
Microsatellite	MSI	26 (26.5%)	15 (22.4%)	0.545
	MSS	72 (73.5%)	52 (77.6%)	
KRAS (rs112445441)	Wild	68 (70.8%)	47 (70.1%)	0.925
	Mutant	28 (29.2%)	20 (29.9%	
BRAFV600E	Wild	89 (92.7%)	62 (93.9%)	0.76
	Mutant	7 (7.3%)	4 (6.1%)	
EGFR	Wild	96 (100%)	67 (100%)	NA
	Mutant	0 (0%)	0 (0%)	
Telomere	Absent	28 (30.8%)	23 (37.7%)	0.375
Shortening	Present	63 (69.2%)	38 (62.3%)	

* Significant at <0.05 level; # *t*-test.

Table 2. LVI and PNI stratified by stage.

Characteristic	Stage	Status	Late-Onset (>40 Years)	Early-Onset (<40 Years)	*Chi-Square Test* *p*-Value
LV Invasion	Stage-1	LVI Absent	26 (86.7%)	6 (85.7%)	0.947
		LVI Present	4 (13.3%)	1 (14.3%)	
	Stage-2	LVI Absent	23 (88.5%)	9 (56.3%)	0.017 *
		LVI Present	3 (11.5%)	7 (43.8%)	
	Stage-3	LVI Absent	22 (52.4%)	21 (47.7%)	0.666
		LVI Present	20 (47.6%)	23 (52.3%)	
PN Invasion	Stage-1	PNI Absent	29 (96.7%)	7 (100%)	0.624
		PNI Present	1 (3.3%)	0 (0%)	
	Stage-2	PNI Absent	24 (92.3%)	15 (93.8%)	0.860
		PNI Present	2 (7.7%)	1 (6.3%)	
	Stage-3	PNI Absent	37 (88.1%)	31 (70.5%)	0.044 *
		PNI Present	5 (11.9%)	13 (29.5%)	

* Significant at <0.05 level.

2.1. Differential Gene Expression in CRC

2.1.1. Age of Onset of CRC and Differential Gene Expression of Cancer Related Gene Sets

In the next step, we asked if the differential expression (CRC tissue vs. paired normal colonic tissue) of some known cancer related gene sets were different in magnitude in EOCRC and LOCRC patients. The detailed list of these gene sets is shown in Supplementary Table S1. Table 3 shows that the "Growth factor receptor" genes were significantly down-regulated in tumor tissue only in LOCRC [fold change −1.12 (95% CI: from −1.16 to −1.08%), $p = 1.21 \times 10^{-8}$], but not in EOCRC patients (see Figure 1). Tumor suppressor genes (TSG) were significantly down-regulated in both LOCRC and EOCRC patients by a similar magnitude. However, anti-TSG (cyclin D group of genes) were significantly up-regulated, both in LOCRC and EOCRC patients, but the magnitude of over-expression was significantly more pronounced in LOCRC than in EOCRC patients [fold change 1.46 (95% CI 1.38–1.55) vs. 1.29 (95% CI 1.20–1.39), ANOVA interaction $p = 0.007$] (see Figure 2). DNA repair genes as a whole were significantly over-expressed in both groups, but to a lesser extent in the young onset group. Therefore, in the next step, we tested the gene sets involved in different DNA repair mechanisms separately.

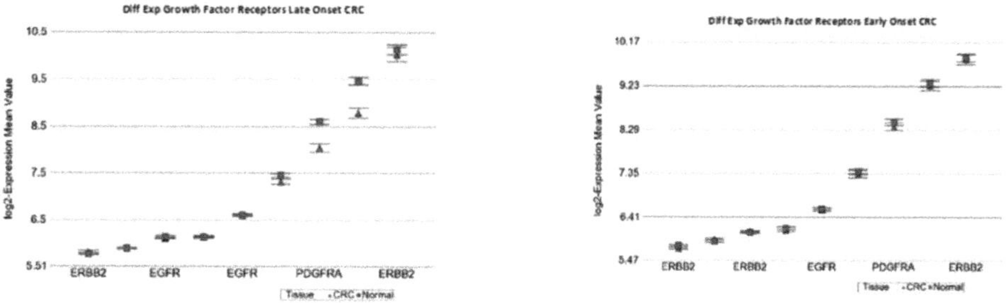

(A) (B)

Figure 1. Differential gene expression of growth factor receptor genes in paired CRC tissue (in blue) and healthy colonic mucosa (in red). Gene probes are arranged on the *x*-axis by expression level, and the mean of $\log_2$ transformed expression value is shown on the *y*-axis. For many genes, there were multiple probes on the chip. Gene symbols for all of the gene probes could not be shown on the *x*-axis. Data from patients from the LOCRC group are shown on the left (**A**) showing down-regulation; and data from EOCRC group are shown on the right (**B**), where there was no differential expression found.

Table 3. Differential expression of cancer-related pathways in LOCRC and EOCRC.

Gene Set	Interaction *p*	Late-Onset (>40 Years)			Early-Onset (<40 Years)		
		Fold Change	(95% CI)	*p*	Fold Change	(95% CI)	*p*
Growth Factor Receptors	6.64×10^{-3}	−1.12	(−1.16 to −1.08)	1.21×10^{-8}	−1.02	(−1.08 to 1.03)	0.34
Anti-TSG	7.59×10^{-3}	1.46	(1.38 to 1.55)	5.23×10^{-35}	1.29	(1.20 to 1.39)	3.8×10^{-11}
DNA Repair	0.06	1.07	(1.05 to 1.08)	2.28×10^{-19}	1.04	(1.02 to 1.06)	5.4×10^{-6}
Pro-Apoptosis	0.10	1.03	(1.01 to 1.05)	7.46×10^{-3}	−1.00	(−1.03 to 1.03)	0.99
Tumor Suppressor Gene	0.10	−1.18	(−1.23 to −1.13)	3.41×10^{-12}	−1.11	(−1.17 to −1.04)	8.1×10^{-4}
Hexokinase	0.32	−1.20	(−1.26 to −1.14)	5.64×10^{-13}	−1.15	(−1.23 to −1.08)	1.6×10^{-5}
Warburg Effect	0.36	1.32	(1.21 to 1.43)	1.22×10^{-10}	1.24	(1.11 to 1.38)	1.2×10^{-4}
Anti-Apoptosis	0.58	−1.10	(−1.15 to −1.05)	1.27×10^{-4}	−1.07	(−1.14 to −1.01)	0.02
Caspases Initiator	0.69	1.03	(1.01 to 1.05)	1.27×10^{-3}	1.02	(−1.00 to 1.05)	0.05
p53 suppressor	0.76	1.00	(−1.01 to 1.02)	0.66	−1.00	(−1.02 to 1.02)	9.6×10^{-1}
Caspases Executor	0.77	−1.18	(−1.23 to −1.14)	4.29×10^{-18}	−1.19	(−1.25 to −1.14)	2.2×10^{-12}
Growth Factors	0.78	−1.18	(−1.26 to −1.10)	3.51×10^{-6}	−1.20	(−1.31 to −1.09)	9.1×10^{-5}

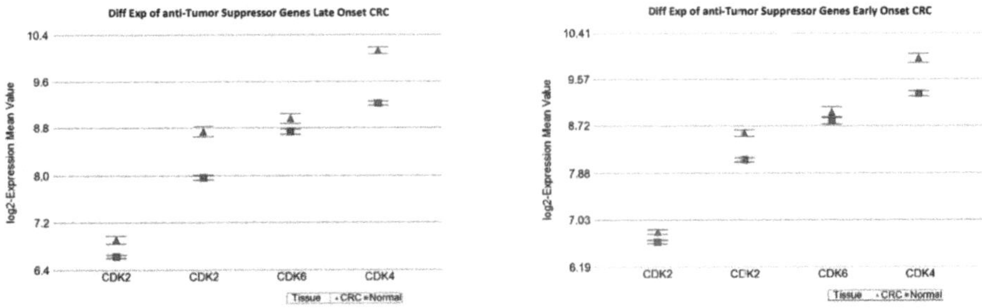

(A) (B)

Figure 2. Differential gene expression of anti-TSG genes in paired CRC tissue (in blue) and healthy colonic mucosa (in red). Gene probes are arranged on the *x*-axis by expression level, and the mean of log2 transformed expression value is shown on the *y*-axis. For many genes, there were multiple probes on the chip. Gene symbols for all the gene probes could not be shown on the *x*-axis. Data from patients from the LOCRC group are shown on the left (**A**), and data from patients of the EOCRC group are shown on the right (**B**). The average magnitude of over-expression was significantly higher ($p = 7.59 \times 10^{-3}$) if the patient had LOCRC [1.46-fold change (95% CI 1.38 to 1.55)] compared to those with EOCRC [1.29-fold change (95% CI 1.20 to 1.39)].

2.1.2. Age of Onset of CRC and Differential Gene Expression of DNA Damage Repair Gene Sets

The detailed gene list is shown in Supplementary Table S2. Except for genes for translesion synthesis (TLS), genes involved in all other DNA repair mechanisms were significantly over-expressed in CRC tissue, irrespective of the age of onset (see Table 4). However, in comparison to the magnitude of over-expression in LOCRC, the over-expression of genes related to "mismatch repair" (see Supplementary Figure S1), "Fanconi anemia" (see Figure 3), "non-homologous end joining (NHEJ)"(see Figure 4), and "microhomology mediated end joining (MMED)" was less pronounced in EOCRC patients (see Table 4). This may suggest a potentially low efficiency of the DNA damage repair mechanism in EOCRC.

Table 4. Differential expression of DNA damage repair genes in LOCRC and EOCRC.

Gene Set	Interaction p	Late-Onset (>40 Years)			Early-Onset (<40 Years)		
		Fold Change	(95% CI)	p	Fold Change	(95% CI)	p
Mismatch Repair (MMR)	1.16×10^{-4}	1.12	(1.10 to 1.14)	2.02×10^{-33}	1.06	(1.03 to 1.08)	7.01×10^{-6}
Fanconi Anemia (FA)	4.86×10^{-4}	1.15	(1.13 to 1.17)	1.50×10^{-46}	1.09	(1.06 to 1.12)	1.42×10^{-11}
Non-homologous end joining (NHEJ)	6.13×10^{-4}	1.12	(1.10 to 1.14)	5.60×10^{-34}	1.06	(1.04 to 1.09)	3.28×10^{-7}
Microhomology mediated end joining (MMEJ)	9.09×10^{-4}	1.26	(1.22 to 1.30)	1.29×10^{-49}	1.16	(1.12 to 1.21)	4.24×10^{-14}
Translesion Synthesis (TLS)	0.03	1.01	(−1.01 to 1.03)	0.30	−1.02	(−1.05 to 1.00)	0.05
Homologous Recombination (HR)	0.15	1.09	(1.07 to 1.10)	5.92×10^{-39}	1.07	(1.05 to 1.09)	1.82×10^{-16}
Nucleotide Excision Repair (NER)	0.19	1.08	(1.07 to 1.09)	1.76×10^{-41}	1.07	(1.05 to 1.09)	2.56×10^{-18}
Checkpoint Signaling	0.21	1.08	(1.06 to 1.09)	6.56×10^{-18}	1.06	(1.03 to 1.08)	4.55×10^{-7}
Base Excision Repair (BER)	0.40	1.06	(1.04 to 1.07)	4.12×10^{-15}	1.05	(1.03 to 1.07)	7.80×10^{-7}
Direct Reversal Repair (DRR)	0.49	1.20	(1.12 to 1.28)	2.96×10^{-7}	1.15	(1.05 to 1.26)	2.01×10^{-3}

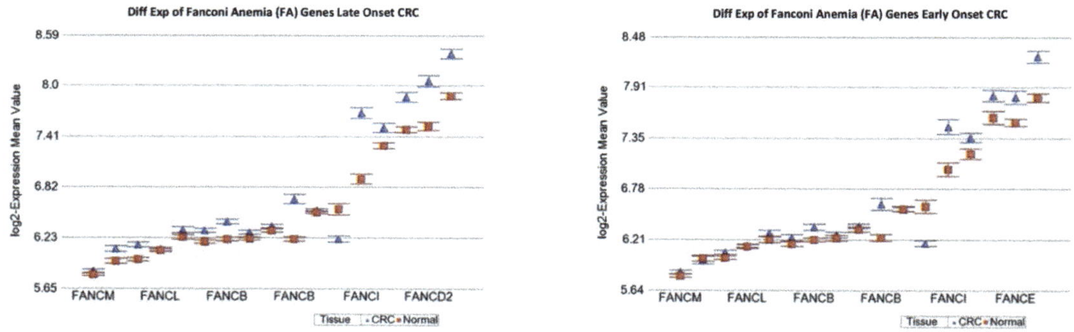

(A) (B)

Figure 3. Differential gene expression of Fanconi anemia genes in paired CRC tissue (in blue) and healthy colonic mucosa (in red). Gene probes are arranged on the x-axis by expression level, and the mean of $\log_2$ transformed expression value is shown on the y-axis. For many genes, there were multiple probes on the chip. Gene symbols for all of the genes could not be shown on the x-axis. Data from patients from the LOCRC group are shown on the left (**A**), and data from patients of the young onset group are shown on the right (**B**). The average magnitude of over-expression was significantly higher ($p = 4.86 \times 10^{-4}$) if the patient had old onset [1.15-fold change (95% CI 1.13 to 1.17)] compared to those with young onset [1.09-fold change (95% CI 1.06 to 1.12)].

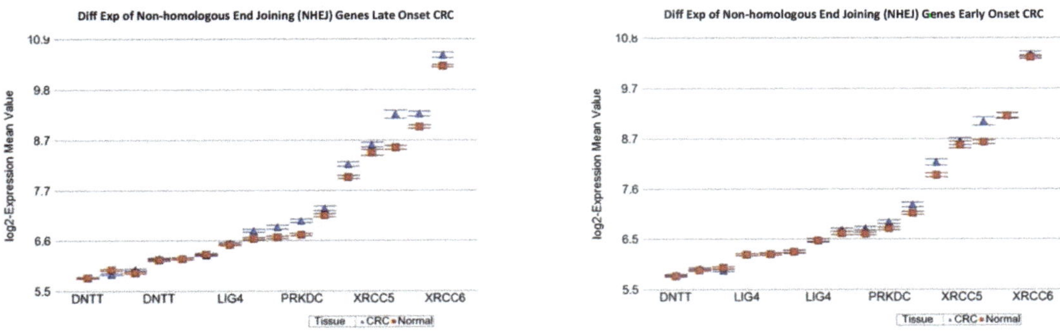

(A) (B)

Figure 4. Differential gene expression of non-homologous end joining genes in paired CRC tissue (in blue) and healthy colonic mucosa (in red). Gene probes are arranged on the x-axis by expression level, and the mean of log_2 transformed expression value is shown on the y-axis. For many genes, there were multiple probes on the chip. Gene symbols for all of the genes could not be shown on the x-axis. Data from patients from the LOCRC group are shown on the left (**A**), and data from patients of the young onset group are shown on the right (**B**). The average magnitude of over-expression was significantly higher ($p = 6.13 \times 10^{-4}$) if the patient had old onset [1.12-fold change (95% CI 1.10 to 1.14)] compared to those with EOCRC [1.06-fold change (95% CI 1.04 to 1.09)].

2.1.3. Age of Onset of CRC and Differential Expression of Replication Stress Gene-Sets

In the light of the above results for the association of the age of onset of CRC and DNA damage repair machinery, we explored if the genes involved in different biological processes in replication stress are different depending on the age of onset of CRC. The total list of genes at the replication stress site and their functional group [19,20] that we tested are shown in Supplementary Table S3. Analyses of our data (see Table 5) suggested that, in response to the cancer, on average, the genes related to "DNA replication repair" were over-expressed in CRC tissue, irrespective of the age of onset. However, it was significantly less pronounced in EOCRC patients compared to LOCRC patients [fold change 1.09 (95%CI 1.08–1.10) vs. 1.14 (95%CI 1.13–1.15), ANOVA interaction $p = 2.54 \times 10^{-8}$] (see Supplementary Figure S2). In the same line, the metabolism related genes were also over-expressed in both, but less pronounced in EOCRC (see Supplementary Figure S3).

Table 5. Differential expression of replication stress genes in LOCRC and EOCRC.

Gene Set	Interaction p	Late-Onset (>40 Years)			Early-Onset (<40 Years)		
		Fold Change	(95% CI)	p	Fold Change	(95% CI)	p
DNA Replication Repair	2.54×10^{-8}	1.14	(1.13 to 1.15)	9.40×10^{-184}	1.09	(1.08 to 1.10)	2.60×10^{-53}
Metabolism	5.96×10^{-6}	1.12	(1.10 to 1.14)	1.34×10^{-24}	1.03	(1.00 to 1.06)	0.03
Cell Movement	6.86×10^{-4}	−1.06	(−1.07 to −1.05)	3.59×10^{-26}	−1.03	(−1.04 to −1.01)	1.25×10^{-4}
RNA Processing	0.11	1.05	(1.03 to 1.06)	6.29×10^{-13}	1.06	(1.05 to 1.08)	5.88×10^{-14}
GF Signaling	0.13	−1.04	(−1.06 to −1.02)	8.99×10^{-5}	−1.01	(−1.04 to 1.01)	0.28
Development Regulation	0.16	1.05	(1.03 to 1.08)	4.04×10^{-7}	1.03	(1.00 to 1.06)	0.04

Table 5. *Cont.*

Gene Set	Interaction p	Late-Onset (>40 Years)			Early-Onset (<40 Years)		
		Fold Change	(95% CI)	p	Fold Change	(95% CI)	p
Cell Survival	0.27	1.02	(1.01 to 1.03)	3.17×10^{-4}	1.01	(−1.00 to 1.02)	0.18
Immune Regulation	0.63	−1.02	(−1.03 to −1.01)	6.66×10^{-3}	−1.02	(−1.04 to −1.01)	7.48×10^{-3}
Protein Translation	0.65	−1.04	(−1.06 to −1.03)	4.68×10^{-9}	−1.05	(−1.07 to −1.03)	4.50×10^{-7}
Stress Responses	0.73	1.04	(1.02 to 1.06)	4.78×10^{-5}	1.04	(1.01 to 1.06)	7.67×10^{-3}
Cell Cycle	0.8	1.04	(1.03 to 1.06)	1.08×10^{-8}	1.04	(1.02 to 1.06)	5.14×10^{-5}
Angiogenesis	0.91	−1.03	(−1.07 to 1.01)	0.18	−1.03	(−1.08 to 1.03)	0.37
Chromatin TF Transcription	0.93	1.01	(−1.00 to 1.02)	0.21	1.01	(−1.01 to 1.02)	0.40

2.1.4. Age of Onset of CRC and Differential Expression of Notch4-GATA4-IRG1 axis Gene-Sets

A recent study suggested the possible association of "NOTCH4-GATA4-IRG1 axis" genes and "leptin and other obesity related genes" with CRC [21,22]. The gene list is shown in Supplementary Table S4. We tested that in our data (see Figure 5). We found that the "Notch4-GATA-IRG" set of genes were over-expressed in tumor tissue by the same magnitude in LOCRC [fold change 1.09 (95% CI 1.07–1.12)] and in EOCRC [fold change 1.08 (95% CI 1.06–1.12)] (ANOVA interaction $p = 0.71$).

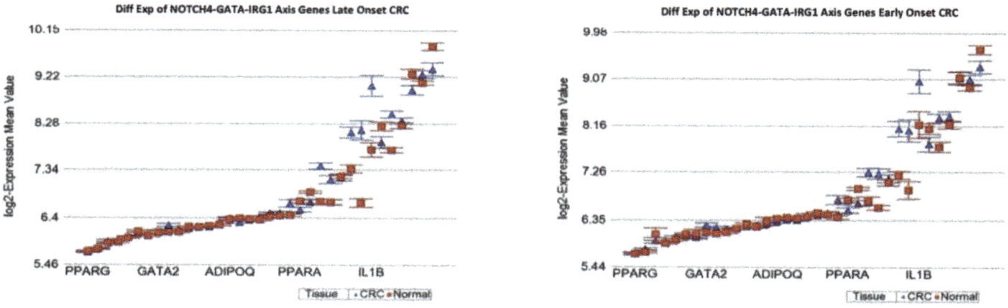

(**A**) (**B**)

Figure 5. Differential gene expression of NOTCH4_GATA4-IRG1 genes in paired CRC tissue (in blue) and healthy colonic mucosa (in red). Gene probes are arranged on the x-axis by expression level, and the mean of $\log_2$ transformed expression value is shown on the y-axis. For many genes, there were multiple probes on the chip. Gene symbols for all the genes could not be shown on the x-axis. Data from patients from the LOCRC group are shown on the left (**A**), and data from patients of the young onset group are shown on the right (**B**). The genes were similarly over-expressed in both of the groups.

We observed the down-regulation of *LEPR* and *GHRL* in CRC, but it was not different by the age of onset (see Supplementary Figure S4). However, we found that the fatty acid synthase (*FASN*) gene was significantly over-expressed in LOCRC [fold change 1.69 (95%CI 1.41–2.03)] compared to a non-significant increase in EOCRC patients [1.25-fold (95% CI: −1.009 to 1.59)] (see Supplementary Figure S5). The over-expression of *FASN* has been reported in CRC [23,24]. The data may suggest *FASN* as a therapeutic target in the LOCRC group, as has been suggested by others [25,26].

2.1.5. Age of Onset of CRC and Differential Expression of KEGG Pathways

In addition to examining the selective biologically relevant gene sets, as shown above, we also tested all of the KEGG pathways to see if the magnitude of the differential expression (CRC vs. normal colon tissue) of any pathway was different in LOCRC and EOCRC patients. The detailed result is shown in Supplementary Table S5. It was interesting to see that, among all of the KEGG pathways, the most significant interaction was seen in the "DNA replication pathway". Genes in the "DNA replication pathway" were over-expressed by 1.27-fold (95%CI 1.25–1.28) in LOCRC patients compared to 1.16-fold (95% CI 1.14–1.18) in EOCRC patients (ANOVA interaction $p = 2.73 \times 10^{-15}$).

2.1.6. Association of Differentially Expressed "Gene Sets" with Histology in CRC

Considering the significance of the tumor stage, LVI and PNI, we tried to see if the gene expression of some of the top most differentially expressed gene sets (CRC vs. normal tissue, EORC and LOCRC combined) were different in magnitude by these histological changes. Assuming that the cancer cells respond to the DNA damage by increasing the DNA damage repair and DNA replication machinery, one can expect that these gene sets will be up-regulated in CRC tissue compared to normal colon tissue. Our data suggests that these cellular responses for repairing the DNA damage are lower in magnitude in the presence of LVI, PNI, and advanced stage. For example, in the absence of LVI, the base excision repair (BER) genes are on average 1.13-fold (95% CI 1.11–1.14) over-expressed, but in the presence of LVI, it is 1.07-fold (95% CI 1.05–1.08). The details are shown in Table 6, which shows that perhaps LVI had the strongest influence on the magnitude of differential expression for these DNA repair machineries. A photomicrograph example of LVI and PNI is shown in Supplementary Figure S6A,B, respectively.

Table 6. Association of differential expression of DNA damage repair gene sets and histological finding of LVI and PNI in CRC tissue.

Stratification	Base Excision Repair Genes		Mismatch Repair Genes		Non-Homologous End-Joining	
	Fold Change	**(95% CI)**	**Fold Change**	**(95% CI)**	**Fold Change**	**(95% CI)**
LV Invasion						
Absent	1.13	(1.11 to 1.14)	1.18	(1.16 to 1.19)	1.11	(1.09 to 1.13)
Present	1.07	(1.05 to 1.08)	1.12	(1.11 to 1.14)	1.07	(1.05 to 1.09)
p	1.39×10^{-8}		1.02×10^{-5}		1.60×10^{-3}	
PN Invasion						
Absent	1.11	(1.09 to 1.12)	1.16	(1.15 to 1.17)	1.09	(1.08 to 1.11)
Present	1.07	(1.05 to 1.09)	1.12	(1.09 to 1.15)	1.07	(1.04 to 1.10)
p	2.93×10^{-3}		6.58×10^{-3}		0.23	
Stage						
Stage-1	1.13	(1.1 to 1.15)	1.18	(1.16 to 1.21)	1.11	(1.08 to 1.15)
Stage-2	1.1	(1.08 to 1.12)	1.15	(1.13 to 1.17)	1.09	(1.06 to 1.11)
Stage-3	1.09	(1.07 to 1.10)	1.14	(1.13 to 1.16)	1.08	(1.06 to 1.10)
p	0.01		0.04		0.18	

2.1.7. Gene Expression Profiling from Therapeutic Point of View

T-cell inflamed gene expression profile (GEP): This group of genes has been used to predict immunotherapy targeting programmed cell death protein-1 (PD-1, also known as CD274)) [27,28]. We looked at the gene expression of these genes in our patients to see if they were different by age of onset (see Figure 6A,B). In both EOCRC and LOCRC groups, these genes were down-regulated to a similar extent. However, stratification by MSI status showed that the down-regulation was more pronounced (ANOVA interaction $p = 1.7 \times 10^{-7}$) in the MSS group [fold change −1.32 (95% CI from −1.28 to −1.36)] than in the MSI group [fold change −1.19 (95% CI from −1.13 to −1.25)].

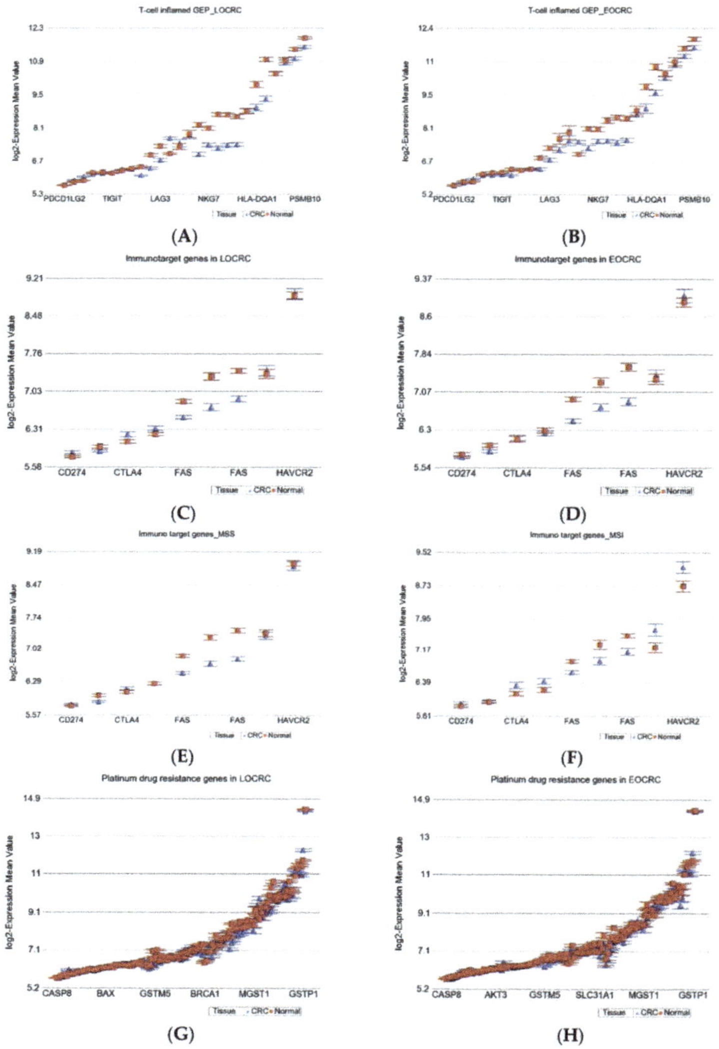

Figure 6. Differential gene expression of few groups of genes related to immune target therapy in paired CRC tissue (in blue) and healthy colonic mucosa (in red). Gene probes are arranged on the *x*-axis by expression level, and the mean of $\log_2$ transformed expression value is shown on the *y*-axis. For many genes, there were multiple probes on the chip. T-cell inflamed GEP was equally down-regulated in both LOCRC and EOCRC and are shown in (**A**) and (**B**), respectively. Differential expression of few target genes for available check-point inhibitors in LOCRC and EOCRC are shown in (**C**) and (**D**), respectively. Note that the magnitude of difference was similar by age of onset. However, when we divided the patients by MSI status, the difference was seen. Differential expression of the same few target genes in MSS and MSI patients are shown in (**E**) and (**F**), respectively. Over-expression of CTLA4 and HAVCR2 was seen only in the MSI patients, but not in the MSS group. Differential expression of "Platinum drug resistance" genes in LOCRC and EOCRC are shown in (**G**) and (**H**), respectively. These were equally down-regulated in LOCRC and EOCRC and are shown in (**G**) and (**H**), respectively.

Other ICI target genes: We also looked at targets of some other immune check-point inhibitors—*LAG3, PDCD1, CD274, CTLA4,* and *HAVCR2.* Again, the age of onset did not influence the differential expression (interaction $p = 0.17$) (see Figure 6C,D), but the MSI status influenced the overall differential expression, as we reported in an earlier publication [11]. *LAG3, PDCD1,* and *CD274 (PDL-1)* were down-regulated in CRC (see Figure 6E,F), but *CTLA4* and *HAVCR2* were up-regulated only in the MSI–CRC group, suggesting a potential benefit from *CTLA4* and *HAVCR2* inhibitors in that subset only [11] (see Figure 6F).

Platinum drug resistance genes: The differential expression was similar in magnitude in EOCRC and LOCRC. The MSI status also did not have any influence either (see Figure 6G,H). The overall slight down-regulation of these genes in CRC suggested a low potential chance of development of cis-platin resistance.

In summary, on the basis of the gene expression profile of genes relevant to therapy (a), it may not be surprising if *PD1* blockers do not show promising results in this CRC population, but some other ICI, such as *CTLA4* or *HAVCR2* blockers, may have promising results in the MSI subgroup; (b) the possibility of platinum drug resistance may be low, irrespective of age of onset in this population.

2.2. DNA Methylation

Age of Onset of CRC and Differential DNA Methylation

First, we carried out the differential methylation analysis (paired CRC tissue vs. corresponding healthy colon tissue) at the individual probe level data. Using paired analysis, we found a large number of differentially methylated loci (DML), both in LOCRC (n = 143,646, at FDR 0.05) and in EOCRC (n = 111,066, at FDR 0.05) (see Figure 7A). A large number of these DML (n = 100,370, which is 69.8% of those found in LOCRC and 90.36% of those found in EOCRC) were common between the two groups (see Figure 7A). These lists of DML do not take the magnitude of the differential methylation (delta beta = beta value of CRC tissue—beta value of normal) into account. Considering our aim to detect the DML, the magnitudes of delta beta of which are significantly different among LOCRC and EOCRC, we used an ANOVA model that included an interaction term "tissue x age of onset of CRC". Our analysis suggested that there were 5607 loci with interaction $p =< 0.05$. These three lists were used in the Venn diagram (Figure 7B). The intersections of the Venn diagram allowed us to identify the following groups of DML:

I. DML common in both EOCRC and LOCRC and the magnitude of delta beta was not different (n = 97,939) as the interaction p was >=0.05. This represents the largest group of DML in CRC. The scatterplot in Figure 7C shows the magnitude of delta beta of these DML in LOCRC and EOCRC and the cut-off lines on both axes show a large number of loci with delta beta > 0.1 or <−0.1, indicating hyper- or hypo-methylation exceeding 10%.

II. DML common in both EOCRC and LOCRC, but the magnitude of delta beta was significantly different between EOCRC and LOCRC patients (n = 2431).

III. DML found only in EOCRC, but the magnitude of the delta beta was not different between the EOCRC and LOCRC (n = 9925). The scatterplot in Figure 7D shows that the magnitudes of delta beta of these DML were low and not different among LOCRC and EOCRC.

IV. DML specific to EOCRC (n = 771): these DML are differentially methylated only in EOCRC, and the magnitude of the delta beta is significantly more pronounced from that seen in LOCRC. The details of these DML are presented in Supplementary Table S6. The scatterplot in Figure 7E shows the magnitudes of delta beta of these DML; only a few exceeded 0.1 or 10% differential methylation. Regardless of the low magnitudes of delta beta of these 771 EOCRC specific DML, because of specificity, these markers were able to separate CRC from normal among the EOCRC patients (see the PCA plot in Figure 7F). The only hypermethylated DML with delta beta >= 0.1 was the *SOX8* gene. Among the three hypomethylated loci with delta beta =< 0.1, one was the *TACC1* gene (see Figure 7E), which is known to be associated with other cancers. The methylation status of the *TACC1* gene in LOCRC and EOCRC is shown in Figure 7G,H, respectively, showing that *TACC1* was hypomethylated only in EOCRC, but not in LOCRC.

V. DML specific to old age onset CRC (1946): these DML are differentially methylated only in LOCRC, and the magnitude of the delta beta is significantly more pronounced from that seen in EOCRC. The details of these DML are presented in Supplementary Table S7. For many of these DML, the magnitudes of delta beta exceeded 0.1 or 10% differential methylation.

In the next step, we tried to identify if the magnitude of the differential methylation of some gene set(s) was different among the patients with LOCRC compared to those with EOCRC patients. We included an interaction term "tissue x age of onset" in the gene set ANOVA model(s); the *p*-value of that interaction term identified pathways/gene sets that were differentially methylated depending on the age of onset.

Among the cancer related gene sets, the genes involved in "Growth factor receptors" and "DNA repair" showed different magnitudes of differential methylation among EOCRC and LOCRC (see Supplementary Table S8). This methylation data for "Growth factor receptor" genes also correlate to our gene expression data in a sense that, for these genes, on average, we observed the hypermethylation of DNA in CRC tissue in general (more in LOCRC than EOCRC) and the down-regulation in mRNA in CRC tissue (more pronounced in LORC than in EOCRC). Among the DNA damage related gene sets, "Direct reversal Repair" genes were differentially methylated depending on the age of onset (see Supplementary Table S9). Among the replication stress site gene sets, "Immune regulation" genes were differentially methylated depending on the age of onset (see Supplementary Table S10). Finally, the top KEGG pathways which showed different magnitudes of differential methylation among EOCRC and LOCRC were "Antigen processing and presentation", "Type I diabetes mellitus", "JAK-STAT signaling pathway" etc. (see Supplementary Table S11).

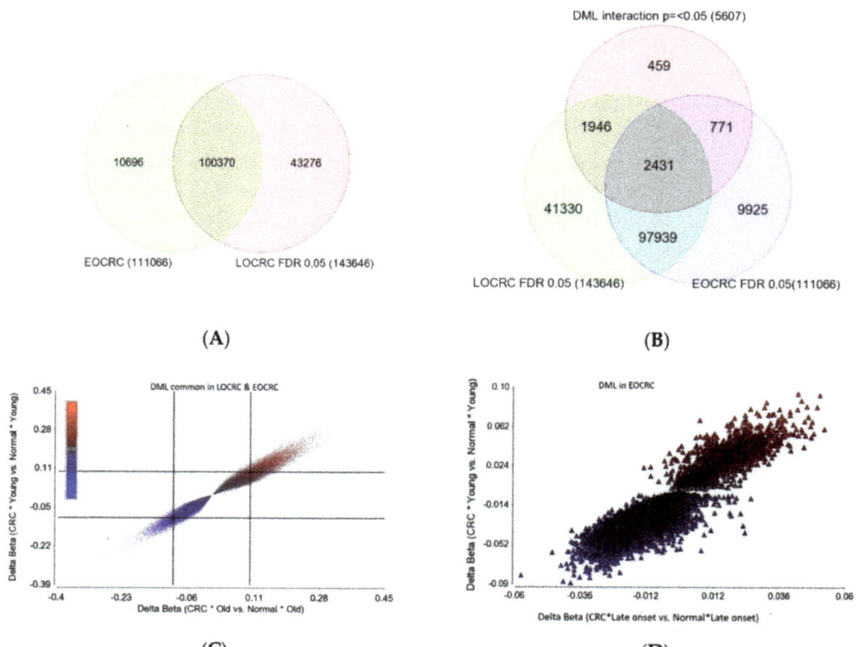

Figure 7. *Cont.*

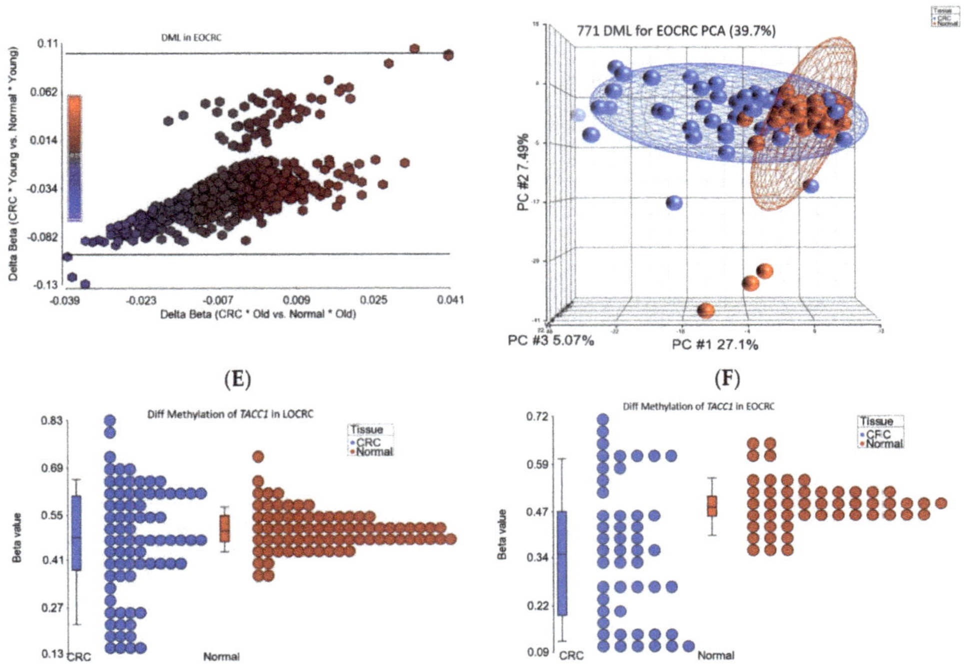

(E) (F)

(G) (H)

Figure 7. Differential methylation analysis of CRC. The overlap of DML (based on FDR <0.05 only without taking the delta beta into account) in LOCRC and EOCRC is shown in (**A**). In (**B**), we also overlapped the methylation markers that showed different magnitude of delta beta in EOCRC and LOCRC. We identified these markers through the "tissue x age of onset" interaction p =< 0.05 in the ANOVA models. (**C**) shows the 97,939 common markers in EOCRC and LOCRC that also had similar magnitude of delta beta in both age groups (because the interaction p was >0.05). The scatterplot shows the delta beta of these markers in LOCRC on $x-$axis and the delta beta of these markers in EOCRC on the $y-$axis. (**D**) shows the 9925 DML that were found only in EOCRC, but the delta beta was equally small; I shows the 771 EOCRC specific DML. (**E**) The scatterplot shows the delta beta of these markers in LOCRC on $x-$axis and the delta beta of these markers in EOCRC on the $y-$axis. Although they are greater than LOCRC, the majority of these DML had small delta beta, even in EOCRC. PCA plot using these 771 EOCRC-specific DML is shown in (**F**). Among the three hypomethylated loci with delta Beta =< 0.1, one was *TACC1* gene. The methylation status of *TACC1* gene in LOCRC and EOCRC are shown in (**G**) and (**H**), respectively, showing *TACC1* was hypomethylated only in EOCRC.

In summary, (a) the vast majority of the DML in CRC were common in EOCRC and LOCRC; (b) the vast majority of the robust DML [the one with the magnitude of delta beta => 0.2 (20%)] were common between EOCRC and LOCRC; (c) although the EOCRC specific DML were relatively small in number and also small in the magnitude of delta beta. These EOCRC specific markers could potentially differentiate cancer from healthy tissue.

3. Discussion

The incidence of CRC among young adults has increased globally [1,29–31]. Siegel RL et al. published a comprehensive overview of current CRC statistics in the United States, including the estimated numbers of new cases and deaths in 2020 by age and incidence, survival, and mortality rates and trends by age and race/ethnicity based on incidence data through 2016 and mortality data through 2017 [31].

In this study from a Bangladeshi population, we attempted to find any histological and transcriptome wide and methylome wide differences in CRC tissue compared to corresponding normal tissue among EOCRC and LOCRC patients. We acknowledge that such a study cannot shed light on the etiology of the increase in incidence in EOCRC. Additionally, we did not have any post-surgical follow-up data to comment on the difference in prognosis of EOCRC, if any. From the histology perspective, we documented that EOCRC is associated with an advanced stage. This association has been seen in other populations as well. However, the greater proportion of patients with advanced stage could not be simply explained by a delay in diagnosis. The other two histological features associated with EOCRC were LVI and PNI. The significance of KRAS and BRAF mutations in CRC is well known [32]. However, we did not see any difference in the frequency of these mutations between EOCRC and LOCRC in our series.

The histological diagnosis of LVI on examination of the hematoxylin & eosin (H&E)-stained slide includes the presence of tumor cells within a vascular space; erythrocytes surrounding the tumor cells; the identification of endothelial cells lining the space; the presence of an elastic lamina surrounding the tumor; and the attachment of tumor cells to the vascular wall [33]. PNI is a pathologic process characterized by a tumor invasion of nervous structures and spread along nerve sheaths. The pathogenesis of PNI likely involves complex signaling between tumor cells, stromal cells, and the nerves. PNI is known to be a marker for a more aggressive tumor phenotype and poor prognosis in several malignancies, most notably head and neck and prostate cancers. PNI was defined as tumor cells within any layer of the nerve sheath or tumor in the perineural space that involved at least one third of the nerve circumference [34]. To our knowledge, our study is the first to show the association of these histological markers of advanced disease (stage, LVI, and PNI) with an underlying transcriptomic profile of impairment in DNA damage repair machinery in CRC tissue.

In one retrospective study among LOCRC, 207 right sided and 207 left sided tumors are compared after curative resection, the authors found that the left-sided tumor exhibited better survival outcomes than the right-sided ones after curative resection [4]. Survival data for EOCRC is conflicting. Survival rates among the young and older groups also show variation depending on the CRC stages. The right sided tumor exhibited a more advanced stage, increased tumor size, more frequently poorly differentiated tumors, more harvested lymph nodes, and more positivity of LVI than left sided ones. They also found better 5-year survival outcomes for the group with left sided CRC [4]. Right sided CRC presents more with iron deficiency anemia [35]. Powell et al. compared 134 (33%) right sided, 125 (30%) left sided, and 152 (37%) rectal tumors. Emergency presentation ($p < 0.001$), anemia ($p < 0.001$), advanced stage ($p < 0.001$), poor differentiation ($p < 0.001$), and older age ($p < 0.05$) were more commonly observed in right sided cancers [36].

There are some differences in the clinicopathological characteristics among EOCRC and LOCRC. The differences between the 94 EOCRC (avg. 27 years) and 275 LOCRC (avg. 67 years) were studied [37]. There were differences in the stage at diagnosis—stage III and IV (76% vs. 46%), signet ring-cell (13% vs. 1%), poorly differentiated (37% vs. 8%), MSI status (27% vs. 13%), and the 5-year disease specific survival (48% vs. 78%) in EOCRC and LOCRC, respectively [37], and no difference was found in KRAS and BRAF mutations. Low BRAF mutations were reported among Bangladeshi immigrants to the UK [38]. Another study also found that, in sporadic CRC, independent of the age, the LVI and PNI were associated with a poor prognosis and recurrence [39]. In our study, we found advanced stage, LVI, and PNI to be significantly associated with EOCRC compared to LOCRC.

Very few studies have compared the gene expression difference between the EOCRC and LOCRC. Tunca B et al. looked at the expression profiles of 114 different genes which were evaluated using mRNA PCR arrays in 39 tumors and 20 surgical margin tissue samples from 39 sporadic CRC patients diagnosed at less than 50 years of age [40]. The expression levels of *IMPDH2, CK20, MAP3K8*, and *EIF5A* were strongly up-regulated in CRC tissues compared with normal colorectal tissues, but not compared to the LOCRC. A similar study was carried out on sporadic EOCRC and healthy controls (<50 years old). Seven genes, *CYR61, UCHL1, FOS, FOS B, EGR1, VIP*, and *KRT24*, were consistently up-regulated in the mucosa of all six patients compared with the mucosa from four healthy controls [41].

Gene expression variation between FFPE tissues from six EOCC patients (<50 years) and six LOCRC patients (>65 years) were examined. Among the 770 genes assayed, changes in expression levels of 88 genes were unique (28 up- and 60 down-regulated) to EOCC (using the cutoff criteria of expression level differences > 2-fold and p value < 0.01) [42]. At the pathway level, *RAS, MAPK, WNT*, and DNA repair pathways were similarly deregulated in both age groups, whereas *PI3K*-AKT signaling was more specific to EOCC and cell cycle pathways to LOCC [42]. Berg M et al. performed an integrated analysis of copy number changes and gene expression in 23 sporadic EOCRC patients with a median age of 44 years and 17 LOCRC patients with median age of 79 years [43]. Tissues were preserved in buffer RLT. In the younger group, *CLC, LTBP4*, and *ZNF574* were up-regulated and *PPAT, RG9MTD2, EIF4E*, and *PLA2G12A* were down-regulated compared to the older group. Agesen TH et al. compared gene expression in EOCRC and LOCRC and found that a good number of genes are over- or under-expressed in EOCRC [44]. Among those, Charcot–Leyden crystal protein (CLC) was 10 times over-expressed and interferon (alpha, beta, and omega) receptor-1 (IFNAR1) was under-expressed.

Gene expression data in EOCRC were analyzed by Mo X et al. [45]. A total of 140 module hub genes were identified and found to be enriched in the 'mitochondrial large ribosomal subunit', 'structural constituent of ribosome', 'poly (A) RNA binding', 'collagen binding', 'protein ubiquitination', and 'ribosome pathway'. Twenty-six module hub genes were found to have a degree score > 5 in the PPI network, seven of which [secreted protein acidic and cysteine rich (*SPARC*), decorin (*DCN*), fibrillin 1 (*FBN1*), WW domain containing transcription regulator 1 (*WWTR1*), transgelin (*TAGLN*), DEAD-box helicase 28 (*DDX28*), and cold shock domain containing C2 (*CSDC2*)], had good prognostic values for patients with early-onset CRC, but not late-onset CRC. Therefore, the authors suggested that *SPARC, DCN, FBN1, WWTR1, TAGLN, DDX28*, and *CSDC2* may contribute to the development of early-onset CRC and may serve as potential diagnostic biomarkers.

Structurally, leptin has similarity with other proteins of the cytokine family, belonging to the group of cytokines commonly called adipocytokines or adipokines. Initially described as an anti-obesity hormone, leptin has subsequently been shown to also influence hematopoiesis, thermogenesis, reproduction, angiogenesis, and immune homeostasis [46]. Some variants of the Leptin gene have been found to be associated with CRC in females [9]. There is an inverse correlation between adiponectin and leptin in obesity [47]. In our study, we observed the down-regulation of LEPR GHRL in both EOCRC and LOCRC without a significant difference; however *FASN* was significantly over-expressed in LOCRC compared to a non-significant increase in EOCRC patients.

In a recent paper, Joo JE et al. compared the DNA methylation of tumor and healthy mucosa in 110 EOCRC (<50 years), 334 intermediate onset CRC (IOCRC) (50–70 years), and 325 LOCRC (>70 years) [48]. They used FFPE samples on Methylation 450K chips using the DNA restoration kit. They found extensive DNA methylation alterations in all CRCs, including EOCRCs. They identified DNA methylation-related changes specific to EOCRC, including *TFAP2A* and *GSX1* genes, and 12 differentially methylated genes associated with the MODY pathway. In our study, we found a relatively smaller number of EOCRC specific DML, but those could cluster the CRC samples well from normal tissue. Some of our methylation findings also correlated with gene expression findings in EOCRC.

We did not have any CRC patients with distal metastasis in this series and we did not have chemotherapy or immunotherapy data. However, we tried to look for the molecular basis in the CRC tissue for the potential use of ICI in these EOCRC and LOCRC patients. The age of onset did not appear to be a factor that can modify the choice of ICI. Rather, the MSI status can be used to select a potential subset of CRC patients for certain ICI. With all of the limitations of our study in mind, some of the strengths of the study may be noted. First, paired tumor-normal samples from the same individual for comparison is the most robust method for detecting any gene expression and methylation changes in cancer. Second, we used tissue samples preserved in RNA later for RNA and fresh frozen samples for DNA, which are the gold standard for such assays. Third, to our knowledge, this is one of the first studies from native Bangladeshi patients with CRC to comprehensively see a difference between EOCRC and LOCRC pathogenesis from the molecular perspective.

4. Materials and Methods

For this study, we included 330 paired (tumor and adjacent normal) samples from 165 CRC patients (m = 96, f = 69). Of them, 33 had right-sided CRC (cecum 8, ascending colon 17, hepatic flexure 4, and transverse colon 4) and 132 had left-sided CRC (descending colon 8, sigmoid colon 15, recto-sigmoid junction 7, and rectum 102). Sixty-seven of the patients (39 male and 28 female) were aged =<40 years. The fresh frozen samples were collected from 165 CRC patients from the department of Pathology, Bangabandhu Sheikh Mujib Medical University (BSMMU), Dhaka, Bangladesh at different times spanning between December 2009 and May 2016. The patients were at different stages of CRC (stage-1: 37, stage 2: 42 stage 3: 86). From each patient, the specimens were collected from the surgically resected tumor and the surrounding unaffected part of the colon about 5–10 cm away from the tumor mass. A surgical pathology fellow collected all samples from the operating room immediately after the surgical resection. A histopathology examination was performed on H&E stained slides in routinely processed paraffin impregnated tissue blocks. The slides were examined independently by two pathologists and there was concordance in all 165 cases. For the staging and grading of the CRC, the World Health Organization Classification of tumors was followed [49]. From each individual, we obtained a pair of tumor and normal tissues, which were frozen immediately and shipped on dry ice to the molecular genomics lab at The University of Chicago for subsequent DNA, RNA extraction, and molecular assay.

For each patient, we also abstracted key demographic and clinical data and tumor characteristics from hospital medical records. Written informed consent was obtained from all participants. The research protocol was approved by the "Ethical Review Committee, Bangabandhu Sheikh Mujib Medical University", Dhaka, Bangladesh (BSMMU/2010/10096), and by the "Biological Sciences Division, University of Chicago Hospital Institutional Review Board", Chicago, IL, USA (10-264-E).

4.1. DNA and RNA Extraction and Quality Control

DNA was extracted from fresh frozen tissue using a Puregene Core kit (Qiagen, MD, USA). The electropherogram from Agilent BioAnalyzer with Agilent DNA 12,000 chips showed the fragment size to be >10,000 bp. RNA was extracted from RNA Later preserved colonic tissue using a Ribopure tissue kit (Ambion, Austin, TX, USA, Cat# AM1924).

4.2. Relative Telomere Length (RTL) Measurement

For RTL measurement, we used a Luminex-based assay using QuantiGene Plex chemistry (Invitrogen, Santa Clara, CA, USA). The details of the assay are described earlier [50,51]. Briefly, the assay requires ~50 ng of DNA, which is hybridized to sequence-specific probes for the telomere repeat sequence (TEL) and reference gene sequence (*ALK*). The TEL and *ALK* gene signals are amplified using branched DNA technology and detected using Luminex technology. We used custom designed probes to measure the abundance of the telomere repeat sequence. The 24-mer probe targeted four repeats–"TTAGGGTTAGGGTTAGGGTTAGGG". As a reference single gene, we used *ALK* which showed very stable copy numbers (CN = 2) in all the DNA samples detected by oligonucleotide-based microarray SNP chips from our previous study. The result is a ratio, and hence there is no unit. The assay precision was good to excellent, with an intra-class correlation coefficient (ICC) of 0.91 (95% CI 0.86–0.94) [50]. The RTL assay failed in 17 samples (nine CRC and eight normal tissue) out of the total 330 samples tested. This failure rate (5.1%) was similar to what we have seen previously in a larger scale study using the same Luminex-based RTL measurement assay [52]. RTL data of this series of CRC patients was recently published [53].

4.3. Genome-Wide Gene Expression Assay

We used microarray data (Illumina HT12 v4 BeadChip) from the first 71 paired tumor and normal tissue RNA (of the same set of 165 patients used for RTL assay in this study). The chip contains a total of 47,231 probes covering 31,335 genes. Therefore, for many genes, there were multiple probes on the chip targeting different genomic regions of the same gene, that also enables the covering of multiple isoforms, if any. We used the probe level data. Paired samples were processed in the same chip (12 samples/chip). One sample from the normal tissue failed on the microarray. Therefore, we had gene expression data from 71 CRC tissues and 70 corresponding normal tissues. Gene expression data was normalized using quantile normalization in the GenomeStudio software.

4.4. Genome-Wide Methylation Assay

We also have methylation data (Illumina HumanMethylation450 DNA analysis BeadChip v1.0 Assay) from the first 125 paired tumor-normal samples (125 pairs out of the same set of 165 patients used for this study) [11]. The DNA samples were subjected to bisulfite conversion using EZ-96 DNA Methylation Kit (Zymo Research, Irvine, CA, USA). The chip presents 485,577 loci of which there are 150,254 in CpG Island, 112,067 in Shore (0–2 kb from island), 47,114 in Shelf (2–4 kb from the island), and 176,112 in deep sea (>4 kb from CpG island). We did not include the markers in the deep sea region in the final differential methylation analysis. Paired samples (CRC and corresponding normal) were processed on the same chip to avoid a batch effect. From this assay, on average 17 loci per gene were interrogated. A Tecan Evo robot was used for automated sample processing and the chips were scanned on a single iScan reader. If the intensity of the methylated loci is X and the intensity of the unmethylated loci is Y, then, the methylation score (beta value) is $X/X + Y$. If all are unmethylated ($X = 0$), then the methylation level is $0/0 + Y = 0$. If all loci are methylated ($Y = 0$), then the beta value is $X/X + 0 = 1$. If 50% of probes are hybridized at the methylated loci and 50% are hybridized at the unmethylated loci, then the methylation score is $50/50 + 50 = 0.5$.

4.5. Microsatellite Instability (MSI) Detection

A high-resolution melting (HRM) analysis method was used for the detection of two mononucleotide MSI markers—BAT25 and BAT26 [54,55]. A tumor was defined as having MSI when it showed instability with at least one of these markers (BAT25 and BAT26), and as MSS when it showed no instability for both of the markers. We used published primer sequences [54]. The amplification conditions included the polymerase activation step at 95 °C for 2 min, followed by five cycles of denaturation at 95 °C for 15 s, annealing starting at 60 °C for 30 s, extension at 72 °C for 30 s, and an additional 33 cycles of denaturation at 95 °C for 15 s, annealing at 53 °C for 30 s, and extension at 72 °C for 30 s. Before the HRM step, the products were heated to 95 °C for 1 min and cooled to 40 °C for 1 min, to allow for the heteroduplex formation. HRM was carried out and the data was collected over the range from 60 to 95 °C, with temperature increments of 0.2 °C/s at each 0.05 s. The BAT25 and BAT26 products were sequenced for validation. In this way, a total of 30 tumor samples showed MSI and all were confirmed by another relatively novel MSI marker CAT25 [56,57].

4.6. KRAS and BRAF Mutation Detection:

Tumor and adjacent healthy colonic tissue from 165 paired (tumor and normal) tissues were tested for KRAS (rs 112445441) and BRAFV600E mutations by high resolution melt analysis, as described previously [55].

4.7. Statistical Analysis

To compare the continuous variables, we used a *t*-test or one-way analysis of variance (ANOVA). The principal component analysis (PCA) and sample histograms were checked as a part of quality control analyses of the microarray data. Mixed-model multi-way ANOVA (which allows more than one ANOVA factor to be entered in each model) was used to compare the individual probe level expression data (for gene expression) or the beta value of CpG loci (for methylation data) across different groups. For statistical analysis, we used Partek Genomics Suite (version 7.0) (https://www.partek.com/partek-genomics-suite/, accessed on 14 November 2022). In general, "tissue" (tumor/adjacent normal), age of onset of CRC (0: =<40 years, 1: >40 years), LVI (0 = no, 1 = yes), PNI (0 = no, 1 = yes), telomere shortening (0 = no, 1 = yes), and MSI status (MSI/MSS) were used as categorical variables with fixed effect. These levels represent all conditions of interest, whereas "person ID#" (as proxy of inter-person variation) was treated as a categorical variable with random effect, since the person ID is only a random sample of all of the levels of that factor. The method of moments estimation was used to obtain estimates of variance components for mixed models [58]. As per the study design, we processed both the CRC tissue and the corresponding adjacent normal sample from one individual in a single chip. In the ANOVA model, the $\log_2$-transformed gene expression or beta-value for the CpG loci were used as the response variable (Y), and "Tumor" (tumor or normal), person ID#, "MSI-status", and "age of onset" were entered as ANOVA factors.

For the paired analysis, we used the following model:

$$Y_{ijk} = \mu + Tumor_i + Person_j + \varepsilon_{ijk}$$

where *Yijk* represents the *k-th* observation on the *i-th* tumor *j-th* person. *M* is the common effect for the whole experiment. *Eijk* represents the random error present in the *k-th* observation on the *i-th* tumor *j-th* person. The errors *εijk* are assumed to be normally and independently distributed with mean 0 and standard deviation *δ* for all measurements. Person is a random effect.

For the detection of the interaction between tumor and age of onset, the following model was used:

$$Y_{ijk} = \mu + Tumor_i + Age\ of\ onset\ of\ CRC_J + Tumor * Age\ of\ onset\ of\ CRC_{ij} + \varepsilon_{ijk}$$

where *Yijk* represents the *k-th* observation on the *i-th* tumor *j-th* age of onset. *M* is the common effect for the whole experiment. *Eijk* represents the random error present in the *k-th* observation on the *i-th* tumor *j-th* age of onset of CRC. The errors *εijk* are assumed to be normally and independently distributed with mean 0 and standard deviation *δ* for all measurements.

Gene ontology (GO) was used to group a set of genes into a category. In GO Enrichment analysis, we tested if the genes found to be differentially expressed or methylated fell into a gene ontology category more often than expected by chance [59]. We used a chi-square test to compare the "number of significant genes from a given category/total number of significant genes" vs. "number of genes on chip in that category/total number of genes on the microarray chip". The negative log of the p-value for this test was used as the enrichment score. Therefore, a GO group with a high enrichment score represents a lead functional group. The enrichment scores were analyzed in a hierarchical visualization and in tabular form.

Gene set ANOVA is a mixed model ANOVA to test the expression or methylation of a set of genes (sharing the same category or functional group) instead of an individual gene in different groups (https://www.partek.com/partek-genomics-suite/, accessed on 14 November 2022). The analysis is performed at the gene level, but the result is expressed at the level of the gene set-category by averaging the member genes' results. The equation for the model was:

$$\text{Model: } Y = \mu + T + P + G + S\,(T * P) + \varepsilon$$

where Y represents the expression or methylation status of a gene set category, μ is the common effect or average expression/methylation of the gene set category, T is the tissue-to-tissue (tumor/normal) effect, P is the patient-to-patient effect, G is the gene-to-gene effect (differential expression or methylation of genes within the gene set category independent of tissue types), S (T*P) is the sample-to-sample effect (this is a random effect, and nested in the tissue and patient), and ε represents the random error. All of the figures were generated using Partek Genomics Suite (version 7.0) (https://www.partek.com/partek-genomics-suite/, accessed on 14 November 2022).

5. Conclusions

We found that a high proportion of CRC patients presented at =<40 years of age. Histologically, EOCRC was more frequently associated with advanced stage, LVI, and PNI. As for molecular mechanisms for pathogenesis, the genome-wide gene expression and methylation suggested that the EOCRC was associated with an impaired DNA damage repair response, DNA replication repair, and immune response. From a therapeutic perspective, the study suggested the potential use of *FASN* targeted therapy, especially in the LOCRC group in this population.

Supplementary Materials: The following supporting information can be downloaded at: https://www.mdpi.com/article/10.3390/ijms232214261/s1, Figure S1: Differential expression of MMR genes in LOCRC and EOCRC; Figure S2: Differential expression of DNA replication repair genes in LOCRC and EOCRC; Figure S3: Differential expression of metabolism genes in LOCRC and EOCRC; Figure S4: Differential expression of obesity related genes in LOCRC and EOCRC; Figure S5: Differential expression of FASN gene in LOCRC and EOCRC; Figure S6: Photo micrograph HE stain at 200× magnification showing LVI (Shown in S6A) and PNI (shown in S6B); Table S1: Cancer related genes; Table S2: DNA damage genes; Table S3: Replicative stress sensitive site genes; Table S4: NOTCH 4 axis leptin related genes; Table S5: KEGG pathway in EOCRC and LOCRC; Table S6: DML early onset CRC; Table S7: DML late onset CRC; Table S8: Methylation cancer genes interaction; Table S9: Methylation DNA damage genes; Table S10: Methylation DNA RSS genes; Table S11: Methylation KEGG genes.

Author Contributions: Conceptualization, M.G.K., F.J., and H.A.; methodology, F.J.; formal analysis, M.G.K. and F.J.; investigation, F.J., M.R., and M.K.; resources, M.K. and H.A.; data curation, M.G.K.; writing—original draft preparation, M.G.K., F.J., and A.Q.; writing—review and editing, M.R. and H.A.; supervision, M.G.K. and H.A.; funding acquisition, H A. All authors have read and agreed to the published version of the manuscript.

Funding: This study was partially supported by NIH funds P20CA210305 and P30ES027792.

Institutional Review Board Statement: The study was conducted according to the guidelines of the Declaration of Helsinki, and approved by the "Biological Sciences Division, University of Chicago Hospital Institutional Review Board", Chicago, IL, USA (10-264-E).

Informed Consent Statement: Informed consent was obtained from all subjects involved in the study.

Data Availability Statement: All supporting data are presented in the tables presented in the main manuscript and as supplemental material.

Acknowledgments: We acknowledge the support and help from all of the patients included in this study. We thank Zahidul Haq, Rupash Paul, and Mustafizur Rahman for their support. We thank the University of Chicago Research, Bangladesh (URB) staff for the handling and shipping of all of the study material to the University of Chicago molecular genomics laboratory.

Conflicts of Interest: The authors declare no conflict of interest. The funders had no role in the design of the study; in the collection, analyses, or interpretation of data; in the writing of the manuscript, or in the decision to publish the results.

References

1. Vuik, F.E.; Nieuwenburg, S.A.; Bardou, M.; Lansdorp-Vogelaar, I.; Dinis-Ribeiro, M.; Bento, M.J.; Zadnik, D.; Pellisé, M.; Esteban, L.; Kaminski, M.F.; et al. Increasing incidence of colorectal cancer in young adults in Europe over the last 25 years. *Gut* **2019**, *68*, 1820–1826. [CrossRef] [PubMed]
2. Burnett-Hartman, A.N.; Powers, J.D.; Chubak, J.; Corley, D.A.; Ghai, N.R.; McMullen, C.K.; Pawloski, P.A.; Sterrett, A.T.; Feigelson, H.S. Treatment patterns and survival differ between early-onset and late-onset colorectal cancer patients: The patient outcomes to advance learning network. *Cancer Causes Control* **2019**, *30*, 747–755. [CrossRef] [PubMed]
3. Crosbie, A.B.; Roche, L.M.; Johnson, L.M.; Pawlish, K.S.; Paddock, L.E.; Stroup, A.M. Trends in colorectal cancer incidence among younger adults-Disparities by age, sex, race, ethnicity, and subsite. *Cancer Med.* **2018**, *7*, 4077–4086. [CrossRef] [PubMed]
4. Lim, D.R.; Kuk, J.K.; Kim, T.; Shin, E.J. Comparison of oncological outcomes of right-sided colon cancer versus left-sided colon cancer after curative resection: Which side is better outcome? *Medicine* **2017**, *96*, e8241. [CrossRef] [PubMed]
5. Perea, J.; García, J.L.; Corchete, L.; Tapial, S.; Olmedillas-López, S.; Vivas, A.; García-Olmo, D.; Urioste, M.; Goel, A.; González-Sarmiento, R. A clinico-pathological and molecular analysis reveals differences between solitary (early and late-onset) and synchronous rectal cancer. *Sci. Rep.* **2021**, *11*, 2202. [CrossRef] [PubMed]
6. Chang, D.T.; Pai, R.K.; Rybicki, L.A.; Dimaio, M.A.; Limaye, M.; Jayachandran, P.; Koong, A.; Kunz, P.A.; Fisher, G.A.; Ford, J.M.; et al. Clinicopathologic and molecular features of sporadic early-onset colorectal adenocarcinoma: An adenocarcinoma with frequent signet ring cell differentiation, rectal and sigmoid involvement, and adverse morphologic features. *Mod. Pathol.* **2012**, *25*, 1128–1139. [CrossRef] [PubMed]
7. Liang, J.T.; Huang, K.C.; Cheng, A.L.; Jeng, Y.M.; Wu, M.S.; Wang, S.M. Clinicopathological and molecular biological features of colorectal cancer in patients less than 40 years of age. *Br. J. Surg.* **2003**, *90*, 205–214. [CrossRef] [PubMed]
8. O'Connell, J.B.; Maggard, M.A.; Liu, J.H.; Etzioni, D.A.; Livingston, E.H.; Ko, C.Y. Do young colon cancer patients have worse outcomes? *World J. Surg.* **2004**, *28*, 558–562. [CrossRef]
9. Li, H.; Boakye, D.; Chen, X.; Hoffmeister, M.; Brenner, H. Association of Body Mass Index With Risk of Early-Onset Colorectal Cancer: Systematic Review and Meta-Analysis. *Am. J. Gastroenterol.* **2021**, *116*, 2173–2183. [CrossRef]
10. Liu, P.-H.; Wu, K.; Ng, K.; Zauber, A.G.; Nguyen, L.; Song, M.; He, X.; Fuchs, C.S.; Ogino, S.; Willett, W.C.; et al. Association of obesity with risk of early-onset colorectal cancer among women. *JAMA Oncol.* **2019**, *5*, 37–44. [CrossRef]
11. Jasmine, F.; Haq, Z.; Kamal, M.; Raza, M.; da Silva, G.; Gorospe, K.; Paul, R.; Strzempek, P.; Ahsan, H.; Kibriya, M.G. Interaction between Microsatellite Instability (MSI) and Tumor DNA methylation in the pathogenesis of colorectal carcinoma. *Cancers* **2021**, *13*, 4956. [CrossRef]
12. Saltzstein, S.L.; Behling, C.A. Age and time as factors in the left-to-right shift of the subsite of colorectal adenocarcinoma: A study of 213,383 cases from the California Cancer Registry. *J. Clin. Gastroenterol.* **2007**, *41*, 173–177. [CrossRef] [PubMed]
13. Bs, A.K.; George, T.; Hughes, S.J.; Delitto, D.; Allegra, C.J.; Hall, W.A.; Chang, G.J.; Tan, S.A.; Shaw, C.M.; Iqbal, A. Rectal cancer patients younger than 50 years lack a survival benefit from NCCN guideline-directed treatment for stage II and III disease. *Cancer* **2018**, *124*, 3510–3519. [CrossRef]
14. Saraste, D.; Järås, J.; Martling, A. Population-based analysis of outcomes with early-age colorectal cancer. *Br. J. Surg.* **2020**, *107*, 301–309. [CrossRef] [PubMed]
15. REACCT Collaborative; Zaborowski, A.M.; Abdile, A.; Adamina, M.; Aigner, F.; D'Allens, L.; Allmer, C.; Álvarez, A.; Anula, R.; Andric, M.; et al. Characteristics of Early-Onset vs Late-Onset Colorectal Cancer: A review. *JAMA Surg.* **2021**, *156*, 865–874. [CrossRef]
16. Zaborowski, A.M.; Murphy, B.; Creavin, B.; Rogers, A.C.; Kennelly, R.; Hanly, A.; Martin, S.T.; O'Connell, P.R.; Sheahan, K.; Winter, D.C. Clinicopathological features and oncological outcomes of patients with young-onset rectal cancer. *Br. J. Surg.* **2020**, *107*, 606–612. [CrossRef] [PubMed]
17. Mauri, G.; Sartore-Bianchi, A.; Russo, A.G.; Marsoni, S.; Bardelli, A.; Siena, S. Early-onset colorectal cancer in young individuals. *Mol. Oncol.* **2019**, *13*, 109–131. [CrossRef] [PubMed]
18. Losi, L.; Di Gregorio, C.; Pedroni, M.; Ponti, G.; Roncucci, L.; Scarselli, A.; Genuardi, M.; Baglioni, S.; Marino, M.; Rossi, G.; et al. Molecular genetic alterations and clinical features in early-onset colorectal carcinomas and their role for the recognition of hereditary cancer syndromes. *Am. J. Gastroenterol.* **2005**, *100*, 2280–2287. [CrossRef]
19. Barlow, J.H.; Faryabi, R.B.; Callén, E.; Wong, N.; Malhowski, A.; Chen, H.T.; Gutierrez-Cruz, G.; Sun, H.-W.; McKinnon, P.; Wright, G.; et al. Identification of early replicating fragile sites that contribute to genome instability. *Cell* **2013**, *152*, 620–632. [CrossRef]
20. Macheret, M.; Bhowmick, R.; Sobkowiak, K.; Padayachy, L.; Mailler, J.; Hickson, I.D.; Halazonetis, T.D. High-resolution mapping of mitotic DNA synthesis regions and common fragile sites in the human genome through direct sequencing. *Cell Res.* **2020**, *30*, 997–1008. [CrossRef]
21. Qiao, L.; Wong, B.C. Role of Notch signaling in colorectal cancer. *Carcinogenesis* **2009**, *30*, 1979–1986. [CrossRef] [PubMed]
22. Scheurlen, K.M.; Chariker, J.H.; Kanaan, Z.; Littlefield, A.B.; George, J.B.; Seraphine, C.; Rochet, A.; Rouchka, E.C.; Galandiuk, S. The NOTCH4-GATA4-IRG1 axis as a novel target in early-onset colorectal cancer. *Cytokine Growth Factor Rev.* **2022**, *67*, 25–34. [CrossRef] [PubMed]

23. Ogino, S.; Kawasaki, T.; Ogawa, A.; Kirkner, G.J.; Loda, M.; Fuchs, C.S. Fatty acid synthase overexpression in colorectal cancer is associated with microsatellite instability, independent of CpG island methylator phenotype. *Hum. Pathol.* **2007**, *38*, 842–849. [CrossRef] [PubMed]
24. Uddin, S.; Hussain, A.R.; Ahmed, M.; Abubaker, J.; Al-Sanea, N.; Abduljabbar, A.; Ashari, L.H.; Alhomoud, S.; Al-Dayel, F.; Bavi, P.; et al. High prevalence of fatty acid synthase expression in colorectal cancers in Middle Eastern patients and its potential role as a therapeutic target. *Am. J. Gastroenterol.* **2009**, *104*, 1790–1801. [CrossRef] [PubMed]
25. Drury, J.; Young, L.E.A.; Scott, T.L.; Kelson, C.O.; He, D.; Liu, J.; Wu, Y.; Wang, C.; Weiss, H.L.; Fan, T.; et al. Tissue-specific downregulation of fatty acid synthase suppresses intestinal adenoma formation via coordinated reprograming of transcriptome and metabolism in the mouse model of Apc-driven colorectal cancer. *Int. J. Mol. Sci.* **2022**, *23*, 6510. [CrossRef]
26. Lu, T.; Sun, L.; Wang, Z.; Zhang, Y.; He, Z.; Xu, C. Fatty acid synthase enhances colorectal cancer cell proliferation and metastasis via regulating AMPK/mTOR pathway. *Onco Targets Ther.* **2019**, *12*, 3339–3347. [CrossRef]
27. Ayers, M.; Lunceford, J.; Nebozhyn, M.; Murphy, E.; Loboda, A.; Kaufman, D.R.; McClanahan, T.K. IFN-γ-related mRNA profile predicts clinical response to PD-1 blockade. *J. Clin. Investig.* **2017**, *127*, 2930–2940. [CrossRef]
28. Cristescu, R.; Mogg, R.; Ayers, M.; Albright, A.; Murphy, E.; Yearley, J.; Sher, X.; Liu, X.Q.; Lu, H.; Nebozhyn, M.; et al. Pan-tumor genomic biomarkers for PD-1 checkpoint blockade-based immunotherapy. *Science* **2018**, *362*, eaar3593. [CrossRef]
29. Lee, J.K.; Merchant, S.A.; Jensen, C.D.; Murphy, C.C.; Udaltsova, N.; Corley, D.A. Rising early-onset colorectal cancer incidence is not an artifact of increased screening colonoscopy use in a large, diverse healthcare system. *Gastroenterology* **2022**, *162*, 325–327. [CrossRef]
30. Nikolic, N.; Spasic, J.; Stanic, N.; Nikolic, V.; Radosavljevic, D. Young-onset colorectal cancer in Serbia: Tertiary cancer center experience. *J. Adolesc. Young Adult Oncol.* **2022**. [CrossRef]
31. Siegel, R.L.; Miller, K.D.; Sauer, A.G.; Fedewa, S.A.; Butterly, L.F.; Anderson, J.C.; Cercek, A.; Smith, R.A.; Jemal, A. Colorectal cancer statistics, 2020. *CA Cancer J. Clin.* **2020**, *70*, 145–164. [CrossRef] [PubMed]
32. Ottaiano, A.; Berretta, M.; Von Arx, C.; Capozzi, M.; Caraglia, M. Editorial: The Treatment of RAS or BRAF Mutated Metastatic Colorectal Cancer: Challenges and Perspectives. *Front. Oncol.* **2022**, *12*, 852445. [CrossRef] [PubMed]
33. Harris, E.I.; Lewin, D.N.; Wang, H.L.; Lauwers, G.Y.; Srivastava, A.; Shyr, Y.; Shakhtour, B.; Revetta, F.; Washington, M.K. Lymphovascular invasion in colorectal cancer: An interobserver variability study. *Am. J. Surg. Pathol.* **2008**, *32*, 1816–1821. [CrossRef] [PubMed]
34. Liebig, C.; Ayala, G.; Wilks, J.; Verstovsek, G.; Liu, H.; Agarwal, N.; Berger, D.H.; Albo, D. Perineural invasion is an independent predictor of outcome in colorectal cancer. *J. Clin. Oncol.* **2009**, *27*, 5131–5137. [CrossRef] [PubMed]
35. Beale, A.L.; Penney, M.D.; Allison, M.C. The prevalence of iron deficiency among patients presenting with colorectal cancer. *Color. Dis.* **2005**, *7*, 398–402. [CrossRef] [PubMed]
36. Powell, A.G.M.T.; Wallace, R.; McKee, R.F.; Anderson, J.H.; Going, J.J.; Edwards, J.; Horgan, P.G. The relationship between tumour site, clinicopathological characteristics and cancer-specific survival in patients undergoing surgery for colorectal cancer. *Color. Dis.* **2012**, *14*, 1493–1499. [CrossRef]
37. Khan, S.A.; Morris, M.; Idrees, K.; Gimbel, M.I.; Rosenberg, S.; Zeng, Z.; Li, F.; Gan, G.; Shia, J.; LaQuaglia, M.P.; et al. Colorectal cancer in the very young: A comparative study of tumor markers, pathology and survival in early onset and adult onset patients. *J. Pediatr. Surg.* **2016**, *51*, 1812–1817. [CrossRef]
38. Sengupta, N.; Yau, C.; Sakthianandeswaren, A.; Mouradov, D.; Gibbs, P.; Suraweera, N.; Cazier, J.-B.; Polanco-Echeverry, G.; Ghosh, A.; Thaha, M.; et al. Analysis of colorectal cancers in British Bangladeshi identifies early onset, frequent mucinous histotype and a high prevalence of RBFOX1 deletion. *Mol. Cancer* **2013**, *12*, 1. [CrossRef]
39. Nikberg, M.; Chabok, A.; Letocha, H.; Kindler, C.; Glimelius, B.; Smedh, K. Lymphovascular and perineural invasion in stage II rectal cancer: A report from the Swedish colorectal cancer registry. *Acta Oncol.* **2016**, *55*, 1418–1424. [CrossRef]
40. Tunca, B.; Tezcan, G.; Cecener, G.; Egeli, U.; Zorluoglu, A.; Yilmazlar, T.; Ak, S.; Yerci, O.; Ozturk, E.; Umut, G.; et al. Overexpression of CK20, MAP3K8 and EIF5A correlates with poor prognosis in early-onset colorectal cancer patients. *J. Cancer Res. Clin. Oncol.* **2013**, *139*, 691–702. [CrossRef]
41. Hong, Y.; Ho, K.S.; Eu, K.W.; Cheah, P.Y. A susceptibility gene set for early onset colorectal cancer that integrates diverse signaling pathways: Implication for tumorigenesis. *Clin. Cancer Res.* **2007**, *13*, 1107–1114. [CrossRef] [PubMed]
42. Jandova, J.; Xu, W.; Nfonsam, V. Sporadic early-onset colon cancer expresses unique molecular features. *J. Surg. Res.* **2016**, *204*, 251–260. [CrossRef] [PubMed]
43. Berg, M.; Ågesen, T.H.; Thiis-Evensen, E.; Merok, M.A.; Teixeira, M.R.; Vatn, M.H.; Nesbakken, A.; Skotheim, R.I.; Lothe, R.A.; the INFAC-Study Group. Distinct high resolution genome profiles of early onset and late onset colorectal cancer integrated with gene expression data identify candidate susceptibility loci. *Mol. Cancer* **2010**, *9*, 100. [CrossRef] [PubMed]
44. Agesen, T.H.; Berg, M.P.V.D.; Clancy, T.; Thiisevensen, E.; Cekaite, L.; Lind, G.E.; Nesland, J.M.; Bakka, A.; Mala, T.; Hauss, H.J.; et al. CLC and IFNAR1 are differentially expressed and a global immunity score is distinct between early- and late-onset colorectal cancer. *Genes Immun.* **2011**, *12*, 653–662. [CrossRef] [PubMed]
45. Mo, X.; Su, Z.; Yang, B.; Zeng, Z.; Lei, S.; Qiao, H. Identification of key genes involved in the development and progression of early-onset colorectal cancer by co-expression network analysis. *Oncol. Lett.* **2020**, *19*, 177–186. [CrossRef] [PubMed]
46. La Cava, A.; Alviggi, C.; Matarese, G. Unraveling the multiple roles of leptin in inflammation and autoimmunity. *J. Mol. Med.* **2004**, *82*, 4–11. [CrossRef] [PubMed]

47. Matsubara, M.; Maruoka, S.; Katayose, S. Inverse relationship between plasma adiponectin and leptin concentrations in normal-weight and obese women. *Eur. J. Endocrinol.* **2002**, *147*, 173–180. [CrossRef]
48. Joo, J.; Clendenning, M.; Wong, E.; Rosty, C.; Mahmood, K.; Georgeson, P.; Winship, I.; Preston, S.; Win, A.; Dugué, P.-A.; et al. DNA methylation signatures and the contribution of age-associated methylomic drift to carcinogenesis in early-onset colorectal cancer. *Cancers* **2021**, *13*, 2589. [CrossRef]
49. Hamilton, S.R. *Tumours of the Colon and Rectum;* World Health Organization Classification of Tumours-Pathology and Genetics of Tumours of the Digestive System; World Health Organization: Lyon, France, 2000.
50. Jasmine, F.; Shinkle, J.; Sabarinathan, M.; Ahsan, H.; Pierce, B.L.; Kibriya, M.G. A novel pooled-sample multiplex luminex assay for high-throughput measurement of relative telomere length. *Am. J. Hum. Biol.* **2018**, *30*, e23118. [CrossRef]
51. Kibriya, M.G.; Jasmine, F.; Roy, S.; Ahsan, H.; Pierce, B. Measurement of telomere length: A new assay using QuantiGene chemistry on a Luminex platform. *Cancer Epidemiol. Biomark. Prev.* **2014**, *23*, 2667–2672. [CrossRef]
52. Demanelis, K.; Jasmine, F.; Chen, L.S.; Chernoff, M.; Tong, L.; Delgado, D.; Zhang, C.; Shinkle, J.; Sabarinathan, M.; Lin, H.; et al. Determinants of telomere length across human tissues. *Science* **2020**, *369*, 1333. [CrossRef] [PubMed]
53. Kibriya, M.G.; Raza, M.; Kamal, M.; Haq, Z.; Paul, R.; Mareczko, A.; Pierce, B.L.; Ahsan, H.; Jasmine, F. Relative telomere length change in colorectal carcinoma and its association with tumor characteristics, gene expression and microsatellite instability. *Cancers* **2022**, *14*, 2250. [CrossRef] [PubMed]
54. Janavicius, R.; Matiukaite, D.; Jakubauskas, A.; Griskevicius, L. Microsatellite instability detection by high-resolution melting analysis. *Clin. Chem.* **2010**, *56*, 1750–1757. [CrossRef] [PubMed]
55. Kibriya, M.G.; Raza, M.; Jasmine, F.; Roy, S.; Paul-Brutus, R.; Rahaman, R.; Dodsworth, C.; Rakibuz-Zaman, M.; Kamal, M.; Ahsan, H. A genome-wide DNA methylation study in colorectal carcinoma. *BMC Med. Genom.* **2011**, *4*, 50. [CrossRef] [PubMed]
56. Deschoolmeester, V.; Baay, M.; Wuyts, W.; Van Marck, E.; Van Damme, N.; Vermeulen, P.; Lukaszuk, K.; Lardon, F.; Vermorken, J.B. Detection of microsatellite instability in colorectal cancer using an alternative multiplex assay of quasi-monomorphic mononucleotide markers. *J. Mol. Diagn.* **2008**, *10*, 154–159. [CrossRef] [PubMed]
57. Findeisen, P.; Kloor, M.; Merx, S.; Sutter, C.; Woerner, S.M.; Dostmann, N.; Benner, A.; Dondog, B.; Pawlita, M.; Dippold, W.; et al. T25 repeat in the 3′ untranslated region of the CASP2 gene: A sensitive and specific marker for microsatellite instability in colorectal cancer. *Cancer Res.* **2005**, *65*, 8072–8078. [CrossRef] [PubMed]
58. Eisenhart, C. The assumptions underlying the analysis of variance. *Biometrics* **1947**, *3*, 1–21. [CrossRef]
59. Downey, T. Analysis of a multifactor microarray study using Partek genomics solution. *Methods Enzymol.* **2006**, *411*, 256–270. [CrossRef]

International Journal of *Molecular Sciences*

MDPI

Review

Deciphering Mechanisms of Action of Sortilin/Neurotensin Receptor-3 in the Proliferation Regulation of Colorectal and Other Cancers

Jean Mazella

CNRS, Institut de Pharmacologie Moléculaire et Cellulaire, UMR 7275, Université Côte d'Azur, OK660 Route des Lucioles, 06560 Valbonne, France; mazella@ipmc.cnrs.fr; Tel.: +33-4-9395-7761; Fax: 33-4-9395-7708

Abstract: The purpose of this review is to decipher the mechanisms of the pathways leading to the complex roles of neurotensin (NTS) receptor-3, also called sortilin, and of its soluble counterpart (sSortilin/NTSR3) in a large amount of physiological and pathological functions, particularly in cancer progression and metastasis. Sortilin/NTSR3 belongs to the family of type I transmembrane proteins that can be shed to release its extracellular domain from all the cells expressing the protein. Since its discovery, extensive investigations into the role of both forms of Sortilin/NTSR3 (membrane-bound and soluble form) have demonstrated their involvement in many pathophysiological processes from cancer development to cardiovascular diseases, Alzheimer's disease, diabetes, and major depression. This review focuses particularly on the implication of membrane-bound and soluble Sortilin/NTSR3 in colorectal cancer tissues and cells depending on its ability to be associated either to neurotrophins (NTs) or to NTS receptors, as well as to other cellular components such as integrins. At the end of the review, some hypotheses are suggested to counteract the deleterious effects of these proteins in order to develop effective anti-cancer treatments.

Keywords: sortilin; neurotensin; neurotensin receptor-3; soluble sortilin; colorectal cancer; cell signaling; cell morphology

Citation: Mazella, J. Deciphering Mechanisms of Action of Sortilin/Neurotensin Receptor-3 in the Proliferation Regulation of Colorectal and Other Cancers. *Int. J. Mol. Sci.* **2022**, *23*, 11888. https://doi.org/10.3390/ijms231911888

Academic Editors: Carmine Stolfi, Donatella Delle Cave and Alessandro Ottaiano

Received: 21 March 2022
Accepted: 30 September 2022
Published: 6 October 2022

1. Introduction

The development of cancerous tumors is known to be the consequence of the overexpression of growth factors. Unfortunately, when treated with radiotherapy or chemical therapy, some tumors can metastasize as a result of the weakening of cancer cell–cell interactions in the tumor tissue, leading to the dissemination of cancer cells in the circulation [1,2]. Both mechanisms of cancer growth and metastasis are regulated by a large panel of circulating activators from several neuropeptides [3,4] to membrane-bound factors released by matrix metalloprotease (MMP)-dependent shedding [5], such as Epidermal Growth Factor Receptor (EGFR) ligands [6,7]. One of the most studied neuropeptides involved in cancer progression is neurotensin (NTS), the three known receptors of which (two G-protein coupled receptors, NTSR1 and NTSR2, and a type I receptor, NTSR3) are expressed in numerous cancers and particularly in digestive cancers [8–11]. Interestingly, NTSR3 [12], also previously identified as Sortilin [13], is shed from the plasma membrane [14], leading to the release of a soluble form of sortilin (sSortilin). However, growing evidence indicates the emerging role of membrane-bound Sortilin/NTSR3 and its soluble counterpart in cancer cell proliferation and dissemination.

The identification of Sortilin/NTSR3 by three different experimental approaches predicted the complexity of the functions of the protein. Chronologically, by using the chaperon protein RAP (receptor-associated protein) affinity column, Petersen and collaborators identified and purified a 95 kDa protein from human brain extract using the detergent CHAPS. Molecular cloning of the encoding gene showed that the protein was a type I receptor with homology to the yeast vacuolar protein sorting 10 protein (Vps10p) of the sorting proteins family [13]. On another hand, by using an affinity column made with antibodies against the glucose transporter Glut4, Kandror's team identified a glycoprotein of 110 kDa as a major component of Glut4-containing vesicles. The molecular cloning of sortilin protein from rat adipocytes indicated a 93% identity to human sortilin [15]. Finally, at the same time, the 100 kDa protein previously identified as the NTS binding site was purified from CHAPS-solubilized mouse and human brain extracts by using an NTS affinity column. The cloning of the human 100 kDa NTS binding site identified the protein as sortilin [12].

Therefore, since its discovery, several cellular functions have been described for Sortilin/NTSR3, including the sorting of proteins to the plasma membrane [16,17] or to lysosomes [18,19]. In addition to its involvement in intracellular trafficking, Sortilin/NTSR3 also displays a receptor function for NTS [20], for a lipoprotein lipase [21], and a co-receptor function to initiate the action of NTS in pancreatic beta cells [22–24], as well as in the HT29 adenocarcinoma colorectal cancer cells [25,26]. Additionally, Sortilin/NTSR3 has been shown to interact with the receptor of Nerve Growth Factor (NGF), the p75 neurotrophin receptor (p75NTR), to trigger neuronal apoptosis induced by the precursor of Nerve Growth Factor (pro-NGF) [27,28] and the precursor of Brain-Derived Neurotrophic Factor (pro-BDNF) [29,30]. Recently, it was identified that Sortilin/NTSR3 acts as a receptor for the brain lipids carrier Apolipoprotein E (apoE), which confers the most important genetic risk factor for Alzheimer's disease (AD), demonstrating the involvement of Sortilin/NTSR3 in the neuroprotective action of apoE in AD pathology [31]. Finally, the role of Sortilin/NTSR3 as a biomarker of risk in cardiovascular disorders in humans has been largely confirmed [32,33].

In the field of cancer, the initial involvement of Sortilin/NTSR3 has been observed by its implication in the NTS-induced proliferation of several cancer cell lines [9]. Subsequently, a series of works were developed to investigate the role of Sortilin/NTSR3 either as an actor or as a biomarker in the development of human cancers [34]. Briefly, the overexpression of Sortilin/NTSR3 is linked to proliferation and migration in neuroendocrine tumors [35], in breast and ovarian carcinomas [36–38], in gliomas [39–41], in thyroid cancers [42], and in chronic lymphocytic leukemias [43].

This review focuses on the role of the two predominant protein forms of sortilin (the membrane-bound and the soluble form), particularly in colorectal cancer, through their interaction with various types of membrane receptors such as NTSR1, epidermal growth factor receptor (EGFR), and tropomyosin receptor kinase B (TrkB).

2. The Membrane-Bound Sortilin/NTSR3

At the cellular level, from its translation to its targeting to the plasma membrane, Sortilin/NTSR3 has been described to be classically associated with the membrane in the endoplasmic reticulum, the trans-Golgi network, and finally, in the plasma membrane after transport in membrane vesicles. Note that the three independent works described above that led to the cloning of Sortilin/NTSR3 indicated that only 10% of the protein was associated with the surface membrane, whilst 90% of the protein remained intracellular [12,13,15]. Once at the cell surface, Sortilin/NTSR3 could be released into the circulation by shedding [14,44] or released in extracellular micro-vesicles termed exosomes [45].

2.1. The Role of Membrane Sortilin/NTSR3 in the Signaling and Trafficking of Neurotrophin Receptors

Neurotrophins (NTs) are growth factors that control a series of functions in the nervous system. The mature forms of NGF and BDNF, as well as those of NT4/5 and NT3, are involved in Trk-dependent neuronal cell survival, whereas their unmatured forms are responsible for cell death through p75NTR [46]. In fact, all the functions described above for NT receptors necessitate their association with Sortilin/NTSR3, which was well described in a previous review [47]. Focusing on colorectal cancer cells, the interaction of Sortilin/NTSR3 with either TrkB or p75NTR, both expressed in colorectal cancer cells, triggers opposite functions. On the one hand, BDNF, the secretion of which is activated by Sortilin/NTSR3 [48], induces cell proliferation and displays anti-apoptotic effects through TrkB [49]. One the other hand, exogenous pro-BDNF induced colorectal cancer cell apoptosis through Sortilin/NTSR3 as a co-receptor of p75NTR, the high-affinity receptor for pro-neurotrophins, suggesting a mechanism of Sortilin/NTSR3 action that can counterbalance cell survival [49] (Figure 1).

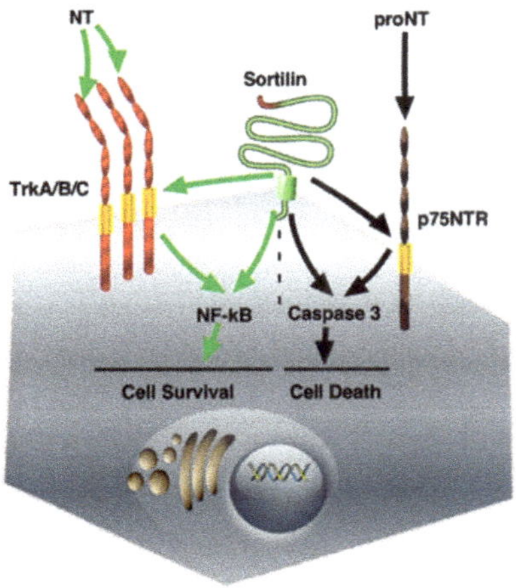

Figure 1. The fate of neurotrophin receptors/Sortilin interactions. The binding of proNT to the complex Sortilin/p75NTR triggers intracellular pathways that activate caspase 3, leading to cell death. On the other hand, the binding of matured NT to Trk receptors is responsible for NF-kB activation, which induces cell survival. The abbreviations used are: Trk, Tropomyosin Receptor Kinase; NT, neurotrophin; proNT, neurotrophin precursor; p75NTR, p75 neurotrophin receptor; NF-kB, nuclear factor kappa-light chain enhancer of activated B cells.

2.2. The Role of Membrane Sortilin/NTSR3 in the Signaling and Trafficking of Neurotensin Receptors

NTS and its receptors NTSR1 and Sortilin/NTSR3 are significantly overexpressed in colorectal cancer cells when compared to the surrounding normal epithelium, an observation that can potentially be used as a prognostic biomarker associated with more advanced colorectal cancer and poorer disease-free survival [50].

In the human colonic adenocarcinoma cell line HT29, Sortilin/NTSR3 is co-expressed with the G-protein coupled receptor NTSR1 (Figure 2). Immunoprecipitation experiments provided evidence for endogenous complex formation between these two receptors. It has also been demonstrated that the NTSR1–Sortilin/NTSR3 complex is internalized on NTS stimulation [25]. More interestingly, the interaction of Sortilin/NTSR3 with NTSR1 modulates both the NTS-induced phosphorylation of mitogen-activated protein (MAP) kinases and the phosphoinositide (PI) turnover mediated by NTSR1 [51], suggesting that Sortilin/NTSR3 may act as a co-receptor to participate in true NTS signaling. To further examine the functionality of Sortilin/NTSR3 trafficking in HT29 cells, the internalization of the Sortilin/NTSR3–NTS complex was followed from the plasma membrane to the trans-Golgi network (TGN), where NTS was bound to a lower molecular form of the receptor compared to the form found at the cell surface or on early endosomes [51]. This result suggested that the signaling and transportation functions of Sortilin/NTSR3 may be mediated through different molecular forms of the protein, a high-molecular-weight membrane form responsible for NTS endocytosis and a low-molecular-weight intracellular form responsible for the sorting of internalized NTS to the TGN. Once again, the role of Sortilin/NTSR3 in HT29 proliferation appears rather essential in the regulation of the action of NTS to modulate cancer cell proliferation.

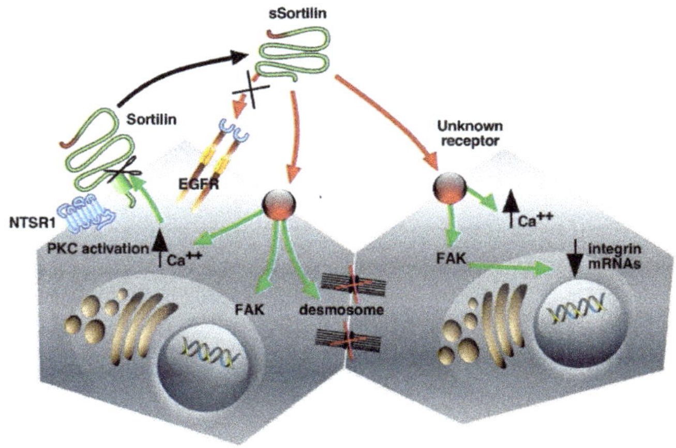

Figure 2. Schematic representation of Sortilin/NTSR3 shedding and signaling cascades in the HT29 cell line. The interaction of the membrane-bound Sortilin/NTSR3 with NTSR1 leads to the modulation of NTS-induced PKC activation. The shedding of Sortilin/NTSR3 releases sSortilin, which can interact with unknown receptors to increase intracellular Ca++, activate FAK, decrease integrin mRNAs, and lead to desmosome disruption. Green, intracellular signaling pathways; red, extracellular interactions. The abbreviations used are: NTSR1, neurotensin receptor-1; Sortilin/NTSR3, neurotensin receptor-3/sortilin; sSortilin/NTSR3, soluble neurotensin receptor-3/sortilin; EGFR, Epidermal Growth Factor Receptor; PKC, Protein Kinase C; FAK, Focal Adhesion Kinase.

In the same colorectal cell line, a study from Navarro et al. demonstrated that NTS-induced proliferation was dependent on the internalization of the Sortilin/NTSR3-NTS complex [52]. Inhibition of the internalization process affected NTS-induced Erk1/2 phosphorylation and cell proliferation, whereas the peptide-induced activation of phospholipase C was unaffected, indicating that the two intracellular pathways activated by NTS in HT29 cells (phospholipase C and MAP kinases) are independent. This can be explained by distinct conformational structures formed by the associated NTSR1 and Sortilin/NTSR3, leading to either G-protein activation or to the process of sequestration. This also indicates that inhibiting the trafficking of surface protein receptors could be an alternative method through which to develop anti-cancer treatments.

3. The Soluble Form of Sortilin/NTSR3

3.1. Shedding of the Cell Surface Sortilin/NTSR3

The shedding of Sortilin/NTSR3 was not stimulated by NTS itself, but the amount of shed protein (sSortilin/NTSR3) recovered in the extracellular medium was enhanced when the internalization process was blocked by hyperosmolar sucrose suggesting an accumulation of the protein at the cell surface and also an increase in the amount of shed protein in these conditions. The shedding process of Sortilin/NTSR3 is activated in a concentration- and time-dependent manner by PMA (Phorbol 12-Myristate 13-Acetate), a protein with a molecular weight of 100 kDa, which is slightly lower than that detected in crude homogenates (110 kDa). PMA acts as an activator of MMPs via the PKC pathway in several types of cells such as neurons, microglial cells, and cancer cells [14]. In the same way, other PKC activators such as carbachol or PGE2 [53] increased the shedding of Sortilin/NTSR3 [54]. Note that other members of the Vps10p receptor family, SorLA and SorCS1-3, are also shed [44,55].

3.2. Binding and Internalization Properties of sSortilin/NTSR3

The shed Vps10p proteins could display their own activities as ligands or could serve as transporters/protectors to avoid the proteolytic degradation of their ligands. Binding experiments performed on HT29 cell homogenates using ^{125}I-radiolabeled proteins showed that sSortilin/NTSR3 specifically bound to HT29 membranes with an affinity of 5 nM but not to the other NTS receptors [54]. Although in numerous cancer cell systems, NTS signaling depends on EGFR activation [56,57], this is not the case in HT29 cells, since sSortilin/NTSR3 is unable to compete with EGF on the EGFR, and reciprocally, EGF is unable to compete with sSortilin/NTSR3 on its binding sites [54]. These results indicate that sSortilin/NTSR3 recognizes a specific receptor in HT29 cells that is neither sortilin nor EGFR (Figure 2).

After binding to a specific receptor, sSortilin/NTSR3 is rapidly and efficiently sequestrated at 37 °C into HT29 cells by a mechanism dependent on hyperosmolar sucrose [54]. Following its internalization, 60–70% of the sequestered protein is recovered into lysosomes and degraded. The remaining non-degraded sSortilin/NTSR3 could be sorted to recycling vesicles or to other cellular compartments to trigger unidentified functions. The intracellular fate of sSortilin/NTSR3 appears to follow the same sorting to lysosomes that the membrane-bound Sortilin/NTSR3 undergoes [58].

3.3. Cell Functions of sSortilin/NTSR3 in HT29 Cells

sSortilin/NTSR3 induces plasma membrane translocation of PKCα and consequently increases the intracellular concentration of calcium at low concentrations (10 nM) [54]. It was shown that the effect of sSortilin/NTSR3 on calcium concentrations can be desensitized, a mechanism frequently observed by the internalization and uncoupling of functional receptors such as G-protein coupled receptors [59] and the low-density lipoprotein lipase receptor family [60].

In HT29 cells, sSortilin/NTSR3 rapidly and transiently activates Akt phosphorylation through the upstream phosphorylation of the complex focal adhesion kinase FAK-Src [54]. The activation of the phosphatidylinositol 3-kinase (PI3 kinase) pathway is an important step to induce calcium release from the intracellular stores (for a review, see [61]), a pathway involved in the development of colorectal cancers [62]. It is important to note that the activation of the FAK pathway is involved in survival mechanisms, and especially in a variety of distinct cancer cell development and metastasis processes [63,64].

3.4. Morphological Changes of HT29 Cells Induced by sSortilin/NTSR3

The activation of the focal adhesion kinase (FAK) pathway is known to be correlated with numerous cellular processes such as cell spreading, adhesion, migration, and survival [65]. In HT29 cells, the shape and the morphology on sSortilin/NTSR3 incubation were investigated to determine the role of the protein in the regulation of cancer cell detachment [66].

The geometric distribution (polygon classes) of cells, assessed using labeling with fluorescent anti-E-cadherin antibodies, illustrates that resting confluent HT29 cells presented a geometric distribution corresponding to 46% hexagons, a distribution in agreement with several other resting cells [67,68]. Interestingly, sSortilin/NTSR3-treated HT29 cells displayed a significant reduction (to 30%) in the proportion of hexagons in favor of pentagons, as well as an increase in the cell surface [66].

The modifications by sSortilin/NTSR3 of the actin cytoskeleton and the cell shape, as well as its ability to activate FAK, were likely linked to the cell–matrix contact weakening, which can lead to cell migration. However, HT29 cells are non-migrating cells [69]; therefore, the role of sSortilin/NTSR3 could correspond to involvement in the first step of a mechanism responsible for cell detachment.

Could the reorganization of the cell shape by sSortilin/NTSR3 be in agreement with the modifications of the architecture of ultra-structural components such as desmosomes and intermediate filaments? [66]. To answer this question, the number and structure of desmosomes have been analyzed. Desmosomes are formed by plaque densities and bundles of intermediate filaments. These structures are involved in cell–cell adhesion by connecting the proteins forming plaque densities to the interfilaments' cytoskeleton. The desmosomes are important to ensure tissue integrity and to maintain homeostasis [70]. In fact, sSortilin/NTSR3 decreases the average number of desmosomes per cell and modifies the architecture of desmosomes, thus weakening the cell–cell and cell–matrix interactions (Figure 2). The disorganization of desmosomes may contribute to the weakening of the cell barrier, which can allow for the crossing of growth factors leading to tissue dysfunction, particularly in the development or progression of human epithelial cancer cells (for reviews see [71,72]).

The marked changes observed in the sSortilin/NTSR3-treated HT29 cell morphology were correlated with the decreased expression of E-cadherin and a series of integrin family members, proteins implicated in cell–cell junctions or cell adhesion [66,73]. A decrease in or loss of integrin couples has already been described in colonic epithelial cells [74,75] in association with a poor prognosis.

Cell detachment from the plates has previously been observed in resting colonic cancer cells including HCT116, HT29, and SW620 cell lines. The action of sSortilin/NTSR3 in the weakening of cell–cell contact and cell–matrix interactions may be part of a mechanism responsible for the initial steps leading to cancer cell detachment and diffusion from primary tumors to healthy non-tumoral cells, thus facilitating metastasis (Figure 2).

4. Another Crucial Function of Sortilin/NTSR3: Possible Role in the Field of Cancer

Involvement of Sortilin/NTSR3 in the Membrane Expression of TREK-1OK

A previous study demonstrated the interaction between the two proteins sortilin and TREK-1, in which mice with deletions of the sortilin (*sort1*) or TREK-1 (*kcnk2*) genes displayed a similar phenotype of resistance to depressive-like behavior during resignation tests such as the forced swimming test (FST) and the tail suspension test (TST) [76]. TREK-1 belongs to the family of two-pore-domain potassium channels, which play important roles in neuroprotectioTablen, pain, analgesia, and depression [77–79]. As one of the first functions identified for Sortilin/NTSR3 was to address numerous proteins from the intracellular compartments to the plasma membrane or lysosomes [17,80], it was crucial to determine whether sortilin and TREK-1 were associated, and if they were, how sortilin was involved in the sorting of TREK-1. This hypothesis was firstly confirmed by demonstrating that the TREK-1 channel expression at the plasma membrane of COS-7 cells was strongly enhanced by the co-expression of Sortilin/NTSR3 [76], and secondly, by observing that the brain of *sort1−/−* mice had an altered TREK-1 function due to a dramatically lower expression of the channel at the plasma membrane of neurons [81]. Therefore, the regulation of the functional expression of TREK-1 could be of importance in a series of human cancers, including prostate [82,83] and endometrial [84] cancers, in which the overexpression of the potassium channel appears to be responsible for tumor development. The ability of spadin, a shorter analog of the pro-peptide (PE) released from the maturation of Sortilin/NTSR3 to block the activity of TREK1, indicates that it could possibly be used as a tool to decrease the proliferation of cancer cells.

5. Conclusions

The initial multiplicity of functions described for Sortilin/NTSR3 has been enhanced by the protein kinase C-dependent shedding of the membrane-bound protein, leading to a soluble extracellular form for which additional actions have been observed. The regulation of both forms involves complex mechanisms. From the various functions of Sortilin/NTSR3 in numerous cell types and tissues, the implications of membrane-bound and soluble Sortilin/NTSR3 in CRC cells are summarized in Table 1.

Table 1. Implication of membrane-bound and soluble Sortilin/NTSR3 in CRC cells.

	Membrane-Bound Sortilin as a Co-Receptor	
Ligand and receptor	Function	Pathways
NTS, NTSR1 BDNF, TrkB Pro-BDNF, p75NTR	Cell proliferation Cell proliferation, anti-apoptotic Cell apoptosis	PKC, ERK1/2, PI3K/Akt PI3K/Akt
	Soluble Sortilin as a Ligand	
Receptor	Function	Pathways
Unknown Unknown Unknown, EGFR- independent	Cell-cell disruption Cell morphological changes Cytoskeleton redistribution Cell proliferation	FAK/Src, PI3K/Akt Integrins expression changes ERK1/2, PKCα

CRC colorectal cancer, NTS neurotensin, NTSR neurotensin receptor, EGFR epidermal growth factor receptor, BDNF brain-derived neurotrophic factor, pro-BDNF precursor of BDNF, Trk tropomyosin receptor kinase, p75NTR p75 neurotrophin receptor, PKC protein kinase C, ERK extracellular signal-regulated kinase, PI3K phosphatidylinositol 3-kinase, FAK, focal adhesion kinase.

Focusing on cancers, and particularly on colorectal cancers, most studies on the role of Sortilin/NTSR3 have been performed on in vitro models (human colorectal cancer cell lines). A recent study carried out on both cell lines and primary cultures from patients demonstrated that the overexpression of Sortilin/NTSR3 was associated with 5-fluorouracil (5-FU) resistance and a poor prognosis in colorectal cancer [85].

At the plasma membrane level, the association of Sortilin/NTSR3 with the neurotrophin receptors TrkA/B/C induced cell proliferation, whereas its association with the neurotrophin receptor p75NTR triggered cell death (Figure 1). In addition, p75NTR can dimerize with other Trk receptors, as well as with NTS receptors 1 and 2 (NTSR1-2) [86]. The observation that p75NTR can undergo ectodomain shedding by γ-secretase and TNFα-convertases, a cleavage that abolishes ligand-induced signaling and produces an active intracellular fragment, increases the complexity of the role of these proteins either in cell survival or in cell death [87].

Although not demonstrated in all cases, the role of the complex Sortilin/NTSR3–NTSR1 in NTS-induced cancer cell proliferation was described to be dependent on the internalization process, at least in the HT29 cell line [52], thus increasing the complexity of the mechanisms of action that regulate cancer cell growth. Further investigations regarding the importance of the internalization process in cancer cell proliferation should assist in the identification of new molecular target(s) to counteract cancer development.

The cleavage of the membrane Sortilin/NTSR3 by a mechanism dependent on the activation of protein kinase C leads to the release of the luminal part of Sortilin/NTSR3. Then, sSortilin/NTSR3 can bind to specific binding sites to participate in several intracellular signaling pathways including the activation of PKCα, an autoregulation process of the expression of the soluble protein in the HT29 cell line. The effective concentrations of sSortilin/NTSR3 that induce PKCα activation and calcium increase are around 10 nM, concentrations that are in accordance with both circulating serum levels of sSortilin/NTSR3 determined in several previous works and with the affinity of sSortilin/NTSR3 to its unidentified receptor [54].

In order to control the expression of Sortilin/NTSR3 and its soluble counterpart, regarding their positive or negative behavior in the development of numerous pathologies, the use of specific inhibitors of the proteins, their unknown receptors, and/or their associated co-receptors would ameliorate treatments, particularly in cancers. However, it is important to avoid targeting sortilin non-selectively due to its physiological expression and function in numerous crucial tissues. In the present case of colorectal cancers, the use of a cytotoxic agent conjugated to NTS, for example, can provide an efficient treatment against tumor development, as shown in breast cancer [88]. Another possibility could be the use of the natural NTS antagonist pro-peptide (PE) released from the maturation of the sortilin precursor, which has been shown to counteract the activation of microglial cells by NTS [20].

The complex formed between Sortilin/NTSR3 and integrin(s) as a receptor for sSortilin/NTSR3 may be responsible for the activation of FAK, as observed in HT29 cells, since integrins are described as being able to stimulate intracellular kinases including FAK-Src (for a review, see [89]). In agreement with this hypothesis, this complex can ensure strong links between cells, which can be weakened by sSortilin/NTSR3 competition with the membrane sortilin, a process leading to the dissociation and dissemination of cancer cells.

Finally, since sSortilin/NTSR3 exhibits important roles in the development of metastasis, and given that its serum levels are deleterious in other important diseases such as cardiovascular disease and depression [90–95], regulation of the soluble protein formation appears to be possible by targeting the activity of MMPs. Indeed, the use of BB3103, the ADAM10 (A Desintegrin And Metalloprotease) inhibitor, was shown to block the formation of sSortilin/NTSR3 by HT29 cells [14]. Although clinical trials targeting MMPs have been canceled in phase I or in phase III for numerous inhibitors, increasing the selectivity for specific MMPs could offer a useful objective for the further development of selective treatments for each type of cancer (for a review see, [96]).

Funding: This work was supported by the Centre National de la Recherche Scientifique.

Institutional Review Board Statement: Not applicable.

Informed Consent Statement: Not applicable.

Data Availability Statement: Not applicable.

Acknowledgments: Not applicable.

Conflicts of Interest: The author declares no conflict of interest.

References

1. Thiery, J.P. Epithelial-mesenchymal transitions in tumour progression. *Nat. Rev. Cancer* **2002**, *2*, 442–454. [CrossRef] [PubMed]
2. Peralta, M.; Osmani, N.; Goetz, J.G. Circulating tumor cells: Towards mechanical phenotyping of metastasis. *eScience* **2022**, *25*, 103969. [CrossRef]
3. Reubi, J.C. Peptide receptors as molecular targets for cancer diagnosis and therapy. *Endocr. Rev.* **2003**, *24*, 389–427. [CrossRef] [PubMed]
4. Gao, Z.; Lei, W.I.; Lee, L.T.O. The Role of Neuropeptide-Stimulated cAMP-EPACs Signalling in Cancer Cells. *Molecules* **2022**, *27*, 311. [CrossRef] [PubMed]
5. Park, K.C.; Dharmasivam, M.; Richardson, D.R. The Role of Extracellular Proteases in Tumor Progression and the Development of Innovative Metal Ion Chelators that Inhibit their Activity. *Int. J. Mol. Sci.* **2020**, *21*, 6805. [CrossRef]
6. Kasina, S.; Scherle, P.A.; Hall, C.L.; Macoska, J.A. ADAM-mediated amphiregulin shedding and EGFR transactivation. *Cell Prolif.* **2009**, *42*, 799–812. [CrossRef] [PubMed]
7. Pavlenko, E.; Cabron, A.S.; Arnold, P.; Dobert, J.P.; Rose-John, S.; Zunke, F. Functional Characterization of Colon Cancer-Associated Mutations in *ADAM17*: Modifications in the Pro-Domain Interfere with Trafficking and Maturation. *Int. J. Mol. Sci.* **2019**, *20*, 2198. [CrossRef] [PubMed]
8. Christou, N.; Blondy, S.; David, V.; Verdier, M.; Lalloué, F.; Jauberteau, M.O.; Mathonnet, M.; Perraud, A. Neurotensin pathway in digestive cancers and clinical applications: An overview. *Cell Death Dis.* **2020**, *11*, 1027. [CrossRef]
9. Dal Farra, C.; Sarret, P.; Navarro, V.; Botto, J.M.; Mazella, J.; Vincent, J.P. Involvement of the neurotensin receptor subtype NTR3 in the growth effect of neurotensin on cancer cell lines. *Int. J. Cancer* **2001**, *92*, 503–509. [CrossRef] [PubMed]
10. Sánchez, M.L.; Coveñas, R. The Neurotensinergic System: A Target for Cancer Treatment. *Curr. Med. Chem.* **2021**, *29*, 3231–3260. [CrossRef] [PubMed]
11. Iyer, M.R.; Kunos, G. Therapeutic approaches targeting the neurotensin receptors. *Expert Opin. Ther. Pat.* **2021**, *31*, 361–386. [CrossRef] [PubMed]
12. Mazella, J.; Zsurger, N.; Navarro, V.; Chabry, J.; Kaghad, M.; Caput, D.; Ferrara, P.; Vita, N.; Gully, D.; Maffrand, J.P.; et al. The 100-kDa neurotensin receptor is gp95/sortilin, a non-G-protein-coupled receptor. *J. Biol. Chem.* **1998**, *273*, 26273–26276. [CrossRef] [PubMed]

13. Petersen, C.M.; Nielsen, M.S.; Nykjaer, A.; Jacobsen, L.; Tommerup, N.; Rasmussen, H.H.; Roigaard, H.; Gliemann, J.; Madsen, P.; Moestrup, S.K. Molecular identification of a novel candidate sorting receptor purified from human brain by receptor-associated protein affinity chromatography. *J. Biol. Chem.* **1997**, *272*, 3599–3605. [CrossRef] [PubMed]
14. Navarro, V.; Vincent, J.P.; Mazella, J. Shedding of the luminal domain of the neurotensin receptor-3/sortilin in the HT29 cell line. *Biochem. Biophys. Res. Commun.* **2002**, *298*, 760–764. [CrossRef]
15. Lin, B.Z.; Pilch, P.F.; Kandror, K.V. Sortilin is a major protein component of Glut4-containing vesicles. *J. Biol. Chem.* **1997**, *272*, 24145–24147. [CrossRef] [PubMed]
16. Willnow, T.E.; Petersen, C.M.; Nykjaer, A. VPS10P-domain receptors—Regulators of neuronal viability and function. *Nat. Rev. Neurosci.* **2008**, *9*, 899–909. [CrossRef]
17. Ouyang, S.; Jia, B.; Xie, W.; Yang, J.; Lv, Y. Mechanism underlying the regulation of sortilin expression and its trafficking function. *J. Cell Physiol.* **2020**, *235*, 8958–8971. [CrossRef] [PubMed]
18. Lefrancois, S.; Zeng, J.; Hassan, A.J.; Canuel, M.; Morales, C.R. The lysosomal trafficking of sphingolipid activator proteins (SAPs) is mediated by sortilin. *EMBO J.* **2003**, *22*, 6430–6437. [CrossRef]
19. Barnes, J.W.; Aarnio-Peterson, M.; Norris, J.; Haskins, M.; Flanagan-Steet, H.; Steet, R. Upregulation of Sortilin, a Lysosomal Sorting Receptor, Corresponds with Reduced Bioavailability of Latent TGFβ in Mucolipidosis II Cells. *Biomolecules* **2020**, *10*, 670. [CrossRef]
20. Martin, S.; Vincent, J.P.; Mazella, J. Involvement of the neurotensin receptor-3 in the neurotensin-induced migration of human microglia. *J. Neurosci.* **2003**, *23*, 1198–1205. [CrossRef] [PubMed]
21. Nielsen, M.S.; Jacobsen, C.; Olivecrona, G.; Gliemann, J.; Petersen, C.M. Sortilin/neurotensin receptor-3 binds and mediates degradation of lipoprotein lipase. *J. Biol. Chem.* **1999**, *274*, 8832–8836. [CrossRef] [PubMed]
22. Beraud-Dufour, S.; Coppola, T.; Massa, F.; Mazella, J. Neurotensin receptor-2 and -3 are crucial for the anti-apoptotic effect of neurotensin on pancreatic beta-TC3 cells. *Int. J. Biochem. Cell Biol.* **2009**, *41*, 2398–2402. [CrossRef] [PubMed]
23. Daziano, G.; Blondeau, N.; Béraud-Dufour, S.; Abderrahmani, A.; Rovère, C.; Heurteaux, C.; Mazella, J.; Lebrun, P.; Coppola, T. Sortilin-derived peptides promote pancreatic beta-cell survival through CREB signaling pathway. *Pharmacol. Res.* **2021**, *167*, 105539. [CrossRef] [PubMed]
24. Blondeau, N.; Béraud-Dufour, S.; Lebrun, P.; Hivelin, C.; Coppola, T. Sortilin in Glucose Homeostasis: From Accessory Protein to Key Player? *Front. Pharmacol.* **2018**, *9*, 1561. [CrossRef] [PubMed]
25. Martin, S.; Navarro, V.; Vincent, J.P.; Mazella, J. Neurotensin receptor-1 and -3 complex modulates the cellular signaling of neurotensin in the HT29 cell line. *Gastroenterology* **2002**, *123*, 1135–1143. [CrossRef] [PubMed]
26. Kim, J.T.; Weiss, H.L.; Evers, B.M. Diverse expression patterns and tumorigenic role of neurotensin signaling components in colorectal cancer cells. *Int. J. Oncol.* **2017**, *50*, 2200–2206. [CrossRef] [PubMed]
27. Nykjaer, A.; Lee, R.; Teng, K.K.; Jansen, P.; Madsen, P.; Nielsen, M.S.; Jacobsen, C.; Kliemannel, M.; Schwarz, E.; Willnow, T.E.; et al. Sortilin is essential for proNGF-induced neuronal cell death. *Nature* **2004**, *427*, 843–848. [CrossRef] [PubMed]
28. Dedoni, S.; Marras, L.; Olianas, M.C.; Ingianni, A.; Onali, P. Valproic acid upregulates the expression of the p75NTR/sortilin receptor complex to induce neuronal apoptosis. *Apoptosis Int. J. Program. Cell Death* **2020**, *25*, 697–714. [CrossRef]
29. Teng, H.K.; Teng, K.K.; Lee, R.; Wright, S.; Tevar, S.; Almeida, R.D.; Kermani, P.; Torkin, R.; Chen, Z.Y.; Lee, F.S.; et al. ProBDNF induces neuronal apoptosis via activation of a receptor complex of p75NTR and sortilin. *J. Neurosci.* **2005**, *25*, 5455–5463. [CrossRef] [PubMed]
30. Eggert, S.; Kins, S.; Endres, K.; Brigadski, T. Brothers in arms: ProBDNF/BDNF and sAPPα/Aβ-signaling and their common interplay with ADAM10, TrkB, p75NTR, sortilin, and sorLA in the progression of Alzheimer's disease. *Biol. Chem.* **2022**, *403*, 43–71. [CrossRef] [PubMed]
31. Asaro, A.; Sinha, R.; Bakun, M.; Kalnytska, O.; Carlo-Spiewok, A.S.; Rubel, T.; Rozeboom, A.; Dadlez, M.; Kaminska, B.; Aronica, E.; et al. ApoE4 disrupts interaction of sortilin with fatty acid-binding protein 7 essential to promote lipid signaling. *J. Cell Sci.* **2021**, *134*, jcs258894. [CrossRef] [PubMed]
32. Di Pietro, P.; Carrizzo, A.; Sommella, E.; Oliveti, M.; Iacoviello, L.; Di Castelnuovo, A.; Acernese, F.; Damato, A.; De Lucia, M.; Merciai, F.; et al. Targeting the ASMase/S1P pathway protects from sortilin-evoked vascular damage in hypertension. *J. Clin. Invest.* **2022**, *132*. [CrossRef]
33. Møller, P.L.; Rohde, P.D.; Winther, S.; Breining, P.; Nissen, L.; Nykjaer, A.; Bøttcher, M.; Nyegaard, M.; Kjolby, M. Sortilin as a Biomarker for Cardiovascular Disease Revisited. *Front. Cardiovasc. Med.* **2021**, *8*, 652584. [CrossRef] [PubMed]
34. Ghaemimanesh, F.; Mehravar, M.; Milani, S.; Poursani, E.M.; Saliminejad, K. The multifaceted role of sortilin/neurotensin receptor 3 in human cancer development. *J. Cell Physiol.* **2021**, *236*, 6271–6281. [CrossRef] [PubMed]
35. Kim, J.T.; Napier, D.L.; Weiss, H.L.; Lee, E.Y.; Townsend, C.M., Jr.; Evers, B.M. Neurotensin Receptor 3/Sortilin Contributes to Tumorigenesis of Neuroendocrine Tumors Through Augmentation of Cell Adhesion and Migration. *Neoplasia* **2018**, *20*, 175–181. [CrossRef]
36. Roselli, S.; Pundavela, J.; Demont, Y.; Faulkner, S.; Keene, S.; Attia, J.; Jiang, C.C.; Zhang, X.D.; Walker, M.M.; Hondermarck, H. Sortilin is associated with breast cancer aggressiveness and contributes to tumor cell adhesion and invasion. *Oncotarget* **2015**, *6*, 10473–10486. [CrossRef] [PubMed]

37. Ghaemimanesh, F.; Ahmadian, G.; Talebi, S.; Zarnani, A.H.; Behmanesh, M.; Hemmati, S.; Hadavi, R.; Jeddi-Tehrani, M.; Farzi, M.; Akhondi, M.M.; et al. The effect of sortilin silencing on ovarian carcinoma cells. *Avicenna J. Med. Biotechnol.* **2014**, *6*, 169–177. [PubMed]
38. Charfi, C.; Demeule, M.; Currie, J.C.; Larocque, A.; Zgheib, A.; Danalache, B.A.; Ouanouki, A.; Béliveau, R.; Marsolais, C.; Annabi, B. New Peptide-Drug Conjugates for Precise Targeting of SORT1-Mediated Vasculogenic Mimicry in the Tumor Microenvironment of TNBC-Derived MDA-MB-231 Breast and Ovarian ES-2 Clear Cell Carcinoma Cells. *Front. Oncol.* **2021**, *11*, 760787. [CrossRef] [PubMed]
39. Xiong, J.; Zhou, L.; Yang, M.; Lim, Y.; Zhu, Y.H.; Fu, D.L.; Li, Z.W.; Zhong, J.H.; Xiao, Z.C.; Zhou, X.F. ProBDNF and its receptors are upregulated in glioma and inhibit the growth of glioma cells in vitro. *Neuro Oncol.* **2013**, *15*, 990–1007. [CrossRef]
40. Pinet, S.; Bessette, B.; Vedrenne, N.; Lacroix, A.; Richard, L.; Jauberteau, M.O.; Battu, S.; Lalloue, F. TrkB-containing exosomes promote the transfer of glioblastoma aggressiveness to YKL-40-inactivated glioblastoma cells. *Oncotarget* **2016**, *7*, 50349. [CrossRef] [PubMed]
41. Yang, W.; Wu, P.F.; Ma, J.X.; Liao, M.J.; Wang, X.H.; Xu, L.S.; Xu, M.H.; Yi, L. Sortilin promotes glioblastoma invasion and mesenchymal transition through GSK-3β/β-catenin/twist pathway. *Cell Death Dis.* **2019**, *10*, 208. [CrossRef] [PubMed]
42. Faulkner, S.; Jobling, P.; Rowe, C.W.; Rodrigues Oliveira, S.M.; Roselli, S.; Thorne, R.F.; Oldmeadow, C.; Attia, J.; Jiang, C.C.; Zhang, X.D.; et al. Neurotrophin Receptors TrkA, p75(NTR), and Sortilin Are Increased and Targetable in Thyroid Cancer. *Am. J. Pathol.* **2018**, *188*, 229–241. [CrossRef] [PubMed]
43. Farahi, L.; Ghaemimanesh, F.; Milani, S.; Razavi, S.M.; Akhondi, M.M.; Rabbani, H. Sortilin as a Novel Diagnostic and Therapeutic Biomarker in Chronic Lymphocytic Leukemia. *Avicenna J. Med. Biotechnol.* **2019**, *11*, 270–276. [PubMed]
44. Hermey, G.; Sjogaard, S.S.; Petersen, C.M.; Nykjaer, A.; Gliemann, J. Tumour necrosis factor alpha-converting enzyme mediates ectodomain shedding of Vps10p-domain receptor family members. *Biochem. J.* **2006**, *395*, 285–293. [CrossRef]
45. Wilson, C.M.; Naves, T.; Vincent, F.; Melloni, B.; Bonnaud, F.; Lalloue, F.; Jauberteau, M.O. Sortilin mediates the release and transfer of exosomes in concert with two tyrosine kinase receptors. *J. Cell Sci.* **2014**, *127 Pt 18*, 3983–3997. [CrossRef]
46. Bartkowska, K.; Turlejski, K.; Djavadian, R.L. Neurotrophins and their receptors in early development of the mammalian nervous system. *Acta Neurobiol. Exp.* **2010**, *70*, 454–467.
47. Blondy, S.; Christou, N.; David, V.; Verdier, M.; Jauberteau, M.O.; Mathonnet, M.; Perraud, A. Neurotrophins and their involvement in digestive cancers. *Cell Death Dis.* **2019**, *10*, 123. [CrossRef]
48. Chen, Z.Y.; Ieraci, A.; Teng, H.; Dall, H.; Meng, C.X.; Herrera, D.G.; Nykjaer, A.; Hempstead, B.L.; Lee, F.S. Sortilin controls intracellular sorting of brain-derived neurotrophic factor to the regulated secretory pathway. *J. Neurosci.* **2005**, *25*, 6156–6166. [CrossRef]
49. Akil, H.; Perraud, A.; Melin, C.; Jauberteau, M.O.; Mathonnet, M. Fine-tuning roles of endogenous brain-derived neurotrophic factor, TrkB and sortilin in colorectal cancer cell survival. *PLoS ONE* **2011**, *6*, e25097. [CrossRef]
50. Qiu, S.; Nikolaou, S.; Zhu, J.; Jeffery, P.; Goldin, R.; Kinross, J.; Alexander, J.L.; Rasheed, S.; Tekkis, P.; Kontovounisios, C. Characterisation of the Expression of Neurotensin and Its Receptors in Human Colorectal Cancer and Its Clinical Implications. *Biomolecules* **2020**, *10*, 1145. [CrossRef]
51. Morinville, A.; Martin, S.; Lavallee, M.; Vincent, J.P.; Beaudet, A.; Mazella, J. Internalization and trafficking of neurotensin via NTS3 receptors in HT29 cells. *Int. J. Biochem. Cell Biol.* **2004**, *36*, 2153–2168. [CrossRef] [PubMed]
52. Navarro, V.; Martin, S.; Mazella, J. Internalization-dependent regulation of HT29 cell proliferation by neurotensin. *Peptides* **2006**, *27*, 2502–2507. [CrossRef] [PubMed]
53. Warhurst, G.; Fogg, K.E.; Higgs, N.B.; Tonge, A.; Grundy, J. Ca(2+)-mobilising agonists potentiate forskolin- and VIP-stimulated cAMP production in human colonic cell line, HT29-cl.19A: Role of [Ca2+]i and protein kinase C. *Cell Calcium* **1994**, *15*, 162–174. [CrossRef]
54. Massa, F.; Devader, C.; Beraud-Dufour, S.; Brau, F.; Coppola, T.; Mazella, J. Focal adhesion kinase dependent activation of the PI3 kinase pathway by the functional soluble form of neurotensin receptor-3 in HT29 cells. *Int. J. Biochem. Cell Biol.* **2013**, *45*, 952–959. [CrossRef]
55. Hampe, W.; Riedel, I.B.; Lintzel, J.; Bader, C.O.; Franke, I.; Schaller, H.C. Ectodomain shedding, translocation and synthesis of SorLA are stimulated by its ligand head activator. *J. Cell Sci.* **2000**, *113*, 4475–4485. [CrossRef]
56. Zhao, D.; Zhan, Y.; Zeng, H.; Koon, H.W.; Moyer, M.P.; Pothoulakis, C. Neurotensin stimulates expression of early growth response gene-1 and EGF receptor through MAP kinase activation in human colonic epithelial cells. *Int. J. Cancer* **2007**, *120*, 1652–1656. [CrossRef]
57. Moody, T.W.; Nuche-Berenguer, B.; Nakamura, T.; Jensen, R.T. EGFR Transactivation by Peptide G Protein-Coupled Receptors in Cancer. *Curr. Drug Targets* **2016**, *17*, 520–528. [CrossRef]
58. Dumaresq-Doiron, K.; Jules, F.; Lefrancois, S. Sortilin turnover is mediated by ubiquitination. *Biochem. Biophys. Res. Commun.* **2013**, *433*, 90–95. [CrossRef]
59. Evron, T.; Daigle, T.L.; Caron, M.G. GRK2: Multiple roles beyond G protein-coupled receptor desensitization. *Trends Pharmacol. Sci.* **2012**, *33*, 154–164. [CrossRef]
60. Hussain, M.M. Structural, biochemical and signaling properties of the low-density lipoprotein receptor gene family. *Front. Biosci.* **2001**, *6*, D417–D428.
61. Toker, A. Phosphoinositide 3-kinases-a historical perspective. *Subcell Biochem.* **2012**, *58*, 95–110. [PubMed]

62. Temraz, S.; Mukherji, D.; Shamseddine, A. Dual Inhibition of MEK and PI3K Pathway in KRAS and BRAF Mutated Colorectal Cancers. *Int. J. Mol. Sci.* **2015**, *16*, 22976–22988. [CrossRef] [PubMed]
63. Buchheit, C.L.; Rayavarapu, R.R.; Schafer, Z.T. The regulation of cancer cell death and metabolism by extracellular matrix attachment. *Semin. Cell Dev. Biol.* **2012**, *23*, 402–411. [CrossRef] [PubMed]
64. Fu, W.; Hall, J.E.; Schaller, M.D. Focal adhesion kinase-regulated signaling events in human cancer. *Biomol. Concepts* **2012**, *3*, 225–240. [CrossRef] [PubMed]
65. Parsons, J.T. Focal adhesion kinase: The first ten years. *J. Cell Sci.* **2003**, *116 Pt 8*, 1409–1416. [CrossRef] [PubMed]
66. Massa, F.; Devader, C.; Lacas-Gervais, S.; Beraud-Dufour, S.; Coppola, T.; Mazella, J. Impairement of HT29 Cancer Cells Cohesion by the Soluble Form of Neurotensin Receptor-3. *Genes Cancer* **2014**, *5*, 240–249. [CrossRef]
67. Kalaji, R.; Wheeler, A.P.; Erasmus, J.C.; Lee, S.Y.; Endres, R.G.; Cramer, L.P.; Braga, V.M. ROCK1 and ROCK2 regulate epithelial polarisation and geometric cell shape. *Biol. Cell* **2012**, *104*, 435–451. [CrossRef]
68. Farhadifar, R.; Roper, J.C.; Aigouy, B.; Eaton, S.; Julicher, F. The influence of cell mechanics, cell-cell interactions, and proliferation on epithelial packing. *Curr. Biol.* **2007**, *17*, 2095–2104. [CrossRef]
69. Stutzmann, J.; Bellissent-Waydelich, A.; Fontao, L.; Launay, J.F.; Simon-Assmann, P. Adhesion complexes implicated in intestinal epithelial cell-matrix interactions. *Microsc. Res. Tech.* **2000**, *51*, 179–190. [CrossRef]
70. Green, K.J.; Gaudry, C.A. Are desmosomes more than tethers for intermediate filaments? *Nat. Rev. Mol. Cell Biol.* **2000**, *1*, 208–216. [CrossRef]
71. Dusek, R.L.; Attardi, L.D. Desmosomes: New perpetrators in tumour suppression. *Nat. Rev. Cancer* **2011**, *11*, 317–323. [CrossRef] [PubMed]
72. Brooke, M.A.; Nitoiu, D.; Kelsell, D.P. Cell-cell connectivity: Desmosomes and disease. *J. Pathol.* **2012**, *226*, 158–171. [CrossRef] [PubMed]
73. Takeichi, M. Dynamic contacts: Rearranging adherens junctions to drive epithelial remodelling. *Nat. Rev. Mol. Cell Biol.* **2014**, *15*, 397–410. [CrossRef]
74. Koretz, K.; Schlag, P.; Boumsell, L.; Moller, P. Expression of VLA-alpha 2, VLA-alpha 6, and VLA-beta 1 chains in normal mucosa and adenomas of the colon, and in colon carcinomas and their liver metastases. *Am. J. Pathol.* **1991**, *138*, 741–750.
75. Stallmach, A.; Riecken, E.O. Colorectal carcinoma—Current pathogenetic concepts. Significance of cell-matrix interaction for invasive growth and metastasis. *Schweiz. Rundsch. Med. Prax.* **1992**, *81*, 847–849. [PubMed]
76. Mazella, J.; Petrault, O.; Lucas, G.; Deval, E.; Beraud-Dufour, S.; Gandin, C.; El-Yacoubi, M.; Widmann, C.; Guyon, A.; Chevet, E.; et al. Spadin, a sortilin-derived peptide, targeting rodent TREK-1 channels: A new concept in the antidepressant drug design. *PLoS Biol.* **2010**, *8*, e1000355. [CrossRef]
77. Borsotto, M.; Veyssiere, J.; Moha Ou Maati, H.; Devader, C.; Mazella, J.; Heurteaux, C. Targeting two-pore domain K(+) channels TREK-1 and TASK-3 for the treatment of depression: A new therapeutic concept. *Br. J. Pharmacol.* **2015**, *172*, 771–784. [CrossRef]
78. Heurteaux, C.; Lucas, G.; Guy, N.; El Yacoubi, M.; Thummler, S.; Peng, X.D.; Noble, F.; Blondeau, N.; Widmann, C.; Borsotto, M.; et al. Deletion of the background potassium channel TREK-1 results in a depression-resistant phenotype. *Nat. Neurosci.* **2006**, *9*, 1134–1141. [CrossRef]
79. Luo, Y.; Huang, L.; Liao, P.; Jiang, R. Contribution of Neuronal and Glial Two-Pore-Domain Potassium Channels in Health and Neurological Disorders. *Neural Plast.* **2021**, *2021*, 8643129. [CrossRef]
80. Xu, S.Y.; Jiang, J.; Pan, A.; Yan, C.; Yan, X.X. Sortilin: A new player in dementia and Alzheimer-type neuropathology. *Biochem. Cell Biol.* **2018**, *96*, 491–497. [CrossRef]
81. Moreno, S.; Devader, C.M.; Pietri, M.; Borsotto, M.; Heurteaux, C.; Mazella, J. Altered Trek-1 Function in Sortilin Deficient Mice Results in Decreased Depressive-Like Behavior. *Front. Pharmacol.* **2018**, *9*, 863. [CrossRef] [PubMed]
82. Voloshyna, I.; Besana, A.; Castillo, M.; Matos, T.; Weinstein, I.B.; Mansukhani, M.; Robinson, R.B.; Cordon-Cardo, C.; Feinmark, S.J. TREK-1 is a novel molecular target in prostate cancer. *Cancer Res.* **2008**, *68*, 1197–1203. [CrossRef]
83. Zhang, G.M.; Wan, F.N.; Qin, X.J.; Cao, D.L.; Zhang, H.L.; Zhu, Y.; Dai, B.; Shi, G.H.; Ye, D.W. Prognostic significance of the TREK-1 K2P potassium channels in prostate cancer. *Oncotarget* **2015**, *6*, 18460–18468. [CrossRef]
84. Patel, S.K.; Jackson, L.; Warren, A.Y.; Arya, P.; Shaw, R.W.; Khan, R.N. A role for two-pore potassium (K2P) channels in endometrial epithelial function. *J. Cell. Mol. Med.* **2013**, *17*, 134–146. [CrossRef]
85. Blondy, S.; Talbot, H.; Saada, S.; Christou, N.; Battu, S.; Pannequin, J.; Jauberteau, M.O.; Lalloué, F.; Verdier, M.; Mathonnet, M.; et al. Overexpression of sortilin is associated with 5-FU resistance and poor prognosis in colorectal cancer. *J. Cell. Mol. Med.* **2021**, *25*, 47–60. [CrossRef]
86. Tomellini, E.; Lagadec, C.; Polakowska, R.; Le Bourhis, X. Role of p75 neurotrophin receptor in stem cell biology: More than just a marker. *Cell Mol. Life Sci.* **2014**, *71*, 2467–2481. [CrossRef]
87. Meldolesi, J. Neurotrophin receptors in the pathogenesis, diagnosis and therapy of neurodegenerative diseases. *Pharmacol. Res.* **2017**, *121*, 129–137. [CrossRef] [PubMed]
88. Demeule, M.; Charfi, C.; Currie, J.C.; Larocque, A.; Zgheib, A.; Kozelko, S.; Béliveau, R.; Marsolais, C.; Annabi, B. TH1902, a new docetaxel-peptide conjugate for the treatment of sortilin-positive triple-negative breast cancer. *Cancer Sci.* **2021**, *112*, 4317–4334. [CrossRef] [PubMed]

89. Mohan, N.; Hosain, S.; Zhao, J.; Shen, Y.; Luo, X.; Jiang, J.; Endo, Y.; Wu, W.J. Atezolizumab potentiates Tcell-mediated cytotoxicity and coordinates with FAK to suppress cell invasion and motility in PD-L1(+) triple negative breast cancer cells. *Oncoimmunology* **2019**, *8*, e1624128. [CrossRef] [PubMed]
90. Benjannet, S.; Rhainds, D.; Essalmani, R.; Mayne, J.; Wickham, L.; Jin, W.; Asselin, M.C.; Hamelin, J.; Varret, M.; Allard, D.; et al. NARC-1/PCSK9 and its natural mutants: Zymogen cleavage and effects on the low density lipoprotein (LDL) receptor and LDL cholesterol. *J. Biol. Chem.* **2004**, *279*, 48865–48875. [CrossRef] [PubMed]
91. Nozue, T.; Hattori, H.; Ogawa, K.; Kujiraoka, T.; Iwasaki, T.; Michishita, I. Effects of Statin Therapy on Plasma Proprotein Convertase Subtilisin/kexin Type 9 and Sortilin Levels in Statin-Naive Patients with Coronary Artery Disease. *J. Atheroscler. Thromb.* **2016**, *23*, 848–856. [CrossRef] [PubMed]
92. Molgaard, S.; Demontis, D.; Nicholson, A.M.; Finch, N.A.; Petersen, R.C.; Petersen, C.M.; Rademakers, R.; Nykjaer, A.; Glerup, S. Soluble sortilin is present in excess and positively correlates with progranulin in CSF of aging individuals. *Exp. Gerontol.* **2016**, *84*, 96–100. [CrossRef] [PubMed]
93. Hu, F.; Padukkavidana, T.; Vaegter, C.B.; Brady, O.A.; Zheng, Y.; Mackenzie, I.R.; Feldman, H.H.; Nykjaer, A.; Strittmatter, S.M. Sortilin-mediated endocytosis determines levels of the frontotemporal dementia protein, progranulin. *Neuron* **2010**, *68*, 654–667. [CrossRef] [PubMed]
94. Tanimoto, R.; Palladino, C.; Xu, S.Q.; Buraschi, S.; Neill, T.; Gomella, L.G.; Peiper, S.C.; Belfiore, A.; Iozzo, R.V.; Morrione, A. The perlecan-interacting growth factor progranulin regulates ubiquitination, sorting, and lysosomal degradation of sortilin. *Matrix Biol. J. Int. Soc. Matrix Biol.* **2017**, *64*, 27–39. [CrossRef] [PubMed]
95. Du, H.; Zhou, X.; Feng, T.; Hu, F. Regulation of lysosomal trafficking of progranulin by sortilin and prosaposin. *Brain Commun.* **2022**, *4*, fcab310. [CrossRef]
96. Bernegger, S.; Jarzab, M.; Wessler, S.; Posselt, G. Proteolytic Landscapes in Gastric Pathology and Cancerogenesis. *Int. J. Mol. Sci.* **2022**, *23*, 2419. [CrossRef] [PubMed]

International Journal of *Molecular Sciences*

MDPI

Article

Targeted Sequencing of Cytokine-Induced PI3K-Related Genes in Ulcerative Colitis, Colorectal Cancer and Colitis-Associated Cancer

Nurul Nadirah Razali [1], Raja Affendi Raja Ali [2,3], Khairul Najmi Muhammad Nawawi [2,3], Azyani Yahaya [4] and Norfilza M. Mokhtar [1,3,*]

[1] Department of Physiology, Faculty of Medicine, Universiti Kebangsaan Malaysia, Cheras, Kuala Lumpur 56000, Malaysia
[2] Gastroenterology Unit, Department of Medicine, Faculty of Medicine, Universiti Kebangsaan Malaysia, Cheras, Kuala Lumpur 56000, Malaysia
[3] GUT Research Group, Faculty of Medicine, Universiti Kebangsaan Malaysia, Cheras, Kuala Lumpur 56000, Malaysia
[4] Department of Pathology, Faculty of Medicine, Universiti Kebangsaan Malaysia, Cheras, Kuala Lumpur 56000, Malaysia
* Correspondence: norfilza@ppukm.ukm.edu.my; Tel.: +60-3-91458610

Abstract: Chronic relapsing inflammatory bowel disease is strongly linked to an increased risk of colitis-associated cancer (CAC). One of the well-known inflammatory carcinogenesis pathways, phosphatidylinositol 3-kinase (PI3K), was identified to be a crucial mechanism in long-standing ulcerative colitis (UC). The goal of this study was to identify somatic variants in the cytokine-induced PI3K-related genes in UC, colorectal cancer (CRC) and CAC. Thirty biopsies (n = 8 long-standing UC, n = 11 CRC, n = 8 paired normal colorectal mucosa and n = 3 CAC) were subjected to targeted sequencing on 13 PI3K-related genes using Illumina sequencing and the SureSelectXT Target Enrichment System. The Genome Analysis Toolkit was used to analyze variants, while ANNOVAR was employed to detect annotations. There were 5116 intronic, 355 exonic, 172 untranslated region (UTR) and 59 noncoding intronic variations detected across all samples. Apart from a very small number of frameshifts, the distribution of missense and synonymous variants was almost equal. We discovered changed levels of *IL23R, IL12Rß1, IL12Rß2, TYK2, JAK2* and *OSMR* in more than 50% of the samples. The *IL23R* variant in the UTR region, rs10889677, was identified to be a possible variant that might potentially connect CAC with UC and CRC. Additional secondary structure prediction using RNAfold revealed that mutant structures were more unstable than wildtype structures. Further functional research on the potential variants is, therefore, highly recommended since it may provide insight on the relationship between inflammation and cancer risk in the cytokine-induced PI3K pathway.

Keywords: targeted sequencing; inflammatory bowel diseases; colorectal cancer; colitis-associated cancer; phosphatidylinositol 3-kinase

Citation: Razali, N.N.; Raja Ali, R.A.; Muhammad Nawawi, K.N.; Yahaya, A.; Mokhtar, N.M. Targeted Sequencing of Cytokine-Induced PI3K-Related Genes in Ulcerative Colitis, Colorectal Cancer and Colitis-Associated Cancer. *Int. J. Mol. Sci.* **2022**, *23*, 11472. https://doi.org/10.3390/ijms231911472

Academic Editors: Alessandro Ottaiano and Donatella Delle Cave

Received: 16 August 2022
Accepted: 20 September 2022
Published: 29 September 2022

1. Introduction

Carcinogenesis is the most severe complication that could arise from prolonged inflammatory bowel disease (IBD). Colitis-associated cancer (CAC) is a type of colorectal cancer (CRC) that develops as a consequence of IBD due to the presence of chronic inflammation in the gastrointestinal tract [1,2]. The likelihood of CAC developing from ulcerative colitis (UC) and Crohn's disease (CD), the two major subtypes of IBD, is 1.4% and 0.8%, respectively [3,4]. Additionally, the prevalence rate of developing CAC is expected to range from 0.6 to 17% in Western nations and from 0.3 to 1.8% in Asia Pacific regions [5]. Moreover, in Malaysia, the mean incidence of IBD doubled to 1.46 per 100,000 person/year in between 2010 and 2018 [6]. The rising urbanization of societies, which includes dietary changes, the use of antibiotics, personal cleanliness standards, microbiological exposures, and pollution, may be the cause of the rising trend in IBD over the past decades [7]. As a result, this increasing trend may ultimately contribute to the dynamic shift of CAC among Asians. Moreover, CAC has contributed to 10 to 15% of IBD fatality cases in Western countries [4,8].

Generally, CAC only accounts for 1 to 2% of CRC cases [8]. There are several characteristics that may distinguish CAC from sporadic CRC. Their clinicopathological characteristics are comparable, but CAC has a higher proportion of numerous cancer lesions, an increased percentage of superficial and invasive type lesions, and a higher proportion of mucinous or signet ring cell carcinomas [9]. Contrary to CRC, CAC develops through the inflammation–dysplasia–carcinoma sequence, where the change from low to high grade dysplasia is triggered by field precursor cells that are present in or close to the dysplastic mucosa [10]. According to a study by Choi et al. (2015), 20% of UC patients with low-grade dysplasia may have developed high-grade dysplasia or CRC within 53 months of their first diagnosis [11].

Interestingly, there are many similarities between the molecular pathogenesis of CAC and CRC. *K-Ras*, *p53*, *APC* and *COX2* are some of the common genes that are altered as CAC and CRC progressed [8]. As p53 is widely distributed in the inflamed mucosal area, this suggests that chronic inflammation has a propensity to become mutagenic [12]. In fact, a study by Claessen et al. (2010) indicated that *p53* staining was found moderately in non-dysplastic tissue and exhibition of stronger expression was seen in the low- and high-grade dysplastic lesion in more than 60% of IBD patients [13].

Immune cells, epithelial cells, stromal cells, cytokines, and chemokines are among the diverse cell types that make up the inflammatory process in CAC development and are similar to those found in the microenvironment of the malignancy [14]. Cytokines have a tendency to control the pro-tumorigenic response in chronic inflammatory conditions by causing cell malignancies and transformation [15,16]. Inflammatory mediators such as tumor necrosis factor alpha (*TNF-α*), interleukin-6 (*IL6*) and *STAT3* played significant roles in pre-neoplastic growth regulation during CAC tumorigenesis as demonstrated in an animal model study [17].

The release of those mediators is more likely to target several signaling pathways that play major roles in carcinogenesis such as NF-κB, PI3K, JAK/STAT and Wnt/B-catenin [18]. Phosphatidylinositol 3-kinases (PI3K) were recognized in promoting cancer progression as they play a key role in the regulation of survival, differentiation, and proliferation of cancer cells. PI3K enzymatic activity was found to be involved in the pathogenesis of various diseases, ranging from chronic inflammation to cancer, for instance CRC [19,20]. A recent microarray study on UC patients with two different durations discovered PI3K as one of the important pathways in the long-duration UC compared to short duration [21]. Nevertheless, there is still a paucity of knowledge about the role of the PI3K signaling pathway in the carcinogenesis progression of colitis-associated cancer. Thus, in the present study, we have performed targeted sequencing on 13 genes that are related to the cytokines-induced PI3K signaling pathway for the identification of driver gene mutations in colitis-associated cancer, long-standing ulcerative colitis, and sporadic colorectal cancer patients.

2. Results

2.1. Information on Clinical Samples

Table 1 displays demographic data for all samples. The median age of all samples was 69 years old (IQR:8.75). Malays made up the majority of the samples (70%) and were followed by Chinese (17%) and Indians (13%). Females had a somewhat larger gender distribution (60%) than men (40%). Most patients' smoking status was non-smoker (93%) as opposed to ex-smoker (7%). The mean disease duration for all long-standing UC was 28.5 ± 6.61 years. Most UC patients had a diagnosis of left-sided colitis or pancolitis, with a Mayo index score of 1 to 3 and a Geboes score of Grade 2A.1 to 2A.2 of. Meanwhile, patients with CAC had a chronic active history of colitis for 22 ± 13 years. The majority of CRC and CAC patients were at stages 1 to 3, and the rectosigmoid and distal colon were the sites of the malignancies. Histologically, the majority of CRC were moderately differentiated, whereas tumors from CAC patients were categorized as poorly and well differentiated. None of the patients had a history of CRC in their families.

Table 1. Clinical and demographic details of the recruited patients. All data are expressed as *n* except where indicated in the table. UC, ulcerative colitis; CRC, colorectal cancer; CAC, colitis-associated cancer; n, number.

	UC (n = 8)	CRC (n = 11)	Normal (n = 8)	CAC (n = 3)
Median age (range)	65.5 (60–69)	60 (36–74)	64 (51–74)	59 (20–69)
Race				
Malay	3	10	7	1
Chinese	3	1	1	-
Indian	2	-	-	2
Gender				
Male	4	5	3	-
Female	4	6	5	3
Smoking status				
Ex-smoker	-	1	1	-
Non-smoker	8	10	7	3
Stage				
I	Not applicable	4	Not applicable	-
II		3		-
III		4		3
Adenocarcinoma types				
Poorly differentiated	Not applicable	1	Not applicable	1
Moderately differentiated		9		-
Well differentiated		1		2
Mayo score (range)	1–3	Not applicable	Not applicable	Data unavailable
Geboes score (range)	2A.1–2A.2	Not applicable	Not applicable	Data unavailable

2.2. The Technical Performance of the Target Enrichment System Panel

The average percentage of clean reads across all raw reads produced by the SureSelectXT Target Enrichment System was 96.4% (range 80.9 to 98.8%). The range of the total reads was 1,200,800 to 6,106,992 reads. For each sample, the percentage of the target base covered by at least of 100× for each sample had reached 100%.

2.3. Summary of Identified Variants in PI3K-Related Genes

Targeted sequencing was performed on 13 genes that are related to PI3K, namely *IL12Rβ1, IL12Rβ2, IL23R, IL31, OSMR, JAK2, TYK2, STAT1, STAT3, STAT4, STAT6, PDK1* and *SGK2*. Long-standing ulcerative colitis (n = 8), colitis-associated cancer (n = 3), colorectal cancer (n = 11) and paired normal colorectal mucosa (n = 8) made up the total 30 samples. In total, we found 5702 variants across all samples in the cytokine-induced PI3K-related genes. Ninety percent (5116) of such variants were intronic mutations, followed by 355 exonic mutations (6%), 172 mutations in the 3′ and 5′ untranslated region (UTR) region (3%) and 59 mutations on the noncoding intronic region (1%) (Figure 1A).

Single-nucleotide polymorphisms (SNPs) made up most of the discovered variants (75% = 4256 variants), followed by 25% (1446 variants) of insertion–deletion (InDel) variants. The top three genes with the most SNPs were *OSMR* (640 variants), *IL12Rβ1* (549 variants) and *IL23R* (515 variants). Meanwhile, *IL12Rβ2* (202 variants), *IL12Rβ1* (174 variants) and *STAT6* (163 variants) showed a markedly high frequency of number of InDel variants (Figure 1B).

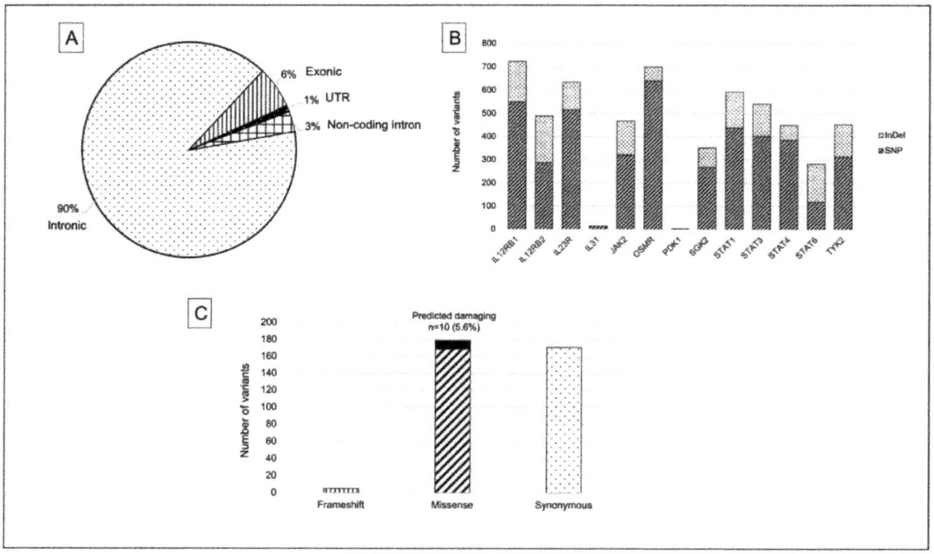

Figure 1. (**A**) Pie chart displaying the overall distribution of the variants identified in PI3K-related genes. (**B**) A bar column showing the total number of variants discovered in each PI3K-related gene and fractionated into Indel and SNPs. (**C**) Bar charts displaying the different somatic alterations found in the exonic region across all samples.

The number of missense and synonymous mutations from a total of 355 exonic variants had somewhat similar distribution, with 179 variants (50.4%) and 171 variants (48.2%), respectively, followed by 5 variants (1.4%) with frameshift mutations (insertion and deletion). Only 5.6% (n = 10) of missense mutations in the four genes *IL12Rβ1, OSMR, JAK2* and *STAT1* were predicted to be damaging or potentially damaging, whereas the remaining 174 variants (94.4%) were predicted to have a benign or neutral function (Figure 1C).

More than 50% of the patients had changed exonic sequences in 6 out of the 13 cytokine-induced PI3K-related genes. *IL23R* missense mutations appeared in all samples including the normal colonic mucosa, which is the paired sample of CRC patients (30/30 samples). This was followed by *IL12Rβ2*, where all paired normal samples had the same mutations as CRC (28/30 samples). In contrast, missense mutations *TYK2* and *OSMR* were found in 27 and 21 out of 30 samples, respectively, even though 6 of those samples had paired normal colonic tissues. A total of 7 out of the 30 samples were matched normal samples where 23 of them contained missense mutations in the *JAK2*. Meanwhile, *IL12Rβ1* missense mutations were present in 19 out of 30 samples, including four matched normal tissues. Apart from *IL12Rβ2* and *JAK2*, where synonymous mutations were frequently observed, practically all top changed genes showed significant amount of missense mutation. Frameshift mutations, however, were infrequently observed and were only found in *OSMR* (2/21 samples) and *TYK2* (1/27 samples) (Figure 2).

2.4. Somatic Variants Distribution in PI3K-Related Genes among All Samples

A total of 634 mutations were found as recurrent mutations, in the most frequently mutated gene, *IL23R*. Almost half of the 314 overall mutations were discovered in two or more samples. Of the 634 total mutations, 57 were recurrent missense mutations, 3 were synonymous mutations, 551 were intronic and 23 were UTR. These mutations were discovered sporadically in exon 2 to 10 (Figure 3A).

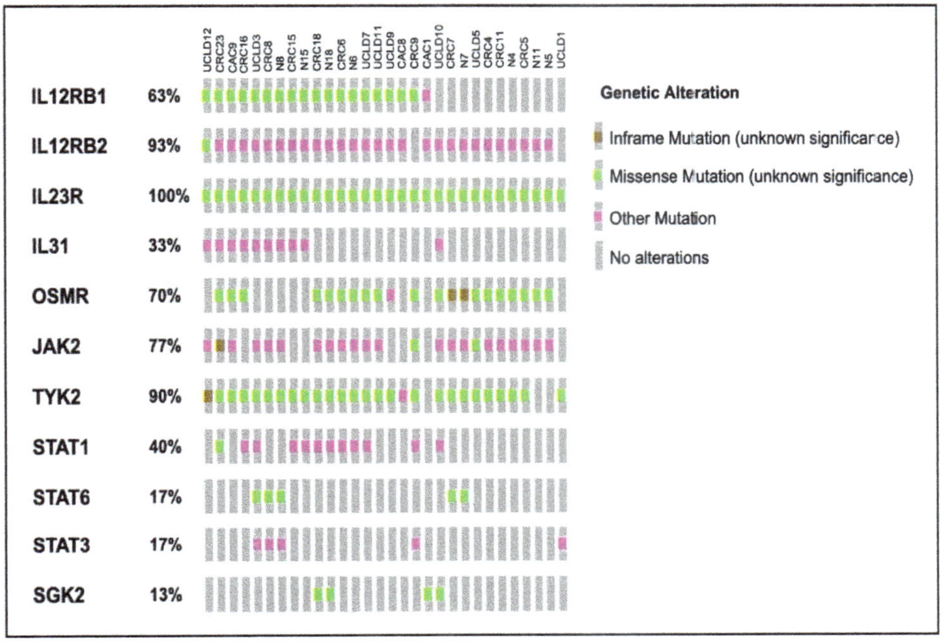

Figure 2. Oncoprint diagram showing the genetic alterations found in the exonic region of PI3K-related genes. Each bar represents the patient's number [22,23].

Only 453 alterations with 102 recurrent mutations were found in the second-most frequently occurring gene, *TYK2*, which had a 90% incidence of mutations. A total of 362 intronic and 3′-5′ flanking regions, 31 missense, and 6 synonymous were discovered throughout exon 1 to 23. *TYK2* had the most splice mutations, 50 in total (Figure 3B).

Meanwhile, *IL12Rβ2* reported 489 alterations with recurring 163 mutations. Among all groups studied, 60 recurrent synonymous mutations were distributed across exons 2 to 15, followed by 416 intronic and 12 splice areas. In addition, exon 10 only contained one missense mutation (Figure 3C).

IL12Rβ1 had the highest number of alterations, with 723 mutations, including 299 recurrences while being affected in just 60% of all cases. Between exon 7 to 15, 55 recurrence missense mutations were discovered, while 28 synonymous mutations were widely distributed around exon 1 to 16, followed by 623 intronic and 8 UTR (Figure 3D).

Another gene with a lot of modification numbers is *OSMR*, with 699 alterations, 281 of which were recurrent mutations. Exon 2 to 18 had 641 intronic mutations, 28 UTR, 20 missense mutations, and 8 synonymous mutations. Two samples contained a single frameshift insertion in exon 10, resulting in the protein's function being truncated (Figure 3E).

On the other hand, *JAK2* has 120 recurring mutations with 461 overall alterations. Throughout exonic region 3 to 25, 34 synonymous, 20 3′UTR, 400 intronic and 2 splice mutations were found in exonic region 3 to 25. Additionally, at exon 9 and 17, only two missenses were discovered. Exon 8 and 13 both had two frameshift deletion mutations that led to truncated proteins (Figure 3F).

However, no mutations were identified in the cancer hotspot locations of *IL23R*, *IL12Rβ1*, *IL12Rβ2*, *OSMR*, *TYK2*, and *JAK2* across all mutant distributions.

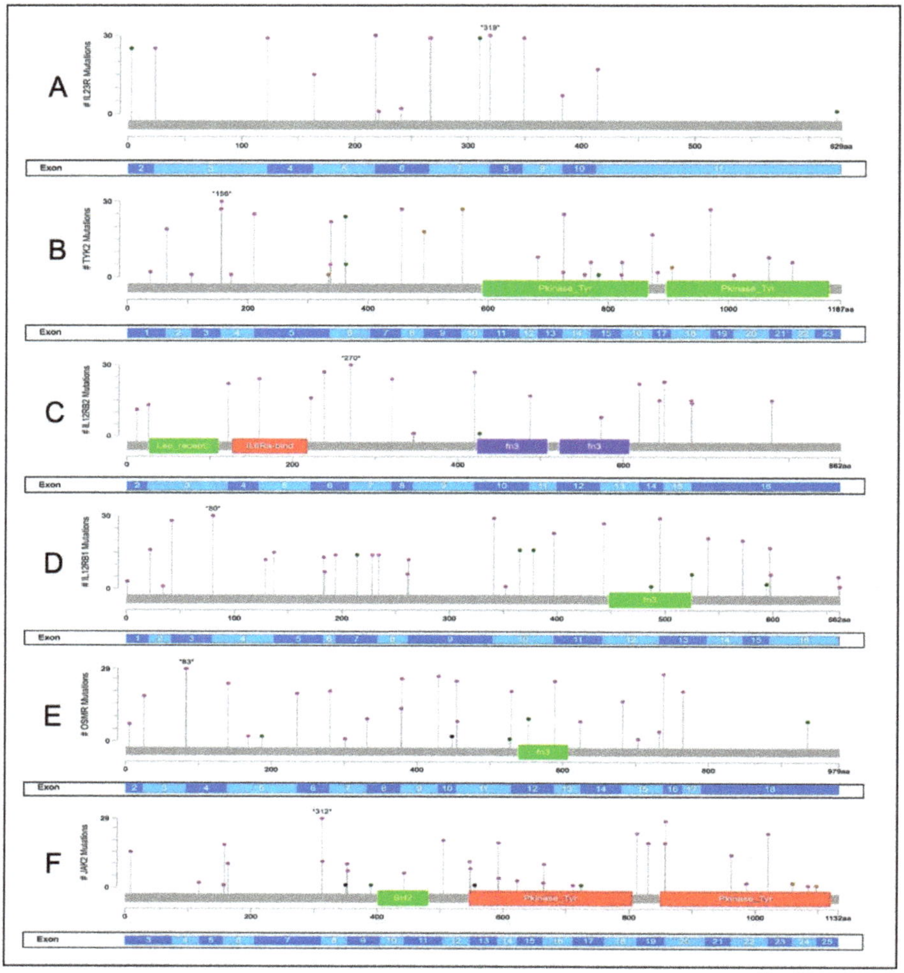

Figure 3. Distribution of somatic mutations within the functional domain of each gene. Circle and the hues green (missense), black (truncating mutation), and purple symbolize the mutations of other genes. The number with asterisk displays the location of protein change. The number of mutations identified in the coding area is shown by the length of the line. (**A**) *IL23R*, (**B**) *TYK2*, (**C**) *IL12Rβ2*, (**D**) *IL12Rβ1*, (**E**) *OSMR*, and (**F**) *JAK2* changes were identified.

2.5. Somatic Variants Distribution of PI3K-Related Genes per Group

The three main groups in this study are ulcerative colitis (UC), colorectal cancer (CRC) and colitis-associated cancer (CAC). As a result, each group's distribution of variants in PI3K-related genes was also examined.

A total 1625 variants, including 51 missense, 46 synonymous, 1 frameshift, 1461 intronic, 48 UTR and 18 non-coding intronic, were observed in PI3K-related genes in the UC group. The top five altered genes were *IL12Rβ2*, *IL23R*, *OSMR*, *STAT1* and *STAT3* (Figure 4A). Additionally, at least two UC samples were the only ones to contain all 26 recurrent mutations in *IL12Rβ1*, *IL12Rβ2*, *IL23R*, *SGK2*, *OSMR*, *STAT4* and *STAT6*. All recurrences, however, were only discovered in the intronic region (Table 2).

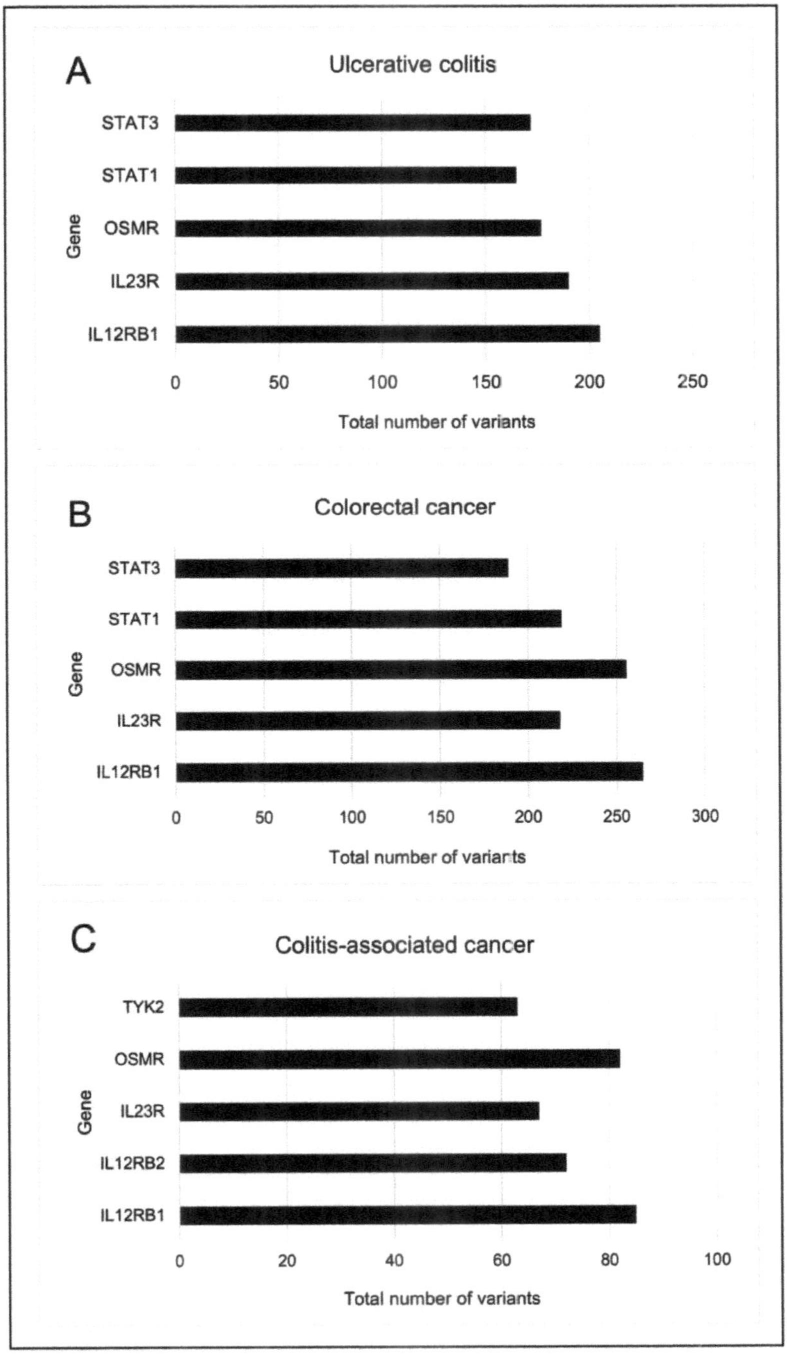

Figure 4. The top five altered genes for each group are shown in a bar graph. (**A**) Ulcerative colitis (**B**) Colorectal cancer, and (**C**) Colitis-associated cancer groups.

Table 2. List of recurrence somatic variants in two samples or more in each UC, CRC and CAC group. UC, ulcerative colitis; CRC, colorectal cancer; CAC, colitis-associated cancer.

Group	Gene	Location	dbSNP	Changes	Prediction
UC	*IL12Rβ1*	Intronic	rs201422056	g.18174947_18174948del	Not applicable
	IL12Rβ2	Intronic	rs17838042	g.67792801G>C	Not applicable
		Intronic	rs17129778	g.67787691A>T	Not applicable
		Intronic	rs17129794	g.67794918A>C	Not applicable
		Intronic	rs147756804	g.67796641_67796646del	Not applicable
	IL23R	Intronic	rs41313260	g.67706309C>T	Not applicable
	SGK2	Intronic	rs73620603	g.42195665C>T	Not applicable
	OSMR	Intronic	rs367864552	g.38881296_38881299del	Not applicable
		Intronic	rs55964556	g.38931022_38931024del	Not applicable
		Intronic	rs757333768	g.38881299ins	Not applicable
	STAT4	Intronic	rs370820216	g.191898876T>C	Not applicable
	STAT6	Intronic	-	g.57494483_57494499del	Not applicable
		Intronic	-	g.57494421ins	Not applicable
CRC	*IL12Rβ1*	Exonic	rs370238890	c.1781G>A; p.G594E	Benign
	IL12Rβ2	Intronic	-	g.67860953_67860954del	Not applicable
	IL23R	Intronic	rs767258696	g.67699612_67699616del	Not applicable
	OSMR	Intronic	rs113727379	g.38885647C>T	Not applicable
		Exonic	rs34675408	c.561T>G; p.H187Q	Benign
	JAK2	Intronic	rs3780378	g.5112288C>T	Not applicable
	STAT4	Intronic	rs35593987	g.191916526_191916527del	Not applicable
		Intronic	rs11272763	g.191992821ins	Not applicable
	STAT6	UTR5	rs71802646	g.57505072_57505076del	Not applicable
CAC	*IL12Rβ2*	Intronic	Not available	g.67795960ins	Not applicable
	TYK2	Intronic	Not available	g.10477409_10477412del	Not applicable

In the CRC group, 2029 variants were found, including 68 missense, 61 synonymous, 3 frameshift, 1809 intronic, 68 UTR and 20 non-coding intronic variants. The top five altered genes in the CRC group comprised *IL12Rβ1*, *IL23R*, *OSMR*, *STAT1* and *STAT3* (Figure 4B). Furthermore, at least two samples from the CRC group only had 19 recurrent mutations in the exonic, intronic, and UTR regions of *IL12Rβ1*, *IL12Rβ2*, *OSMR*, *JAK2*, *IL23R*, *STAT4* and *STAT6*. Both missense mutations that affected the exonic region were predicted to be benign or neutral (Table 2).

In all, 614 variants were found in the CAC group, comprising 16 missense, 20 synonymous, 561 intronic, 13 UTR and 4 non-coding intronic mutations. Most alterations were found in *IL12Rβ1*, *IL12Rβ2*, *IL23R*, *OSMR* and *TYK2* (Figure 4C). Only two recurrence frameshift mutations in *IL12Rβ2* and *TYK2* were discovered in at least two samples of CAC, in contrast to the UC and CRC groups (Table 2).

We were able to identify variants that co-occurred in the CAC group with the UC and CRC, respectively. Data filtration was applied on the selection of only CAC group, paired with either the UC or CRC group which have yielded a total of 27 intronic variants. For CAC with UC, these co-existing variants were scattered from *IL12Rβ1*, *IL12Rβ2*, *IL23R*, *OSMR*, *JAK2*, *TYK2*, *STAT1*, *STAT3* and *STAT6*, but for CAC with CRC, only two genes were involved (*JAK2* and *STAT4*) (Table 3).

Table 3. List of somatic variants from UC and CRC that co-exist with CAC group. UC, ulcerative colitis; CRC, colorectal cancer; CAC, colitis-associated cancer.

Group	Gene	Location	dbSNP	Change
CAC with UC	*IL12Rβ1*	Intronic	rs372889	g.18173603T>C
		Intronic	rs439409	g.18193613A>G

Table 3. *Cont.*

Group	Gene	Location	dbSNP	Change
		Intronic	rs382634	g.18187562G>A
		Intronic	rs17878594	g.18173513C>T
		Intronic	Not available	g.18179560_18179562del
	IL12Rβ2	Intronic	rs12410480	g.67803994G>T
		Intronic	rs145598332	g.67833145ins
		Intronic	rs66726768	g.67795956_67795960del
	IL23R	Intronic	Not available	g.67672557_67672569del
	OSMR	Intronic	rs79215370	g.38382285C>T
		Intronic	rs137968159	g.38919257_38919270del
	JAK2	Intronic	rs10283730	g.5073289G>A
		Intronic	rs7865719	g.5082333A>G
CAC with UC		Intronic	rs138377711	g.5111358_5111359del
	TYK2	Intronic	rs12720294	g.10469699A>G
		Intronic	rs12720293	g.10470293A>G
		Intronic	rs143429818	g.10469743ins
	STAT1	Intronic	rs2066803	g.191839459C>A
		Intronic	rs41371944	g.191844745T>C
		Intronic	rs376961322	g.191844269_191844270
	STAT3	Intronic	rs9909659	g.40473835G>A
		Intronic	rs8081037	g.40499158C>T
	STAT6	Intronic	Not available	g.57494483ins
		Intronic	Not available	g.57494421ins
		Intronic	rs398019756	g.57494880_57494881del
CAC with CRC	*JAK2*	Intronic	rs9987451	g.5113452C>T
	STAT4	Intronic	Not available	g.191940749_191940750del

2.6. Identification of Potential Variants That Link UC, CRC and CAC

Additional research was also carried out to identify possible polymorphisms that link CAC with UC and CRC. In particular, we were looking for variants that should be present in the majority of CAC samples. Data were filtered with a high CAC number (minimum n = 2), and low paired normal number (maximum is three out of eight) for better analysis. Data on UC and CRC numbers, however, were not filtered. In addition, because the analysis's objective was to detect in the coding region, data on the variants' function were also sorted by 'exonic' and 'UTR' exclusively.

With the help of data filtering, we identified six potential variants that were present in at least two CAC samples, three in matched normal samples and half of the CRC and UC samples. In *IL12Rβ1*, *IL12Rβ2* and *IL23R*, one missense, four synonymous and one UTR variant were found. Sanger sequencing was used to confirm these possible variants further. Only the variant rs10889677 (c.*309C>A), located at the UTR region of *IL23R*, was validated, and found in practically all CAC samples.

Other than missense mutations, it was not possible to apply protein function prediction methods such as SIFT and PolyPhen-2. Nevertheless, based on their mRNA secondary structure, predictions about the effects of those synonymous and UTR variations are still possible. In terms of the enhanced presence of hairpins, stem-loops, bulge loops, multi-branch loops, and stacking, changes in the secondary structure of the mRNA could be seen.

In silico prediction analysis showed obvious changes in the secondary structure *IL23R* variant (rs10889677) (Figure 5). In comparison to the wildtype, the presence of variant rs10889677 significantly altered the structure close to the terminal branch by adding additional branches and loops. The minimum free energy (MFE) value predicted from the created structure was also affected because of the changing of the color coding on the base structure.

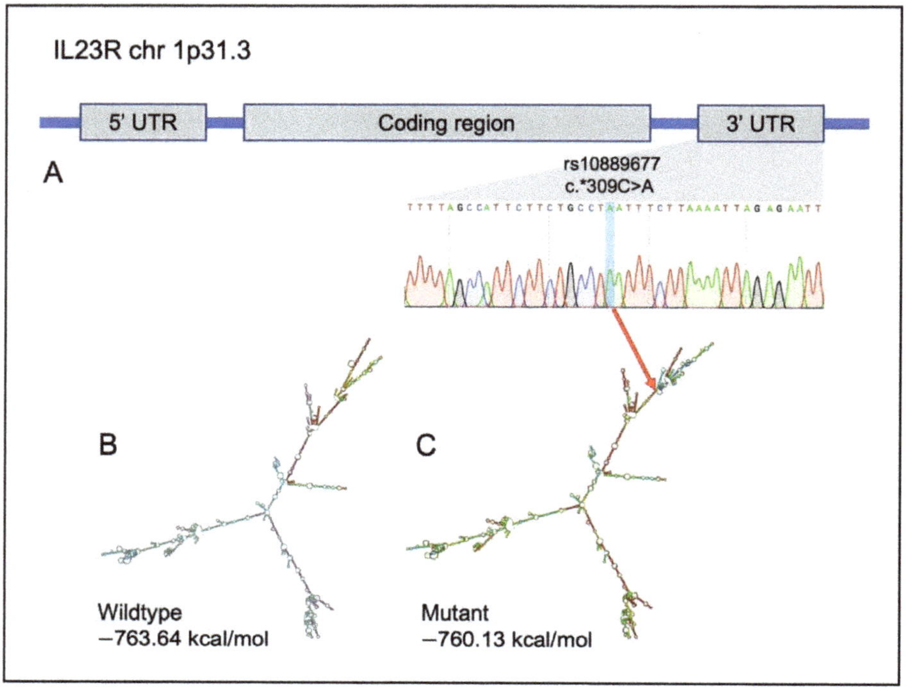

Figure 5. Details on *IL23R* variant rs10889677 and the in silico prediction of the mRNA secondary structure by RNAfold. (**A**) Location of the variant in the *IL23R*. (**B**) Wildtype (**C**) Mutant. Red arrow showing the location of nucleotide change [24].

3. Discussion

The risk of contracting colitis-related cancer rose in UC patients who had chronic inflammation that persisted for an extended period of time. The development of CAC was recently studied in relation to *p53*, *APC* and *K-Ras* [25]. However, the underlying role of the inflammation–carcinogenesis pathway in the pathogenesis of CAC is still poorly understood. We are interested in investigating how the PI3K signaling pathway contributes to the pathophysiology of CAC. In this study, we used the targeted sequencing approach via the SureSelectXT Target Enrichment System to test for somatic mutations in 13 cytokine-induced PI3K-related genes in long-standing UC, CAC and CRC patients, instead of choosing the common cancer-related genes.

Screening on the somatic alterations in our group samples showed that the majority of the variants were found on the intron, not the exonic region. In fact, silent mutations and neutral predicted variants predominated in the coding region. Few insertions and deletions that could result in frameshift mutations were seen. Nevertheless, the SureSelectXT Target Enrichment System is professed as an accurate genome analysis method from small-scale research to large sample cohorts. It has demonstrated high performance, as measured by capture efficiency, sensitivity, reproducibility, and SNP detection [26]. In this study, this application has successfully sequenced more than 95% of the clean reads coverage with an average error rate of less than 0.15% across all bases. In fact, the high precision of genome sequencing was explained by the fact that there are at least 100× as many reads per given nucleotides in the genome.

In each of our sample populations, interleukin-23 receptor (*IL23R*) was found to be the gene that was most frequently altered. A previous study has reported *IL23R* as a gene associated with inflammatory bowel disease (IBD), whereby nine SNPs in *IL23R* at various locations such as intronic, exonic and UTR have shown significant associations with Crohn's disease [27]. Moreover, it has been shown that *IL23R* variants in IBD may operate as a protective variant or contribute to the development of inflammation [28,29]. It has been demonstrated that *IL23R* polymorphisms also may increase the risk of CRC [30].

There are several *IL23R* variants that we have discovered, but just two of them, rs7530511 (c.929T>C: p.L310P) and rs1884444 (c.9G>T: p.Q3H), have been linked with colorectal-related diseases. In contrast, both variants were found in intracerebral hemorrhage and cancer-related disease (bladder and esophageal) [31–33]. Despite this, one interesting variant in *IL23R*, rs10889677 (c.*309C>A), was primarily discovered in UC, CRC and CAC. This is in line with recent studies that found this polymorphism to be a strong predictor of CRC in Asians and a risk factor for IBD [34,35]. In fact, according to a prior genome-wide association study, rs10889677 demonstrated a favorable link with IBD and might subsequently make the condition worse clinically [27]. Additionally, *IL23R* has just recently been discussed as a potential IBD treatment [36,37]. The activity of the IL23 signaling pathway could be inhibited by specific binding of the oral peptide (PTG-200) to the *IL23R*, which would subsequently influence *JAK2* and *TYK2* activation and perhaps result in abnormal *STAT3* and *STAT4* expression.

The next frequently altered gene in our study was interleukin-12 receptor beta 1 (*IL12Rß1*). Twenty percent of our samples included the *IL12Rß1* variant (rs11575935; c.1573G>A: p.A525T) that may be harmful. This variant has not yet been connected to any illnesses. Only 1.7% and 0.4% of the population in Asian and European regions, respectively, had this variant [38]. The next frequently altered gene in our study was tyrosine kinase-2 (*TYK2*). *TYK2* has lately gained attention as a potential therapeutic for IBD; its connection to gastrointestinal disorders has long been established [39,40]. *TYK2* mutation rs2304256 (c.1084G>T: p.V362F), which was discovered in almost 80% of our samples, was an intriguing finding. This variant has been linked to autoimmune and inflammatory diseases, including IBD [41,42]. The mutation rs2304256 was initially thought to be benign, but subsequent studies showed that it might encourage exon 8 inclusion and subtly boost *TYK2* expression in whole blood [43]. In fact, the Genotype-Tissue Expression (GTEx) database shows that rs2304256 is indeed linked to a slight increase in *TYK2* in several tissues, including colonic tissue. The Oncostatin M receptor (*OSMR*) is a different gene that has drawn attention. Our finding on the *OSMR* variant, rs2278329 (c.1657G>A: p.D553N), which predominately affects CRC and UC, is analogous to prior studies on *OSMR* that looked at the role of *OSMR* in inflammation and its potential link with other cancers such as bladder and thyroid cancer [44–46].

Apart from that, there was only one variant in the untranslated region (UTR) region of *IL23R*, rs10889677, which was found in 85% of CAC samples, and validated as the potential variant that correlates CAC with UC and CRC. Although, the selected potential variant was not located in the coding region and predicted as damaging, yet the effect of that variant on the translation efficacy still can be predicted via the construction of the mRNA secondary structure. The terminal branch of the *IL23R* variant rs10889677's mRNA secondary structure was clearly altered. Compared to the wildtype structure, prominent additional branches and loops had impacted the minimum free energy (MFE) value. MFE demonstrates the stability of a structure. In a stable structure, there should be more negative values. Stability is essential in the mRNA secondary structure because stable complexes may increase translation efficiency [47]. This finding was also supported by another studies that showed the presence of a non-damaging intronic variant had caused alteration in the mRNA secondary structure and stability, thus affecting the mRNA expression level in the brain tissue [48]. Moreover, results of additional studies also corroborated this assumption, where the mutant rs10889677 variant that had aberrant translation efficiency showed lower rates of T-cell proliferation, subsequently increasing susceptibility of the IBD and elevating the risk for developing cancers, such as breast, lung and nasopharyngeal [49,50].

There are a few limitations of this study. This study only involved a single center. Most likely, it was due to our stringent inclusion and exclusion criteria. Moreover, we were cautious in selecting the potential CAC variants, as data filtering was conducted based only exonic and UTR variants. Therefore, further studies with a larger cohort would be highly recommended to corroborate the results. In addition, exploring the underlying mechanism of more variety of potential variants in CAC via in vitro functional study would be highly beneficial in discovering the linkage of those gene variants with tumorigenesis.

Through our findings, we successfully identified somatic variants in cytokine-induced PI3K-related genes in long-standing UC, CAC and CRC samples. Targeted therapies for CRC now focus on a few common pathways, including EGFR (cetuximab and panitumumab) and VEGF (bevacizumab), which can stimulate a number of downstream intracellular signaling pathways including the PI3K signaling pathway [51]. So far, therapeutics interventions based on cytokine-induced PI3K-related genes such as *IL23R* have primarily been studied in inflammatory illnesses [37,52]. Hence, introducing *IL23R* as the next cancer small-molecule inhibitor may be advantageous for future therapies.

4. Materials and Methods

4.1. Sample Collection

A total of 30 fresh frozen and archive samples from long-standing UC, CAC, CRC and the corresponding adjacent normal colorectal mucosa tissue were collected from patients that were attending the Endoscopy Unit, Universiti Kebangsaan Malaysia Medical Centre (UKMMC), Kuala Lumpur, Malaysia. Upon admission, informed consent was obtained from each patient. Basic clinical and demographic data were gathered and analyzed while reviewing the patient's medical records. Prior to further processing, the tissues were collected in RNAlater (Sigma Aldrich, St. Louis, MO, USA) and kept frozen at −80 °C. An experienced pathologist examined the confirmation of the diagnosis, the level of inflammation, and the presence of metastases based on the hematoxylin and eosin (H&E)-stained sections. Only cancer tissues with more than 80% tumor cell content were used in this study for cancer samples. The normal samples were confirmed to be free from tumor or inflammatory cells. The Universiti Kebangsaan Malaysia Research Ethics Committee (UKM/PPI/111/8/JEP-2019-572) granted approval for this study.

4.2. Nucleic Acid Extraction and Quality Assessment

DNA extraction from the fresh frozen tissues was performed using the AllPrep DNA/RNA/miRNA Universal Kit (Qiagen, Valencia, CA, USA) in accordance with the manufacturer's protocol. Meanwhile, using GENEREAD DNA FFPE Kit (Qiagen, Valencia, CA, USA), DNA was extracted from formalin-fixed paraffin-embedded (FFPE) blocks of archival samples. DNA concentration was measured using DeNovix DS11+ Spectrophotometer (DeNovix Inc., Wilmington, DE, USA) and Qubit® DNA Assay Kit in Qubit® 2.0 Fluorometer (Life Technologies, Carlsbad, CA, USA). The agarose gel electrophoresis was used to evaluate the extracted DNA's purity. Targeted sequencing was performed on the genomic DNA that was largely undamaged and free of RNA.

4.3. Targeted Sequencing

Library preparation was conducted using DNA random fragmentation by sonication (Covaris, MA, USA) to the size of 180–280 bp fragments, followed by PCR enrichment and purification with AMPure XP system (Beckman Coulter, Beverly, USA). The library was then quantified using the high-sensitivity DNA assay on the Agilent Bioanalyzer 2100 (Agilent Technologies Inc, Santa Clara, CA, USA). Targeted sequencing was carried out using Illumina sequencing and SureSelectXT Target Enrichment System (Agilent Technologies Inc., Santa Clara, CA, USA).

4.4. Sequence Alignment and Variant Annotation

Burrows–Wheeler Aligner (BWA) was utilized to map the paired-end clean reads to the human reference genome (hg19). After the discovery of the genomic variant, the program ANNOVAR [53] was used to annotate the variants in a variety of ways, including the genomic regions impacted by the variants (RefSeq and Genecode), protein-coding changes and deleteriousness prediction (SIFT [54], PolyPhen [55] and MutationAssessor [56]), mRNA secondary structure (RNAfold) [24], allele frequency (1000 Human Genome) [57], disease associations (dbSNP [58], COSMIC (cancer.sanger.ac.uk) [59], OMIM [60], GWAS Catalog [61] and HGMD [62]) and pathway annotation (Gene Ontology [63], KEGG [64] and Reactome [65]).

4.5. Validation of Genomic Variants

The discovered somatic variants were then validated using the Sanger sequencing method. Primers were designed using the NCBI Primer Tool (National Center for Biotechnology Information, Bethesda MD, USA) and Primer3Plus [66]. The sequencing results were analyzed using the SnapGene Viewer 5.3.2. The primers used for validation were *IL12Rß2* rs2229546 5′-GCTGAGAGCAGACAACTGGT-3′ (forward), 5′-CCATCATGGGTGGGAAGGTC-3′ (reverse), rs2228420 5′-GGGCGCATACACCAATCAG-3′ (forward), 5′-TTTCCCTGACCCATGGCAG-3′, *IL23R* rs10889677 5′-TCTGTGCTCCTACCATCACC-3′ (forward), 5′-TGTGCCTGTATGTGTGACCA-3′ (reverse) and *JAK2* rs2230722 5′-GAGATCTTGCCATGTTGCCC-3′ (forward), and 5′-ACACTGCCATCCCAAGACAT-3′ (reverse).

4.6. Statistical Analysis

Normally distributed variables are presented as the mean ± standard deviation, and non-normally distributed variables as the median (25th and 75th percentiles). Statistical analyses were performed using SPSS 26 software (SPSS Inc., Chicago, IL, USA).

5. Conclusions

We were able to identify somatic variants in the PI3K-related genes among UC, CRC and CAC, and it was discovered that most of these variants were found in the *IL23R, IL12Rß1* and *IL12Rß2* genes, followed by *TYK2, JAK2* and *OSMR*. The discovery of *IL23R* variant rs10889677 as a possible mutation may help to give an insight on how the cytokine-induced PI3K pathway links inflammation with a higher risk of developing cancer, opening the door to improved care for CAC patients in the near future.

Author Contributions: Conceptualization, N.M.M. and R.A.R.A.; methodology and participant recruitments, N.M.M., R.A.R.A., K.N.M.N. and N.N.R.; validation, N.N.R., A.Y. and N.M.M.; data analysis, N.N.R. and N.M.M.; writing—original draft preparation, N.N.R.; writing—review and editing, N.M.M., R.A.R.A. and K.N.M.N.; supervision, N.M.M., R.A.R.A. and K.N.M.N. All authors have read and agreed to the published version of the manuscript.

Funding: This work was supported by the Fundamental Research Grant Scheme, Ministry of Higher Education, Malaysia (FRGS/1/2018/SKK06/UKM/02/4).

Institutional Review Board Statement: The study was conducted in accordance with the Declaration of Helsinki and approved by the Universiti Kebangsaan Malaysia Research Ethics Committee (UKM/PPI/111/8/JEP-2019-572).

Informed Consent Statement: Informed consent was obtained from all subjects involved in the study.

Data Availability Statement: The raw data in this study are available upon request.

Acknowledgments: We would like to thank the staff from the Gastroenterology Unit, UKMMC and the Department of Physiology, UKM for their assistance and coordination with biospecimen collection.

Conflicts of Interest: The authors declare no conflict of interest.

References

1. Grivennikov, S.I.; Cominelli, F. Colitis-Associated and Sporadic Colon Cancers: Different Diseases, Different Mutations? *Gastroenterology* **2016**, *150*, 808–810. [CrossRef]
2. Liverani, E.; Scaioli, E.; John Digby, R.; Bellanova, M.; Belluzi, A. How to predict clinical relapse in inflammatory bowel disease patients. *World J. Gastroenterol.* **2016**, *22*, 1017–1033. [CrossRef] [PubMed]

3. Laukoetter, M.G.; Mennigen, R.; Hannig, C.M.; Osada, N.; Rijcken, E.; Vowinkel, T.; Krieglstein, C.F.; Senniger, N.; Anthoni, C.; Bruewer, M. Intestinal Cancer Risk in Crohn's Disease: A Meta-Analysis. *J. Gastrointest. Surg.* **2011**, *15*, 576–583. [CrossRef] [PubMed]
4. Zhou, Q.; Shen, Z.F.; Wu, B.S.; Xu, C.B.; He, Z.Q.; Chen, T.; Shang, H.T.; Xie, C.F.; Huang, S.Y.; Chen, Y.G.; et al. Risk of Colorectal Cancer in Ulcerative Colitis Patients: A Systematic Review and Meta-Analysis. *Gastroenterol. Res. Pract.* **2019**, *2019*, 5363261. [CrossRef] [PubMed]
5. Zhiqin, W.; Palaniappan, S.; Raja Ali, R.A. Inflammatory Bowel Disease-related Colorectal Cancer in the Asia-Pacific Region: Past, Present, and Future. *Intest. Res.* **2014**, *12*, 194. [CrossRef]
6. Mokhtar, N.M.; Nawawi, K.N.M.; Verasingam, J.; Zhiqin, W.; Sagap, I.; Azman, Z.A.M.; Mazlan, L.; Hamid, H.A.; Yaacob, N.Y.; Rose, I.M.; et al. A four-decade analysis of the incidence trends, sociodemographic and clinical characteristics of inflammatory bowel disease patients at single tertiary centre, Kuala Lumpur, Malaysia. *BMC Public Health* **2019**, *19*, 1–10. [CrossRef]
7. Ananthakrishnan, A.N.; Bernstein, C.N.; Iliopoulos, D.; Macpherson, A.; Neurath, M.F.; Raja Ali, R.A.; Vavricka, S.R.; Fiocchi, C. Environmental triggers in IBD: A review of progress and evidence. *Nat. Rev. Gastroenterol. Hepatol.* **2018**, *15*, 39–49. [CrossRef]
8. Kameyama, H.; Nagahashi, M.; Shimada, Y.; Tajima, Y.; Ichikawa, H.; Nakano, M.; Sakata, J.; Kobayashi, T.; Narayanan, S.; Takabe, K.; et al. Genomic characterization of colitis-associated colorectal cancer. *World J. Surg. Oncol.* **2018**, *16*, 4–9. [CrossRef]
9. Watanabe, T.; Konishi, T.; Kishimoto, J.; Kotake, K.; Muto, T.; Sugihara, K. Ulcerative Colitis-associated Colorectal Cancer Shows a Poorer Survival than sporadic colorectal cancer: A Nationwide Japanese Study. *Inflamm. Bowel Dis.* **2011**, *17*, 1–7. [CrossRef]
10. Hartnett, L.; Egan, L.J. Inflammation, DNA methylation and colitis-associated cancer. *Carcinogenesis* **2012**, *33*, 723–731. [CrossRef]
11. Choi, C.R.; Ignjatovic-Wilson, A.; Askari, A.; Lee, G.H.; Warusavitarne, J.; Moorghen, M.; Thomas-Gibson, S.; Saunders, B.P.; Rutter, M.D.; Graham, T.A.; et al. Low-Grade Dysplasia in Ulcerative Colitis: Risk Factors for Developing High-Grade Dysplasia or Colorectal Cancer. *Am. J. Gastroenterol.* **2015**, *110*, 1461–1471. [CrossRef] [PubMed]
12. Ullman, T.A.; Itzkowitz, S.H. Intestinal inflammation and cancer. *Gastroenterology* **2011**, *140*, 1807–1816.e1. [CrossRef] [PubMed]
13. Claessen, M.M.H.; Schipper, M.E.I.; Oldenburg, B.; Siersema, P.D.; Offerhaus, G.J.A.; Vleggaar, F.P. WNT-pathway activation in IBD-associated colorectal carcinogenesis: Potential biomarkers for colonic surveillance. *Cell Oncol.* **2010**, *32*, 303–310. [CrossRef] [PubMed]
14. Axelrad, J.E.; Lichtiger, S.; Yajnik, V. Inflammatory Bowel Disease: Global view Inflammatory bowel disease and cancer: The role of inflammation, immunosuppression, and cancer treatment. *World J. Gastroenterol.* **2016**, *22*, 4794–4801. [CrossRef]
15. Shrihari, T.G. Dual role of inflammatory mediators in cancer. *Ecancermedicalscience* **2017**, *11*, 1–9. [CrossRef]
16. Galdiero, M.R.; Marone, G.; Mantovani, A. Cancer inflammation and cytokines. *Cold Spring Harb. Perspect. Biol.* **2018**, *10*, a028662. [CrossRef]
17. Grivennikov, S.I.; Cominello, F. IL-6 and Stat3 Are Required for Survival of Intestinal Epithelial Cells and Development of Colitis-Associated Cancer. *Cancer Cell* **2009**, *15*, 103–113. [CrossRef]
18. Qu, X.; Tang, Y.; Hua, S. Immunological approaches towards cancer and inflammation: A cross talk. *Front. Immunol.* **2018**, *9*, 699. [CrossRef]
19. Ciraolo, E.; Gulluni, F.; Hirsch, E. Methods to measure the enzymatic activity of PI3Ks. *Methods Enzym.* **2014**, *543*, 115–140.
20. Abdul, S.N.; Ab Mutalib, N.S.; Sean, K.S.; Syafruddin, S.E.; Ishak, M.; Sagap, I.; Mazlan, L.; Rose, I.M.; Abu, N.; Mokhtar, N.M.; et al. Molecular characterization of somatic alterations in Dukes' B and C colorectal cancers by targeted sequencing. *Front. Pharm.* **2017**, *8*, 465. [CrossRef]
21. Low, E.N.D.; Mokhtar, N.M.; Wong, Z.; Raja Ali, R.A. Colonic mucosal transcriptomic changes in patients with long-duration ulcerative colitis revealed colitis-associated cancer pathways. *J. Crohns Colitis* **2017**, *13*, 755–763. [CrossRef] [PubMed]
22. Gao, J.; Aksoy, B.A.; Dogrusoz, U.; Dresdner, G.; Gross, B.; Sumer, S.O.; Sun, Y.; Jacobsen, A.; Sinha, R.; Larsson, E.; et al. Integrative Analysis of Complex Cancer Genomics and Clinical Profiles Using the cBioPortal Complementary Data Sources and Analysis Options. *Sci. Signal.* **2014**, *6*, 1–20. [CrossRef]
23. Cerami, E.; Gao, J.; Dogrusoz, U.; Gross, B.E.; Sumer, S.O.; Aksoy, B.A.; Jacobsen, A.; Byrne, C.J.; Heuer, M.L.; Larsson, E.; et al. The cBio Cancer Genomics Portal: An Open Platform for Exploring Multidimensional Cancer Genomics Data. *Cancer Discov.* **2012**, *2*, 401–404. [CrossRef] [PubMed]
24. Gruber, A.R.; Lorenz, R.; Bernhart, S.H.; Neubock, R.; Hofacker, I.L. The Vienna RNA websuite. *Nucleic Acids Res.* **2008**, *36*, W70–W74. [CrossRef]
25. Du, L.; Kim, J.J.; Shen, J.; Chen, B.; Dai, N. KRAS and TP53 mutations in inflammatory bowel disease associated colorectal cancer: A meta-analysis. *Oncotarget* **2017**, *8*, 22175–22186. [CrossRef]
26. Ong, J.; Giuffre, A.; Joshi, S.; Ravi, H.; Pabon-Pena, C.; Novak, B.; Visitacion, M.; Hamady, M.; Useche, F.; Arezi, B.; et al. Overview of the Agilent Technologies SureSelectTM Target Enrichment System. *J. Mol. Tech.* **2011**, *22*, S30–S31.
27. Duerr, R.H.; Taylor, K.D.; Brant, S.R.; Rioux, J.D.; Silverberg, M.S.; Daly, M.J.; Steinhart, A.H.; Abraham, C.; Regueiro, M.; Griffiths, A.; et al. A genome-wide association study identifies IL23R as an inflammatory bowel disease gene. *Science* **2006**, *314*, 1461–1463. [CrossRef]
28. Sivanesan, D.; Beauchamp, C.; Quinou, C.; Lee, J.; Lesage, S.; Chemtob, S.; Rioux, J.D.; Michnick, S.W. IL23R (Interleukin 23 Receptor) variants protective against inflammatory bowel diseases (IBD) display loss of function due to impaired protein stability and intracellular trafficking. *J. Biol. Chem.* **2016**, *291*, 8673–8685. [CrossRef]

29. Zhu, Y.; Jiang, H.G.; Chen, Z.H.; Lu, B.H.; Li, J.; Shen, X.N. Genetic association between IL23R rs11209026 and rs10889677 polymorphisms and risk of Crohn's disease and ulcerative colitis: Evidence from 41 studies. *Inflamm. Res.* **2020**, *69*, 87–103. [CrossRef]
30. Poole, E.M.; Curtin, K.; Hsu, L.; Duggan, D.J.; Makar, K.W.; Xiao, L.; Carlson, C.S.; Caan, B.J.; Potter, J.D.; Slattery, M.L.; et al. Genetic variability in IL23R and risk of colorectal adenoma and colorectal cancer. *Cancer Epidemiol.* **2012**, *36*, e104–e110. [CrossRef]
31. Park, H.J.; Kim, S.K.; Park, H.K.; Chung, J.H. Association of IL23R polymorphism (rs7530511) with intracerebral hemorrhage in Korean population. *Neurol. Sci.* **2016**, *37*, 983–985. [CrossRef] [PubMed]
32. El-Gedamy, M.; El-khayat, Z.; Abol-Enein, H.; El-said, A. Rs-1884444 G/T variant in IL-23 receptor is likely to modify risk of bladder urothelial carcinoma by regulating IL-23/IL-17 inflammatory pathway. *Cytokine* **2021**, *138*, 155355. [CrossRef] [PubMed]
33. Li, M.; Yue, C.; Jin, G.; Guo, H.; Ma, H.; Wang, G.; Huangm, S.; Wu, F.; Zhao, X. Rs1884444 variant in IL23R gene is associated with a decreased risk in esophageal cancer in Chinese population. *Mol. Carcinog.* **2019**, *58*, 1822–1831. [CrossRef]
34. Peng, L.L.; Wang, Y.; Zhu, F.L.; Xu, W.D.; Ji, X.L.; Ni, J. IL-23R mutation is associated with ulcerative colitis: A systemic review and meta-analysis. *Oncotarget* **2017**, *8*, 4849–4863. [CrossRef] [PubMed]
35. Mosallaei, M.; Simonian, M.; Esmaeilzadeh, E.; Bagheri, H.; Miraghajani, M.; Salehi, A.R.; Mehrzad, V.; Salehi, R. Single nucleotide polymorphism rs10889677 in miRNAs Let-7e and Let-7f binding site of IL23R gene is a strong colorectal cancer determinant: Report and meta-analysis. *Cancer Genet.* **2019**, *239*, 46–53. [CrossRef]
36. Cheng, X.; Taranath, R.; Mattheakis, L.; Bhandari, A.; Liu, D. The biomarker profile of PTG-200, an oral peptide antagonist of IL-23 receptor, tracks with efficacy in a preclinical model of IBD. *J. Crohn's Colitis* **2017**, *11*, S502–S540. [CrossRef]
37. Cheng, X.; Lee, T.Y.; Ledet, G.; Zemade, G.; Tovera, M.; Campbell, R.; Purro, N.; Annamali, T.; Masjedizadeh, M.; Liu, D.; et al. Safety, Tolerability, and Pharmacokinetics of PTG-200, an Oral GI-Restricted Peptide Antagonist of IL-23 Receptor, in Normal Healthy Volunteers. *Am. J. Gastroenterol.* **2019**, *114*, S439–S440. [CrossRef]
38. van de Vosse, E.; Haverkamp, M.H.; Ramirez-Alejo, N.; Martinez-Gallo, M.; Blancas-Galicia, L.; Metin, A.; Garty, B.Z.; Sun-Tan, C.; Broides, A.; de Paus, R.A.; et al. IL-12Rβ1 deficiency: Mutation update and description of the IL12RB1 variation database. *Hum. Mutat.* **2013**, *34*, 1329–1339. [CrossRef]
39. Danese, S.; Peyrin-Biroulet, L. Selective tyrosine kinase 2 inhibition for treatment of inflammatory bowel disease: New hope on the rise. *Inflamm. Bowel Dis.* **2021**, *27*, 2023–2030. [CrossRef]
40. Villanueva, M.T. TYK2 inhibition shows promise. *Nature* **2019**, *18*, 668–669. [CrossRef]
41. Tao, J.H.; Zou, Y.F.; Feng, X.L.; Li, J.; Wang, F.; Pan, F.M.; Ye, D.Q. Meta-analysis of TYK2 gene polymorphisms association with susceptibility to autoimmune and inflammatory diseases. *Mol. Biol. Rep.* **2010**, *38*, 4663–4672. [CrossRef] [PubMed]
42. Can, G.; Tezel, A.; Gürkan, H.; Can, H.; Yilmaz, B.; Unsal, G.; Soylu, A.R.; Ummit, H.C. Tyrosine kinase-2 gene polymorphisms are associated with ulcerative colitis and Crohn's disease in Turkish Population. *Clin. Res. Hepatol. Gastroenterol.* **2015**, *39*, 489–498. [CrossRef] [PubMed]
43. Li, Z.; Rotival, M.; Patin, E.; Michel, F.; Pellegrini, S. Two common disease-associated TYK2 variants impact exon splicing and TYK2 dosage. *PLoS ONE* **2020**, *15*, e0225289. [CrossRef] [PubMed]
44. West, N.R.; Hegazy, A.N.; Owens, B.M.J.; Bullers, S.J.; Linggi, B.; Buonocore, S.; Coccia, M.; Gortz, D.; This, S.; Stockenhuber, K.; et al. Oncostatin M drives intestinal inflammation and predicts response to tumor necrosis factor–neutralizing therapy in patients with inflammatory bowel disease. *Nat. Med.* **2017**, *23*, 579–589. [CrossRef] [PubMed]
45. Hong, I.K.; Eun, Y.G.; Chung, D.H.; Kwon, K.H.; Kim, D.Y. Association of the oncostatin M receptor gene polymorphisms with papillary thyroid cancer in the Korean population. *Clin. Exp. Otorhinolaryngol.* **2011**, *4*, 193–198. [CrossRef]
46. Deng, S.; He, S.Y.; Zhao, P.; Zhang, P. The role of oncostatin M receptor gene polymorphisms in bladder cancer. *World J. Surg. Oncol* **2019**, *17*, 1–9. [CrossRef]
47. Gaspar, P.; Moura, G.; Santos, M.A.S.; Oliveira, J.L. mRNA secondary structure optimization using a correlated stem-loop prediction. *Nucleic Acids Res.* **2012**, *41*, e73. [CrossRef]
48. Chen, M.H.; Fang, C.; Wu, N.Y.; Xia, Y.H.; Zeng, Y.J.; Ouyang, W. Genetic variation of rs12918566 affects GRIN2A expression and is associated with spontaneous movement response during sevoflurane anesthesia induction. *Brain Behav.* **2021**, *11*, 1–9. [CrossRef]
49. Zwiers, A.; Kraal, L.; van de Pouw Kraan, T.C.; Wurdinger, T.; Bouma, G.; Kraal, G. Cutting edge: A variant of the IL-23R gene associated with inflammatory bowel disease induces loss of microRNA regulation and enhanced protein production. *J. Immunol.* **2012**, *188*, 1573–1577. [CrossRef]
50. Zheng, J.; Jiang, L.; Zhang, L.; Yang, L.; Deng, J.; You, Y.; Li, N.; Wu, H.; Li, W.; Lu, J.; et al. Functional genetic variations in the IL-23 receptor gene are associated with risk of breast, lung and nasopharyngeal cancer in Chinese populations. *Carcinogenesis* **2012**, *33*, 2409–2416. [CrossRef]
51. Xie, Y.H.; Chen, Y.X.; Fang, J.Y. Comprehensive review of targeted therapy for colorectal cancer. *Signal. Transduct. Target* **2020**, *5*, 22. [CrossRef] [PubMed]
52. Quiniou, C.; Domínguez-Punaro, M.; Cloutier, F.; Erfani, A.; Ennaciri, J.; Sivanesan, D.; Sanchez, M.; Chognard, G.; Hou, X.; Rivera, J.C.; et al. Specific targeting of the IL-23 receptor, using a novel small peptide noncompetitive antagonist, decreases the inflammatory response. *Am. J. Physiol. Regul. Integr. Comp. Physiol.* **2014**, *307*, R1216–R1230. [CrossRef] [PubMed]
53. Wang, K.; Li, M.; Hakonarson, H. ANNOVAR: Functional annotation of genetic variants from high-throughput sequencing data. *Nucleic Acids Res.* **2010**, *38*, e164. [CrossRef]

54. Ng, P.C.; Henikoff, S. SIFT: Predicting amino acid changes that affect protein function. *Nucleic Acids Res.* **2003**, *31*, 3812–3814. [CrossRef] [PubMed]
55. Adzhubei, I.A.; Jordan, D.M.; Sunyaev, S.R. Predicting functional effect of human missense mutations using polyphen-2. In *Current Protocols in Human Genetics*; John Wiley & Sons: Hoboken, NJ, USA, 2013. [CrossRef]
56. Reva, B.; Antipin, Y.; Sander, C. Predicting the functional impact of protein mutations: Application to cancer genomics. *Nucleic Acids Res.* **2011**, *39*, 37–43. [CrossRef]
57. Auton, A.; Brooks, L.D.; Durbin, R.M.; Garrison, E.P.; Kang, H.M.; Korbel, J.O.; Marchini, J.L.; McCarthy, S.; McVean, G.A.; Abecasis, G.R.; et al. A global reference for human genetic variation. *Nature* **2015**, *526*, 68–74. [CrossRef]
58. Sherry, S.T.; Ward, M.; Sirotkin, K. dbSNP—Database for single nucleotide polymorphisms and other classes of minor genetic variation. *Genome Res.* **1999**, *9*, 677–679. [CrossRef]
59. Bamford, S.; Dawson, E.; Forbes, S.; Clements, J.; Pettett, R.; Dogan, A.; Flanagan, A.; Teague, J.; Futreal, P.A.; Stratton, M.R.; et al. The COSMIC (Catalogue of Somatic Mutations in Cancer) database and website. *Br. J. Cancer* **2004**, *91*, 355–358. [CrossRef]
60. McKusick, V.A. *Mendelian Inheritance in Man: A catalog of Human Genes and Genetic Disorders*; JHU Press: Baltimor, MA, USA, 1998.
61. Buniello, A.; MacArthur, J.A.L.; Cerezo, M.; Harris, L.W.; Hayhurst, J.; Malangone, C.; McMahon, A.; Morales, J.; Mountjoy, E.; Sollis, E.; et al. The NHGRI-EBI GWAS Catalog of published genome-wide association studies, targeted arrays and summary statistics 2019. *Nucleic Acids Res.* **2019**, *47*, D1005–D1012. [CrossRef]
62. Stenson, P.D.; Ball, E.V.; Mort, M.; Phillips, A.D.; Shiel, J.A.; Thomas, N.S.; Abeysinghe, S.; Krawczak, M.; Cooper, D.N. Human Gene Mutation Database (HGMD): 2003 update. *Hum. Mutat.* **2003**, *21*, 577–581. [CrossRef]
63. Gene Ontology Consortium. The Gene Ontology resource: Enriching a GOld mine. *Nucleic Acids Res.* **2021**, *49*, D325–D334. [CrossRef] [PubMed]
64. Kanehisa, M.; Goto, S. KEGG: Kyoto Encyclopedia of Genes and Genomes. *Nucleic Acids Res.* **2000**, *28*, 27–30. [CrossRef] [PubMed]
65. Wu, G.; Haw, R. Functional Interaction Network Construction and Analysis for Disease Discovery. *Methods Mol. Biol.* **2017**, *1558*, 235–253. [CrossRef] [PubMed]
66. Untergasser, A.; Nijveen, H.; Rao, X.; Bisseling, T.; Geurts, R.; Leunissen, J.A.M. Primer3Plus, an enhanced web interface to Primer3. *Nucleic Acids Res.* **2007**, *35*, 71–74. [CrossRef]

International Journal of *Molecular Sciences*

MDPI

Article

S-Adenosylmethionine Inhibits Colorectal Cancer Cell Migration through Mirna-Mediated Targeting of Notch Signaling Pathway †

Luigi Borzacchiello ‡, Roberta Veglia Tranchese ‡, Roberta Grillo, Roberta Arpino, Laura Mosca *, Giovanna Cacciapuoti § and Marina Porcelli §

Department of Precision Medicine, University of Campania "Luigi Vanvitelli", Via Luigi De Crecchio 7, 80138 Naples, Italy; luigi.borzacchiello@unicampania.it (L.B.); roberta.vegliatranchese@unicampania.it (R.V.T.); roberta.grillo@unicampania.it (R.G.); roberta.arpino@unicampania.it (R.A.); giovanna.cacciapuoti@unicampania.it (G.C.); marina.porcelli@unicampania.it (M.P.)

* Correspondence: laura.mosca@unicampania.it

† This article is dedicated to the memory of Professor Vincenzo Zappia, who passed away six months ago.
‡ These authors contributed equally to this work.
§ These authors contributed equally to this study as co-last authors.

Abstract: Metastasis is a leading cause of mortality and poor prognosis in colorectal cancer (CRC). Thus, the identification of new compounds targeting cell migration represents a major clinical challenge. Recent findings evidenced a central role for dysregulated Notch in CRC and a correlation between Notch overexpression and tumor metastasis. MicroRNAs (miRNAs) have been reported to cross-talk with Notch for its regulation. Therefore, restoring underexpressed miRNAs targeting Notch could represent an encouraging therapeutic approach against CRC. In this context, S-adenosyl-L-methionine (AdoMet), the universal biological methyl donor, being able to modulate the expression of oncogenic miRNAs could act as a potential antimetastatic agent. Here, we showed that AdoMet upregulated the onco-suppressor miRNAs-34a/-34c/-449a and inhibited HCT-116 and Caco-2 CRC cell migration. This effect was associated with reduced expression of migration-/EMT-related protein markers. We also found that, in colorectal and triple-negative breast cancer cells, AdoMet inhibited the expression of Notch gene, which, by luciferase assay, resulted the direct target of miRNAs-34a/-34c/-449a. Gain- and loss-of-function experiments with miRNAs mimics and inhibitors demonstrated that AdoMet exerted its inhibitory effects by upregulating miRNAs-34a/-34c/-449a. Overall, these data highlighted AdoMet as a novel Notch inhibitor and suggested that the antimetastatic effects of AdoMet involve the miRNA-mediated targeting of Notch signaling pathway.

Keywords: S-adenosylmethionine; colorectal cancer; breast cancer; Notch; miRNA; metastasis; EMT

Citation: Borzacchiello, L.; Veglia Tranchese, R.; Grillo, R.; Arpino, R.; Mosca, L.; Cacciapuoti, G.; Porcelli, M. S-Adenosylmethionine Inhibits Colorectal Cancer Cell Migration through Mirna-Mediated Targeting of Notch Signaling Pathway. *Int. J. Mol. Sci.* **2022**, *23*, 7673. https://doi.org/10.3390/ijms23147673

Academic Editors: Alessandro Ottaiano and Donatella Delle Cave

Received: 24 June 2022
Accepted: 11 July 2022
Published: 12 July 2022

1. Introduction

Colorectal cancer (CRC) is the third most diagnosed cancer and the second leading cause of cancer associated mortality worldwide according to a Globocan 2020 survey [1]. Despite advances in early diagnosis and treatments including chemotherapy, immunotherapy, antiangiogenics, and surgical treatment, many of CRC patients still undergo high risks of tumor recurrence and metastasis. Several studies indicated that CRC aggressiveness and potential for metastatic spread are associated with the activation of epithelial-mesenchymal transition (EMT), a process playing a crucial role in driving carcinoma invasion and metastasis [2]. Thus, identifying chemical compounds targeting cancer cell migration is highly advantageous.

CRC is the result of dysregulated cellular pathways that promote inappropriate stem-cell-like phenotype, apoptotic resistance, uncontrolled proliferation, and metastatic spread [3]. Recent studies indicated that Notch signaling activation is responsible for the induction of aggressive phenotypic and functional changes in tumor cells consistent with mesenchymal transformation [4]. Notch signaling is an evolutionarily conserved pathway in multicellular organisms that, through cell-to-cell contacts, influences cell-fate decisions during embryonic and postnatal development and plays a critical role in maintaining the balance between cell proliferation, differentiation, and apoptosis and is also involved in metastasis, angiogenesis, and self-renewal [5]. Dysregulated Notch signaling has been found in a variety of neoplastic diseases, where it plays a complex oncogenic or tumor-suppressive role depending on tissue and cellular context [6]. In addition, the overexpression of Notch signaling has been found to be associated with poor prognosis or poor response to treatment of some solid tumors [5,7]. Therefore, therapeutic strategies have first been developed in the preclinical phase and in early-phase clinical trials to target oncogenic functions of Notch in tumor cells [8]. Recent findings highlighted a central role for abnormal Notch in CRC and indicated a correlation between the overexpression of Notch signaling components and CRC progression and metastasis [9].

Accumulating evidence suggests that microRNAs (miRNAs) play a crucial role in the regulation of genes driving CRC initiation, progression, and metastasis [10]. MiRNAs significantly affect numerous cellular processes, including cell development and differentiation, DNA damage repair, cell death, and intercellular communication. To date, more than 2500 miRNAs have been identified in humans, and nearly a third of human genes closely associated with many physiological processes are regulated by miRNAs [11]. Notably, miRNAs do not need perfect complementarity for target recognition, and therefore, a single miRNA can regulate up to one hundred target genes, showing pleiotropic effects and providing opportunities in the field of cancer therapy [12–15]. MiRNAs are crucially involved in cancer, acting as oncogenes or tumor suppressors depending on the regulatory effects exerted on the expression of their target genes [16]. Generally, oncogenic miRNAs are overexpressed, while tumor suppressive miRNAs are downregulated or completely lost in tumorigenesis, resulting in enhanced tumor progression, invasion, and metastasis [17]. Several lines of evidence suggest that miRNAs play a crucial role in the regulation of genes driving CRC initiation, progression, and metastasis [10]. Moreover, miRNAs have recently been reported to cross-talk with Notch pathway for its regulation [18]. Therefore, based on the remarkable role exerted by Notch in the promotion of CRC metastasis [4], restoring underexpressed miRNAs that target this signaling pathway could represent a promising therapeutic approach against CRC progression.

The new structures, the potential pharmacological activities, and the few harmful side effects on normal cells make natural compounds and their structural analogues effective tools that have been widely used in different clinical settings. Recent findings have documented that natural compounds exert anti-carcinogenic activities by interfering with the initiation, development, and progression of cancer through regulating epigenetic modifications and affecting various signaling pathways [19]. Beyond targeting protein functions, more and more evidence has demonstrated that natural agents exert antitumor activities by altering miRNA expression, providing a new approach to develop innovative and more efficient anticancer strategies based on synergistic combinatorial therapies [20]. In this context, S-adenosyl-L-methionine (AdoMet), a multitargeted and safe FDA-approved natural compound and the universal biological methyl donor in transmethylation reactions, has emerged, over the past two decades, as a promising anticancer therapeutic agent [21,22]. Recently, the antiproliferative properties of AdoMet and its implication in multiple cellular processes including proliferation, differentiation, cell cycle regulation, and apoptosis in various tumor cells have been thoroughly examined in the literature [21–31], and several findings have highlighted the therapeutical potential of AdoMet as an effective adjuvant to chemotherapeutic agents to be used in combined therapy to overcome drug resistance [28–31]. More and more evidence has also shown that the epigenetic modulation of miRNAs involved in oncogenic functions represents one of the main mechanisms underlying the anticancer activity of AdoMet. The regulation of miRNA's expression profile by AdoMet has been recently evaluated in breast and in head and neck cancer cells, suggesting that the ability of this natural compound to inhibit proliferation and cell migration, as well as to induce apoptosis in these tumor cells, is mediated by miRNAs [32–35].

Growing evidence accumulating in the literature in recent years on the anticancer activity exerted by AdoMet in colon cancer cells highlighted the pleiotropic effects of this eclectic multi-target sulfonium compound, evidencing its ability to overcome 5-FU chemoresistance by targeting multiple pathways such as autophagy, P-gp expression, and NF-kB signaling activation and to influence tumor progression by modulating gene expression [36–38].

Here, we demonstrated that AdoMet suppressed CRC progression through the inhibition of EMT and migration of HCT-116 and Caco-2 cells. We found that AdoMet upregulated key tumor-suppressive miRNAs as miRNA34a, miRNA34c, and miRNA449a in these tumor cells and that combined treatment AdoMet/miRNA inhibitors partially reversed the antimigratory effect of AdoMet. Finally, we provided novel evidence that AdoMet-induced inhibition of CRC cell migration involves miRNA-mediated targeting of Notch signaling pathway and that the same mechanism is utilized by AdoMet to inhibit cell migration in MDA-MB-231 and MDA-MB-468 triple negative breast cancer (TNBC) cell lines. The findings highlighted AdoMet as a new Notch inhibitor and a promising candidate for the treatment of Notch-dependent highly invasive cancers such as CRC and TNBC.

2. Results

2.1. AdoMet Upregulated miR-34a, miR-34c, and miR-449a Expression in HCT-116 and Caco-2 CRC Cell Lines

MiR-34a, miR-34c, and miR-449a, belonging to the miRNA-34/449 superfamily, are downregulated in many types of human cancers, including CRC, and play critical roles in tumor development and progression [39–41]. The involvement of miRNA-34/449 superfamily in the regulation of oncogenic pathways such as cell proliferation, metastasis, and apoptosis, proposes their potential role as tumor suppressors [41–43]. Recently, AdoMet-induced modulation of miR-34c and miR-449a expression has been evaluated in MDA-MB-231 and MDA-MB-468 breast cancer cell lines [33], providing evidence that the inhibition exerted by AdoMet on TNBC cell migration is mediated by AdoMet-induced upregulation of these tumor suppressor miRNAs.

To gain new information into the molecular mechanisms underlying AdoMet's antitumor activity in CRC cells and to confirm the ability of AdoMet to act as epigenetic regulator of miRNAs, the expression profile of miR-34a, miR-34c, and miR-449a was analyzed in HCT-116 and Caco-2 cells by quantitative real-time PCR (qRT-PCR) analysis with pre-designed probe-primer sets, after cell treatment with 500 μM AdoMet (Figure 1). The results achieved showed that after 48 h, the relative expression of the three miRNAs appeared remarkably upregulated by AdoMet in both CRC cell lines when compared to untreated cells with fold-change values of 3.9-fold and 1.9-fold for miR-34a, 2.0-fold and 5.2-fold for miR-34c, and 1.9-fold and 4.2-fold for miR-449a, in HCT-116 (Figure 1A) and Caco-2 (Figure 1B) cells, respectively. The ability to reprogram the expression of miRNA 34/449 superfamily in HCT-116 and Caco-2 cells, in good agreement with the results obtained in MDA-MB-231 and MDA-MB-468 cells [33], is indicative of a generalized miRNA regulation mechanism exerted by AdoMet in highly invasive cancers such as CRC and TNBC.

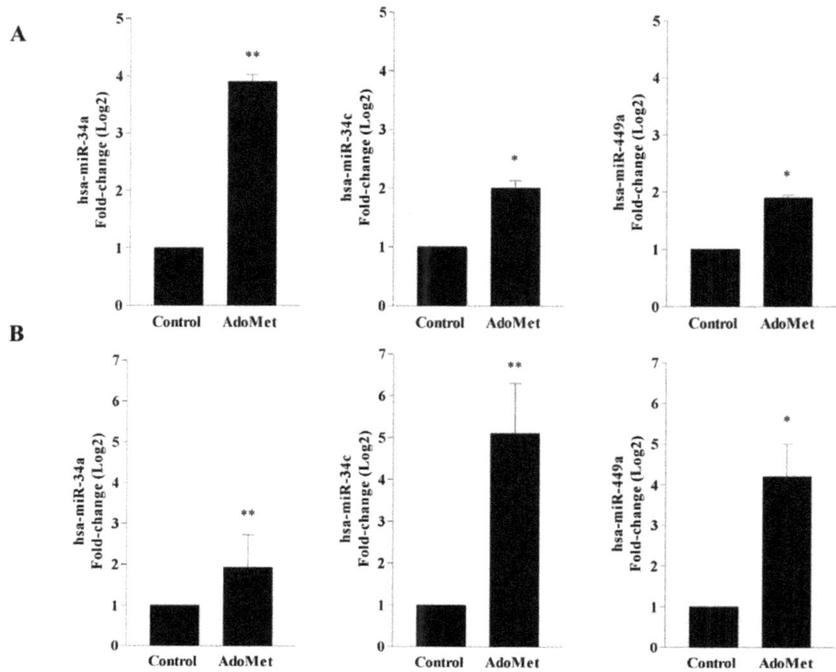

Figure 1. Effect of AdoMet on miR-34a, miR-34c, and miR-449a expression in CRC cells. The relative expression of miR-34a, miR-34c, and miR-449a in HCT-116 (**A**) and Caco-2 (**B**) cells treated with 500 µM AdoMet for 48 h was analyzed by qRT-PCR, following normalization with U6 endogenous control. The analysis was carried out by triplicate determination of at least 3 separate experiments. The results are expressed as fold-change (Log_2) ± SD. The means were compared using Student's *t* test, * $p < 0.05$, ** $p < 0.01$.

2.2. *miR-34a, miR-34c, and miR-449a Mediated the AdoMet-Induced Inhibition of HCT-116 and Caco-2 Cell Migration*

In order to evaluate the involvement of AdoMet-induced miRNA-34/449 superfamily up-regulation in the migration process of HCT-116 and Caco-2 cells, we investigated the effect of AdoMet, alone or in combination with miR-34a, miR-34c, and miR-449a mimics or inhibitors by wound healing assay monitored for 24 h (Figure 2). In both cell lines, we evidenced that compared to untreated cells, the treatment with 500 µM AdoMet for 48 h inhibited CRC cell motility causing approximately 21.1% and 42.0% wound closure in HCT-116 (Figure 2A) and Caco-2 (Figure 2B) cells, respectively. The inhibitory effect of AdoMet was significantly increased following the combined treatments with the sulfonium compound and miR-34a, miR-34c, and miR-449a, leading to wound closure values of about 12.1% and 27.3% for miR-34a, 5.6% and 28.5% for miR-34c and 11.5% and 13.5% for miR-449a in HCT-116 (Figure 2A) and Caco-2 (Figure 2B) cells, respectively, strongly indicating the role of miR-34a, miR-34c, and miR-449a as mediators of the process. Confirming evidence of these results came from loss-of-function experiments showing that treatment with AdoMet in combination with miR-34a, miR-34c, and miR-449a inhibitors reverted the miRNAs- and AdoMet-induced cellular effects and restored the migratory capacity of either HCT-116 (Figure 2A) and Caco-2 (Figure 2B) cells, as revealed by the wound size. Notably, the inhibitory effect exerted by miR-34a, miR-34c, and miR-449a on CRC cell migration is in line with other literature reports on the antimetastatic effects exerted by these miRNAs in various tumors [43–45]. Overall, these data indicated that epigenetic regulation of miRNA-34/449 superfamily by AdoMet played a crucial role in regulating HCT-116 and Caco-2 cell motility and represented the likely mechanism underlying AdoMet-induced inhibition of CRC cell migration.

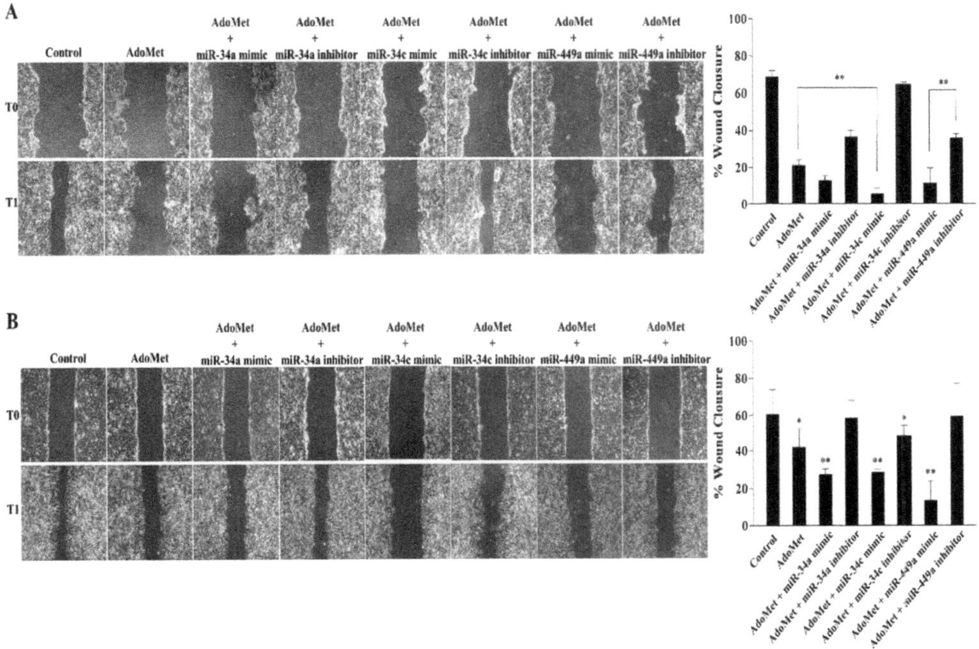

Figure 2. Effect of AdoMet alone and in combination with miR-34a, miR-34c, and miR-449a mimics or inhibitors on CRC cell migration. Confluent monolayers of HCT-116 (**A**) and Caco-2 (**B**) cells treated or not (Control) with 500 μM AdoMet alone or in combination with miR-34a, miR-34c, and miR-449a mimics or inhibitors for 48 h, were scratched with a micropipette tip and snapshot pictures were captured by microscope to check for wound closure. Pictures of the wounds corresponding to time zero (T0) and 24 h (T1) from the scrape in both cell lines are reported. Histograms show the quantification of wound area calculated as a percentage of the control using ImageJ software 1.48v (U.S. National Institutes of Health, Bethesda, MD, USA). Data represent the average of three independent experiments. The means and SD are shown. The means were compared using analysis of variance (ANOVA) plus Bonferroni's *t*-test. * $p < 0.05$, ** $p < 0.01$ versus untreated cells (Control).

2.3. *miR-34a, miR-34c, and miR-449a Mediated AdoMet-Induced Inhibition of Migration- and EMT-Related Protein Expression in HCT-116 and Caco-2 Cells*

To confirm the results of the wound healing assay and to investigate whether the epigenetic regulation of miRNAs by AdoMet was involved in the inhibition of cell migration-associated EMT process, we analyzed by Western blot the effect of AdoMet and miR-34a, miR-34c, and miR-449a, alone and in combination, on the expression levels of the main markers characterizing cell migration and EMT in HCT-116 and Caco-2 cells. First, we examined metalloproteinase-2 (MMP-2) and metalloproteinase-9 (MMP-9), key proteolytic enzymes involved in the degradation of the basement membrane and extracellular matrix [46] whose overexpression has been recently correlated with poor survival outcome in CRC patients [47]. We found that AdoMet and miR-34a, miR-34c, and miR-449a individually reduced the expression of MMP-2 and MMP-9 in HCT-116 (Figure 3A) and Caco-2 (Figure 3B) cells as compared to control and that their combined treatment was more effective than single treatment indicating a functional relationship between AdoMet and these miRNAs and suggesting that the antimigratory properties of AdoMet in CRC cells are mediated by the upregulation of tumor suppressors miR-34a, miR-34c, and miR-449a.

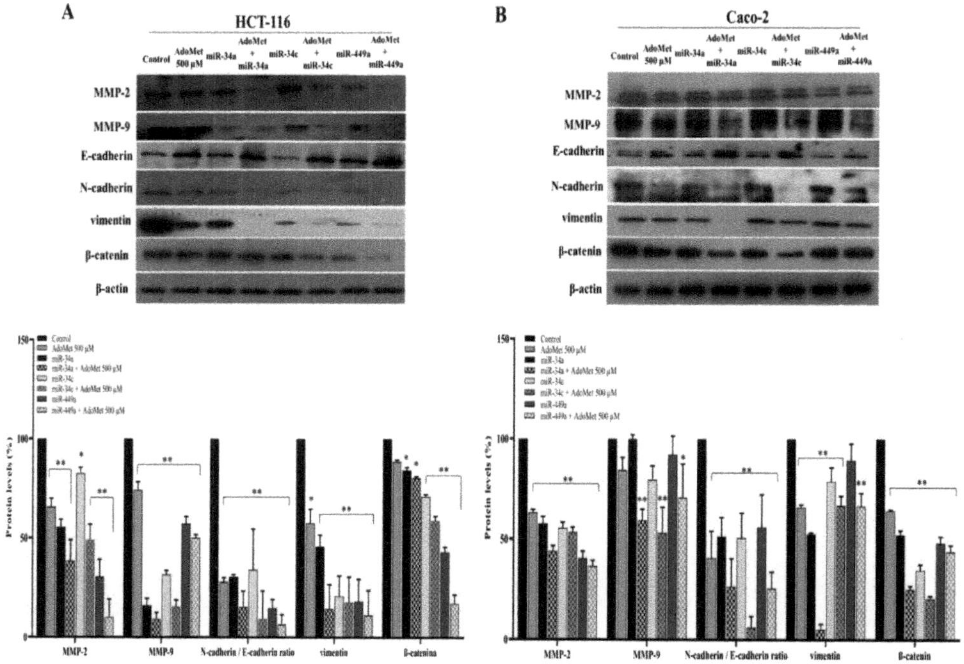

Figure 3. Effect of AdoMet alone and in combination with miR-34a, miR-34c, and miR-449a mimics on the levels of migration- and EMT-related proteins in CRC cells. Cells were transfected with 100 nM miR-34a, miR-34c, and miR-449a in the presence or not (control) of 500 µM AdoMet for 48 h. The expression levels of MMP2, MMP9, E-cadherin, N-cadherin, vimentin, and β-catenin, were detected by Western blot using the total cell lysate of HCT-116 (**A**) and Caco-2 (**B**) cells. The densitometric analysis was reported as the percentage of protein expression of untreated control (100%). For the equal loading of proteins in the lanes, β-actin was used as a standard. The images are representative of three immunoblotting analyses obtained from at least three independent experiments. The means were compared using analysis of variance (ANOVA) plus Bonferroni's t-test. * $p < 0.05$, ** $p < 0.01$ versus untreated cells (Control). Uncropped images of Western blots are reported in Figures S1 and S2.

In invasive tumors, the upregulation of the mesenchymal marker N-cadherin and downregulation of the epithelial marker E-cadherin represent the hallmark of migratory and invasive traits during EMT [48]. We therefore analyzed the expression patterns of EMT-associated proteins, including E-cadherin, N-cadherin, and vimentin. We found that, after single treatment of HCT-116 (Figure 3A) and Caco-2 (Figure 3B) cells with AdoMet or with miR-34a, miR-34c, and miR-449a, the intensity of E-cadherin protein band compared to untreated cells became more evident in treated cells, while N-cadherin behaved in the opposite way, resulting in an N- to E-cadherin switch [48] indicative of AdoMet-induced inhibition of EMT. The enhanced effect observed in both CRC cell lines following combined AdoMet/miRNAs treatment clearly indicated that AdoMet-dependent upregulation of miR-34a, miR-34c, and miR-449a is involved in the AdoMet-induced transition from N- to E-cadherin expression, in turn, responsible for the suppression of cell migration. We then evaluated vimentin, the major cytoskeletal component of motile mesenchymal cells including metastatic tumor cells of epithelial origin [49]. Additionally, in this case, a strong decrease in the levels of this protein was observable following either single and, more markedly, combined treatment with AdoMet and miR-34a, miR-34c, and miR-449a, furnishing evidence that the modulation of vimentin expression by AdoMet is dependent on these miRNAs.

Finally, we examined β-catenin, the most important mediator of the canonical Wnt signaling pathway, whose dysregulation has been recently demonstrated to play a primary role in driving colon cancer cell growth, invasion, and survival [50]. The results showed, in comparison with the control, a downregulation of β-catenin in HCT-116 (Figure 3A) and Caco-2 (Figure 3B) cells treated with AdoMet or with miR-34a, miR-34c, and miR-449a, which became more evident following a combined AdoMet/miRNA treatment, thus providing evidence that AdoMet negatively regulates β-catenin signaling through modulating miR-34a, miR-34c, and miR-449a expression.

In conclusion, our findings indicated that the combined treatment of AdoMet and miR-34a, miR-34c, and miR-449a was more effective compared to the single agents in inhibiting the main cell migration- and EMT-related protein markers, further confirming the crucial role of these tumor suppressive miRNAs as mediators of the antimigratory and antimetastatic properties of AdoMet in CRC cells.

2.4. Notch Was Directly Targeted by miR-34a, miR 34c, and miR 449a in Colorectal HCT-116 and Caco-2 and in Triple Negative MDA-MB-231 and MDA-MB-468 Breast Cancer Cells

To unravel the molecular mechanisms of how AdoMet suppresses cell migration and EMT in HCT-116 and Caco-2 CRC cells, the downstream mediators or effectors of AdoMet/miR-34a/34c/449a cascade were predicted and analyzed by two distinct computational algorithms, available online in silico tools, TargetScan 7.1 [51] and miRBase software [52]. Among the putative targets of the miR34/449 superfamily, we have identified and selected Notch gene for further studies on in vitro cell lines, since, in CRC, as well documented in the literature, Notch signaling is dysregulated, and Notch overexpression has been correlated with poor survival, cancer stem cell-like phenotype, EMT, and metastasis resulting in tumor progression [4–9,53]. As indicated in Figure 4A, showing the results of the miRNAs/Notch alignments, the 3′-UTR of Notch mRNA contains a perfectly complementary binding site for the seed region of miR-34a/c and miR-449a.

To validate the binding prediction and to find out whether miR-34a/c and miR-449a were able to establish a direct interaction with Notch mRNA, a dual-luciferase reporter assay was performed. The 3′-UTR of Notch gene containing the putative miRNAs binding sites was cloned downstream of luciferase reporter vector to generate the construct shown in Figure 4B. HCT-116 and Caco-2 cells were co-transfected with 3′-UTR reporter construct of Notch along with miR-34a/c or miR-449a mimics prior to cell harvest and luciferase assay. As expected, co-transfected miRNA mimics substantially reduced in both cell lines the luciferase activity of 3′-UTR reporter compared to the control cells (Figure 4C), indicating that the miRNAs of interest target Notch, leading to its downregulation. Notably, the transfection of cells with a luciferase reporter construct carrying mutation in the 3′-UTR binding region abrogated the repression of luciferase expression exerted by miRNAs, confirming the specificity of the interaction (data not shown). In addition, based on the finding of our previous study demonstrating that miR-34c and miR-449a mediated the AdoMet-induced inhibition of MDA-MB-231 and MDA-MB-468 cell migration [33], we decided to carry out the same experiments also in these TNBC cell lines. Figure 4D shows that the results obtained are indicative of a functional interaction between miRNAs and their target site at Notch-3′-UTR, confirming the results obtained in CRC cells and demonstrating that Notch represents the target of miR34/449 superfamily in highly invasive cancer cells such as CRC and TNBC cells.

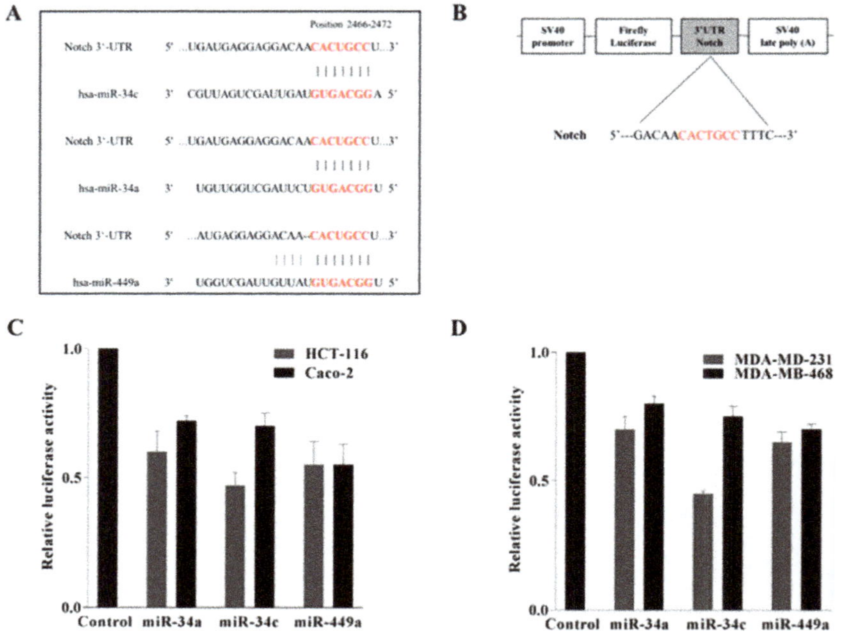

Figure 4. AdoMet-upregulated miR-34a, miR-34c, and miR-449a directly bind to 3′-UTR of Notch in CRC and TNBC cells. Alignments of miR-34a, miR-34c, and miR-449a with Notch 3′-UTR obtained from miRNA-mRNA integration analysis using the microRNA target prediction software TargetScan (**A**). Schematic diagram of luciferase reporter plasmid containing 3′-UTR target sequence of Notch for miRNAs (**B**). The relative luciferase activity was determined using co-transfected CRC HCT-116 and Caco-2 (**C**) and TNBC MDA-MB-231 and MDA-MB-468 (**D**) cells with miR-34a, miR-34c, and miR-449a mimics and the corresponding luciferase reporter plasmid. Each sample was run in triplicate. The means were compared using analysis of variance (ANOVA) plus Bonferroni's *t*-test. Error bars show mean ± SD.

2.5. AdoMet Inhibited Notch Expression in CRC and TNBC Cells through miR-34a, miR-34c, and miR-449a

To investigate the functional relevance of miRNAs/Notch-3′-UTR interaction in colorectal and in breast cancer cells and to explore whether AdoMet might be able to target Notch expression through miRNAs upregulation, CRC and TNBC cells were transfected with miR-34a, miR-34c, and miR-449a mimics in the presence and in absence of 500 µM AdoMet and Notch expression was then analyzed by RT-qPCR and Western blot. The results indicated that after treatment with AdoMet and miRNA mimics alone, a reduced amount of Notch mRNA (Figure 5A,B, left) and its encoded protein (Figure 6A,B) was detected in both cancer cell types, as compared to the control, demonstrating that AdoMet and miRNAs individually suppressed Notch expression at both translational and transcriptional levels and highlighting Notch as an important target of AdoMet and miR-34a/c/449a superfamily in these tumor cell lines. Interestingly, AdoMet/miRNA mimics co-treatment significantly potentiated the effect of AdoMet (Figure 5A,B left and Figure 6A,B), providing evidence that AdoMet negatively regulated Notch expression through miR-34a, miR-34c, and miR-449a. Notably, loss-of-function studies carried out in HCT-116 cells evidenced that transfection with miRNAs inhibitors was able to revert the inhibitory effect of AdoMet and to restore the levels of Notch mRNA (Figure 5A,B right), further confirming that miRNAs are important mediators of AdoMet-induced inhibition of Notch expression. Taken together, these data highlighted AdoMet as a novel Notch inhibitor and strongly suggested that the antimetastatic effects exerted by AdoMet in colorectal and breast cancer cells are probably mediated by miRNA 34a-34c-449a/Notch axis.

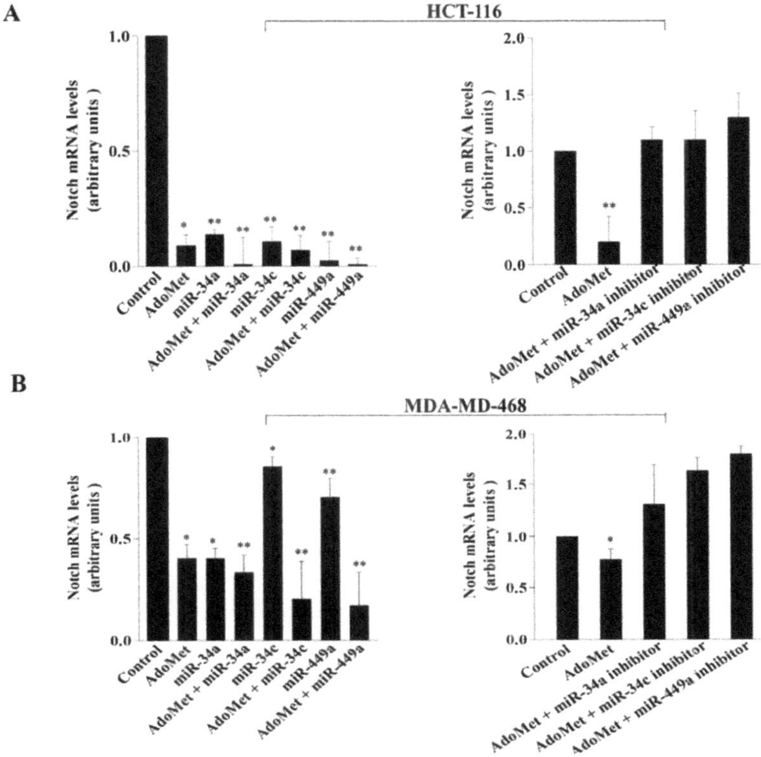

Figure 5. Effect of AdoMet and miR-34a, miR-34c, and miR-449a on Notch mRNA levels in CRC and TNBC cells. HCT-116 (**A**) and MDA-MB-468 (**B**) cells were transfected with miR-34a, miR-34c, and miR-449a mimics (**left**) and inhibitors (**right**) in the presence or not (Control) of 500 μM AdoMet for 48 h. Total-RNA of HCT-116 and MDA-MB-468 cells was extracted, and cDNA was synthesized by qRT-PCR to analyze the transcriptional level of the predicted target. The graphs show the fold-change of Notch in the different experimental conditions normalized to GAPDH mRNA and compared to untreated cells. Data represent the average of three independent experiments. The means were compared using analysis of variance (ANOVA) plus Bonferroni's t-test. Error bars show mean ± SD. * $p < 0.05$, ** $p < 0.01$ versus control untreated cells.

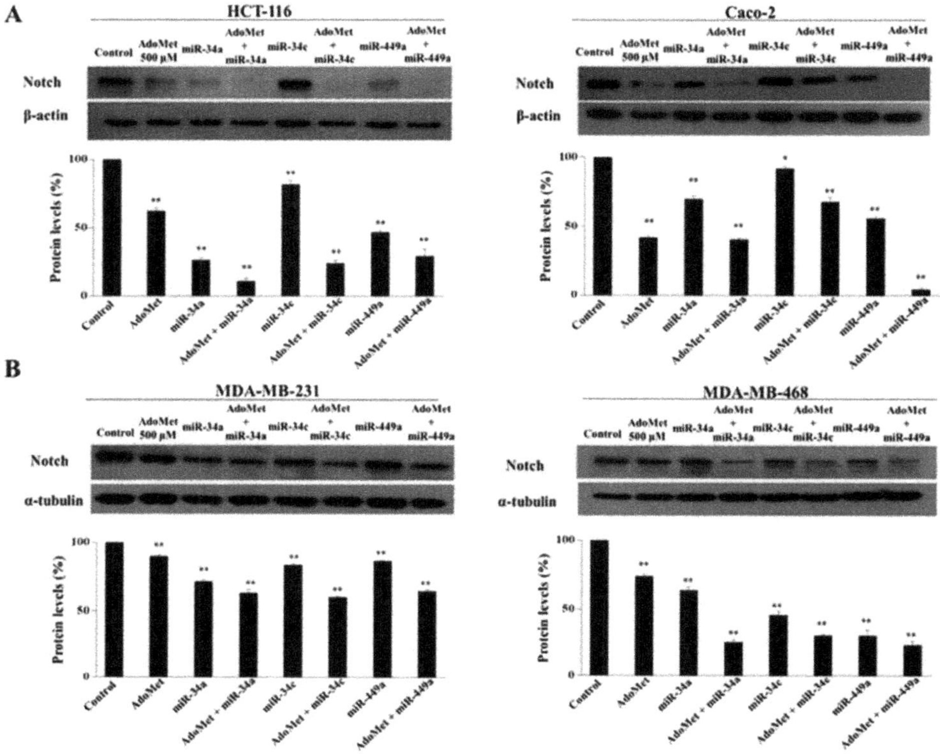

Figure 6. Effect of AdoMet and miR-34a, miR-34c, and miR-449a mimics on Notch protein levels in CRC and TNBC cells. CRC (**A**) and TNBC (**B**) cells were transfected with miR-34a, miR-34c, and miR-449a mimics, in the presence or not (Control), of 500 µM AdoMet for 48 h. Then, cell lysates were subjected to SDS-PAGE, incubated with antibodies against the indicated proteins and analyzed by Western blot. Housekeeping proteins, β-actin and α-tubulin, were used as a loading control. The densitometric analysis was reported as a percentage of protein expression of untreated control (100%). Data represent the average of three independent experiments. The means were compared using analysis of variance (ANOVA) plus Bonferroni's *t*-test. Error bars show mean ± SD. * $p < 0.05$, ** $p < 0.01$ versus control untreated cells. Uncropped images of Western blots are reported in Figure S3.

3. Discussion

The metastatic spread of cancer cells is a leading cause of mortality in CRC patients. Many studies have pointed out that tumor cell migration and invasion are prerequisites for subsequent metastasis to distant organs and that activation of EMT, a process in which epithelial cells lose their adhesions and gain a mesenchymal migratory and aggressive phenotype, representing the early and crucial step of cancer metastasis and playing a major role in CRC progression [2]. Therefore, it is important to elaborate therapeutic approaches able to target metastasis and its preliminary stages such as cell migration and EMT. Although the pathogenesis of CRC has not been fully elucidated, accumulating evidence has indicated that the dysregulation of miRNAs contributes to the metastatic process and influences CRC development. Thus, restoring the expression of underexpressed tumor suppressive miRNAs is considered a promising and advantageous strategy to control cancer metastasis and to improve the outcome in CRC patients [54].

AdoMet, identified in recent decades as a potent anticancer molecule, is one of the most studied epigenetic regulators. It plays a primary role in cellular metabolism and represents the major methyl donor required in numerous methylation reactions [55–57].

Notably, AdoMet has been found to regulate the expression of miRNAs and lncRNAs through epigenetic mechanisms [35]. Recently, the effect of AdoMet on miRNA expression profile has been evaluated in breast [32,33] and in head and neck cancer [34], providing evidence that the ability of AdoMet to inhibit cell proliferation and migration, as well as to induce apoptosis in these tumor cells, is mediated by miRNAs.

In the present study, we demonstrated that AdoMet exhibited antimetastatic activity in CRC cells and provided experimental evidence on the underlying mechanism. The novel finding of this work indicated that AdoMet suppressed EMT and the migration of HCT-116 and Caco-2 CRC cells by activating the expression of miR-34a/c/miR-449a, which directly targeted the Notch gene, leading to its post-transcriptional repression. Comparable results obtained by expanding the study to MDA-MB-231 and MDA-MB-468 cell lines allow us to propose this mechanism as a general strategy utilized by AdoMet to target EMT and cell migration in invasive cancers such as CRC and TNBC (Table S1).

In agreement with the findings previously obtained by our laboratory in the highly invasive breast cancer cells MDA-MB-231 and MDA-MB-468, we found that AdoMet was able to up-regulate the expression of Mir34a/c and miR-449a in HCT-116 and Caco-2 cells, evidencing the ability of the sulfonium compound to reprogram non-coding RNA expression in CRC cells.

The miR-34/449 family is conserved in mammals and includes six homologous genes, miR-34a, miR-34b, miR-34c, miR-449a, miR-449b, and miR-449c, located at three genomic loci. The miR-34/449 family is widely distributed in vertebrate organisms and is important for the regulation of virus–host interactions by modulating immune responses and virus replication [58]. In mammals, miR-34 includes three miRNAs with miR-34a having its own transcript located at chromosome 1p36.22, whereas miR-34b and miR-34c share a primary transcript located at chromosome 11q23.1 [43]. The seed sequence of miR-34a and miR-34c is identical, while miR-34b is slightly different. The mature miR-34a sequence shares 86% and 82% homology with miR-34b and miR-34c, respectively, suggesting that miR-34s could recognize similar mRNA targets and thus could be functionally redundant [59]. Notably, miR-34s show p53-responsive elements in their promoter regions and have been identified as one of the first p53-regulated miRNAs [60]. Converging evidence demonstrated that the overexpression of miR-34s triggers various p53-downstream effects in a context-dependent manner, highlighting these miRNAs as important mediators of tumor-suppressive activities of p53 [61]. MiR-34s play a crucial role in repressing tumor progression by involving in EMT via p53, EMT-transcription factors, and some important signaling pathways, including Notch and Wnt/β-catenin [59]. Consistently with their suppressive role in tumorigenesis, miR-34s are frequently expressed at reduced levels in a broad range of tumors due to aberrant transcriptional regulation, genomic deletions, and promoter hypermethylation [62]. Epigenetic silencing of miR-34s is associated with the proliferation, migration, and invasion of CRC [63]. In analogy with miR-34s, miR-449a is a member of a cluster (miR-449a, miR-449b, and miR-449c) characterized by sequences and secondary structures similar to those of miR-34 members. MiR-449a is considered a tumor-suppressive miRNA, and its downregulated expression has been associated with advanced clinical stage and the poor histological differentiation of CRC [40,64]. Recently, it has been reported that the upregulation of autophagy-dependent miR-449a suppresses CRC tumorigenesis both in vitro and in vivo [65]. Based on these findings, the ability to reactivate the expression of the tumor-suppressive miR-34a/c/miR-449a, thereby restoring their levels in CRC cells, makes AdoMet a potential therapeutic agent for colorectal cancer.

We found that AdoMet caused a decrease in the migratory rate of HCT-116 and Caco-2 cells indicative of its ability to lower the aggressiveness of these cancer cells and potentially reduce their metastatic power.

MMPs have a key function in extracellular matrix remodeling and degradation and play roles in all stages of carcinogenesis [46,47]. Among MMPs, MMP-2 and MMP-9 have drawn much attention for their implication in tumor invasion and metastasis. These enzymes have a unique ability to break down type IV collagen, which is a major component of the basement membrane. Elevated expression of MMP-2/9 has been associated with increased metastatic potential in many tumor cells, including CRC [46]. Moreover, many studies revealed a correlation between increased MMP-2/9 expression and worst outcome of CRC suggesting that the control of the expression and/or the activity of these proteolytic enzymes may offer a new therapeutic opportunity for treating this highly invasive cancer [66]. In line with this view, we found that the decrease in cell migration induced by AdoMet in HCT-116 and Caco-2 cells was associated with decreased protein levels of MMP-2 and MMP-9.

The switch from E-cadherin to N-cadherin expression is considered an important marker of ongoing EMT and has been shown to promote cancer cell motility and invasion [48]. Furthermore, tumor cells subjected to EMT significantly modify the composition of cytoskeletal intermediate filaments with the repression of keratin and expression of vimentin that is believed to be responsible for the adoption of a mesenchymal shape and increased motility and is regarded as the main and canonical marker of EMT [67]. Vimentin is highly expressed in CRC cells, where it plays a critical role in metastasis and prognosis [68]. Notably, our results showed that in HCT-116 and Caco-2 cells, AdoMet significantly decreased vimentin levels and promoted the switch from N- to-E-cadherin expression, as evidenced by comparing the intensities of the corresponding protein bands in AdoMet-treated and untreated cells highlighting the ability of AdoMet to slow down CRC cell migration by inhibiting the transition from the epithelial to the mesenchymal state.

Wnt/β-catenin signaling is one of the most frequently dysregulated pathways in CRC [69]. Aberrant activation of this signaling causes the accumulation of β-catenin in the nucleus and promotes the transcription of many Wnt target genes involved in the initiation, progression, metastasis, drug resistance, and immune evasion of cancer [70]. Notably, β-catenin has been recently recognized as a direct target of miR-34b in human colon cancer cells and it has been reported that miR-34b may inhibit migration and invasion of Caco-2 cells by regulating Wnt/β-catenin signaling [71]. Several inhibitors of Wnt/β-catenin signaling pathway have been developed for CRC treatment [71]. Interestingly, many of these inhibitors currently under investigation include natural compounds, drugs, small molecules, and biological agents [72]. In line with this evidence, we found that AdoMet caused a decrease in β-catenin level in HCT-116 and Caco-2 cells, allowing us to consider this physiological sulfonium compound a promising agent for the selective targeting of Wnt/β-catenin signaling pathway.

The activation of Notch signaling has been shown to play a key role in the induction of the aggressive and metastatic phenotype of CRC cells [73]. In recent literature, a complex network of mutual interconnections between miR-34s and Notch signaling has been reported and discussed indicating that this interplay is implicated in cancer initiation/progression, metastasis, and chemoresistance [73]. In ovarian cancer, the overexpression of miR-34 mimic induced cell death and autophagy through downregulating Notch1, whereas Notch1 transfection reverted anti-proliferative effects of miR-34 [74]. In CRC, it has been reported that miR-34 inversely correlated with metastasis and that its overexpression suppressed cell invasiveness and migration by targeting Notch1 and JAG1 [75]. Moreover, a natural compound, genistein, inhibited cell growth and induced apoptosis in pancreatic cancer cells through the upregulation of miR-34a, leading to decreased Notch1 [76]. Regarding miR-449a, it has been recently reported that its overexpression in vitro led to the silencing of genes associated with Notch signaling in prostate cancer cells [77]. Consistent with these reports, we identified Notch as a direct target of miR-34a/c/miR-449a and provided new evidence that AdoMet was able to inhibit Notch expression through upregulation of miR-34a/c/miR-449a.

Members of miR-34/miR-449 family sharing similar sequences allow to expect similar biological roles and target genes [78]. Notably, we found that: (i) the overexpression of miR-34a/c/miR-449a induced by AdoMet or through transfecting the cells with the corresponding miRNA mimics, caused comparable inhibitory effects on the migratory power of HCT-116 and Caco-2 cells and on the expression levels of relevant proteins associated with cell migration and EMT such as MMP-2, MMP-9, E-cadherin, N-cadherin, vimentin, β-catenin, and Notch; (ii) the ectopic expression of miR-34a/c/miR-449a in cells transfected with miRNA mimics significantly enhanced the inhibitory effects induced by AdoMet resulting in a gain-of-function; (iii) the loss-of-function studies performed by transfecting CRC cells with miRNA inhibitors reverted the AdoMet-induced inhibition on cell migration and on the levels of Notch mRNA and its encoded protein. Consistent with these findings, the key primary step of the antimetastatic activity of AdoMet in HCT-116 and Caco-2 cells is represented by the AdoMet-induced upregulation of miR-34a/c/miR-449a. In fact, transfection with miRNA inhibitors significantly diminished the observed inhibitory effects of AdoMet on EMT and cell migration. In turn, miR-34a/c/miR-449a act as crucial players downstream of AdoMet through the directly targeting of Notch signaling pathway.

4. Materials and Methods

4.1. Materials

Tissue culture dishes were purchased from Corning (Corning, NY, USA). Bovine serum albumin (BSA), fetal bovine serum (FBS), Dulbecco's modified Eagle's medium (DMEM), phosphate-buffered saline (PBS), and trypsin-EDTA were obtained from Gibco (Grand Island, NY, USA). AdoMet was provided from New England Biolabs, prepared in a solution of 5 mM H_2SO_4 and 10% ethanol, filtered, and stored at −20 °C until use. Opti-minimal essential medium (Opti-MEM), mirVANA PARIS Kit, Lipofectamine 2000, TaqMan-MiRNA Reverse Transcription Kit, High-capacity cDNA reverse transcription kit, TaqMan Universal PCR Master Mix, small-nuclear-U6, miRNA-34a, miRNA-34c and miRNA-449a mimics and inhibitors, SYBR™ Green PCR Master Mix and custom DNA Oligos: neurogenic locus notch homolog protein (Notch), glyceraldehyde 3-phosphate dehydrogenase (GAPDH) were obtained from Thermo Fisher Scientific (Waltham, MA, USA). The pEZX-MT06 plasmid containing 3′-UTR of Notch were obtained from GeneCopoeia (Rockville, MD, USA), and Dual-Luciferase Reporter Assay System was purchased from Promega (Madison, WI, USA). Radioimmunoprecipitation assay buffer (RIPA buffer) was purchased from Sigma-Aldrich (St. Louis, MO, USA). Monoclonal antibodies (mAbs) to β-actin (#3700, dilution: 1:5000), α-tubulin (#2125, dilution: 1:5000), MMP9 (#13667, dilution: 1:1000), vimentin (#5741, dilution: 1:1000), β-catenin (#8480, dilution: 1:1000), N-cadherin (#13116, dilution: 1:1000), E-cadherin (#14472, dilution: 1:1000), and Notch (#5732, dilution: 1:1000) and polyclonal antibodies (pAbs) to MMP2 (#4022, dilution: 1:1000) were purchased from Cell Signaling Technology (Danvers, MA, USA). Horseradish peroxidase (HRP)-conjugated goat anti-mouse (GxMu-003-DHRPX) and HRP-conjugated goat anti-rabbit (GtxRb-003-DHRPX) secondary antibodies were obtained from ImmunoReagents Inc. (Raleigh, NC, USA). All buffers and solutions were prepared with ultra-high-quality water. All reagents were of the purest commercial grade.

4.2. Cell Cultures and Treatments

CRC cell lines HCT-116 and Caco-2 and TNBC cell lines MDA-MB-468 and MDA-MB-231 provided from the American Type Culture Collection (ATCC, Manassas, VA, USA) were cultured at 37 °C in a 5% CO_2 humidified atmosphere and grown in DMEM supplemented with 10% FBS, 2 mM L-glutamine and 50 U/mL penicillin-streptomycin. Typically, subconfluent cells were seeded in 6-well plates at a density of 2.5×10^5 cells/well and 1.5×10^5 cells/well for CRC and TNBC cells, respectively, to achieve 80% confluence. After 24 h, the cells were treated with 10% FBS fresh medium containing 500 μM AdoMet for 48 h. Subsequently, fluctuating cells were recovered from culture medium by centrifugation, whereas adherent cells were collected by trypsinization.

4.3. Cell Transfections

CRC and TNBC cells, at 80% confluence, were transfected with 100 nM miR-34a, miR-34c and miR-449a mimics or inhibitors diluted in Opti-MEM medium added or not (Control) with 500 μM AdoMet, using Lipofectamine 2000 according to the manufacturer's protocol. Lipofectamine was also used alone as a negative control. After 48 h from transfection, the cells were harvested and then processed to carry out the appropriate analyses.

4.4. qRT-PCR for miRNAs Detection and mRNA Expression

Total RNA from cultured HCT-116 and Caco-2 cells and from MDA-MB-468 cells transfected with 100 nM miR-34a, miR-34c and miR-449a mimics, treated or not with AdoMet 500 μM for 48 h, was purified using the mirVANA PARIS kit, according to the manufacturer's protocol. RNA concentration was determined using a NanoDrop 1000 spectrophotometer (Thermo Fisher Scientific, Waltham, MA, USA). Subsequently, using the TaqMan MiRNA Reverse Transcription Kit or high-capacity cDNA reverse transcription kit, single-stranded cDNA was synthesized from total RNA samples.

The expression of individual miRNAs was determined using pre-designed probe-primer sets from Life Technologies and TaqMan Universal PCR Master Mix by quantitative real-time PCR (qRT-PCR) performed on a ViiA7™ Real-time PCR system (Applied Biosystems, Darmstadt, Germany), as previously reported [79]. To normalize total RNA samples, the small-nuclear-U6 was selected as endogenous control.

The expression of Notch transcript was determined independently by qRT-PCR, using SYBR Green PCR Master Mix. To normalize total RNA samples, GAPDH was selected as an appropriate constitutively expressed endogenous control. The qRT-PCR primer sequences are as follows: Notch, (Forward 5′-GGAGTCAGGGAGAGGTTCTAT-3′ and Reverse 5′-GGAGGTGTGACTAATTGGATGT-3′); GAPDH, (Forward 5′-GGAGTCAACGGATTTGGTCG-3′ and Reverse 5′-CTTCCCGTTCTCAGCCTTGA-3′). The relative expression of the transcripts was measured by using ViiA7™ Real-Time PCR software (Applied Biosystems, Darmstadt, Germany).

4.5. Protein Extraction and Western Blot Analysis

CRC and TNBC cell lines transfected with 100 nM miR-34a, miR-34c, and miR-449a mimics or inhibitors, treated or not with AdoMet 500 μM, after 48 h were collected by centrifugation, washed twice with ice-cold PBS, then lysed using 100 μL of RIPA buffer as previously reported [80]. Protein concentration was performed by Bradford method as previously reported [81]. Equal amounts of cell proteins were separated by sodium dodecyl sulfate-polyacrylamide gel electrophoresis (SDS-PAGE) and electrotransferred to nitrocellulose membranes by Trans-Blot turbo Transfer System (Bio-Rad Laboratories, Hercules, CA, USA). The membranes were washed in 10 mM Tris-HCl, pH 8.0, 150 mM NaCl, 0.05% Tween 20 (TBST), and blocked with TBST supplemented with 5% nonfat dry milk. Then, membranes were incubated first with specific primary antibodies at 4 °C overnight in TBST and 5% nonfat dry milk, washed, and then incubated 1 h with HRP-conjugated secondary antibodies. All primary antibodies were used at a dilution of 1:1000; all secondary antibodies were used at a dilution of 1:5000. Blots were then developed using enhanced chemiluminescence detection reagents ECL (Westar, Cyanagen, Bologna, Italy) and exposed to X-ray film. All films were scanned by using Image J software 1.48v (U.S. National Institutes of Health, Bethesda, MD, USA).

4.6. Migration Process Evaluated by Scratch-Wound Assay

HCT-116 and Caco-2 cells were seeded in a serum-containing media in 6-well culture plates in the appropriate number and cultured overnight until 100% confluence was reached. The cells were then transfected with 100 nM miR-34a, miR-34c and miR-449a mimics or inhibitors diluted in Opti-MEM free medium integrated or not (Control) with 500 μM AdoMet, by using Lipofectamine 2000 according to manufacturer's protocol. The wounds were obtained by manually scratching, in a sterile environment, a confluent cell monolayer with a sterile 200 μL pipette tip, rapidly washed twice with medium to remove cell debris and replaced with 2 mL of complete medium. Initial images of the wounds were captured using a microscope (Leica Microsystems GmbH) corresponding to time zero (T0). After 24 h

(T1) of treatment, snapshot images were taken to examine the wound closure. Wound areas of control and treated cells were quantified using ImageJ software 1.48v (U.S. National Institutes of Health, Bethesda, MD, USA).

4.7. Luciferase Reporter Assay

CRC and TNBC cells were seeded into 24-well plates at 20,000 cells/well and 15,000 cells/well, respectively, and cultured overnight. Thereafter, cells were co-transfected with luciferase pEZX-MT06 plasmids containing wild-type or mutated (5′-GACAATGTCATTTTTC-3′) 3′-UTR of Notch and miR-34a, miR-34c and miR-449a mimics using Lipofectamine 2000, following the provided manual. After 24 h incubation, cell medium was exchanged with fresh DMEM containing 10% FBS. Dual-luciferase reporter assay was conducted using a Dual-Luciferase reporter assay system according to the manufacturer's instructions. Luciferase activity was detected under the control of Tecan Infinite M200 (Tecan, Männedorf, Switzerland). For the normalization of Firefly luciferase activity, the luminescence intensity of Renilla luciferase was used as an internal control of transfected cells.

4.8. Statistical Analysis

Experiments were performed at least three times with replicate samples. Data are expressed as mean ± standard deviation (SD). For the comparisons of mean values between two groups, unpaired Student's *t*-test was used. For the comparisons of mean values among three or more groups, the means were compared using analysis of variance (ANOVA) plus Bonferroni's *t*-test. A *p*-value < 0.05 was considered to indicate a statistically significant result, $p < 0.01$ was considered highly significant.

5. Conclusions

As a promising tumor suppressor, miR-34s have attracted considerable attention in recent years for their ability to modulate a variety of oncogenic functions in different cancers, and miR-34 mimetics are currently tested for treatment of advanced cancers [82]. However, delivery systems for miR-34 replacement therapy were not without toxicity and side-effects. Notably, the first phase I clinical trial of MRX34, a miR-34a mimic encapsulated in lipid nanoparticles, was stopped because of the occurrence of severe immune-related adverse events [59]. The selective targeting of Notch could also represent an ideal strategy for prevention and treatment of colon cancer owing to the crucial role played by Notch signaling in CRC progression and metastasis. However, despite promising preclinical results and early-phase clinical trials, Notch inhibitors such as anti-Notch receptors/ligands antibodies or γ-secretase inhibitors have failed in clinical translation due to their poor selectivity, dose-limiting toxicity, and low efficacy, encouraging to investigate dietary and natural compounds with multi-target activities to achieve better outcomes [83].

Our study for the first time highlighted the ability of the physiological methyl donor AdoMet to control EMT and the migratory power of CRC cells through the miR-34a/c/miR-449a/Notch axis, thus offering a promising approach for the treatment of this highly invasive cancer. Worthy of interest in this context is to mention that AdoMet is an approved dietary supplement which can therefore be used for therapeutic purposes without the common contraindications of chemotherapy. Moreover, as documented by several clinical studies, AdoMet, at pharmacological doses, shows a low incidence of side-effects with an excellent tolerability record and no toxic or antiproliferative effects in normal, non-tumorigenic cells [21]. The ability of AdoMet to modulate miRNA expression in different cancer cell types and the well-documented potential of miRNA mimics and miRNA inhibitors to restore tumor suppressor miRNAs or to downregulate oncogenic miRNAs, respectively, greatly stimulates the design of adjuvant therapeutic approaches based on combined AdoMet/miRNAs treatments, which could allow synergistic effects resulting in increased cellular chemosensitivity and in reduced drug toxicity.

Supplementary Materials: The following supporting information can be downloaded at: https://www.mdpi.com/article/10.3390/ijms23147673/s1.

Author Contributions: L.M., G.C. and M.P. conceived the idea, designed experiments, analyzed the data and wrote the manuscript. L.B. and R.V.T. contributed to the design of the study and to the data analysis and suggested improvement for organization and style in the revision of the manuscript; R.G. and R.A. analyzed the data, contributed to the design and accomplishment of the experiments based on Scratch-Wound assay and prepared the figures. All authors have read and agreed to the published version of the manuscript.

Funding: This research received no external funding.

Data Availability Statement: The data presented in this study are available on request.

Acknowledgments: The work was partially supported by Intradepartmental Projects, Department of Precision Medicine, Università della Campania "Luigi Vanvitelli".

Conflicts of Interest: The authors declare no conflict of interest.

References

1. Sung, H.; Ferlay, J.; Siegel, R.L.; Laversanne, M.; Soerjomataram, I.; Jemal, A.; Bray, F. Global cancer statistics 2020: GLOBOCAN estimates of incidence and mortality worldwide for 36 cancers in 185 countries. *CA A Cancer J. Clin.* **2021**, *71*, 209–249. [CrossRef] [PubMed]
2. Vu, T.; Datta, P.K. Regulation of EMT in colorectal cancer: A culprit in metastasis. *Cancers* **2017**, *9*, 171. [CrossRef] [PubMed]
3. Malki, A.; ElRuz, R.A.; Gupta, I.; Allouch, A.; Vranic, S.; Al Moustafa, A.E. Molecular mechanisms of colon cancer progression and metastasis: Recent insights and advancements. *Int. J. Mol. Sci.* **2020**, *22*, 130. [CrossRef] [PubMed]
4. Tyagi, A.; Sharma, A.K.; Damodaran, C. A review on Notch signaling and colorectal cancer. *Cells* **2020**, *9*, 1549. [CrossRef] [PubMed]
5. Anusewicz, D.; Orzechowska, M.; Bednarek, A.K. Notch signaling pathway in cancer-review with bioinformatic analysis. *Cancers* **2021**, *13*, 768. [CrossRef]
6. Aster, J.C.; Pear, W.S.; Blacklow, S.C. The varied roles of Notch in cancer. *Annu. Rev. Pathol. Mech. Dis.* **2017**, *12*, 245–275. [CrossRef] [PubMed]
7. Shaik, J.P.; Alanazi, I.O.; Pathan, A.; Parine, N.R.; Almadi, M.A.; Azzam, N.A.; Aljebreen, A.M.; Alharbi, O.; Alanazi, M.S.; Khan, Z. Frequent activation of Notch signaling pathway in colorectal cancers and its implication in patient survival outcome. *J. Oncol.* **2020**, *2020*, 6768942. [CrossRef]
8. Allen, F.; Maillard, I. Therapeutic targeting of Notch signaling: From cancer to inflammatory disorders. *Front. Cell. Dev. Biol.* **2021**, *9*, 649205. [CrossRef]
9. Vinson, K.E.; George, D.C.; Fender, A.W.; Bertrand, F.E.; Sigounas, G. The Notch pathway in colorectal cancer. *Int. J. Cancer* **2016**, *138*, 1835–1842. [CrossRef]
10. To, K.K.; Tong, C.W.; Wu, M.; Cho, W.C. MicroRNAs in the prognosis and therapy of colorectal cancer: From bench to bedside. *World J. Gastroenterol.* **2018**, *24*, 2949–2973. [CrossRef]
11. Lewis, B.P.; Burge, C.B.; Bartel, D.P. Conserved seed pairing, often flanked by adenosines, indicates that thousands of human genes are microRNA targets. *Cell* **2005**, *120*, 15–20. [CrossRef] [PubMed]
12. Wang, W.T.; Han, C.; Sun, Y.M.; Chen, T.Q.; Chen, Y.Q. Noncoding RNAs in cancer therapy resistance and targeted drug development. *J. Hematol. Oncol.* **2019**, *12*, 55–69. [CrossRef] [PubMed]
13. Si, W.; Shen, J.; Zheng, H.; Fan, W. The role and mechanisms of action of microRNAs in cancer drug resistance. *Clin. Epigenetics* **2019**, *11*, 25–48. [CrossRef] [PubMed]
14. Mollaei, H.; Safaralizadeh, R.; Rostami, Z. MicroRNA replacement therapy in cancer. *J. Cell. Physiol.* **2019**, *234*, 12369–12384. [CrossRef] [PubMed]
15. Lai, X.; Eberhardt, M.; Schmitz, U.; Vera, J. Systems biology-based investigation of cooperating microRNAs as monotherapy or adjuvant therapy in cancer. *Nucleic Acids Res.* **2019**, *47*, 7753–7766. [CrossRef]
16. Liu, B.; Shyr, Y.; Cai, J.; Liu, Q. Interplay between miRNAs and host genes and their role in cancer. *Brief. Funct. Genom.* **2018**, *18*, 255–266. [CrossRef]
17. Gebert, L.; MacRae, I.J. Regulation of microRNA function in animals. *Nat. Rev. Mol. Cell. Biol.* **2019**, *20*, 21–37. [CrossRef]
18. Ghafouri-Fard, S.; Glassy, M.C.; Abak, A.; Hussen, B.M.; Niazi, V.; Taheri, M. The interaction between miRNAs/lncRNAs and Notch pathway in human disorders. *Biomed. Pharmacother.* **2021**, *138*, 111496–111508. [CrossRef]
19. Shen, B. A new golden age of natural products drug discovery. *Cell* **2015**, *163*, 1297–1300. [CrossRef]
20. Alnuqaydan, A.M. Targeting micro-RNAs by natural products: A novel future therapeutic strategy to combat cancer. *Am. J. Transl. Res.* **2020**, *12*, 3531–3556.
21. Mosca, L.; Vitiello, F.; Pagano, M.; Coppola, A.; Veglia Tranchese, R.; Grillo, R.; Cacciapuoti, G.; Porcelli, M. S-adenosylmethionine, a promising antitumor agent in oral and laryngeal cancer. *Appl. Sci.* **2022**, *12*, 1746. [CrossRef]

22. Pascale, R.M.; Simile, M.M.; Calvisi, D.F.; Feo, C.F.; Feo, F. S-adenosylmethionine: From the discovery of its inhibition of tumorigenesis to its use as a therapeutic agent. *Cells* **2022**, *11*, 409. [CrossRef] [PubMed]
23. Ilisso, C.P.; Sapio, L.; Delle Cave, D.; Illiano, M.; Spina, A.; Cacciapuoti, G.; Porcelli, M. S-Adenosylmethionine affects ERK1/2 and Stat3 pathways and induces apoptosis in osteosarcoma cells. *J. Cell. Physiol.* **2016**, *231*, 428–435. [CrossRef] [PubMed]
24. Mahmood, N.; Cheishvili, D.; Arakelian, A.; Tanvir, I.; Khan, H.A.; Pépin, A.S.; Szyf, M.; Rabbani, S.A. Methyl donor S-adenosylmethionine (SAM) supplementation attenuates breast cancer growth, invasion and metastasis in vivo; therapeutic and chemopreventive applications. *Oncotarget* **2018**, *9*, 5169–5183. [CrossRef] [PubMed]
25. Yan, L.; Liang, X.; Huang, H.; Zhang, G.; Liu, T.; Zhang, J.; Chen, Z.; Zhang, Z.; Chen, Y. S-adenosylmethionine affects cell cycle pathways and suppresses proliferation in liver cells. *J. Cancer* **2019**, *10*, 4368–4379. [CrossRef]
26. Mosca, L.; Minopoli, M.; Pagano, M.; Vitiello, F.; Carriero, M.V.; Cacciapuoti, G.; Porcelli, M. Effects of S-adenosyl-L-methionine on the invasion and migration of head and neck squamous cancer cells and analysis of the underlying mechanisms. *Int. J. Oncol.* **2020**, *56*, 1212–1224. [CrossRef]
27. Mahmood, N.; Arakelian, A.; Muller, W.J.; Szyf, M.; Rabbani, S.A. An enhanced chemopreventive effect of methyl donor S-adenosylmethionine in combination with 25-hydroxyvitamin D in blocking mammary tumor growth and metastasis. *Bone Res.* **2020**, *8*, 28–31. [CrossRef]
28. Mahmood, N.; Arakelian, A.; Cheishvili, D.; Szyf, M.; Rabbani, S.A. S-adenosylmethionine in combination with decitabine shows enhanced anti-cancer effects in repressing breast cancer growth and metastasis. *J. Cell. Mol. Med.* **2020**, *24*, 10322–10337. [CrossRef]
29. Mosca, L.; Pagano, M.; Ilisso, C.P.; Delle Cave, D.; Desiderio, V.; Mele, L.; Caraglia, M.; Cacciapuoti, G.; Porcelli, M. AdoMet triggers apoptosis in head and neck squamous cancer by inducing ER-stress and potentiates cell sensitivity to cisplatin. *J. Cell. Physiol.* **2019**, *234*, 13277–13291. [CrossRef]
30. Cave, D.D.; Desiderio, V.; Mosca, L.; Ilisso, C.P.; Mele, L.; Caraglia, M.; Cacciapuoti, G.; Porcelli, M. S-Adenosylmethionine-mediated apoptosis is potentiated by autophagy inhibition induced by chloroquine in human breast cancer cells. *J. Cell. Physiol.* **2017**, *233*, 1370–1383. [CrossRef]
31. Mosca, L.; Vitiello, F.; Coppola, A.; Borzacchiello, L.; Ilisso, C.P.; Pagano, M.; Caraglia, M.; Cacciapuoti, G.; Porcelli, M. Therapeutic potential of the natural compound S-adenosylmethionine as a chemoprotective synergistic agent in breast, and head and neck cancer treatment: Current status of research. *Int. J. Mol. Sci.* **2020**, *21*, 8547. [CrossRef] [PubMed]
32. Ilisso, C.P.; Delle Cave, D.; Mosca, L.; Pagano, M.; Coppola, A.; Mele, L.; Caraglia, M.; Cacciapuoti, G.; Porcelli, M. Adenosylmethionine regulates apoptosis and autophagy in MCF-7 breast cancer cells through the modulation of specific microRNAs. *Cancer Cell Int.* **2018**, *18*, 197–209. [CrossRef] [PubMed]
33. Coppola, A.; Ilisso, C.P.; Stellavato, A.; Schiraldi, C.; Caraglia, M.; Mosca, L.; Cacciapuoti, G.; Porcelli, M. S-Adenosylmethionine inhibits cell growth and migration of triple negative breast cancer cells through upregulating MiRNA-34c and MiRNA-449a. *Int. J. Mol. Sci.* **2020**, *22*, 286. [CrossRef]
34. Pagano, M.; Mosca, L.; Vitiello, F.; Ilisso, C.P.; Coppola, A.; Borzacchiello, L.; Mele, L.; Caruso, F.P.; Ceccarelli, M.; Caraglia, M.; et al. Mi-RNA-888-5p is involved in S-adenosylmethionine antitumor effects in laryngeal squamous cancer cells. *Cancers* **2020**, *12*, 3665. [CrossRef]
35. Mosca, L.; Vitiello, F.; Borzacchiello, L.; Coppola, A.; Tranchese, R.V.; Pagano, M.; Caraglia, M.; Cacciapuoti, G.; Porcelli, M. Mutual correlation between non-coding RNA and S-adenosylmethionine in human cancer: Roles and therapeutic opportunities. *Cancers* **2021**, *13*, 3264. [CrossRef] [PubMed]
36. Mosca, L.; Pagano, M.; Pecoraro, A.; Borzacchiello, L.; Mele, L.; Cacciapuoti, G.; Porcelli, M.; Russo, G.; Russo, A. S-adenosyl-L-methionine overcomes uL3-mediated drug resistance in p53 deleted colon cancercells. *Int. J. Mol. Sci.* **2020**, *22*, 103. [CrossRef]
37. Mosca, L.; Pagano, M.; Borzacchiello, L.; Mele, L.; Russo, A.; Russo, G.; Cacciapuoti, G.; Porcelli, M. S-adenosylmethionine increases the sensitivity of human colorectal cancer cells to 5-fluorouracil by inhibiting P-glycoprotein expression and NF-B activation. *Int. J. Mol. Sci.* **2021**, *22*, 9286. [CrossRef]
38. Zsigrai, S.; Kalmár, A.; Nagy, Z.B.; Barták, B.K.; Valcz, G.; Szigeti, K.A.; Galamb, O.; Dankó, T.; Sebestyén, A.; Barna, G.; et al. S-adenosylmethionine treatment of colorectal cancer cell lines alters DNA methylation, DNA repair and tumor progression-related gene expression. *Cells* **2020**, *9*, 1864. [CrossRef]
39. Zhang, N.; Hu, X.; Du, Y.; Du, J. The role of miRNAs in colorectal cancer progression and chemoradiotherapy. *Biomed. Pharmacother.* **2021**, *134*, 111099–111109. [CrossRef]
40. Niki, M.; Nakajima, K.; Ishikawa, D.; Nishida, J.; Ishifune, C.; Tsukumo, S.I.; Shimada, M.; Nagahiro, S.; Mitamura, Y.; Yasutomo, K. MicroRNA-449a deficiency promotes colon carcinogenesis. *Sci. Rep.* **2017**, *7*, 10696–10706. [CrossRef]
41. Ishikawa, D.; Takasu, C.; Kashihara, H.; Nishi, M.; Tokunaga, T.; Higashijima, J.; Yoshikawa, K.; Yasutomo, K.; Shimada, M. The significance of MicroRNA-449a and its potential target HDAC1 in patients with colorectal cancer. *Anticancer Res.* **2019**, *39*, 2855–2860. [CrossRef] [PubMed]
42. Veena, M.S.; Raychaudhuri, S.; Basak, S.K.; Venkatesan, N.; Kumar, P.; Biswas, R.; Chakrabarti, R.; Lu, J.; Su, T.; Gallagher-Jones, M.; et al. Dysregulation of hsa-miR-34a and hsa-miR-449a leads to overexpression of PACS-1 and loss of DNA damage response (DDR) in cervical cancer. *J. Biol. Chem.* **2020**, *295*, 17169–17186. [CrossRef] [PubMed]
43. Zhang, L.; Liao, Y.; Tang, L. MicroRNA-34 family: A potential tumor suppressor and therapeutic candidate in cancer. *Exp. Clin. Cancer Res.* **2019**, *38*, 53–65. [CrossRef]

44. Luo, W.; Huang, B.; Li, Z.; Li, H.; Sun, L.; Zhang, Q.; Qiu, X.; Wang, E. MicroRNA-449a is downregulated in non-small cell lung cancer and inhibits migration and invasion by targeting c-Met. *PLoS ONE* **2013**, *8*, e64759. [CrossRef] [PubMed]
45. Sandbothe, M.; Buurman, R.; Reich, N.; Greiwe, L.; Vajen, B.; Gürlevik, E.; Schäffer, V.; Eilers, M.; Kühnel, F.; Vaquero, A.; et al. The microRNA-449 family inhibits TGF-β-mediated liver cancer cell migration by targeting SOX4. *J. Hepatol.* **2017**, *66*, 1012–1021. [CrossRef] [PubMed]
46. Gialeli, C.; Theocharis, A.D.; Karamanos, N.K. Roles of matrix metalloproteinases in cancer progression and their pharmacological targeting. *FEBS J.* **2011**, *278*, 16–27. [CrossRef] [PubMed]
47. Salem, N.; Kamal, I.; Al-Maghrabi, J.; Abuzenadah, A.; Peer-Zada, A.A.; Qari, Y.; Al-Ahwal, M.; Al-Qahtani, M.; Buhmeida, A. High expression of matrix metalloproteinases: MMP-2 and MMP-9 predicts poor survival outcome in colorectal carcinoma. *Future Oncol.* **2016**, *12*, 323–331. [CrossRef]
48. Loh, C.Y.; Chai, J.Y.; Tang, T.F.; Wong, W.F.; Sethi, G.; Shanmugam, M.K.; Chong, P.P.; Looi, C.Y. The E-cadherin and N-cadherin switch in epithelial-to-mesenchymal transition: Signaling, therapeutic implications, and challenges. *Cells* **2019**, *8*, 1118. [CrossRef]
49. Strouhalova, K.; Přechová, M.; Gandalovičová, A.; Brábek, J.; Gregor, M.; Rosel, D. Vimentin intermediate filaments as potential target for cancer treatment. *Cancers* **2020**, *12*, 184. [CrossRef]
50. Bian, J.; Dannappel, M.; Wan, C.; Firestein, R. Transcriptional regulation of Wnt/β-catenin pathway in colorectal cancer. *Cells* **2020**, *9*, 2125. [CrossRef]
51. Agarwal, V.; Bell, G.W.; Nam, J.; Bartel, D.P. Predicting effective microRNA target sites in mammalianm RNAs. *eLife* **2015**, *4*, e05005. [CrossRef] [PubMed]
52. Kozomara, A.; Birgaoanu, M.; Griffths-Jones, S. miRBase: From microRNA sequences to function. *Nucleic Acids Res.* **2019**, *47*, D155–D162. [CrossRef] [PubMed]
53. Xiu, M.X.; Liu, Y.M. The role of oncogenic Notch2 signaling in cancer: A novel therapeutic target. *Am. J. Cancer Res.* **2019**, *9*, 837–854. [PubMed]
54. Niu, L.; Yang, W.; Duan, L.; Wang, X.; Li, Y.; Xu, C.; Liu, C.; Zhang, Y.; Zhou, W.; Liu, J.; et al. Biological implications and clinical potential of metastasis-related miRNA in colorectal cancer. *Mol. Ther.-Nucleic Acids* **2020**, *23*, 42–54. [CrossRef]
55. Lu, S.C. S-adenosylmethionine. *Int. J. Biochem. Cell Biol.* **2000**, *32*, 391–395. [CrossRef]
56. Lu, S.C.; Mato, J.M. S-adenosylmethionine in cell growth, apoptosis and liver cancer. *J. Gastroenterol. Hepatol.* **2008**, *23* (Suppl. 1), S73–S77. [CrossRef]
57. Lu, S.C.; Mato, J.M. S-adenosylmethionine in liver health, injury, and cancer. *Physiol. Rev.* **2012**, *92*, 1515–1542. [CrossRef]
58. Lv, J.; Zhang, Z.; Pan, L.; Zhang, Y. MicroRNA-34/449 family and viral infections. *Virus Res.* **2019**, *260*, 1–6. [CrossRef]
59. Li, W.J.; Wang, Y.; Liu, R.; Kasinski, A.L.; Shen, H.; Slack, F.J.; Tang, D.G. MicroRNA-34a: Potent tumor suppressor, cancer stem cell inhibitor, and potential anticancer therapeutic. *Front. Cell Dev. Biol.* **2021**, *9*, 640587–640607. [CrossRef]
60. Bommer, G.T.; Gerin, I.; Feng, Y.; Kaczorowski, A.J.; Kuick, R.; Love, R.E.; Zhai, Y.; Giordano, T.J.; Qin, Z.S.; Moore, B.B.; et al. p53-mediated activation of miRNA34 candidate tumor-suppressor genes. *Curr. Biol.* **2007**, *17*, 1298–1307. [CrossRef]
61. Rokavec, M.; Li, H.; Jiang, L.; Hermeking, H. The p53/miR-34 axis in development and disease. *J. Mol. Cell Biol.* **2014**, *6*, 214–230. [CrossRef] [PubMed]
62. Okada, N.; Lin, C.P.; Ribeiro, M.C.; Biton, A.; Lai, G.; He, X.; Bu, P.; Vogel, H.; Jablons, D.M.; Keller, A.C.; et al. A positive feedback between p53 and miR-34 miRNAs mediates tumor suppression. *Genes Dev.* **2014**, *28*, 438–450. [CrossRef] [PubMed]
63. Toyota, M.; Suzuki, H.; Sasaki, Y.; Maruyama, R.; Imai, K.; Shinomura, Y.; Tokino, T. Epigenetic silencing of microRNA-34b/c and B-cell translocation gene 4 is associated with CpG island methylation in colorectal cancer. *Cancer Res.* **2008**, *68*, 4123–4132. [CrossRef] [PubMed]
64. Sun, X.; Liu, S.; Chen, P.; Fu, D.; Hou, Y.; Hu, J.; Liu, Z.; Jiang, Y.; Cao, X.; Cheng, C.; et al. miR-449a inhibits colorectal cancer progression by targeting SATB2. *Oncotarget* **2016**, *8*, 100975–100988. [CrossRef]
65. Lan, S.H.; Lin, S.C.; Wang, W.C.; Yang, Y.C.; Lee, J.C.; Lin, P.W.; Chu, M.L.; Lan, K.Y.; Zuchini, R.; Liu, H.S.; et al. Autophagy upregulates miR-449a expression to suppress progression of colorectal cancer. *Front. Oncol.* **2021**, *11*, 738144–738156. [CrossRef]
66. Herszényi, L.; Hritz, I.; Lakatos, G.; Varga, M.Z.; Tulassay, Z. The behavior of matrix metalloproteinases and their inhibitors in colorectal cancer. *Int. J. Mol. Sci.* **2012**, *13*, 13240–13263. [CrossRef]
67. Wang, Q.; Zhu, G.; Lin, C.; Lin, P.; Chen, H.; He, R.; Huang, Y.; Yang, S.; Ye, J. Vimentin affects colorectal cancer proliferation, invasion, and migration via regulated by activator protein 1. *J. Cell. Phys.* **2021**, *236*, 7591–7604. [CrossRef]
68. Cancer Genome Atlas Network. Comprehensive molecular characterization of human colon and rectal cancer. *Nature* **2012**, *487*, 330–337. [CrossRef]
69. Zhan, T.; Rindtorff, N.; Boutros, M. Wnt signaling in cancer. *Oncogene* **2017**, *36*, 1461–1473. [CrossRef]
70. Ye, K.; Xu, C.; Hui, T. MiR-34b inhibits the proliferation and promotes apoptosis in colon cancer cells by targeting Wnt/β-catenin signaling pathway. *Biosci. Rep.* **2019**, *39*, BSR20191799. [CrossRef]
71. Cheng, X.; Xu, X.; Chen, D.; Zhao, F.; Wang, W. Therapeutic potential of targeting the Wnt/β-catenin signaling pathway in colorectal cancer. *Biomed. Pharmacother.* **2019**, *110*, 473–481. [CrossRef] [PubMed]
72. Yu, W.K.; Xu, Z.Y.; Yuan, L.; Mo, S.; Xu, B.; Cheng, X.D.; Qin, J.J. Targeting β-catenin signaling by natural products for cancer prevention and therapy. *Front. Pharmacol.* **2020**, *11*, 984. [CrossRef] [PubMed]
73. Majidinia, M.; Darband, S.G.; Kaviani, M.; Nabavi, S.M.; Jahanban-Esfahlan, R.; Yousefi, B. Cross-regulation between Notch signaling pathway and miRNA machinery in cancer. *DNA Repair* **2018**, *66–67*, 30–41. [CrossRef] [PubMed]

74. Jia, Y.; Lin, R.; Jin, H.; Si, L.; Jian, W.; Yu, Q.; Yang, S. MicroRNA-34 suppresses proliferation of human ovarian cancer cells by triggering autophagy and apoptosis and inhibits cell invasion by targeting Notch 1. *Biochimie* **2019**, *160*, 193–199. [CrossRef] [PubMed]
75. Zhang, X.; Ai, F.; Li, X.; Tian, L.; Wang, X.; Shen, S.; Liu, F. MicroRNA-34a suppresses colorectal cancer metastasis by regulating Notch signaling. *Oncol. Lett.* **2017**, *14*, 2325–2333. [CrossRef]
76. Xia, J.; Duan, Q.; Ahmad, A.; Bao, B.; Banerjee, S.; Shi, Y.; Ma, J.; Geng, J.; Chen, Z.; Rahman, K.M.; et al. Genistein inhibits cell growth and induces apoptosis through up-regulation of miR-34a in pancreatic cancer cells. *Curr. Drug Targets* **2012**, *13*, 1750–1756. [CrossRef]
77. Bauer, S.; Ratz, L.; Heckmann-Nötzel, D.; Kaczorowski, A.; Hohenfellner, M.; Kristiansen, G.; Duensing, S.; Altevogt, P.; Klauck, S.M.; Sültmann, H. miR-449a repression leads to enhanced Notch signaling in *TMPRSS2: ERG* fusion positive prostate cancer cells. *Cancers* **2021**, *13*, 964. [CrossRef]
78. Mercey, O.; Popa, A.; Cavard, A.; Paquet, A.; Chevalier, B.; Pons, N.; Magnone, V.; Zangari, J.; Brest, P.; Zaragosi, L.E.; et al. Characterizing isomiR variants within the microRNA-34/449 family. *FEBS Lett.* **2017**, *591*, 693–705. [CrossRef]
79. Stasio, D.D.; Mosca, L.; Lucchese, A.; Cave, D.D.; Kawasaki, H.; Lombardi, A.; Porcelli, M.; Caraglia, M. Salivary mir-27b expression in oral lichen planus patients: A series of cases and a narrative review of literature. *Curr. Top. Med. Chem.* **2019**, *19*, 2816–2823. [CrossRef]
80. Mele, L.; Del Vecchio, V.; Marampon, F.; Regad, T.; Wagner, S.; Mosca, L.; Bimonte, S.; Giudice, A.; Liccardo, D.; Prisco, C.; et al. β2-AR blockade potentiates MEK1/2 inhibitor effect on HNSCC by regulating the Nrf2-mediated defense mechanism. *Cell Death Dis.* **2020**, *11*, 850–863. [CrossRef]
81. Minici, C.; Mosca, L.; Ilisso, C.P.; Cacciapuoti, G.; Porcelli, M.; Degano, M. Structures of catalytic cycle intermediates of the *Pyrococcus furiosus* methionine adenosyltransferase demonstrate negative cooperativity in the archaeal orthologues. *J. Struct. Biol.* **2020**, *210*, 107462–107472. [CrossRef] [PubMed]
82. Adams, B.D.; Parsons, C.; Slack, F.J. The tumor-suppressive and potential therapeutic functions of miR-34a in epithelial carcinomas. *Expert Opin. Ther. Targets* **2016**, *20*, 737–753. [CrossRef] [PubMed]
83. Kiesel, V.A.; Stan, S.D. Modulation of Notch signaling pathway by bioactive dietary agents. *Int. J. Mol. Sci.* **2022**, *23*, 3532. [CrossRef] [PubMed]

International Journal of *Molecular Sciences*

MDPI

Review

MicroRNA Methylome Signature and Their Functional Roles in Colorectal Cancer Diagnosis, Prognosis, and Chemoresistance

Rashidah Baharudin [1], Nurul Qistina Rus Bakarurraini [2], Imilia Ismail [3], Learn-Han Lee [4,*] and Nurul Syakima Ab Mutalib [1,4,5,*]

1 UKM Medical Molecular Biology Institute (UMBI), Universiti Kebangsaan Malaysia, Kuala Lumpur 56000, Wilayah Persekutuan Kuala Lumpur, Malaysia; ieda_baharudin@yahoo.com
2 Faculty of Applied Sciences, Universiti Teknologi Mara (UiTM), Shah Alam 40450, Selangor, Malaysia; nurulqistina2312@gmail.com
3 Faculty of Health Sciences, School of Biomedicine, Universiti Sultan Zainal Abidin (UniSZA), Kuala Nerus 21300, Terengganu, Malaysia; imilia@unisza.edu.my
4 Novel Bacteria and Drug Discovery Research Group, Microbiome and Bioresource Research Strength, Jeffrey Cheah School of Medicine and Health Sciences, Monash University Malaysia, Subang Jaya 47500, Selangor, Malaysia
5 Faculty of Health Sciences, Universiti Kebangsaan Malaysia, Kuala Lumpur 50300, Wilayah Persekutuan Kuala Lumpur, Malaysia
* Correspondence: lee.learn.han@monash.edu (L.-H.L.); syakima@ppukm.ukm.edu.my (N.S.A.M.); Tel.: +60-3-5514-5887 (L.-H.L.); +60-3-9145-9073 (N.S.A.M.)

Abstract: Colorectal cancer (CRC) is one of the leading causes of cancer-related deaths worldwide. Despite significant advances in the diagnostic services and patient care, several gaps remain to be addressed, from early detection, to identifying prognostic variables, effective treatment for the metastatic disease, and the implementation of tailored treatment strategies. MicroRNAs, the short non-coding RNA species, are deregulated in CRC and play a significant role in the occurrence and progression. Nevertheless, microRNA research has historically been based on expression levels to determine its biological significance. The exact mechanism underpinning microRNA deregulation in cancer has yet to be elucidated, but several studies have demonstrated that epigenetic mechanisms play important roles in the regulation of microRNA expression, particularly DNA methylation. However, the methylation profiles of microRNAs remain unknown in CRC patients. Methylation is the next major paradigm shift in cancer detection since large-scale epigenetic alterations are potentially better in identifying and classifying cancers at an earlier stage than somatic mutations. This review aims to provide insight into the current state of understanding of microRNA methylation in CRC. The new knowledge from this study can be utilized for personalized health diagnostics, disease prediction, and monitoring of treatment.

Keywords: microRNA; colorectal cancer; epigenetics; methylation; biomarker

Citation: Baharudin, R.; Rus Bakarurraini, N.Q.; Ismail, I.; Lee, L.-H.; Ab Mutalib, N.S. MicroRNA Methylome Signature and Their Functional Roles in Colorectal Cancer Diagnosis, Prognosis, and Chemoresistance. *Int. J. Mol. Sci.* **2022**, *23*, 7281. https://doi.org/10.3390/ijms23137281

Academic Editors: Alessandro Ottaiano and Donatella Delle Cave

Received: 13 May 2022
Accepted: 27 June 2022
Published: 30 June 2022

1. Introduction

Colorectal cancer (CRC) is one of the most commonly diagnosed cancers worldwide. The economic burden of CRC management of new cases in Malaysia is estimated at MYR 62 million per year [1]. While there have been significant advancements in diagnostic services and patient care, several gaps remain, from early detection to the identification of prognostic variables, the effective treatment of metastatic disease, and the implementation of customized treatment strategies. A quite recent study concludes that CRC is an expensive disease, with provider costs ranging from MYR 13,672 for stage I to MYR 27,972 for stage IV [2].

Cancer is a global burden with over 14.1 million new incidences in 2012 and is projected to increase in the next decade. In Malaysia, CRC is the most common cancer, with an overall incidence rate of 21.3 cases per 100,000 population [3]. Many CRC studies have revealed molecular alterations involved in its pathogenesis [4–6], yet the prognosis of advanced CRC is still dismal and the search continues for biomarkers which could accurately guide the medical practitioners in the management and treatment of CRC. Therefore, robust prognostic and predictive biomarkers are undoubtedly an important goal. One of the candidates for the biomarkers could be discovered by analyzing the epigenome of the tumors.

MicroRNAs are small (~22 nucleotides), non-coding RNAs that modulate gene expression in various eukaryotes [7]. These single-stranded RNAs exert their roles by interacting with specific target mRNAs through partial complementarity with sequences located mainly in the 3′UTR, subsequently causing mRNA degradation or translational inhibition [7]. MicroRNAs perform critical roles in a variety of cellular processes, including apoptosis, cell cycle, proliferation, differentiation, and angiogenesis, by simultaneously regulating the expression levels of several genes. MicroRNAs are found in all tissues and play a function in every cell type [7]. A large number of studies in CRC have also revealed that microRNA expression profiles change remarkably between normal tissues and tumors, were associated with drug resistance, as well as possess diagnostic, prognostic, and theranostic values [8–11]. Moreover, microRNAs play dual roles as oncogenes and tumor suppressors, which is the key function in tumorigenesis [12].

Although the specific mechanism underlying microRNA deregulation in cancer has yet to be determined, multiple studies have demonstrated that epigenetic mechanisms play a significant role in the regulation of microRNA expression in cancer cells [13–15], particularly DNA methylation, which is a biological process that adds methyl groups (CH_3) to the cytosine ring, thus producing 5-methylcytosine (5mC) (Figure 1a). Expression of microRNAs might be epigenetically regulated via DNA methylation of CpG islands located at promoter regions [13] (Figure 1b,c). Alterations in those mechanisms might perturb microRNAs expression, subsequently altering gene and protein expression, leading to cancer progression. The study of DNA methylation in microRNA genes is not entirely new. Publications on this topic began to emerge around a decade ago, yet the gap in knowledge remains, particularly in CRC. Most of the published data have been derived from CRC cell lines and not from clinical specimens [16,17]. In addition, to the best of our knowledge, there has been a limited number of publications on genome-wide microRNA methylome profiling in this cancer [18]. Most of the studies were focused on the selected microRNAs known to be hypermethylated in CRC, such as miR-34b/c, miR-124, miR-133b, and miR-324, etc. [19–21]. While reports on epigenome-wide microRNA methylation profiles have already been published in other cancers, such as pancreatic, breast, and oral cancers [22–24], studies on CRC are severely lacking.

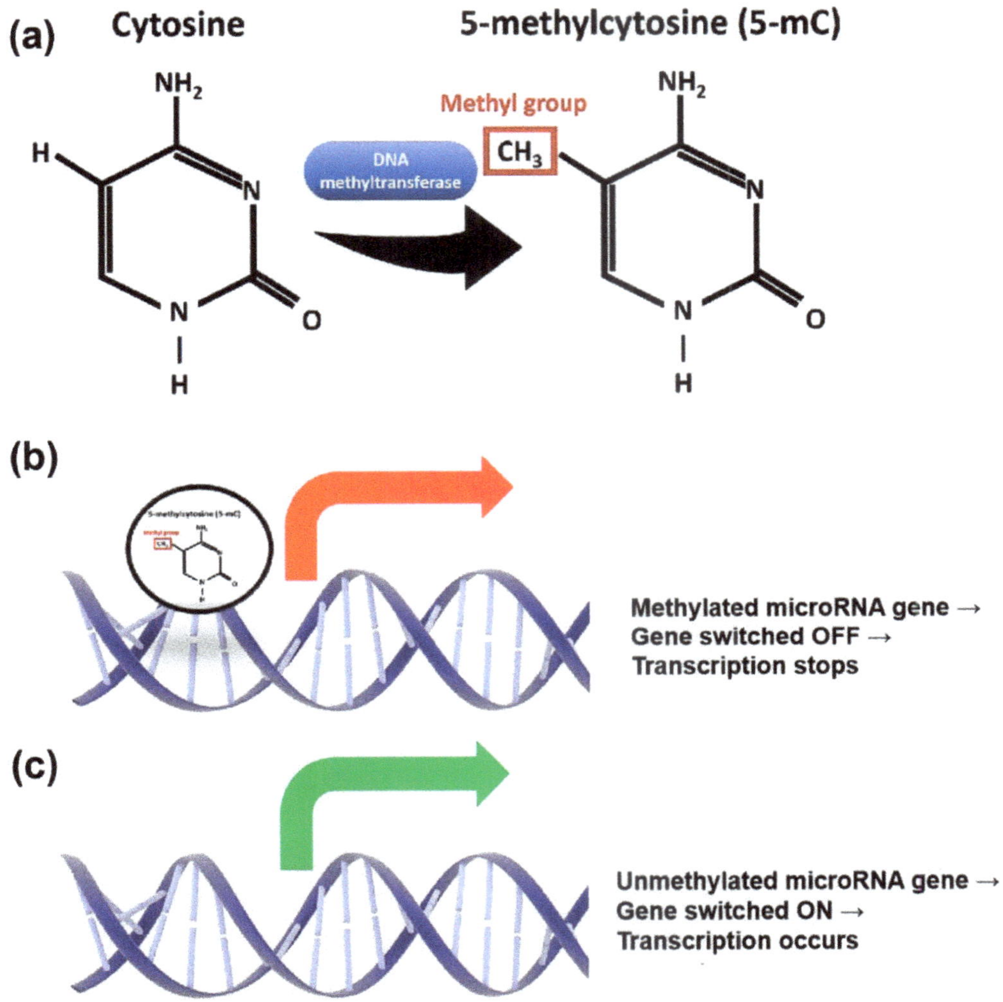

Figure 1. Overview of the methylation process. (**a**) Methylation is defined as an addition of a methyl group to the cytosine ring. (**b**) When the microRNA gene is methylated, it will be switched off and the transcription process will stop. No mature microRNA will be expressed. (**c**) When there is absence of methylation, the microRNA gene will be switched on, therefore transcription will occur and mature microRNA will be expressed.

2. MicroRNA Methylation with Diagnostic and Prognostic Markers

MicroRNAs may function as tumor suppressor genes, and their downregulation is commonly detected in CRCs. Epigenetic regulation, including DNA methylation, is one of the mechanisms associated with microRNA silencing. Since epigenetic silencing is a reversible process, aberrant methylation of microRNA emerges as a novel class of biomarkers, with strong potential as diagnostic and prognostic markers in CRCs.

2.1. MiRNA-124a

MiR-124a is among the first microRNAs in CRC that have been shown to be silenced via an epigenetic mechanism [17,25]. It was discovered through a genetic unmasking experiment from a cell line model with disrupted DNA methyltransferases [17]. DNA methylation of miR-124a may serve as an epigenetic biomarker for CRC, since this gene is more frequently methylated in CRCs than in other cancers [26]. Lujambio et al. have demonstrated that the epigenetic silencing of miR-124a via CpG island hypermethylation leads to the activation of cyclin D kinase 6 (CDK6), an oncogene, and the phosphorylation of the retinoblastoma (Rb) tumor suppressor gene [17]. The authors proved that miR-124a was specifically methylated in cancer cells, suggesting a tumor suppressive role [17]. Encoded by three independent loci (miR-124a-1, -2, and -3), miR-124a is associated with various CpG islands [13]. Aberrant DNA methylation of miR-124a-1, -2, and -3 was detected in bowel lavage fluid (BLF) specimen in CRC patients. Among these three genes, methylated miR-124a-3 showed the greatest sensitivity for CRC detection, highlighting the potential of this microRNA as a non-invasive diagnostic marker for CRC screening [27].

Ueda et al. [28] also reported that three miR-124a genes were methylated during carcinogenesis in patients with ulcerative colitis (UC). Nonetheless, methylation of miR-124a-3 was frequently detected in the early stage of colitis-associated cancer (CAC), indicating the importance of this microRNA in estimating the individual risk of developing CAC. Moreover, another related study found that miR-124a was methylated in UC patients with CRC. The authors identified higher methylation level in rectal tissues in an age-dependent manner [29]. Considered together, these two studies suggest that methylation of miR-124a is a potential marker for identifying UC patients with high risk of developing CRC.

Numerous studies have demonstrated the role of miR-124a as a prognostic biomarker in CRC patients. The miR-124a expression level varied depending on the tumor differentiation grades, while the low miR-124a was measured in tissues with moderate to poor differentiation. In addition, survival analysis of 96 CRC patients showed that the group with downregulated miR-124a exhibited worse prognosis in overall survival (OS) and disease-free survival (DFS) [30]. Similarly, Jinushi et al. [31] discovered the low expression of miR-124a in plasma samples of CRC patients with poor OS. Moreover, another study found that downregulation of miR-124a could induce cell proliferation, migration, invasion, and metastasis in CRC by negatively regulated ROCK1 expression [32]. In the future, miR-124a may constitute an effective new prognostic biomarker for CRC patients with advanced disease or metastasis.

2.2. MiRNA-137

MiR-137 is located on the chromosome 1p22 within the gene sequence of MIR137HG [33]. This microRNA is embedded in a CpG island and is often downregulated in several tumors, including CRC, due to the promoter hypermethylation [34–36]. Several studies have reported the potential use of miR-137 methylation as a diagnostic biomarker. In a study by Balaguer et al., miR-137 was epigenetically silenced in CRC cell lines. In addition, the authors investigated the methylation status of miR-137 in CRC tissues and its adjacent normal. They discovered that the methylation of miR-137 is tumor-specific, considering that higher methylation was significantly detected in CRCs compared with normal tissues. Interestingly, a similar methylation frequency of miR-137 was observed in CRCs and adenomas, indicating that the methylation of miR-137 may occur in the early event of colorectal carcinogenesis [35]. This finding is in agreement with Kashani et al. [36], in which methylation of miR-137 occurred in CRCs and no methylation was observed in normal tissues. In addition, this study revealed increased hypermethylation of miR-137 in patients with a family history of CRC or other gastrointestinal-related cancers. These encapsulate the crucial role of methylation in miR-137 as a diagnostic biomarker in CRC.

CRC arises from abnormal growth of colon epithelium and subsequently transforms to adenomatous polyps, which over time progress to cancer. A study by Huang et al. observed a gradual decrease in miR-137 expression during the process of colorectal carcinogenesis. Therefore, they postulated that DNA methylation subsequently downregulates miR-137 in polyps is an early event in the development of CRC [34]. As discussed earlier, methylation of miR-124 could be a valuable marker in identifying UC patients with high risk of developing CRC. In addition, the authors discovered the potential of methylation in miR-137 as an independent risk factor in differentiating UC patients with high risk of developing CRC. Moreover, methylation of this microRNA showed a substantial AUC value in discriminating UC patients with high or low risk of developing cancer, further demonstrating its importance in CRC screening [29].

Dysregulation of miR-137 is associated with prognosis of CRC. The decline of miR-137 expression is able to predict recurrence and survival of stage II CRC patients [37]. Furthermore, altered miR-137 expression has been shown to be associated with the progression of CRC. Through an in vitro model, downregulation of this microRNA induces cell proliferation, migration, and invasion in CRC by hindering the expression of TCF4. However, miR-137 could also target other downstream genes in addition to TCF4 to promote tumor progression. A study by Sakaguchi et al. demonstrated the capability of ectopic expression of miR-137 to suppress the tumorigenicity of colon cancer stem cells without affecting normal cells. Furthermore, they discovered that the presence of miR-137 restrained the colon cancer metastasis through the downregulation of DCLK1 expression [38]. Interestingly, research by Chen et al. suggests that miR-137 expression in CRC is subject to epigenetic silencing mediated by Mecp2, a DNA methyl CpG binding protein. Mecp2 can directly bind to the promoter region of miR-137 and lead to a decrease in expression. Restoring the expression of miR-137 led to the inhibition of the colorectal tumor growth in a xenograft model, as well as in vivo hepatic metastasis [39].

2.3. MiRNA-34

MiR-34, a tumor suppressive microRNA family, has been observed to be directly regulated by the tumor suppressor p53 [40]. The miR-34 family consists of three members, including miR-34a, miR-34b, and miR-34c. Interestingly, three miR-34 family members are produced by two different transcriptional units [41]. Human miR-34a is located at chromosome 1p36.22, whereas miR-34b and miR-34c reside on chromosome 11q23.1 [41]. MiR-34 is frequently methylated in CRC tissues and to a lesser extent in adjacent normal tissues. Notably, Wu et al. [42] discovered that methylation of miR-34a was observed in 76.8% of CRCs and 5% of healthy volunteer stool samples. Intriguingly, miR-34b/c methylation was displayed in 93.6% of CRC stool samples and no methylation was observed in the healthy samples. This finding is consistent with those of Kalimutho et al., whereby they found that 75% of fecal CRC patients exhibited aberrant methylation of miR-34b/c [43]. High sensitivity detection of methylation miR-34b/c in stool samples may be an effective non-invasive screening method for the diagnosis of CRC.

MiR-34a expression is useful for CRC prognosis. Gao et al. evaluated the expression of miR-34a-5p in recurrence and non-recurrence groups of stage II and stage III CRC patients. Their results revealed that miR-34a-5p was downregulated in the recurrence group despite the TNM stage. In addition, the elevated expression of this microRNA was directly proportional to DFS. This suggests that miR-34a-5p is a potential prognostic marker to predict the aggressiveness of cancer in stage II and III CRCs. Moreover, the authors discovered that the inhibition of metastatic properties in CRCs is a p53-dependent manner [44]. The downregulation of miR-34a in CRC is presumably caused by the aberrant methylation at the promoter region. High methylation frequency of miR-34a has been observed in primary tumors that have developed liver and lymph node metastases. Furthermore, silencing of this microRNA was associated with an increased expression of c-Met and β-catenin, which exhibited pro-metastatic function. Therefore, the epigenetic silencing of miR-34a together with upregulation of c-Met and β-catenin in primary colon cancer may have a prognostic value to identify patients with a high risk of liver metastases [45].

However, two studies presented the opposite findings. Rapti et al. showed that miR-34a was overexpressed in poorly differentiated CRC, which is highest in grade III tumors as compared with the lower grades. Deregulation of this microRNA leads to worsened DFS and OS, independently of clinicopathological factors, such as tumor size, histological grade, tumor invasion, and nodal status apart from distant metastasis. Therefore, elevated miR-34a expression is a potential unfavorable prognosis marker in CRC [46]. Another study by Hasakova et al. found that miR-34a-5p was upregulated in CRCs as compared with the adjacent tissues. Nevertheless, they found that the expression of miR-34a-5p varied in accordance with the sex, whereby downregulation of this microRNA was ascertained in male patients rather than females. In addition, a better survival rate was observed in male patients who exhibited high miR-34a-5p, and was unlikely associated with advanced stages [47]. A possible explanation of these contrasting results is the tumor microenvironment heterogeneity of CRCs.

2.4. Other microRNA Genes

MiR-133b is a tumor suppressor gene and is often silenced in CRC [48]. Silencing of this microRNA is correlated with CpG methylation in the promoter region. In addition, miR-133b was reported to be downregulated in the primary CRC and metastatic hepatic tissues. Remarkably, miR-133b negatively regulates the HOXA9/ZEB1 pathway, which then promotes tumor metastases and poor outcomes in CRCs [49]. DNA hypermethylation of miR-1 was first observed in hepatocellular carcinoma (HCC) primary tissues and cells [50]. Later, Chen et al. discovered the methylation of miR-1 in primary CRC tissues [51]. In addition, the DNA methylation-mediated downregulation of miR-1 was observed in 12 out of 14 colon cancer metastases. Interestingly, miR-1 was shown to interact with miR-133a in CRC and concurrent silencing of these microRNAs negatively regulate TAGLN2 expression. Therefore, miR-1-133a interaction with upregulation of TAGLN2 has a significant role in CRC metastasis.

The expression of miR-9 may be regulated by DNA methylation and histone modification in CRC. Methylation of this microRNA was detected in 56% of primary CRC. However, high methylation frequency was observed in advanced stages of CRC with regional nodal and vascular invasion aside from metastasis. The finding of this study showed that miR-9 silencing is crucially involved in CRC progression [52]. Moreover, deregulation of miR-9 has been reported to promote proliferation and tumor cell survival in CRC [53].

In addition to the microRNAs mentioned above, miR-345 and miR-342 are highly methylated, with low expression in CRCs in comparison with non-cancerous tissues [54,55]. Ectopic expression of these microRNAs is able to suppress colon cancer proliferation and invasiveness. Tang et al. discovered that miR-345 inhibits tumor growth by targeting BCL2-associated athanogene 3 (BAG3), a molecule that regulates the apoptosis process [54]. In contrast, restoration of miR-342 has been found to reduce the expression of DNMT1, which subsequently demethylates tumor suppressor genes, such as ADAM23, HINT1, RASSF1A, and RECK in CRC [55].

A non-exhaustive compiled summary of microRNA methylation implicated in CRC is presented in Table 1. Clearly, epigenome-wide profiling of microRNA methylation using high-throughput approaches, such as microarray or whole-genome bisulfide sequencing has not been performed, further highlighting the importance of our study. Finally, an illustration on the involvement of microRNA methylation in CRC progression is provided in Figure 2a,b.

Table 1. Snapshot of methylation-sensitive microRNAs implicated in CRC.

MicroRNA(s) and Reference	MicroRNA Methylation Detection Method	Types of Specimens	Key Findings
miR-124a [17] Known targets: STAT3, IASPP PRRX1, KITENIN, PRPS1, RPIA PTB1/PKM1/PKM2, DNMT3B, DNMT1, ROCK1, PRRX1, PLCB1	Methylation-specific PCR (MSP) and bisulfite sequencing	Cell line model with disrupted DNA methyltransferase	• Epigenetic silencing of miR-124a via CpG island hypermethylation leads to CDK6 oncogene activation and Rb phosphorylation

Table 1. *Cont.*

MicroRNA(s) and Reference	MicroRNA Methylation Detection Method	Types of Specimens	Key Findings
miR-34b/c [19] Known targets: SATB2	Methylation-specific PCR (MSP) and bisulfite sequencing	CRC cell lines	• miR-34b/c and NTG4 are novel tumor suppressors in CRC • miR-34b/c CpG island is a frequent target of epigenetic silencing in CRC
miR-133b [20] Known targets: CXCR4, HOXA9	Methylation-specific PCR (MSP) and combined bisulfite restriction analysis (COBRA)	Screening using CRC cell lines and validation in the tissues (6 CRCs, 2 adjacent non-tumors, and 2 healthy colorectal tissues)	• miR-133b promoter hypermethylation is upregulated in CRC tissues • The regulation of miR-133b methylation has potential therapeutic utility for CRC treatment
miR-324 [21] Known targets: ELAVL1	Methylation-specific PCR (MSP) and bisulfite sequencing	42 CRCs, 9 colorectal adenomas, and 16 normal mucosae in patients with and without CRC	• Methylation at the EVL/miR-342 locus was identified in 86% CRCs and in 67% adenomas, suggesting that it is an early event in CRC carcinogenesis
miR-137, miR- 342 [36] Known targets: miR-137: TCF4, FMNL2, Aurora-A miR-342: DNMT1, FOXM1, FOXQ1	Methylation-specific PCR (MSP)	Fresh-frozen tissues (51 polyps, 8 tumors, and 14 normal mucosa)	• miR-137 hypermethylation is higher in male patients • miR-342 hypermethylation is associated with patients' age
miR-9, miR-129, miR-137 [52] Known targets: miR-9: TM4SF1, FOXP2, ANO1 miR-129: MALAT1 miR-137: TCF4, FMNL2, Aurora-A	Methylation-specific PCR (MSP) and bisulfite sequencing	CRC cell lines and 50 primary CRCs with adjacent normal tissues	• miR-9-1, miR-129-2, and miR-137 methylation occurred commonly in CRC cell lines and primary CRC tumors, but not in normal colonic mucosa • miR-9-1 methylation was associated with lymph node metastasis
miR-345 [54] No known target	Methylation-specific PCR (MSP) and bisulfite sequencing	CRC cell lines and 31 CRC patients	• miR-345 hypermethylation was detected in tumor vs. normal tissues and is associated with its low expression, lymph node metastasis, and worse histological type
miR-129-2, miR-345, miR-132 [56] Known targets: miR-129: MALAT1 miR-345: No known target miR-132: ZEB2, ERK1	Bisulfite sequencing and Methylation-Specific Multiplex Ligation-Dependent Probe Amplification (MS-MLPA)	CRC cell lines treated with 5-aza-2′deoxycytidine followed by validation in 205 CRCs	• miR-345 and miR-132 hypermethylation is associated with a mismatch-repair deficiency in CRC • miR-132 hypermethylation distinguished sporadic MMR-deficient CRC from Lynch-CRC
miR-132 [57] Known targets: miR-132: ZEB2, ERK1	Methylation-specific PCR (MSP) and bisulfite sequencing	CRC cell lines and 36 CRCs with adjacent normal tissues	• miR-132 is epigenetically silenced in CRC cell lines and implies a poor prognosis in CRC

Table 1. *Cont.*

MicroRNA(s) and Reference	MicroRNA Methylation Detection Method	Types of Specimens	Key Findings
miR-1, miR-9, miR-124, miR-137 [29] Known targets: miR-1: SMAD3 miR-9: TM4SF1, FOXP2, ANO1 miR-124: STAT3, IASPP PRRX1, KITENIN, PRPS1, RPIA. PTB1/PKM1/PKM2, DNMT3B, DNMT1, ROCK1, PRRX1, PLCB1. miR-137: TCF4, FMNL2, Aurora-A	Quantitative bisulfite pyrosequencing	387 colorectal epithelial specimens (362 non-neoplastic and 25 neoplastic tissues)	• Among patients with ulcerative colitis without neoplasia, the rectal tissues had significantly higher levels of microRNA methylation • Methylation level was associated with age and duration of ulcerative colitis
miR-125 [58] Known targets: BCL2, BCL2L12, MCL1, SMURF1, VEGFA, TAZ, CXCL12/CXCR4	Bisulfite sequencing PCR	CRC tissues and adjacent normal tissues from 68 CRC patients	• Patients with hypermethylation of miR-125a and miR-125b had a shorter life expectancy than those with normal levels
miR-941 [59] No known target	Bisulfite sequencing	CRC cell lines	• Hypermethylated in HCT116 cells • Suppresses cell growth and migration in CRC cells
miR-1237 [59] No known target	Bisulfite sequencing	CRC cell lines	• Hypermethylated in HCT116 cells • Transcriptionally independent from the host gene
miR-1247 [60] No known target	Methylation-specific PCR (MSP) and bisulfite sequencing	CRC cell lines and patients (hypermethylated and non-methylated CRCs)	• Downregulated in methylated CRC and hypermethylated cell lines (RKO, HCT116) • Novel tumor suppressor by targeting MYCBP2 in methylated CRC
miR-128 [61] Known targets: IRS1, Galectin-3	Bisulfite sequencing PCR	CRC cell lines and patients	• miR-128 was epigenetically silenced by DNA methylation, implies a poor prognosis in CRC • Restoration of miR-128 could inhibit cell proliferation by inducing cell cycle arrest
miR-148a [62] Known targets: BCL2, ERBB3	Bisulfite pyrosequencing	273 CRC patients (76 stage II, 125 stage III, 72 stage IV)	• miR-148a was significantly downregulated in tumor stage III/IV and correlated with promoter hypermethylation • Low miR-148a expression leads to poor therapeutic response and patients' overall survival

Table 1. *Cont.*

MicroRNA(s) and Reference	MicroRNA Methylation Detection Method	Types of Specimens	Key Findings
miR-126 [63] Known targets: CXCR4	Methylation-specific PCR (MSP) and bisulfite sequencing	CRC cell lines and patients	• Silencing of miR-126 in CRC tissue and cell lines was due to the promoter methylation • Restoration of miR-126 inhibits VEGF expression, thus hindering tumor progression
miR-27b [64] Known targets: RAB3D	Methylation-specific PCR (MSP)	CRC cell lines	• DNA hypermethylation of miR-27b CpG island decreases miR-27b expression • Targets VEGFC to inhibit tumor growth and angiogenesis in vivo
miR-149 [65] Known targets: FOXM1, EPHB3	Methylation-specific PCR (MSP)	CRC cell lines	• Treatment using polyphenol (BPIS) induces hypomethylation of miR-149 CpG island in HCT-8/Fu cells • Upregulation of miR-149 improved chemosensitivity of CRC through miR-149/Akt-mediated cell cycle arrest
miR-497/195 [66] Known targets: IGF1R, NRDP1, KSR1, FRA-1, PTPN3, CARMA3, FGF2	Combined bisulfite restriction analysis (COBRA) and bisulfite genomic sequencing (BGS)	CRC cell lines and patients	• Both miRNAs were hypermethylated and under expressed in precancerous lesion • Pri-miR-497/195 was monoallelic methylated at CpG island in normal colorectal and biallelic methylated in most colorectal adenomas
miR-212 [67] Known targets: MnSOD	Methylation-specific PCR (MSP) and bisulfite sequencing	CRC cell lines and tissues	• miR-212 was hypermethylated at upstream promoter region in CRC tissues and cell lines, but not in FHC cells • Low miR-212 level associated with aggressive tumor phenotype and poor disease prognosis
miR-200c/141 [68] Known targets: ZEB1, DLC1, TRAF5	Methylation-specific PCR (MSP)	CRC tissues	• miR-200c/141 cluster promoter region was significantly hypermethylated in colorectal tumors and adenomatous polyps, but not in hyperplastic polyp tissues

Table 1. *Cont.*

MicroRNA(s) and Reference	MicroRNA Methylation Detection Method	Types of Specimens	Key Findings
miR-373 [69] No known targets	Methylation-specific PCR (MSP) and bisulfite sequencing	CRC cell lines and 40 CRC patients	• CpG island at promoter region of miR-373 was significantly hypermethylated in CRC tissues and cell lines • May inhibit cell viability in CRC cell lines by targeting oncogene RAB22A

Known targets in CRC were identified using miRCancer database [70].

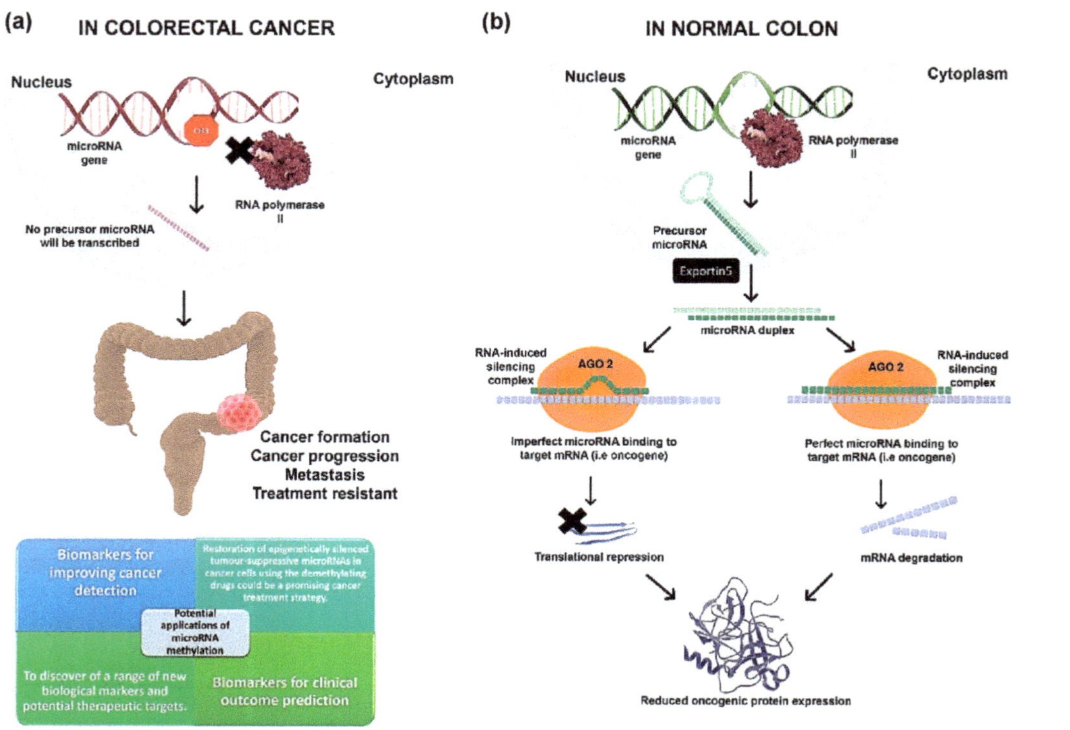

Figure 2. The involvement of methylated microRNA in CRC. (**a**) Simplified illustration of microRNA and its potential involvement in CRC. DNA methylation is a key epigenetic mechanism for silencing RNA polymerase II-transcribed genes [71]. When the microRNA gene is methylated, no precursor microRNA will be transcribed, thus reducing its mature microRNA expression [72]. This in turn could lead to cancer formation, progression, and treatment resistance. (**b**) Simplified illustration of microRNA and its potential involvement and application in normal colon. The unmethylated microRNA gene will lead to transcription of microRNA precursor by RNA polymerase II, which will then be exported into the cytoplasm by Exportin 5, followed by processing with the RISC, which will result in target gene translation repression or mRNA degradation. As a result, oncogenic protein expression will be reduced.

As previously mentioned, the methylome profiles of microRNAs in CRC patients have not been extensively characterized. While there are several published findings from our research group on DNA methylation profiles in CRC [6,73], none have focused on microRNA methylome in detail. MicroRNAs are considered the master regulators that control gene expression [74]. Therefore, research on the elements controlling microRNA is indispensable.

3. MicroRNA Methylation in CRC Chemoresistance

Emerging evidence has revealed that abnormal expression of microRNAs also plays a vital role in chemotherapeutic drug resistance. FOLFOX, which is a mixture of folic acid (FOL), 5-fluorouracil (F), and oxaliplatin (OX) [75], is one of the most extensively used chemotherapy regimens for the treatment of cancer, mainly CRC. While cancer treatment is progressing, the formation of chemoresistance clones have emerged as a significant obstacle in the clinic. Finding prospective biomarkers and therapeutic targets that could lead to an increase in the success rate of suggested therapies is critical to achieving a successful outcome. Since it has been established that microRNAs are significant participants in the biological system, researchers have become increasingly interested in understanding their functional activities. When it comes to overcoming chemoresistance to FOLFOX, microRNAs as post-transcriptional regulators have the potential to be extremely beneficial. A review on differentially expressed microRNAs involved in CRC chemoresistance was previously published by our group and should serve as complementary reading [76]. In this section, we will focus primarily on the methylated microRNAs and their roles in CRC chemoresistance.

MiR-26b expression was analyzed in 5-fluorouracil (5-FU) resistant CRC cell lines and parental cells. The results showed that miR-26b was significantly downregulated in the 5-FU resistant cell lines, and thus, it is probably involved in CRC chemoresistance. Importantly, the downregulation of miR-26b was associated with promoter methylation and treatment with a demethylating drug (5-aza-2′-deoxycytidine was able to restore the expression of miR-26b in resistant cell lines). Upregulation of miR-26b conferred 5-FU chemosensitivity by repressing PGP expression and further activating caspase-9 and caspase-3 [77].

Takahashi et al. have provided evidence that miR-148a is frequently downregulated through the promoter hypermethylation in the advanced CRC. Moreover, downregulation of miR-148a was significantly associated with a poor outcome in patients with stage III CRC treated with adjuvant 5-FU. In addition, low expression of this microRNA is associated with worse therapeutic response and survival rate in stage IV CRC patients treated with 5-FU and oxaliplatin chemotherapy [62].

Low expression of miR-181a, 135a, and 302c is mediated by DNA methylation in colon cancer. Shi et al. proved that dysregulation of these microRNAs promotes 5-FU resistance in microsatellite instable (MSI) CRC. Restoration of microRNAs expression attenuates PLAG1 expression and was shown to re-sensitize 5-FU resistant MSI CRC cell lines [78].

Another microRNA associated drug resistance is miR-149. A previous study showed that aberrant methylation is the main mechanism that is responsible for the silencing of miR-149 in CRC [79]. The expression of miR-149 is downregulated in 5-FU resistant cells as compared with their parental cells. Re-expression of this microRNA was able to enhance the 5-FU sensitivity of CRC cells by suppressing FOXM1 gene [80]. In addition, another recent study demonstrated that the upregulation of miR-149 expression together with the DNA de-methylation (5-aza-dc) therapy could positively elevate the chemosensitivity of CRC [65]. Similarly, the co-administration of dichloroacetate (DCA) and overexpression of miR-149 in CRC was shown to not only improve 5-Fu apoptosis, but also to help in minimizing glucose metabolism [81].

Other downregulated microRNAs, such as miR-200 [82], miR-17-5p [83], miR-124, miR-506 [84], miR-143 [85], and miR-340 [86] were associated with chemoresistance of multi-drugs in CRCs. The downregulation of these microRNAs was correlated with DNA methylation [87–90]. Considered together, these data suggest that epigenetic silencing of microRNAs has strong potential as a marker to predict chemotherapy response in CRC.

4. Conclusions

The methylation of a subset of microRNA genes could serve as a useful biomarker for the improvement of cancer detection and/or clinical outcome prediction. In addition, the restoration of epigenetically silenced tumor-suppressive microRNAs in cancer cells using the demethylating drugs could be a promising cancer treatment strategy. In the future, we envisage that further cancer epigenome and microRNA studies will lead to the discovery of a range of new biological markers and potential therapeutic targets.

Author Contributions: Conceptualization, N.S.A.M.; writing—original draft preparation, R.B. and N.Q.R.B.; writing—review and editing, N.S.A.M., I.I. and L.-H.L.; visualization, N.S.A.M.; supervision, N.S.A.M.; project administration, N.S.A.M.; funding acquisition, N.S.A.M. and L.-H.L. All authors have read and agreed to the published version of the manuscript.

Funding: This research was funded by the Ministry of Higher Education Malaysia, grant number FRGS/1/2020/SKK0/UKM/02/11. The APC was funded by Monash University Malaysia.

Institutional Review Board Statement: Not applicable.

Informed Consent Statement: Not applicable.

Data Availability Statement: Not applicable.

Conflicts of Interest: The authors declare no conflict of interest.

References

1. Ezat, S.W.; Natrah, M.S.; Aljunid, S.; Rizal, M.A.; Saperi, S.; Ismail, S.; Fuad, I.; Azrif, M.A. Economic evaluation of monoclonal antibody in the management of colorectal cancer. *J. Cancer Res. Ther.* **2013**, *1*, 34–39.
2. Azzani, M.; Dahlui, M.; Ishak, W.Z.W.; Roslani, A.C.; Su, T.T. Provider costs of treating colorectal cancer in government hospital of Malaysia. *Malays J. Med. Sci.* **2019**, *26*, 73–86. [CrossRef] [PubMed]
3. Azizah, A.M.; Nor Saleha, I.T.; Noor Hashimah, A.; Asmah, Z.A.; Mastulu, W. *Malaysian National Cancer Registry Report 2007–2011*; National Cancer Institute: Putrajaya, Malaysia, 2016.
4. The Cancer Genome Atlas (TCGA) Research Network. Comprehensive molecular characterization of human colon and rectal cancer. *Nature* **2012**, *487*, 330–337. [CrossRef]
5. Naumov, V.A.; Generozov, E.V.; Zaharjevskaya, N.B.; Matushkina, D.S.; Larin, A.K.; Chernyshov, S.V.; Alekseev, M.V.; Shelygin, Y.A.; Govorun, V.M. Genome-scale analysis of DNA methylation in colorectal cancer using infinium humanmethylation450 beadchips. *Epigenetics* **2013**, *8*, 921–934. [CrossRef] [PubMed]
6. Baharudin, R.; Ab Mutalib, N.-S.; Othman, S.N.; Sagap, I.; Rose, I.M.; Mohd Mokhtar, N.; Jamal, R. Identification of predictive DNA methylation biomarkers for chemotherapy response in colorectal cancer. *Front. Pharmacol.* **2017**, *8*, 47. [CrossRef]
7. Bartel, D.P. MicroRNAs: Genomics, biogenesis, mechanism, and function. *Cell* **2004**, *116*, 281–297. [CrossRef]
8. Zhang, Y.; Wang, J. MicroRNAs are important regulators of drug resistance in colorectal cancer. *Biol. Chem.* **2017**, *398*, 929–938. [CrossRef]
9. Al-Akhrass, H.; Christou, N. The clinical assessment of microRNA diagnostic, prognostic, and theranostic value in colorectal cancer. *Cancers* **2021**, *13*, 2916. [CrossRef]
10. Imedio, L.; Cristóbal, I.; Rubio, J.; Santos, A.; Rojo, F.; García-Foncillas, J. MicroRNAs in rectal cancer: Functional significance and promising therapeutic value. *Cancers* **2020**, *12*, 2040. [CrossRef]
11. Pidíková, P.; Herichová, I. MiRNA clusters with up-regulated expression in colorectal cancer. *Cancers* **2021**, *13*, 2979. [CrossRef]
12. Svoronos, A.A.; Engelman, D.M.; Slack, F.J. OncomiR or tumor suppressor? The duplicity of MicroRNAs in cancer. *Cancer Res.* **2016**, *76*, 3666–3670. [CrossRef] [PubMed]
13. Kaur, S.; Lotsari-Salomaa, J.E.; Seppänen-Kaijansinkko, R.; Peltomäki, P. MicroRNA methylation in colorectal cancer. In *Non-Coding RNAs in Colorectal Cancer*; Slaby, O., Calin, G.A., Eds.; Springer International Publishing: Cham, Switzerland, 2016; Volume 937, pp. 109–122. ISBN 978-3-319-42057-8.
14. Wang, S.; Wu, W.; Claret, F.X. Mutual regulation of micrornas and DNA methylation in human cancers. *Epigenetics* **2017**, *12*, 187–197. [CrossRef] [PubMed]
15. Stark, V.A.; Facey, C.O.B.; Viswanathan, V.; Boman, B.M. The role of MiRNAs, MiRNA clusters, and isomirs in development of cancer stem cell populations in colorectal cancer. *Int. J. Mol. Sci.* **2021**, *22*, 1424. [CrossRef] [PubMed]
16. Lujambio, A.; Calin, G.A.; Villanueva, A.; Ropero, S.; Sanchez-Cespedes, M.; Blanco, D.; Montuenga, L.M.; Rossi, S.; Nicoloso, M.S.; Faller, W.J.; et al. A MicroRNA DNA methylation signature for human cancer metastasis. *Proc. Natl. Acad. Sci. USA* **2008**, *105*, 13556–13561. [CrossRef] [PubMed]

17. Lujambio, A.; Ropero, S.; Ballestar, E.; Fraga, M.F.; Cerrato, C.; Setién, F.; Casado, S.; Suarez-Gauthier, A.; Sanchez-Cespedes, M.; Git, A.; et al. Genetic unmasking of an epigenetically silenced microrna in human cancer cells. *Cancer Res.* **2007**, *67*, 1424–1429. [CrossRef] [PubMed]
18. Patil, N.; Abba, M.L.; Zhou, C.; Chang, S.; Gaiser, T.; Leupold, J.H.; Allgayer, H. Changes in methylation across structural and MicroRNA genes relevant for progression and metastasis in colorectal cancer. *Cancers* **2021**, *13*, 5951. [CrossRef]
19. Toyota, M.; Suzuki, H.; Sasaki, Y.; Maruyama, R.; Imai, K.; Shinomura, Y.; Tokino, T. Epigenetic silencing of MicroRNA-34b/c and B-cell translocation gene 4 is associated with cpg island methylation in colorectal cancer. *Cancer Res.* **2008**, *68*, 4123–4132. [CrossRef]
20. Lv, L.V.; Zhou, J.; Lin, C.; Hu, G.; Yi, L.U.; Du, J.; Gao, K.; Li, X. DNA methylation is involved in the aberrant expression of mir-133b in colorectal cancer cells. *Oncol. Lett.* **2015**, *10*, 907–912. [CrossRef]
21. Grady, W.M.; Parkin, R.K.; Mitchell, P.S.; Lee, J.H.; Kim, Y.-H.; Tsuchiya, K.D.; Washington, M.K.; Paraskeva, C.; Willson, J.K.V.; Kaz, A.M.; et al. Epigenetic silencing of the intronic MicroRNA Hsa-MiR-342 and Its host gene EVL in colorectal cancer. *Oncogene* **2008**, *27*, 3880–3888. [CrossRef]
22. Oltra, S.S.; Peña-Chilet, M.; Vidal-Tomas, V.; Flower, K.; Martinez, M.T.; Alonso, E.; Burgues, O.; Lluch, A.; Flanagan, J.M.; Ribas, G. Methylation deregulation of MiRNA promoters identifies MiR124-2 as a survival biomarker in breast cancer in very young women. *Sci. Rep.* **2018**, *8*, 1–12. [CrossRef]
23. Zhang, J.; Shi, K.; Huang, W.; Weng, W.; Zhang, Z.; Guo, Y.; Deng, T.; Xiang, Y.; Ni, X.; Chen, B.; et al. The DNA methylation profile of non-coding RNAs improves prognosis prediction for pancreatic adenocarcinoma. *Cancer Cell Int.* **2019**, *19*, 107. [CrossRef] [PubMed]
24. Roy, R.; Chatterjee, A.; Das, D.; Ray, A.; Singh, R.; Chattopadhyay, E.; Sarkar, N.D.; Eccles, M.; Pal, M.; Maitra, A.; et al. Genome-wide MiRNA methylome analysis in oral cancer: Possible biomarkers associated with patient survival. *Epigenomics* **2019**, *11*, 473–487. [CrossRef] [PubMed]
25. Lujambio, A.; Esteller, M. CpG island hypermethylation of tumor suppressor MicroRNAs in human cancer. *Cell Cycle* **2007**, *6*, 1455–1459. [CrossRef]
26. Hibner, G.; Kimsa-Furdzik, M.; Francuz, T. Relevance of MicroRNAs as potential diagnostic and prognostic markers in colorectal cancer. *Int. J. Mol. Sci.* **2018**, *19*, 2944. [CrossRef] [PubMed]
27. Harada, T.; Yamamoto, E.; Yamano, H.; Nojima, M.; Maruyama, R.; Kumegawa, K.; Ashida, M.; Yoshikawa, K.; Kimura, T.; Harada, E.; et al. Analysis of DNA methylation in bowel lavage fluid for detection of colorectal cancer. *Cancer Prev. Res.* **2014**, *7*, 1002–1010. [CrossRef]
28. Ueda, Y.; Ando, T.; Nanjo, S.; Ushijima, T.; Sugiyama, T. DNA methylation of MicroRNA-124a is a potential risk marker of colitis-associated cancer in patients with ulcerative colitis. *Dig. Dis. Sci.* **2014**, *59*, 2444–2451. [CrossRef]
29. Toiyama, Y.; Okugawa, Y.; Tanaka, K.; Araki, T.; Uchida, K.; Hishida, A.; Uchino, M.; Ikeuchi, H.; Hirota, S.; Kusunoki, M.; et al. A panel of methylated MicroRNA biomarkers for identifying high-risk patients with ulcerative colitis-associated colorectal cancer. *Gastroenterology* **2017**, *153*, 1634.e8–1646.e8. [CrossRef]
30. Wang, M.-J.; Li, Y.; Wang, R.; Wang, C.; Yu, Y.-Y.; Yang, L. Downregulation of MicroRNA-124 is an independent prognostic factor in patients with colorectal cancer. *Int. J. Colorectal. Dis.* **2013**, *28*, 183–189. [CrossRef]
31. Jinushi, T.; Shibayama, Y.; Kinoshita, I.; Oizumi, S.; Jinushi, M.; Aota, T.; Takahashi, T.; Horita, S.; Dosaka-Akita, H.; Iseki, K. Low expression levels of MicroRNA-124-5p correlated with poor prognosis in colorectal cancer via targeting of SMC4. *Cancer Med.* **2014**, *3*, 1544–1552. [CrossRef]
32. Zhou, L.; Xu, Z.; Ren, X.; Chen, K.; Xin, S. MicroRNA-124 (MiR-124) inhibits cell proliferation, metastasis and invasion in colorectal cancer by downregulating rho-associated protein kinase 1(ROCK1). *CPB* **2016**, *38*, 1785–1795. [CrossRef]
33. Mahmoudi, E.; Cairns, M.J. MiR-137: An important player in neural development and neoplastic transformation. *Mol. Psychiatry* **2017**, *22*, 44–55. [CrossRef] [PubMed]
34. Huang, Y.-C.; Lee, C.-T.; Lee, J.-C.; Liu, Y.-W.; Chen, Y.-J.; Tseng, J.T.; Kang, J.-W.; Sheu, B.-S.; Lin, B.-W.; Hung, L.-Y. Epigenetic silencing of MiR-137 contributes to early colorectal carcinogenesis by impaired aurora—A inhibition. *Oncotarget* **2016**, *7*, 76852–76866. [CrossRef] [PubMed]
35. Balaguer, F.; Link, A.; Lozano, J.J.; Cuatrecasas, M.; Nagasaka, T.; Boland, C.R.; Goel, A. Epigenetic silencing of MiR-137 is an early event in colorectal carcinogenesis. *Cancer Res.* **2010**, *70*, 6609–6618. [CrossRef] [PubMed]
36. Kashani, E.; Hadizadeh, M.; Chaleshi, V.; Mirfakhraie, R.; Young, C.; Savabkar, S. The differential DNA hypermethylation patterns of MicroRNA-137 and MicroRNA-342 locus in early colorectal lesions and tumours. *Biomolecules* **2019**, *9*, 519. [CrossRef]
37. Bahnassy, A.A.; El-Sayed, M.; Ali, N.M.; Khorshid, O.; Hussein, M.M.; Yousef, H.F.; Mohanad, M.A.; Zekri, A.-R.N.; Salem, S.E. Aberrant expression of MiRNAs predicts recurrence and survival in stage-II colorectal cancer patients from Egypt. *Appl. Cancer Res.* **2017**, *37*, 39. [CrossRef]
38. Sakaguchi, M.; Hisamori, S.; Oshima, N.; Sato, F.; Shimono, Y.; Sakai, Y. MiR-137 regulates the tumorigenicity of colon cancer stem cells through the inhibition of DCLK1. *Mol. Cancer Res.* **2016**, *14*, 354–362. [CrossRef]
39. Chen, T.; Cai, S.-L.; Li, J.; Qi, Z.-P.; Li, X.-Q.; Ye, L.-C.; Xie, X.-F.; Hou, Y.-Y.; Yao, L.-Q.; Xu, M.-D.; et al. Mecp2-Mediated epigenetic silencing of MiR-137 contributes to colorectal adenoma-carcinoma sequence and tumor progression via relieving the suppression of c-met. *Sci. Rep.* **2017**, *7*, 44543. [CrossRef]

40. Navarro, F.; Lieberman, J. MiR-34 and P53: New insights into a complex functional relationship. *PLoS ONE* **2015**, *10*, e0132767. [CrossRef]
41. Zhang, L.; Liao, Y.; Tang, L. MicroRNA-34 family: A potential tumor suppressor and therapeutic candidate in cancer. *J. Exp. Clin. Cancer Res.* **2019**, *38*, 53. [CrossRef]
42. Wu, X.; Song, Y.-C.; Cao, P.-L.; Zhang, H.; Guo, Q.; Yan, R.; Diao, D.-M.; Cheng, Y.; Dang, C.-X. Detection of MiR-34a and MiR-34b/c in stool sample as potential screening biomarkers for noninvasive diagnosis of colorectal cancer. *Med. Oncol.* **2014**, *31*, 894. [CrossRef]
43. Kalimutho, M.; Di Cecilia, S.; Del Vecchio Blanco, G.; Roviello, F.; Sileri, P.; Cretella, M.; Formosa, A.; Corso, G.; Marrelli, D.; Pallone, F.; et al. Epigenetically silenced MiR-34b/c as a novel faecal-based screening marker for colorectal cancer. *Br. J. Cancer* **2011**, *104*, 1770–1778. [CrossRef] [PubMed]
44. Gao, J.; Li, N.; Dong, Y.; Li, S.; Xu, L.; Li, X.; Li, Y.; Li, Z.; Ng, S.S.; Sung, J.J.; et al. MiR-34a-5p suppresses colorectal cancer metastasis and predicts recurrence in patients with stage II/III colorectal cancer. *Oncogene* **2015**, *34*, 4142–4152. [CrossRef] [PubMed]
45. Siemens, H.; Neumann, J.; Jackstadt, R.; Mansmann, U.; Horst, D.; Kirchner, T.; Hermeking, H. Detection of MiR-34a promoter methylation in combination with elevated expression of c-met and β-catenin predicts distant metastasis of colon cancer. *Clin. Cancer Res.* **2013**, *19*, 710–720. [CrossRef] [PubMed]
46. Rapti, S.-M.; Kontos, C.K.; Christodoulou, S.; Papadopoulos, I.N.; Scorilas, A. MiR-34a overexpression predicts poor prognostic outcome in colorectal adenocarcinoma, independently of clinicopathological factors with established prognostic value. *Clin. Biochem.* **2017**, *50*, 918–924. [CrossRef]
47. Hasakova, K.; Reis, R.; Vician, M.; Zeman, M.; Herichova, I. Expression of MiR-34a-5p is up-regulated in human colorectal cancer and correlates with survival and clock gene PER2 expression. *PLoS ONE* **2019**, *14*, e0224396. [CrossRef]
48. Li, D.; Xia, L.; Chen, M.; Lin, C.; Wu, H.; Zhang, Y.; Pan, S.; Li, X. MiR-133b, a particular member of myomirs, coming into playing its unique pathological role in human cancer. *Oncotarget* **2017**, *8*, 50193–50208. [CrossRef]
49. Wang, X.; Bu, J.; Liu, X.; Wang, W.; Mai, W.; Lv, B.; Zou, J.; Mo, X.; Li, X.; Wang, J.; et al. MiR-133b suppresses metastasis by targeting HOXA9 in human colorectal cancer. *Oncotarget* **2017**, *8*, 63935–63948. [CrossRef]
50. Datta, J.; Kutay, H.; Nasser, M.W.; Nuovo, G.J.; Wang, B.; Majumder, S.; Liu, C.-G.; Volinia, S.; Croce, C.M.; Schmittgen, T.D.; et al. Methylation mediated silencing of MicroRNA-1 gene and its role in hepatocellular carcinogenesis. *Cancer Res.* **2008**, *68*, 5049–5058. [CrossRef]
51. Chen, W.-S.; Leung, C.-M.; Pan, H.-W.; Hu, L.-Y.; Li, S.-C.; Ho, M.-R.; Tsai, K.-W. Silencing of MiR-1-1 and MiR-133a-2 cluster expression by DNA hypermethylation in colorectal cancer. *Oncol. Rep.* **2012**, *28*, 1069–1076. [CrossRef]
52. Bandres, E.; Agirre, X.; Bitarte, N.; Ramirez, N.; Zarate, R.; Roman-Gomez, J. Epigenetic regulation of MicroRNA expression in colorectal cancer. *Int. J. Cancer* **2009**, *125*, 2737–2743. [CrossRef]
53. Cekaite, L.; Rantala, J.K.; Bruun, J.; Guriby, M.; Ågesen, T.H.; Danielsen, S.A.; Lind, G.E.; Nesbakken, A.; Kallioniemi, O.; Lothe, R.A.; et al. MiR-9, -31, and -182 deregulation promote proliferation and tumor cell survival in colon cancer. *Neoplasia* **2012**, *14*, 868–881. [CrossRef] [PubMed]
54. Tang, J.-T.; Wang, J.-L.; Du, W.; Hong, J.; Zhao, S.-L.; Wang, Y.-C.; Xiong, H.; Chen, H.-M.; Fang, J.-Y. MicroRNA 345, a methylation-sensitive MicroRNA is involved in cell proliferation and invasion in human colorectal cancer. *Carcinogenesis* **2011**, *32*, 1207–1215. [CrossRef] [PubMed]
55. Wang, H.; Wu, J.; Meng, X.; Ying, X.; Zuo, Y.; Liu, R.; Pan, Z.; Kang, T.; Huang, W. MicroRNA-342 inhibits colorectal cancer cell proliferation and invasion by directly targeting DNA methyltransferase 1. *Carcinogenesis* **2011**, *32*, 1033–1042. [CrossRef] [PubMed]
56. Kaur, S.; Lotsari, J.E.; Al-Sohaily, S.; Warusavitarne, J.; Kohonen-Corish, M.R.; Peltomäki, P. Identification of subgroup-specific MiRNA patterns by epigenetic profiling of sporadic and lynch syndrome-associated colorectal and endometrial carcinoma. *Clin. Epigenetics* **2015**, *7*, 20. [CrossRef]
57. Wang, Z.; Qin, J.; Ke, J.; Wang, F.; Zhou, Y.; Jiang, Y.; Xu, J. Downregulation of MicroRNA-132 by DNA hypermethylation is associated with cell invasion in colorectal cancer. *OncoTargets Ther.* **2015**, 3639. [CrossRef]
58. Chen, H.; Xu, Z. Hypermethylation-associated silencing of MiR-125a and MiR-125b: A potential marker in colorectal cancer. *Dis. Markers* **2015**, *2015*, 345080. [CrossRef]
59. Yan, H.; Choi, A.; Lee, B.H.; Ting, A.H. Identification and functional analysis of epigenetically silenced microRNAs in colorectal cancer cells. *PLoS ONE* **2011**, *6*, e20628. [CrossRef]
60. Liang, J.; Zhou, W.; Sakre, N.; DeVecchio, J.; Ferrandon, S.; Ting, A.H.; Bao, S.; Bissett, I.; Church, J.; Kalady, M.F. Epigenetically regulated MiR-1247 functions as a novel tumour suppressor via MYCBP2 in methylator colon cancers. *Br. J. Cancer* **2018**, *119*, 1267–1277. [CrossRef]
61. Takahashi, Y.; Iwaya, T.; Sawada, G.; Kurashige, J.; Matsumura, T.; Uchi, R.; Ueo, H.; Takano, Y.; Eguchi, H.; Sudo, T.; et al. Up-regulation of NEK2 by MicroRNA-128 methylation is associated with poor prognosis in colorectal cancer. *Ann. Surg. Oncol.* **2014**, *21*, 205–212. [CrossRef]
62. Takahashi, M.; Cuatrecasas, M.; Balaguer, F.; Hur, K.; Toiyama, Y.; Castells, A.; Boland, C.R.; Goel, A. The clinical significance of MiR-148a as a predictive biomarker in patients with advanced colorectal cancer. *PLoS ONE* **2012**, *7*, e46684. [CrossRef]

63. Zhang, Y.; Wang, X.; Xu, B.; Wang, B.; Wang, Z.; Liang, Y.; Zhou, J.; Hu, J.; Jiang, B. Epigenetic silencing of MiR-126 contributes to tumor invasion and angiogenesis in colorectal cancer. *Oncol. Rep.* **2013**, *30*, 1976–1984. [CrossRef] [PubMed]
64. Ye, J.; Wu, X.; Wu, D.; Wu, P.; Ni, C.; Zhang, Z.; Chen, Z.; Qiu, F.; Xu, J.; Huang, J. MiRNA-27b targets vascular endothelial growth factor C to inhibit tumor progression and angiogenesis in colorectal cancer. *PLoS ONE* **2013**, *8*, e60687. [CrossRef] [PubMed]
65. Shan, S.; Lu, Y.; Zhang, X.; Shi, J.; Li, H.; Li, Z. Inhibitory effect of bound polyphenol from foxtail millet bran on MiR-149 methylation increases the chemosensitivity of human colorectal cancer HCT-8/Fu cells. *Mol. Cell Biochem.* **2021**, *476*, 513–523. [CrossRef] [PubMed]
66. Menigatti, M.; Staiano, T.; Manser, C.; Bauerfeind, P.; Komljenovic, A.; Robinson, M.; Jiricny, J.; Buffoli, F.; Marra, G. Epigenetic silencing of monoallelically methylated MiRNA loci in precancerous colorectal lesions. *Oncogenesis* **2013**, *2*, e56. [CrossRef]
67. Meng, X.; Wu, J.; Pan, C.; Wang, H.; Ying, X.; Zhou, Y.; Yu, H.; Zuo, Y.; Pan, Z.; Liu, R.-Y.; et al. Genetic and epigenetic down-regulation of MicroRNA-212 promotes colorectal tumor metastasis via dysregulation of MnSOD. *Gastroenterology* **2013**, *145*, 426.e1–6–436.e1–6. [CrossRef]
68. Taheri, Z.; Asadzadeh Aghdaei, H.; Irani, S.; Modarressi, M.H.; Noormohammadi, Z. Clinical correlation of MiR-200c/141 cluster DNA methylation and MiR-141 expression with the clinicopathological features of colorectal primary lesions/tumors. *Rep. Biochem. Mol. Biol.* **2019**, *8*, 208–215.
69. Tanaka, T.; Arai, M.; Wu, S.; Kanda, T.; Miyauchi, H.; Imazeki, F.; Matsubara, H.; Yokosuka, O. Epigenetic silencing of MicroRNA-373 plays an important role in regulating cell proliferation in colon cancer. *Oncol. Rep.* **2011**, *26*, 1329–1335. [CrossRef]
70. Xie, B.; Ding, Q.; Han, H.; Wu, D. MiRCancer: A microRNA–Cancer association database constructed by text mining on literature. *Bioinformatics* **2013**, *29*, 638–644. [CrossRef]
71. Glaich, O.; Parikh, S.; Bell, R.E.; Mekahel, K.; Donyo, M.; Leader, Y.; Shayevitch, R.; Sheinboim, D.; Yannai, S.; Hollander, D.; et al. DNA methylation directs MicroRNA biogenesis in mammalian cells. *Nat. Commun.* **2019**, *10*, 5657. [CrossRef]
72. Suzuki, H.; Maruyama, R.; Yamamoto, E.; Kai, M. DNA methylation and microRNA dysregulation in cancer. *Mol. Oncol.* **2012**, *6*, 567–578. [CrossRef]
73. Kok-Sin, T.; Mohktar, N.M.; Hassan, N.Z.A.; Sagap, I.; Rose, I.M.; Harun, R.; Jamal, R. Identification of diagnostic markers in colorectal cancer via integrative epigenomics and genomics data. *Oncol. Rep.* **2015**, *34*, 22–32. [CrossRef] [PubMed]
74. Garofalo, M.; Croce, C.M. MicroRNAs: Master regulators as potential therapeutics in cancer. *Annu. Rev. Pharmacol. Toxicol.* **2011**, *51*, 25–43. [CrossRef] [PubMed]
75. André, T.; Boni, C.; Mounedji-Boudiaf, L.; Navarro, M.; Tabernero, J.; Hickish, T.; Topham, C.; Zaninelli, M.; Clingan, P.; Bridgewater, J.; et al. Oxaliplatin, fluorouracil, and leucovorin as adjuvant treatment for colon cancer. *N. Engl J. Med.* **2004**, *350*, 2343–2351. [CrossRef]
76. Hon, K.W.; Abu, N.; Ab Mutalib, N.-S.; Jamal, R. MiRNAs and LncRNAs as predictive biomarkers of response to FOLFOX therapy in colorectal cancer. *Front. Pharmacol.* **2018**, *9*, 846. [CrossRef] [PubMed]
77. Wang, B.; Lu, F.-Y.; Shi, R.-H.; Feng, Y.-D.; Zhao, X.-D.; Lu, Z.-P.; Xiao, L.; Zhou, G.-Q.; Qiu, J.-M.; Cheng, C.-E. MiR-26b regulates 5-FU-resistance in human colorectal cancer via down-regulation of pgp. *Am. J. Cancer Res.* **2018**, *8*, 2518–2527.
78. Shi, L.; Li, X.; Wu, Z.; Li, X.; Nie, J.; Guo, M.; Mei, Q.; Han, W. DNA methylation-mediated repression of MiR-181a/135a/302c expression promotes the microsatellite-unstable colorectal cancer development and 5-FU resistance via targeting PLAG1. *J. Genet. Genom.* **2018**, *45*, 205–214. [CrossRef]
79. Wang, F.; Ma, Y.-L.; Zhang, P.; Shen, T.-Y.; Shi, C.-Z.; Yang, Y.-Z.; Moyer, M.-P.; Zhang, H.-Z.; Chen, H.-Q.; Liang, Y.; et al. SP1 mediates the link between methylation of the tumour suppressor MiR-149 and outcome in colorectal cancer. *J. Pathol.* **2013**, *229*, 12–24. [CrossRef]
80. Liu, X.; Xie, T.; Mao, X.; Xue, L.; Chu, X.; Chen, L. MicroRNA-149 increases the sensitivity of colorectal cancer cells to 5-fluorouracil by targeting forkhead box transcription factor FOXM1. *CPB* **2016**, *39*, 617–629. [CrossRef]
81. Liang, Y.; Hou, L.; Li, L.; Li, L.; Zhu, L.; Wang, Y.; Huang, X.; Hou, Y.; Zhu, D.; Zou, H.; et al. Dichloroacetate restores colorectal cancer chemosensitivity through the P53/MiR-149-3p/PDK2-mediated glucose metabolic pathway. *Oncogene* **2020**, *39*, 469–485. [CrossRef]
82. Senfter, D.; Holzner, S.; Kalipciyan, M.; Staribacher, A.; Walzl, A.; Huttary, N.; Krieger, S.; Brenner, S.; Jäger, W.; Krupitza, G.; et al. Loss of MiR-200 family in 5-fluorouracil resistant colon cancer drives lymphendothelial invasiveness in vitro. *Hum. Mol. Genet.* **2015**, *24*, 3689–3698. [CrossRef]
83. Fang, L.; Li, H.; Wang, L.; Hu, J.; Jin, T.; Wang, J.; Yang, B.B. MicroRNA-17-5p promotes chemotherapeutic drug resistance and tumour metastasis of colorectal cancer by repressing PTEN expression. *Oncotarget* **2014**, *5*, 2974–2987. [CrossRef] [PubMed]
84. Chen, Z.; Liu, S.; Tian, L.; Wu, M.; Ai, F.; Tang, W.; Zhao, L.; Ding, J.; Zhang, L.; Tang, A. MiR-124 and MiR-506 inhibit colorectal cancer progression by targeting DNMT3B and DNMT1. *Oncotarget* **2015**, *6*, 38139–38150. [CrossRef] [PubMed]
85. Qian, X.; Yu, J.; Yin, Y.; He, J.; Wang, L.; Li, Q.; Zhang, L.-Q.; Li, C.-Y.; Shi, Z.-M.; Xu, Q.; et al. MicroRNA-143 inhibits tumor growth and angiogenesis and sensitizes chemosensitivity to oxaliplatin in colorectal cancers. *Cell Cycle* **2013**, *12*, 1385–1394. [CrossRef] [PubMed]
86. Zhang, L.-L.; Xie, F.-J.; Tang, C.-H.; Xu, W.-R.; Ding, X.-S.; Liang, J. MiR-340 suppresses tumor growth and enhances chemosensitivity of colorectal cancer by targeting RLIP76. *Eur. Rev. Med. Pharmacol. Sci* **2017**, *21*, 2875–2886.

87. O'Brien, S.J.; Carter, J.V.; Burton, J.F.; Oxford, B.G.; Schmidt, M.N.; Hallion, J.C.; Galandiuk, S. The role of the MiR-200 family in epithelial–mesenchymal transition in colorectal cancer: A systematic review. *Int. J. Cancer* **2018**, *142*, 2501–2511. [CrossRef] [PubMed]
88. Konno, M.; Koseki, J.; Asai, A.; Yamagata, A.; Shimamura, T.; Motooka, D.; Okuzaki, D.; Kawamoto, K.; Mizushima, T.; Eguchi, H.; et al. Distinct methylation levels of mature microRNAs in gastrointestinal cancers. *Nat. Commun.* **2019**, *10*, 1–7. [CrossRef]
89. Ng, E.K.O.; Tsang, W.P.; Ng, S.S.M.; Jin, H.C.; Yu, J.; Li, J.J.; Röcken, C.; Ebert, M.P.A.; Kwok, T.T.; Sung, J.J.Y. MicroRNA-143 targets DNA methyltransferases 3A in colorectal cancer. *Br. J. Cancer* **2009**, *101*, 699–706. [CrossRef]
90. Huang, Z.; Li, Q.; Luo, K.; Zhang, Q.; Geng, J.; Zhou, X.; Xu, Y.; Qian, M.; Zhang, J.; Ji, L.; et al. MiR-340-FHL2 axis inhibits cell growth and metastasis in ovarian cancer. *Cell Death Dis.* **2019**, *10*, 372. [CrossRef]

International Journal of *Molecular Sciences*

MDPI

Article

Tissue-Specific Downregulation of Fatty Acid Synthase Suppresses Intestinal Adenoma Formation via Coordinated Reprograming of Transcriptome and Metabolism in the Mouse Model of Apc-Driven Colorectal Cancer

James Drury [1], Lyndsay E. A. Young [2], Timothy L. Scott [3,4], Courtney O. Kelson [1], Daheng He [5], Jinpeng Liu [5], Yuanyan Wu [5], Chi Wang [5], Heidi L. Weiss [5], Teresa Fan [1,3,4], Matthew S. Gentry [2,3], Ramon Sun [3,6] and Yekaterina Y. Zaytseva [1,3,*]

1 Department of Toxicology and Cancer Biology, University of Kentucky, Lexington, KY 40536, USA; james.drury12@uky.edu (J.D.); courtney.kelson@uky.edu (C.O.K.); teresa.fan@uky.edu (T.F.)
2 Department of Molecular and Cellular Biochemistry, University of Kentucky, Lexington, KY 40536, USA; lyndsay.young@uky.edu (L.E.A.Y.); matthew.gentry@uky.edu (M.S.G.)
3 Markey Cancer Center, University of Kentucky, Lexington, KY 40536, USA; tim.scott@uky.edu (T.L.S.); ramon.sun@uky.edu (R.S.)
4 Center for Environmental and Systems Biochemistry, University of Kentucky, Lexington, KY 40536, USA
5 Markey Cancer Center Biostatistics and Bioinformatics Shared Resource Facility, University of Kentucky, Lexington, KY 40536, USA; daheng.he@uky.edu (D.H.); jinpeng.liu@uky.edu (J.L.); ywu244@g.uky.edu (Y.W.); chi.wang@uky.edu (C.W.); heidi.weiss@uky.edu (H.L.W.)
6 Department of Neuroscience, University of Kentucky, Lexington, KY 40536, USA
* Correspondence: yyzayt2@uky.edu; Tel.: +1-859-218-0134

Citation: Drury, J.; Young, L.E.A.; Scott, T.L.; Kelson, C.O.; He, D.; Liu, J.; Wu, Y.; Wang, C.; Weiss, H.L.; Fan, T.; et al. Tissue-Specific Downregulation of Fatty Acid Synthase Suppresses Intestinal Adenoma Formation via Coordinated Reprograming of Transcriptome and Metabolism in the Mouse Model of Apc-Driven Colorectal Cancer. *Int. J. Mol. Sci.* **2022**, *23*, 6510. https://doi.org/10.3390/ijms23126510

Academic Editor: Stephen Bustin

Received: 27 May 2022
Accepted: 8 June 2022
Published: 10 June 2022

Publisher's Note: MDPI stays neutral with regard to jurisdictional claims in published maps and institutional affiliations.

Abstract: Altered lipid metabolism is a potential target for therapeutic intervention in cancer. Overexpression of Fatty Acid Synthase (FASN) correlates with poor prognosis in colorectal cancer (CRC). While multiple studies show that upregulation of lipogenesis is critically important for CRC progression, the contribution of FASN to CRC initiation is poorly understood. We utilize a C57BL/6-Apc/Villin-Cre mouse model with knockout of FASN in intestinal epithelial cells to show that the heterozygous deletion of FASN increases mouse survival and decreases the number of intestinal adenomas. Using RNA-Seq and gene set enrichment analysis, we demonstrate that a decrease in FASN expression is associated with inhibition of pathways involved in cellular proliferation, energy production, and CRC progression. Metabolic and reverse phase protein array analyses demonstrate consistent changes in alteration of metabolic pathways involved in both anabolism and energy production. Downregulation of FASN expression reduces the levels of metabolites within glycolysis and tricarboxylic acid cycle with the most significant reduction in the level of citrate, a master metabolite, which enhances ATP production and fuels anabolic pathways. In summary, we demonstrate the critical importance of FASN during CRC initiation. These findings suggest that targeting FASN is a potential therapeutic approach for early stages of CRC or as a preventive strategy for this disease.

Keywords: colorectal cancer; fatty acid synthase; lipid metabolism; colorectal cancer initiation; Apc mutation

1. Introduction

Currently ranked as the second leading cause of cancer-related deaths in the United States, colorectal cancer (CRC) remains a substantial public health problem with an estimated 149,500 new cases and 52,980 deaths during 2021 (https://www.cancer.org/cancer/colon-rectal-cancer/about/key-statistics.html, accessed on 31 July 2021). Abnormally elevated lipid synthesis provides cancer cells with membrane building blocks, signaling lipid molecules, posttranslational modifications of proteins, and energy supply to support rapid cell proliferation [1–3]. Fatty Acid Synthase (FASN), a key enzyme of de novo lipid synthesis, has been actively investigated as a therapeutic target in cancer. FASN is the most targetable among lipogenesis genes due to its high degree of overexpression in cancer cells [1,4,5]. Multiple studies, including reports from our laboratory [1,6–8], have found that elevated expression of FASN is associated with advanced stages of CRC and CRC metastasis [6,7]. Pre-clinical studies show significant anti-cancer effects when lipid synthesis is inhibited by genetic and pharmacological inhibition of FASN [9–11]. Currently, a novel FASN inhibitor, TVB-2640, is being tested in one Phase I and two Phase II clinical trials [12]. Even though numerous studies show the benefit of targeting FASN in cancer including CRC, knowledge about the contribution of lipid synthesis to CRC initiation is very limited, and the utility of this pathway as a therapeutic target for the early stages of this disease is unclear.

Multiomics-based analyses of paired normal and tumor tissues from 275 patients with colorectal cancer revealed that metabolic alterations occur at the adenoma stage of carcinogenesis [13]. Several studies have shown that FASN is significantly upregulated in the early stages of CRC [6,13–15]. Increased expression of FASN in 86% of aberrant crypt foci from patients with sporadic CRC or familial adenomatous polyposis also suggests its importance in the early stages of colonic neoplasm development [16]. Consistently, FASN is significantly overexpressed in rectal biopsies from patients harboring adenomas compared with those with no adenomas [17].

A recent study using mouse embryonic fibroblasts transfected with polyomavirus middle T antigen (PyMT), a breast cancer oncoprotein, provides strong evidence that FASN plays an important role in the initial step of cell transformation and is required for cancer cells to acquire 3D growth properties during transformation [18]. FASN deletion results in low rates of glycolysis and mitochondrial respiration as well as an accumulation in reactive oxygen species [18], suggesting that upregulation of fatty acid synthesis promotes tumorigenesis via alteration of metabolic pathways and redox status. This study further emphasizes the potential importance of utilizing FASN as a therapeutic target for the prevention and treatment of early stages of cancer. However, the feasibility of targeting FASN as a preventive strategy or early-stage treatment in CRC has not been explored.

Epidemiological studies demonstrate that diet and dietary fatty acids can contribute to CRC initiation and development [19,20]. Whereas most tumors are dependent on fatty acid synthesis, they can also scavenge lipids from their environment [1]. We have shown that upregulation of CD36, a fatty acid transporter, and an increase in exogenous lipid uptake can compensate for the effect of pharmacological or genetic inhibition of FASN [21]. Therefore, to identify FASN-mediated vulnerabilities which can be efficiently targeted in cancer, it is very important to understand the contribution of FASN to CRC initiation in the context of the in vivo in a mouse model with ad libitum feeding. Mouse models carrying mutations in the *Apc* (adenomatosis polyposis coli) gene are genetically parallel to familial and sporadic colon adenoma development in humans [22]. The alteration of FASN in this model provides a unique opportunity to better understand the contribution of this enzyme and de novo lipid synthesis to Apc-driven carcinogenesis.

We discovered that heterozygous deletion of FASN significantly increases mouse survival and decreases the number of intestinal adenomas formed. Analysis of gene set enrichment data revealed a decrease in FASN expression leading to a significant decrease in the enrichment of genes associated with pathways involved in cellular proliferation, energy production, and CRC progression. In agreement with these data, a decrease in FASN expression reduces the levels of metabolites involved in glycolysis and the tricarboxylic acid (TCA) cycle with the most significant reduction in the level of cellular citrate, a metabolite involved in ATP production and fueling of anabolic pathways. Using the reverse phase protein array (RPPA), we demonstrate alterations in the protein levels of multiple metabolic enzymes. Interestingly, the levels of diglycerides in adenomas and free fatty acids in adenomas and plasma are not affected by the changes in FASN expression in the intestine.

In summary, this study provides strong evidence that FASN is critically important in CRC initiation by orchestrating changes in the transcriptome and metabolic pathways consistent with an increase in proliferation, ATP production, and anabolism. Therefore, targeting FASN should be further explored as a potential preventive strategy or early-stage treatment for CRC.

2. Results

2.1. Heterozygous Deletion of FASN Increases Survival and Decreases the Number of Intestinal Adenomas during Apc-Driven Carcinogenesis

We utilized a C57BL/6-*Apc*^{tm1Tyj} mouse model in which the 15 coding exons of the *Apc* gene are flanked by LoxP sites. Germline heterozygous deletion of the floxed region results in a mouse highly suspectable to spontaneous intestinal adenoma formation and serves as a mouse model of CRC. Immunohistochemistry staining of intestinal tissues from Apc/Villin-Cre (Apc/Cre) mice demonstrates high expression of FASN in adenomas as compared to surrounding tissues (Figure 1A). These mice were bred with mice that have LoxP-flanked FASN alleles [23] to establish mouse colonies with heterozygous (FASN$^{+/\Delta}$/Apc/Cre) and homozygous (FASN$^{\Delta/\Delta}$/Apc/Cre) deletion of FASN in the intestinal epithelium (Figure S1A) [21]. Consistent with a previously published study, homozygous deletion of FASN in the intestine leads to a smaller size litter and premature death of approximately 70–80% of animals within 2 months after birth due to disruptions in the intestinal mucus barrier [24]. However, approximately 20% of FASN$^{\Delta/\Delta}$/Apc/Cre mice survive and have a phenotype similar to FASN$^{+/\Delta}$/Apc/Cre mice (Figure S1B–D). Immunohistochemistry staining demonstrates residual expression of FASN in these mice (Figure S1E,F), potentially due to the inefficiency of the *Cre* transgene [25], which would explain the phenotype similar to FASN$^{+/\Delta}$/Apc/Cre mice. We previously demonstrated that hetero- and homozygous deletion of FASN in Apc/Cre mice lead to the upregulation of CD36, a fatty acid transporter [21]. Indeed, immunohistochemistry staining reveals a high level of CD36 expression in FASN$^{\Delta/\Delta}$/Apc/Cre intestinal mucosa, suggesting that a potential compensation of inhibited lipid synthesis is increased fatty acid uptake (Figure S1G). Due to the variability in FASN expression and survival in FASN$^{\Delta/\Delta}$/Apc/Cre group, these animals were omitted from the survival analysis and studies evaluating the number of intestinal adenomas.

To investigate the contribution of FASN to survival and adenoma formation, we used Apc/Cre and FASN$^{+/\Delta}$/Apc/Cre mice kept on standard laboratory chow. The heterozygous deletion of FASN significantly increases mouse survival (Figure 1B) and decreases the number of intestinal adenomas in both male and female mice (Figure 1C,D). There were no significant changes in animal size due to differences in FASN expression (Figure S1B,C). We noted that female mice develop a higher number of adenomas and have shorter survival as compared to male mice, but the differences between genders were not statistically significant (Figure 1B,C). Even though we excluded FASN$^{\Delta/\Delta}$/Apc/Cre mice from analysis, we noted that the higher degree of FASN inhibition in this group is associated with a much lower number of intestinal adenomas. Several mice in this group were adenoma-free and survived over a year until they were sacrificed for analysis (Figure S1B,C).

Ki-67 expression has been widely used in clinical practice as an index to evaluate the proliferative activity of tumor cells [26]. The sections of the intestine from Apc/Cre, FASN$^{+/\Delta}$/Apc/Cre, and FASN$^{\Delta/\Delta}$/Apc/Cre mice were prepared using the swiss roll technique, and immunohistochemistry staining for Ki67 was performed. The analysis of Ki67 staining using the HALO digital pathology analysis platform revealed high expression of FASN in Apc/Cre adenomas is associated with a higher percentage of Ki67 positive cells and higher intensity staining as compared to FASN$^{+/\Delta}$/Apc/Cre and FASN$^{\Delta/\Delta}$/Apc/Cre adenomas (Figure 1E,F), suggesting a higher proliferative capability of these adenomas.

Together, these data suggest that FASN promotes CRC initiation by increasing the number of adenomas formed and the proliferation of CRC cells, thus decreasing mouse survival.

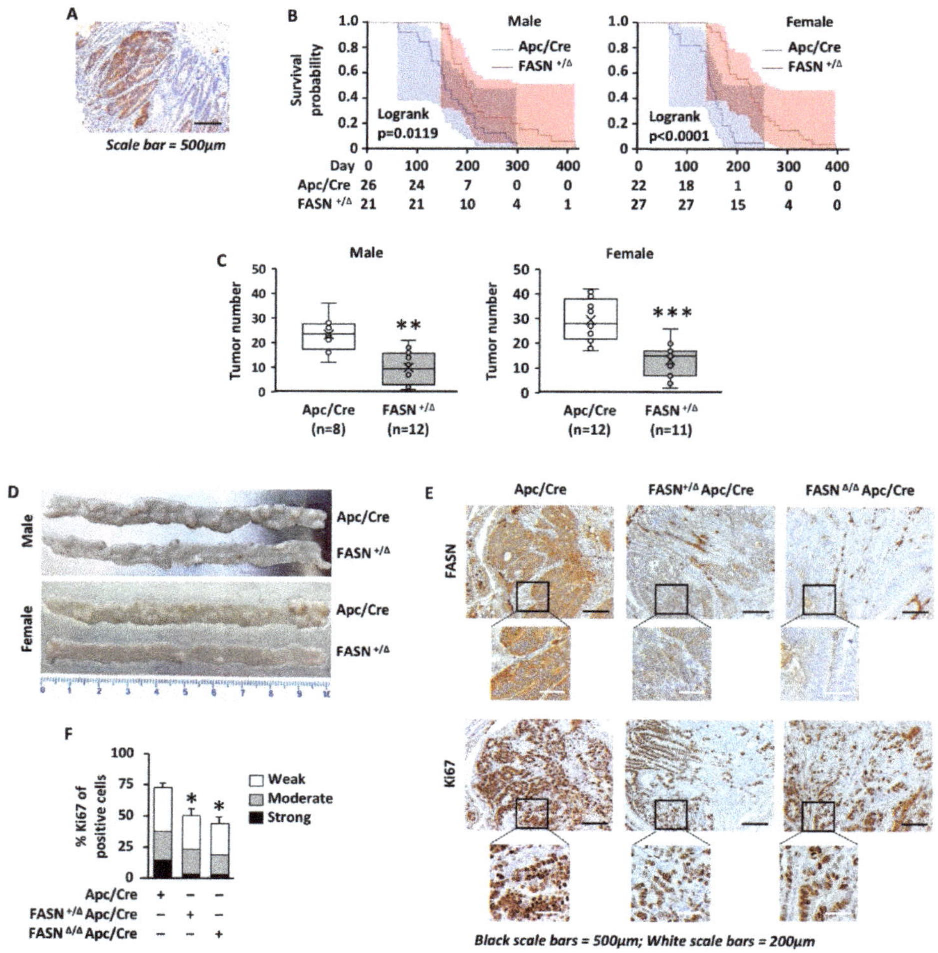

Figure 1. Heterozygous deletion of FASN increases mice survival and decreases the number of adenomas formed in the Apc-driven carcinogenesis mouse model. (**A**) Immunohistochemistry staining for FASN in intestinal tissues from Apc/Cre mice. (**B**) The effect of Villin-Cre-mediated heterozygous deletion of FASN in mouse intestinal tissues on mice survival. (**C**) The effect of Villin-Cre-mediated heterozygous deletion of FASN in mouse intestinal tissues on formation of mouse adenomas. Number of adenomas was quantified within 10 cm sections of distal intestine from Apc/Cre mice and mice with heterozygous deletion of FASN. (**D**) Representative images of intestinal tissues from male and female mice, Apc/Cre vs. Apc/Cre with heterozygous deletion of FASN. (**E**) Representative images of immunochemistry staining for FASN and Ki67 in Apc/Cre mice and Apc/Cre mice with hetero- and homozygous deletion of FASN. (**F**) Quantification of Ki67 staining in mouse adenomas with the different levels of FASN expression. (* $p < 0.05$, ** $p < 0.01$, *** $p < 0.001$).

2.2. A Decrease in FASN Expression Is Associated with Downregulation of Pathways Linked to Cellular Proliferation, Energy Production, and Cancer-Associated Signaling

To determine the effect of FASN on gene expression profile during Apc-driven carcinogenesis, we performed RNA-Seq analysis on adenomas collected from Apc/Cre, $FASN^{+/\Delta}$/Apc/Cre and $FASN^{\Delta/\Delta}$/Apc/Cre mice (Table S1A,B). Venn diagram shows numbers of overlapping and non-overlapping genes differentially expressed among three different genotypes (Figure 2A).

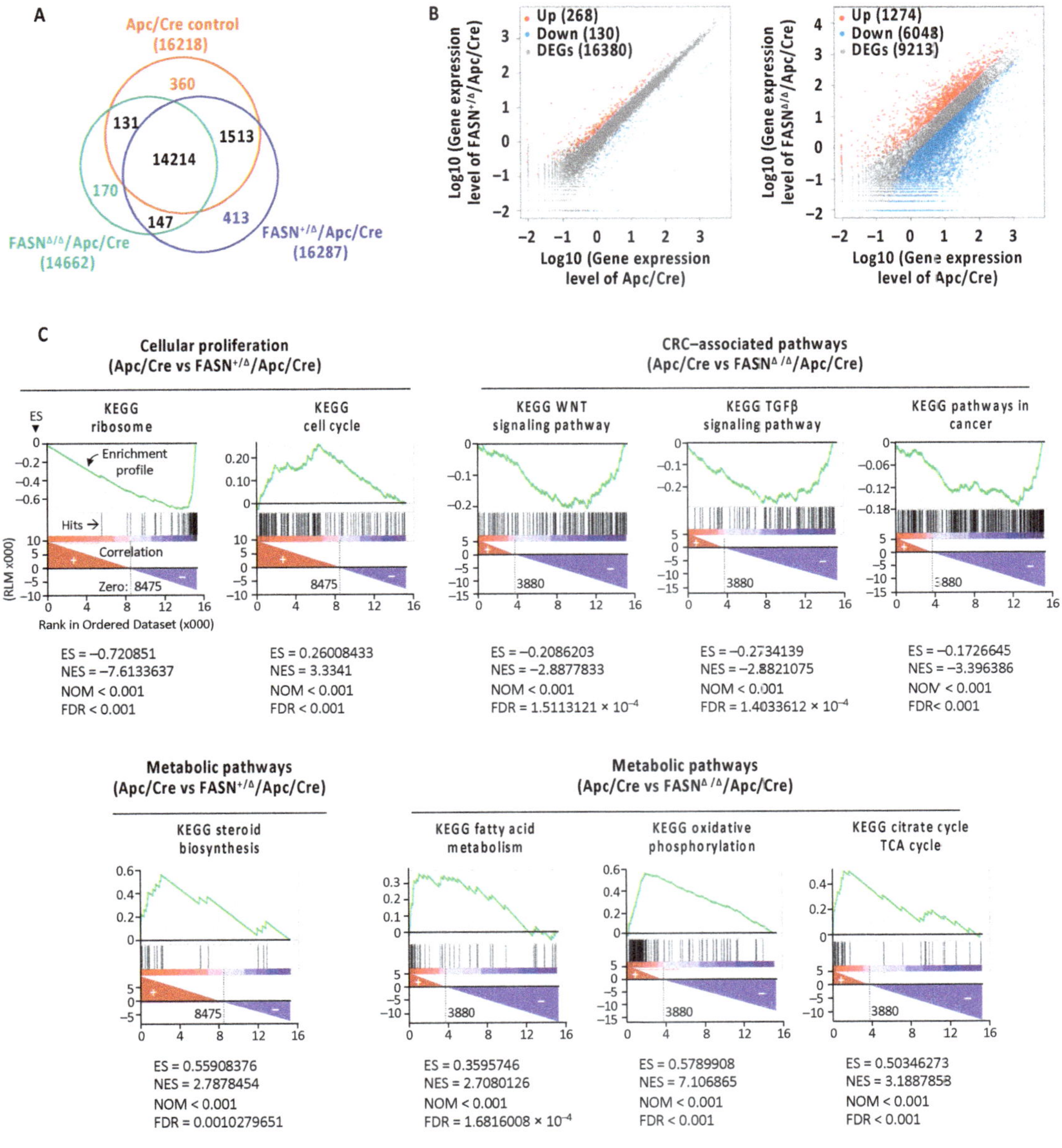

Figure 2. FASN promotes adenoma formation via pathways upregulation of pathways involved in cell growth and energy metabolism. (**A**) Venn diagram displaying the overlapping genes identified in tumors from Apc/Cre mice and Apc/Cre mice with heterozygous and homozygous expression of FASN. (**B**) Scatter plots showing the numbers of differentially expressed genes in Apc/Cre mice and Apc/Cre mice with heterozygous and homozygous expression of FASN. Significantly changed DEGs are indicated in colors. Red and blue dots are up- and downregulated genes, respectively. The detailed lists of differentially expressed genes are provided in Table S2A. (**C**) Representative gene set enrichment analysis plots generated from RNA-Seq expression data of Apc/Cre and FASN knockout mice. The bar codes indicate the location of the members of the gene set in the ranked list of all genes. ES, enrichment score; NES, normalized enrichment score; NOM, nominal *p*-value; FDR, false discovery rate adjusted *p*-value.

Deletion of FASN resulted in significant transcriptome changes with exacerbated changes observed in $FASN^{\Delta/\Delta}$/Apc/Cre mice compared to $FASN^{\Delta/-}$/Apc/Cre mice (Figure 2B, Table S1C-1,C-2). Gene set enrichment analysis shows that high FASN expression in adenomas from Apc/Cre mice is associated with enrichment of genes associated with cellular proliferation, energy production, and oncogenic signaling as compared to adenomas collected from mice with hetero- and homozygous deletion of FASN (Figure 2C). Lists of the top 20 positively and negatively enriched pathways and genes associated with downregulation of FASN in $FASN^{+/\Delta}$/Apc/Cre and $FASN^{\Delta/\Delta}$/Apc/Cre mice are included in Table S1D-1–D-4.

In summary, these data further confirm that FASN promotes adenoma formation via altered expression of genes involved in proliferation, energy production, and CRC progression.

2.3. Heterozygous Deletion of FASN Alters the Levels of Diglycerides, but Does Not Change the Total Levels of Free Fatty Acid and Sphingolipid Species in Mouse Adenomas

We have shown that the shRNA-mediated knockdown of FASN abolishes de novo lipid synthesis but does not affect the total palmitate level in established CRC cell lines [6]. Adenoma tissues and plasma from Apc/Cre and $FASN^{+/\Delta}$/Apc/Cre mice were analyzed to determine the levels of free fatty acids. Due to the limited number of $FASN^{\Delta/\Delta}$/Apc/Cre mice available and the low number of adenomas formed in this genotype group, we did not perform lipid analysis for this genotype. As shown in Figure 3A,B and Table S2A,B, we did not find any significant differences in the levels of free fatty acids between Apc/Cre and $FASN^{+/\Delta}$/Apc/Cre mice. Our previous studies suggest that FASN activity regulates lipid storage and de novo sphingolipid synthesis [8,27]. Therefore, we measured the total levels of diglycerides and sphingolipids in adenomas from Apc/Cre and $FASN^{+/\Delta}$/Apc/Cre mice. We observed a significant decrease in the levels of some diglycerides (C16:1:20:0-DAG; Di-C14-DAG; C14:0:18:0-DAG; C16:1:18:0-DAG; C18:0:18:1-DAG, p-value < 0.05) (Figure 3C and Table S2C). However, we would like to note that further statistical analysis and adjustments for the false discovery rate (q-value) show no statistical significance. No significant changes were seen in the total levels of sphingolipids (Figure 3D and Table S2D). We have also analyzed the levels of triglycerides in adenoma tissue and plasma of Apc/Cre, $FASN^{+/\Delta}$/Apc/Cre, and $FASN^{\Delta/\Delta}$/Apc/Cre using a triglyceride quantification kit. As shown in Figure S2A,B, no significant differences were observed between Apc/Cre mice and Apc/Cre mice with hetero- and homozygous deletion of FASN. Together, these data suggest that the heterozygous deletion of FASN primarily alters the levels of diglycerides but does not significantly affect the total level of free fatty acids and sphingolipids in mouse adenomas.

2.4. Downregulation of FASN in Mouse Adenomas Alters the Levels of Cellular Metabolites

Our previous studies demonstrated that FASN regulates glycolysis and mitochondrial respiration in vitro [8]. To assess FASN-mediated changes of metabolites in glycolysis and the TCA cycle, we performed a metabolic analysis of adenomas from Apc/Cre, $FASN^{+/\Delta}$/Apc/Cre, and $FASN^{\Delta/\Delta}$/Apc/Cre mice using GC-MS. Polar metabolites were extracted from pulverized tumors of the three genotypes and metabolites were identified. Metabolites from all major pathways were detected, including glycolysis and TCA cycle intermediates, lipids, sugars, amino acids, and others. Using this information-rich dataset, supervised clustering analysis was performed to assess overall metabolic profiles for each cohort. The heat map of these data demonstrated significant changes in the metabolite levels among Apc/Cre, $FASN^{+/\Delta}$/Apc/Cre, and $FASN^{\Delta/\Delta}$/Apc/Cre (Figure 4A). The PLS-DA further demonstrates a significant difference in the metabolic profiles of adenomas among Apc/Cre, $FASN^{+/\Delta}$/Apc/Cre, and $FASN^{\Delta/\Delta}$/Apc/Cre mice (Figure 4B). Based on VIP score, citric acid, cholesterol, alanine, uridine, glutamate, 6-phosphoglucose, and palmitate were identified as the top metabolic intermediates that contribute to FASN-driven differences observed among mice genotypes (Figure 4C).

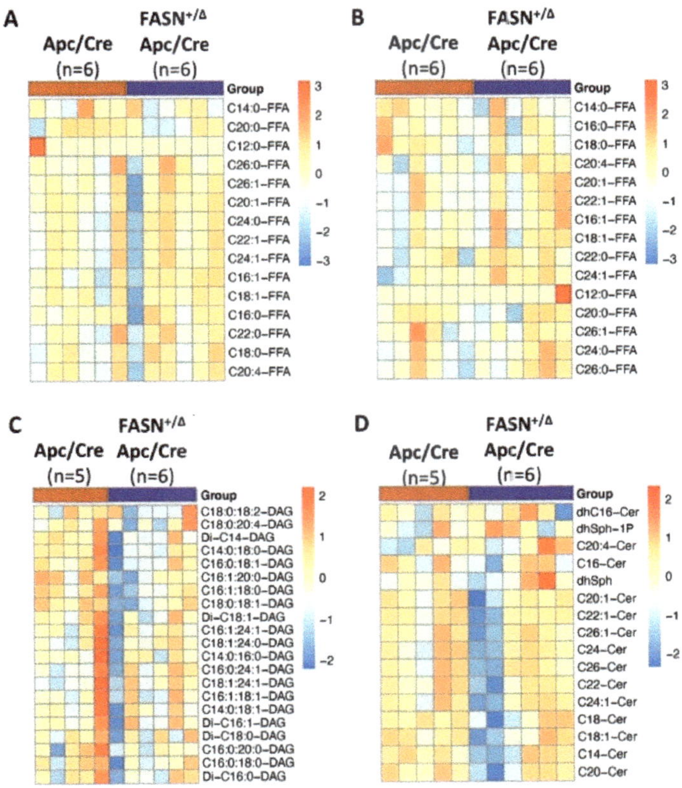

Figure 3. The effect of heterozygous deletion of FASN on lipid composition. Heat maps showing the composition of free fatty acids in (**A**) adenoma tissues and (**B**) plasma, and the composition of (**C**) diglycerides and (**D**) sphingolipids in mouse adenomas from Apc/Cre and FASN$^{+/\Delta}$ Apc/Cre mice.

Taken together, these data demonstrate that FASN upregulation is associated with metabolic pathways involved in the turnover of citrate, palmitate, cholesterol, and 6-phosphoglucose, as well as in the synthesis of uridine and amino acids such as alanine and glutamate.

2.5. Deletion of FASN Alters Expression of Metabolites and Their Metabolizing Enzymes in Adenomas

To profile the expression of metabolic enzymes within glycolysis and the TCA cycle, we performed reverse phase protein array (RPPA) [28]. The heat map of the RPPA analysis shows the levels of metabolic enzyme expression in adenomas collected from Apc/Cre, FASN$^{+/\Delta}$/Apc/Cre, and FASN$^{\Delta/\Delta}$/Apc/Cre mice (Figure 5A). The expression of FASN in samples analyzed by RPPA is also shown by western blot (Figure 5B). Interestingly, we observed an increase in the expression of glucose transporter 1 and glycolytic enzyme hexokinases 2 and 3. The expression of glucose-6-phosphate dehydrogenase (G6PD), a rate-limiting enzyme of the pentose phosphate pathway is also increased (Figure 5A,C and Table S3). In contrast, the enzymes glutamine-fructose-6-phosphate transaminase 1 and O-linked N-acetylglucosamine transferase, which are involved in hexosamine synthesis, a branch of glycolysis and O-linked glycosylation, significantly decreased (Figure 5A and Table S3). The decrease in the levels of glucose in adenomas from FASN$^{+/\Delta}$/Apc/Cre and FASN$^{\Delta/\Delta}$/Apc/Cre mice suggests that glycolytic enzymes may be upregulated due to the limited substrate availability (Figure 5C).

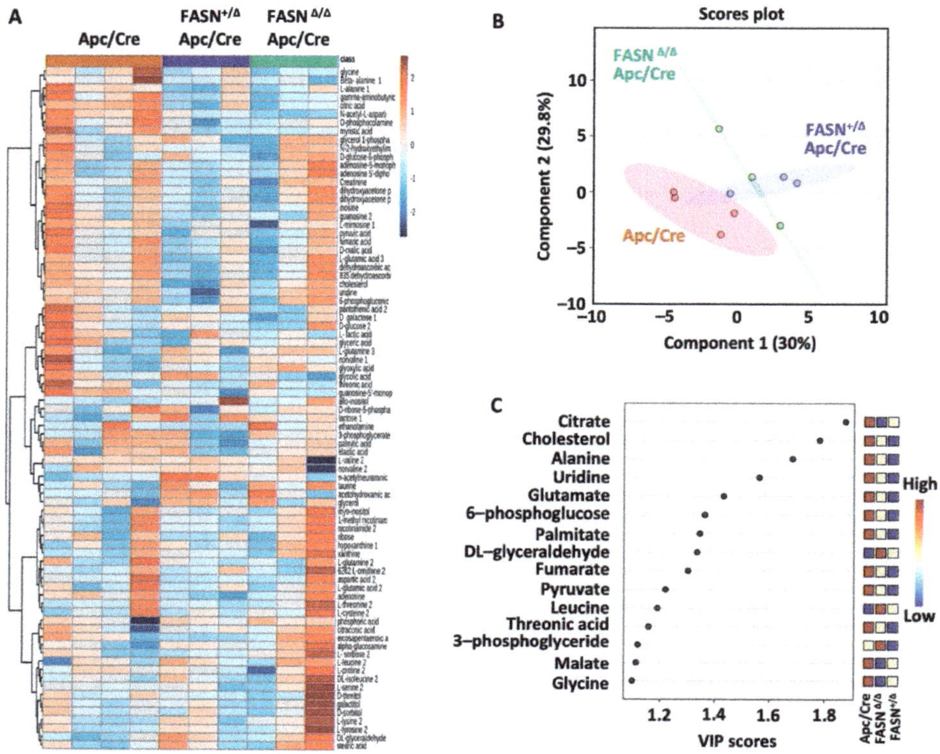

Figure 4. Downregulation of FASN is associated with alteration in multiple metabolic intermediates in mouse adenomas. (**A**) Heat map demonstrating the levels of metabolites (log normalization) identified in adenoma tissues from Apc/Cre (n = 4) and Apc/Cre mice with hetero-(n = 3) and homozygous (n = 3) deletion of FASN. (**B**) Partial least-squares discriminant analysis (PLS-DA) on metabolic data from Apc/Cre, FASN$^{+/\Delta}$ Apc/Cre, and FASN $^{\Delta/\Delta}$ Apc/Cre mice. (**C**) Variable importance in projection (VIP) values from PLS-DA. The VIP score of a metabolite is calculated as a weighted sum of the squared correlations between the PLS-DA components and the original variable (FASN expression). The x-axis indicates the VIP scores corresponding to each metabolite on the y-axis.

Interestingly, even though we did not observe a change in the level of lactate, levels of pyruvate and glucose trended lower in FASN$^{+/\Delta}$/Apc/Cre and FASN$^{\Delta/\Delta}$/Apc/Cre mice as compared to Apc/Cre mice (Figure 5C). Pyruvate dehydrogenase E1 subunit alpha 1, a component of the pyruvate dehydrogenase enzyme complex, links glycolysis and the TCA cycle and is important for cancer metabolic shift [29]. Strikingly, we observed significant downregulation of this enzyme in adenomas from FASN$^{+/\Delta}$/Apc/Cre and FASN$^{\Delta/\Delta}$/Apc/Cre mice as compared to Apc/Cre mice (Figure 5A,C and Table S3).

Nicotinamide adenine dinucleotide phosphate (NADPH) is produced by metabolic enzymes such as G6PD and 6-phosphogluconate dehydrogenase (6PGD) of the pentose phosphate pathway, malic enzymes (MEs), isocitrate dehydrogenases, and enzymes in one-carbon-tetrahydrofolate oxidation pathways [30]. Intriguingly, we found the ME1, a cytosolic NADP^{+}-dependent isoform, is upregulated in FASN$^{+/\Delta}$/Apc/Cre and FASN$^{\Delta/\Delta}$/Apc/Cre mice as compared to Apc/Cre mice, and ME2, a mitochondrial NAD^{+}-dependent isoform, is significantly downregulated (Figure 5A and Table S3). ME1 plays important role in generating NADPH for lipid and cholesterol synthesis and increases FASN expression in the intestine, suggesting that upregulation of ME1 may be a potential compensation mechanism due to a decrease in lipid synthesis.

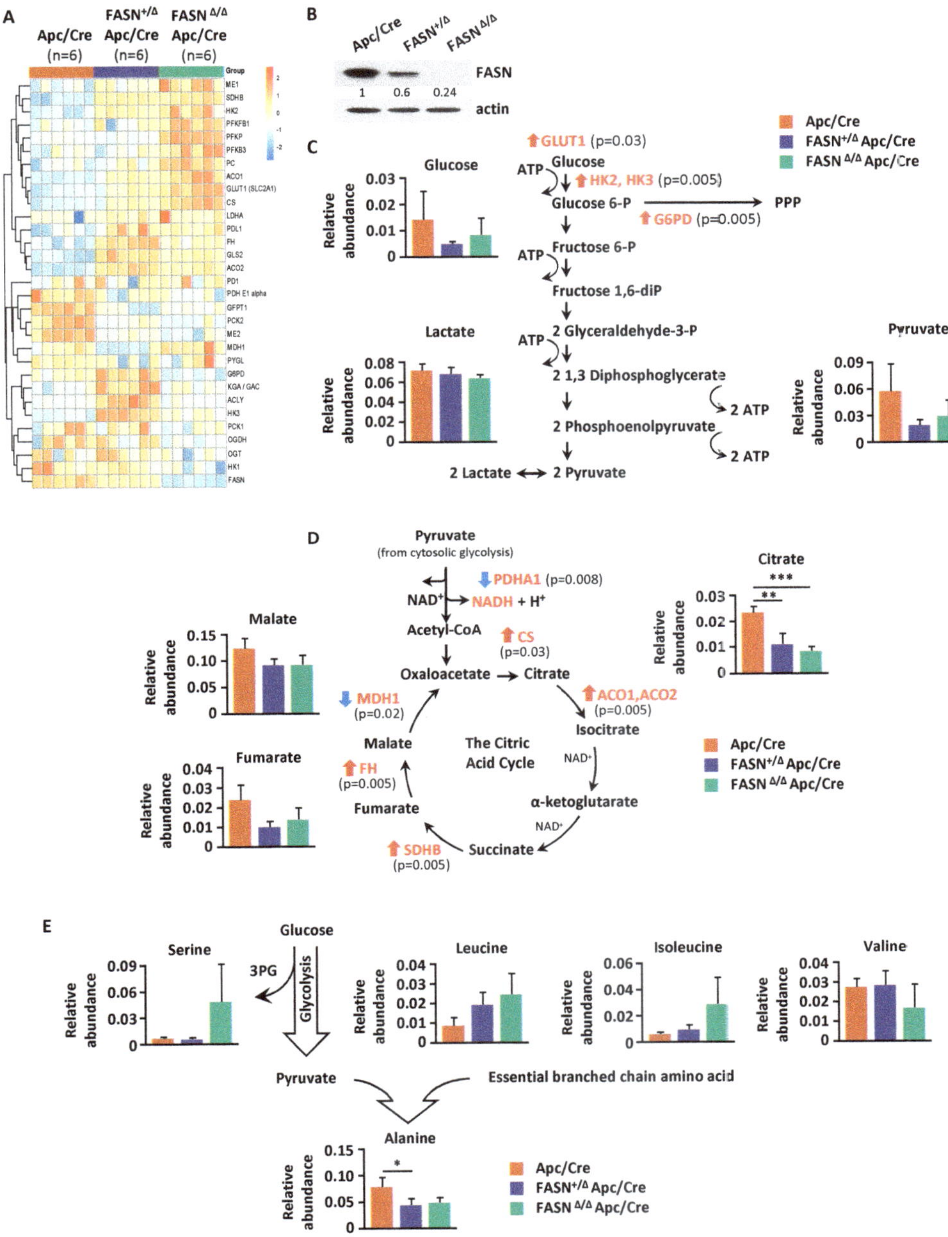

Figure 5. FASN knockdown alters the levels of metabolites and metabolic enzymes involved in glycolysis, TCA cycle, and amino acid metabolism. (**A**) Heat map demonstrating expression of metabolic enzymes as determined by Reverse Phase Protein Array (RPPA) analysis. (**B**) The level of FASN expression by western blot of mouse adenomas used for RPPA analysis. (**C**) FASN-mediated changes in metabolites and metabolizing enzymes within glycolysis. (**D**) FASN-mediated changes in metabolites and metabolizing enzymes within the TCA cycle. (**E**) Hetero- and homozygous knockdown of FASN alters the levels of branched-chain amino acids and decreases synthesis of alanine. (* $p < 0.05$, ** $p < 0.01$, *** $p < 0.001$).

The TCA cycle constitutes the epicenter of cell metabolism because it oxidizes multiple substrates while providing precursors for the synthesis of lipids, nucleotides, and proteins [31]. The TCA cycle begins with the reaction that combines the two-carbon acetyl-CoA with a four-carbon oxaloacetate to generate the six-carbon citrate [31]. We found that the level of citrate is significantly decreased in both FASN$^{+/\Delta}$/Apc/Cre and FASN$^{\Delta/\Delta}$/Apc/Cre mice as compared to Apc/Cre mice (Figure 5D). Furthermore, according to the PLS-VIP analysis, citrate has the highest VIP score (see Figure 4C), suggesting its significance in FASN-mediated metabolic changes in our genetically modified mouse models. We noted that other TCA cycle substrates such as malate and fumarate are also decreased in FASN$^{+/\Delta}$/Apc/Cre and FASN$^{\Delta/\Delta}$/Apc/Cre mice as compared to Apc/Cre mice, but these changes did not reach statistical significance (Figure 5D).Even though the level of citrate significantly decreased, the expression of CS increased, potentially due to feedback regulation by the availability of the precursor pyruvate (thus acetyl CoA) (Figure 5A,C and Table S3). We confirmed that knockout of FASN leads to upregulation of CS in mouse tissues using western blot analysis (Figure S3). Intriguingly, the level of mitochondrial isoform phosphoenolpyruvate carboxykinase 2, which converts oxaloacetate into phosphoenolpyruvate, is significantly decreased due to FASN downregulation, suggesting a decrease in utilization of non-glucose substrates such as glutamine to fuel the pentose phosphate pathway and serine/glycine metabolism [32].

Amino acids play diverse roles in cancer cells, supporting biosynthetic pathways, redox balance, epigenetic regulation, and immune responses [33,34]. Alanine can be synthesized from pyruvate and branched-chain amino acids (BCAA) such as valine, leucine, and isoleucine [35] and plays an important role in the synthesis of proteins, amino acids, and other biomolecules as well as energy fuel for cancer cells [36]. We found that both the hetero- and homozygous deletion of FASN increase the levels of leucine and isoleucine, and significantly decrease the level of alanine in intestinal adenomas (Figure 5E), suggesting the potential impairment in BCAA metabolism.

Cumulatively, these data suggest that a decrease in FASN expression is associated with a decrease in metabolic intermediates of the TCA cycle with the most significant change in the level of intracellular citrate, which is a key metabolite supporting mitochondrial ATP production and anabolic reactions.

2.6. *Expression of CS Is Upregulated and Correlates with Expression of FASN in Colorectal Cancer*

CS activation drives flux toward lipid and triglyceride synthesis in cancer [37]. Since FASN knockout upregulates CS expression in transgenic mouse models, we sought to evaluate the levels of FASN and CS expression in mouse tissues during Apc-driven carcinogenesis. Immunohistochemistry staining of FASN and CS revealed that both are significantly upregulated in intestinal epithelium and adenomas in Apc/Cre mice as compared to intestinal tissues of wild-type C57BL/6J mice (Figure 6A). We noted that even though the expression pattern of CS does not always recapitulate that of FASN, their expression seems to localize in the same areas of developing adenomas (Figure 6A). Interestingly, the inactivation of *Apc* gene using tamoxifen-inducible Villin-Cre-ERT2 resulted in upregulation of both FASN and CS expression in the intestinal mucosa (Figure 6B). To translate our findings to human cancer, we analyzed The Cancer Genome Atlas RNA-Seq data and identified a highly significant positive correlation between FASN and CS gene expression in human colon adenocarcinomas (Figure 6C). In agreement with these data, western blot analysis of fresh normal human colon mucosa and matched primary tumors demonstrates an increase in FASN and CS in tumors as compared to normal tissues (Figure 6D).

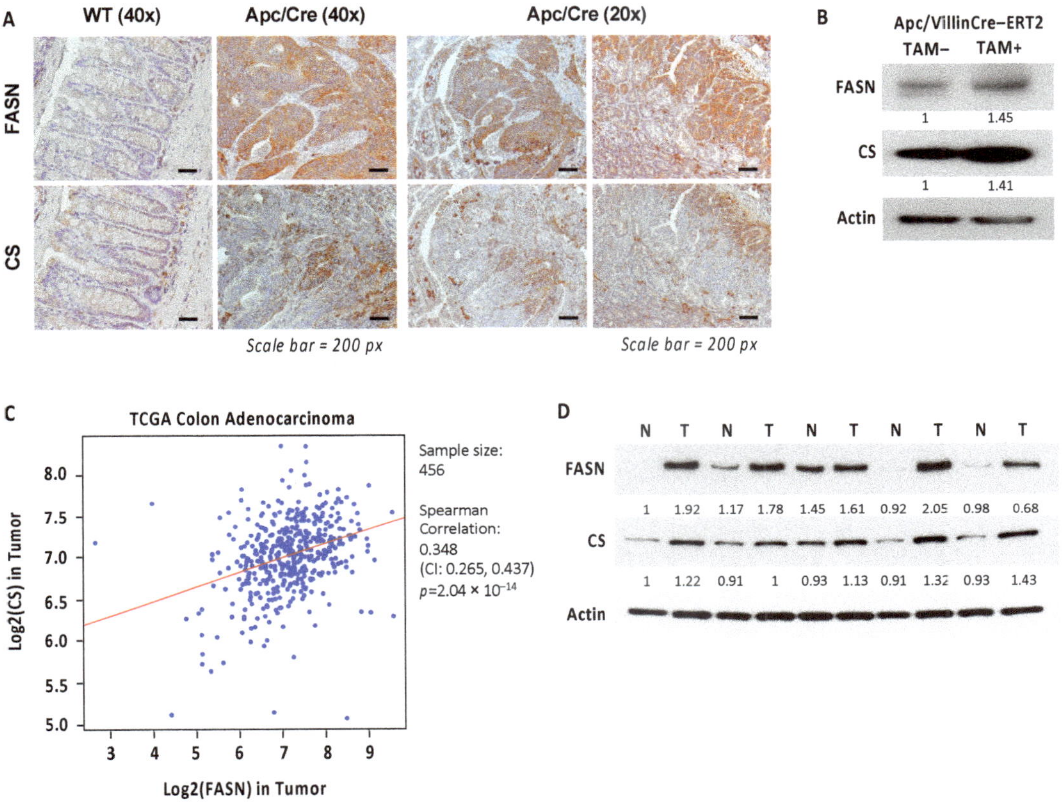

Figure 6. Expression of FASN correlates with expression of CS in CRC. (**A**) Expression of FASN and CS in intestinal tissues of wild type (WT) C57BL/6J mice and in intestinal tissues and adenomas in Apc/Cre mice. (**B**) Heterozygous deletion of *Apc* gene using tamoxifen (TAM) inducible Villin-Cre recombinase leads to upregulation of FASN and CS in mouse intestinal tissues. Tamoxifen was injected for 5 days, and intestinal tissues were collected on day 10 after the last injection. (**C**) Correlations between FASN and CS were determined based on RNA-Seq data of CRC patient tumor tissues (n = 456) from The Cancer Genome Atlas. (**D**) Expression of FASN and CS in human normal colon mucosa and matched tumor tissues. N-normal mucosa, T-primary tumor.

In summary, these data suggest that CS is highly expressed in CRC, and there is a positive correlation between CS and FASN expression at mRNA and protein levels in colon adenocarcinomas.

3. Discussion

Aberrant lipid synthesis via upregulation of FASN is crucial for cancer cells, and targeting FASN can be a potential therapeutic strategy in many cancers, including CRC [1,3]. Even though multiple studies demonstrate that inhibition of FASN decreases CRC cell growth and survival in vitro, pre-clinical studies demonstrate much less efficacy of FASN inhibition on tumor growth in vivo, potentially due to compensation mechanisms such as dietary fatty acid uptake [9,21]. A better understanding of the timing and conditions for using FASN-targeted therapy is critical for developing successful therapeutic strategies [4,9,10]. A recently published study suggests that FASN activity could be essential during the initial steps of the transformation process and, thus, can be a target for cancer prevention [18]. However, the contribution of FASN to CRC initiation has not been extensively studied. Therefore, the goal of this study was to understand the role of FASN in Apc-driven carcinogenesis and to evaluate it as a potential target for CRC prevention using transgenic mouse models.

Our study is the first to elucidate the effect of Villin-Cre-mediated downregulation of FASN expression in intestinal epithelial cells on mice survival, adenoma formation, and transcriptome and metabolome of adenomas in the transgenic model of Apc-driven CRC. The results of our study show that heterozygous deletion of FASN in Apc/Cre mice significantly increases mouse survival and decreases the number of intestinal adenomas. These results are consistent with the study showing that inhibition of FASN by orlistat, a drug used for treating obesity [38], increases survival rates in Apc^{Min} mice, a commonly used model for Apc-driven CRC [39]. Another study using pharmacological inhibitors in Apc^{Min} mice showed that orlistat and lovastatin, inhibitors of cholesterol biosynthesis, significantly reduced FASN enzymatic activities and gene expression in colonic tissues. However, they did not affect the number of intestinal polyps and there was a statistically significant reduction in polyp volume only in the mouse group treated with lovastatin [40]. The discrepancy in outcomes of these studies is potentially due to different diets and different doses of orlistat used [39,40]. Our results showing that FASN promotes carcinogenesis are also in agreement with studies on other types of cancer, showing that transgenic expression of FASN results in a significant increase in prostate intraepithelial neoplasia [41] and that pharmacological inhibition of FASN with Fasnall [42] or C75 [43] significantly delays tumor progression in *neu*-N mice, a model of mammary cancer.

The meta-analysis of 34 studies and 6180 CRC patients demonstrates that high expression of Ki67, a proliferation marker, is significantly correlated with poor overall survival and disease-free survival [44]. Indeed, our results demonstrate that high expression of FASN in Apc/Cre mice is associated with a higher percentage of Ki67 positive cells. Conversely, downregulation of FASN expression in $FASN^{+/\Delta}$/Apc/Cre and $FASN^{\Delta/\Delta}$/Apc/Cre is associated with a decrease in the percentage of Ki67 positive cells and the intensity of Ki67 staining, suggesting that high expression of FASN is associated with a higher proliferative activity of CRC cells. These findings are further supported by the GSEA analysis of RNA-Seq data on adenomas from Apc/Cre, $FASN^{+/\Delta}$/Apc/Cre, and $FASN^{\Delta/\Delta}$/Apc/Cre mice showing the significant enrichment of genes involved in cell cycle progression in Apc/Cre mice as compared to adenomas from mice with hetero- and homozygous deletion of FASN.

Our data from transgenic mice demonstrate that FASN significantly upregulates the set of genes associated with the pathways involved in the cell cycle, steroid biosynthesis, and metabolism. These results are in agreement with data obtained from in vitro studies on human CRC cells showing that similar pathways are modulated by TVB-3166, a FASN inhibitor and an analog of TVB-2640, which is currently used in clinical trials [10,45]. Together, the data further confirm the specificity of TVB inhibitors in targeting FASN in cancer cells. Interestingly, we also found that gene expression of several stem cell markers implicated in CRC, such as LGR5, ALDH, CD44, and CD166, is significantly downregulated in FASN knockout adenomas, suggesting that FASN may promote APC-driven carcinogenesis via an increase in stemness of intestinal epithelial cells. This mechanism is currently under investigation in our laboratory.

Our previous studies demonstrated that genetic and pharmacological inhibition of FASN is associated with the inhibition of glycolysis, TCA cycle activity, and beta-oxidation in vitro [8]. Consistently, low respiration and glycolytic capacity were observed when FASN was deleted in MEFs infected with retroviral particles expressing the PyMT breast cancer oncogene ($FASN^{\Delta/\Delta}$-PyMT) as compared to control $FASN^{lox/lox}$-PyMT MEFs [18]. In agreement with these studies, our GSEA results show reduced expression of genes involved in the TCA cycle and beta-oxidation in $FASN^{+/\Delta}$/Apc/Cre and $FASN^{\Delta/\Delta}$/Apc/Cre as compared to Apc/Cre mice. Interestingly, heterozygous deletion of FASN leads to a drastic and significant decrease in the expression of genes associated with the cell cycle and ribosome pathways, suggesting that an approximate 50% decrease in FASN expression in these mice is sufficient to significantly inhibit cellular proliferation and protein synthesis.

Steroid biosynthesis is also significantly downregulated in these mice as compared to Apc/Cre control mice. The higher extent of FASN downregulation in mice with homozygous deletion leads to additional alterations in gene expression associated with inhibition of energy production, fatty acid biosynthesis, and CRC-promoting pathways, suggesting that a higher degree of FASN inhibition elicits a more efficient and global antitumor effect.

Our previous work demonstrated that shRNA-mediated deletion of FASN significantly decreases the incorporation of ^{13}C sodium acetate into palmitate but does not affect the total palmitate level [6]. Similarly, the relative independence from FASN activity to maintain stable intracellular lipid levels was observed in FASN$^{lox/lox}$-PyMT and FASN$^{\Delta/\Delta}$-PyMT MEFs [18]. In this study, we noted that heterozygous deletion of FASN in intestinal epithelial cells decreased the levels of several diglycerides in adenomas. However, we did not observe any unequivocal changes in the total levels of free fatty acids or sphingolipids in adenomas from mice with heterozygous deletion of FASN. We did not perform the comprehensive lipid analysis on adenomas from FASN$^{\Delta/\Delta}$/Apc/Cre mice due to the limited number of mice of this genotype. However, analysis of the triglyceride levels in intestinal tissue and plasma from these mice shows a similar, statistically non-significant decrease in triglycerides for FASN$^{+/\Delta}$/Apc/Cre mice compared to Apc/Cre. These results have several potential explanations. We have previously shown that both shRNA-mediated and pharmacological inhibition of FASN led to increased FA uptake [9,21], suggesting that the cellular lipid pool may be replenished by dietary FAs. Alternatively, in the Villin-Cre mouse model we used, Cre recombinase is expressed in villus and crypt epithelial cells of the small and large intestines, but the expression of FASN is intact in other cell types within intestinal tissues and in other organs. FASN is highly expressed in liver and adipose tissues [1]. It has been shown that the contribution of liver fatty acid synthesis appears to be less than that of fats derived from peripheral tissues or dietary fat [46] but FAs from all these sources can contribute to the total level of circulating lipids. The components of the tumor microenvironment and the intestinal microbiota can also alter the levels of lipid and metabolites [47]. FASN is highly expressed in endothelial cells [48,49], immune cells [50,51], and fibroblasts [52] associated with cancer, suggesting the potential impact of FASN expression in these cells on the levels of FAs observed during adenoma formation. To better understand the effect of FASN on lipid synthesis and lipid uptake and utilization, we plan to perform stable-isotope tracing studies to identify lipid species that drive Apc/FASN-driven carcinogenesis in a better-controlled environment such as organoid culture and primary CRC cell lines.

In current studies, the metabolic analysis of adenomas demonstrates changes in the level of several metabolites in glycolysis and the TCA cycle, including a decrease in D-glucose, pyruvate, citrate, malate, and fumarate. The changes in abundance of some metabolites did not reach statistical significance which can be explained by tissue- and cell-specific deletion of FASN in our model and the potential contribution of stromal compartment and microbiota to their levels. Based on PLS-DA analysis, the highest VIP was assigned to citrate, suggesting that a reduction in the level of citrate is the most significant change due to the reduced expression of FASN in our transgenic mouse models. This conclusion is supported by previously published work showing that the lack of FASN impairs glycolysis and the anaplerotic shift of the TCA cycle. This study also shows a diminished incorporation of carbon derived from glucose into the TCA cycle intermediates including citrate [18]. Citrate is an intermediate in the TCA cycle, which is produced in mitochondria by the action of CS, which combines acetyl-CoA and oxaloacetate to generate citrate for the TCA cycle [53]. Since CS catalyzes the first reaction of the TCA cycle, it is generally assumed to be the rate-limiting enzyme of the cycle [54]. Using RPPA and western blot analysis, we show that the level of CS is significantly increased in FASN$^{+/\Delta}$/Apc/Cre and FASN$^{\Delta/\Delta}$/Apc/Cre mice as compared to Apc/Cre, potentially, due to a decrease in the substrate availability. Several mechanisms contribute to mitochondrial citrate synthesis, including the serine/glycine pathway, truncated or reversed TCA cycle, and lactate uptake [55]. Citrate is important for ATP production, lipid synthesis, and epigenetic regulation [56]. Moreover, cytosolic citrate is obligatory for the promotion of cancer cell growth and proliferation [55], thus supporting our results that a decrease in FASN expression and the level of citrate are associated with less proliferative properties of intestinal adenomas and an increase of survival of mice. In addition, citrate is a key regulatory molecule, which targets (directly or indirectly) catabolic and anabolic pathways in a manner such that when one pathway is activated, the other is inhibited [53]. Indeed, administration of high doses of citrate inhibits the proliferation of various cancer cells via inhibition of glycolysis and other anti-cancer effects [53,57]. The complexity of citrate synthesis and utilization in cancer cells warrants the use of stable isotope tracing to better understand the metabolic adaptations associated with the downregulation of FASN expression in transgenic mouse models.

Consistent with the published study on CRC [58], our data show that CS is overexpressed in CRC as compared to normal mucosa. Our data shows that the heterozygous deletion of *Apc* in normal intestinal epithelium leads to the upregulation of both CS and FASN, suggesting that an increase in citrate and lipid synthesis are metabolic futures required for Apc-driven carcinogenesis. Even though we did not see significant changes in CS mRNA expression between normal mucosa and tumor tissues, the TCGA data demonstrate a significant correlation between the expression of CS and FASN in human colorectal cancer. We further confirmed this correlation by analyzing FASN and CS protein levels in matched normal colon and tumor tissues. Our data warrant in-depth studies to further delineate the mechanisms of how the level of citrate is regulated by FASN and better understand the functional consequences of these changes in CRC.

Another significant change identified due to a downregulation in FASN expression is a decrease in the beta-alanine level. Interestingly, beta-alanine was found to be the most upregulated metabolite in colon carcinoma tissues as compared to normal mucosa [59], suggesting it is potentially important for metabolic alterations in CRC. The synthesis of alanine from pyruvate is thought to lie in the mitochondria matrix [60]. Even though the role of alanine in cancer is poorly understood, emerging evidence suggests that alanine plays a role in the proliferation and survival of cancer cells [61]. Interestingly, alanine contributes significantly to bioenergetic and anabolic pathways including de novo synthesis of fatty acids in pancreatic cancer [62].

Limitations of the study. The rigorous analysis did not identify any statistically significant changes in free fatty acids, sphingolipids, and triglycerides between Apc/Cre mice and mice with altered expression of FASN. Moreover, even though the abundance of several metabolites decreased, statistical significance was not reached on all of them. There are several explanations for these results. We did not address the potential contribution of diet, adipose tissue, or stromal compartment (where the expression of FASN is intact in our mouse model) to the level of FAs or the contribution of metabolites to adenoma tissues and circulation [47]. Furthermore, we could not account for the effect of high heterogeneity of collected tissues and individual diversity among mice in our model, and this could greatly contribute to the outcome of our study. Indeed, it has been shown that principal component analysis does not show an unequivocal separation between cancer tissue and normal mucosa in CRC patients; paired comparison of cancer tissue and normal mucosa obtained from the same subject must be done to identify the difference [63]. Another potential explanation is that, in many cases, the patterns of non-significant differences in gene/protein expression or in the levels of metabolic intermediates as identified in the current study can lead to significant differences in the development of disease and, therefore, analysis of these patterns is as important as the identification of significant differences. Therefore, follow-up studies using different models need to be performed to confirm our findings and further delineate the effect of FASN in the development of CRC.

Use of stable isotope tracers in a controlled environment such as organoid cultures would address some issues described above and help advance understanding of mechanisms of how FASN contributes to carcinogenesis in CRC. ^{13}C glucose and ^{13}C acetate stable isotope tracing would facilitate a better understanding of the mechanisms of how changes in FASN expression alter glucose utilization and contribute to TCA cycle intermediates and lipid metabolism.

Our previous study demonstrates that pharmacological inhibition of FASN leads to a significant decrease in the levels of lipid species in several patient-derived xenografts treated with TVB-3664, and significant changes seem to be mostly associated with patient-derived xenografts established from metastatic tissues [9]. These data suggest that mutations other than in the *Apc* gene, stage of cancer, and aggressiveness of the tumor can contribute to diversity in lipid utilization and uptake. Therefore, analysis of other models beyond the Apc-driven model should be utilized to better understand the role of de novo lipid synthesis in CRC carcinogenesis.

In summary, despite some limitations, our study provides compelling evidence that FASN plays an important role in CRC initiation by promoting the expression of genes supporting cellular proliferation and upregulating metabolic pathways involved in catabolic reactions and energy production. Therefore, this work warrants further investigation of FASN as a potential target for CRC prevention in the setting of more complex models when CRC is driven by mutations other than the *Apc* gene to further confirm the potential for use of FASN-targeted therapy in individuals who have a high risk based on either genetics or screening colonoscopy results.

4. Materials and Methods

4.1. Mouse Colonies

Mice were housed at the facility supervised by the Division of Laboratory Animal Resources, University of Kentucky in accordance with the NIH Guide for the Care and Use of Lab Animals (https://www.ncbi.nlm.nih.gov/books/NBK54050/, accessed on 31 July 2021). All animal experimental procedures were carried out under approval from the University Committee on Use and Care of Animals, University of Kentucky, protocol # 2016-2521. Mice were fed 2018 Teklad global 18% protein rodent diets from ENVIGO during breeding, strain maintenance, and experimental procedures. C57BL/6J mice with LoxP-flanked FASN alleles (FASN$^{f/f}$) were obtained from Clay Semenkovich, MD, at Washington University [23]. Apc/Villin-Cre mouse colonies with hetero- and homozygous deletion of FASN were established by mating these mice with Villin-Cre mice (B6.Cg-Tg(Vil1-cre)1000 Gum/J, stock #021504) and with Apc mice (C57BL/6-Apc^{tm1Tyj}/J, stock #00945).

4.2. Survival Analysis and Tumor Number Studies

For survival studies, mice were observed daily for signs associated with adenoma development such as weight loss, lethargy/cachexia, paleness of the paws, hinged posture, obstruction, bloody stool, and anal bleeding. Animals were euthanized when they reached the endpoint (recumbent and unable to drink and eat due to symptoms associated with disease progression). For survival studies, we observed 26 Apc/Cre male mice, 22 Apc/Cre females, 21 FASN$^{+/\Delta}$/Apc/Cre males, and 27 FASN$^{+/\Delta}$/Apc/Cre females.

For adenoma count, intestines were removed. The section of 10 cm starting 1 cm from cecum was used for determination of adenoma number. The intestine was washed with PBS, placed on an ice-cold metal platform, opened, cleaned, and the number of visual adenomas was counted. The tumor numbers are reported as the average number of tumors for male and female mice. Tissues from 8 Apc/Cre male mice, 12 Apc/Cre females, 12 FASN$^{+/\Delta}$/Apc/Cre males, and 11 FASN$^{+/\Delta}$/Apc/Cre females were used for analysis.

The tissues were processed for immunohistochemistry analysis using the swiss-roll techniques or adenomas were collected for further quantitative Reverse Transcription Polymerase Chain Reaction and western blot analysis.

4.3. Histologic Analysis and Immunohistochemical (IHC) Staining

Paraffin-embedded tissue section slides were prepared from transgenic mouse tissues using the Biospecimen Procurement and Translational Pathology Shared Resource Facility services. Tissue slides were stained with hematoxylin and eosin. For IHC staining, paraffin-embedded tissue sections were deparaffinized, rehydrated, and antigen retrieval was performed using Antigen Retriever Buffer #T6455 (Sigma-Aldrich Inc., St. Louis, MO, USA). IHC staining was performed using ImmPRESS® HRP Universal (Horse Anti-Mouse/Rabbit IgG) PLUS Polymer Kit, Peroxidase, MP-7800 (Vector Laboratories Inc., Burlingame, CA, USA) according to manufacturer instructions. The stained sections were visualized and imaged using a Nikon Eclipse 80i upright microscope (Melville, NY, USA). The HALO digital pathology analysis platform at the Biospecimen Procurement and Translational Pathology Shared Resource Facility was used to quantify the percentage of Ki67 positive cells and intensity of staining in intestinal tissues.

4.4. Western Blot Analysis

Mouse adenoma tissues were harvested and homogenized using metal beads in Cell Lysis Buffer #9803 (Cell Signaling, Danvers, MA, USA) supplemented with additional protease inhibitors. Equal amounts of cell lysates were resolved by SDS-PAGE and subjected to western blot analysis.

4.5. Antibodies for Western Blot and IHC Staining

Antibodies were purchased from Cell Signaling (Danvers, MA): Fatty Acid Synthase (C20G5) Rabbit mAb (#3180), Citrate Synthase (D7V8B) Rabbit mAb (#14309), Ki-67 (D3B5) Rabbit mAb (Mouse Preferred; IHC Formulated) #12202. All antibodies were used at a concentration of 1:1000 for western blot and 1:100 for IHC.

4.6. RNA-Sequencing and Gene Set Enrichment Analysis

RNA samples from pulled adenomas (n = 3) collected from Apc/Villin-Cre, $FASN^{+/\Delta}$/Apc/Villin-Cre and $FASN^{\Delta/\Delta}$/Apc/Villin-Cre were prepared using a QIAGEN RNeasy kit and library preparation, sequencing, and standard bioinformatics analysis were performed by BGI Genomics, Cambridge, MA, USA (https://www.bgi.com/global/, accessed on 31 July 2021). The quality control assessment of samples and detailed summary of the sequencing coverage, quality statistics, and data analysis report are included in Supplemental Table S1A,B. The gene set enrichment analysis was performed by the Biostatistics and Bioinformatics Shared Resource Facility, University of Kentucky (Lexington, KY, USA) using gene set enrichment analysis (GSEA) software (version 4.0.3) and the KEGG pathways in the Molecular Signature Database (MSigDB) [64,65]. Hypergeometric tests were used to test enrichment of KEGG pathways based on the R package clusterProfiler (version 3.18.1) [66].

4.7. Metabolite Extraction

For metabolic analysis, the adenomas were removed from intestine, rinse with ice cold PBS and immediately cryopreserved to minimize any further changes in metabolite levels. Isolated adenomas were removed from cryostorage and transferred to a micro vial set for use with a Freezer/Mill Cryogenic Grinder (SPEX SamplePrep model 6875D, Cole-Parmer North America, Vernon Hills, IL, USA). Tissue was pulverized to 5 µm particles. Metabolites were extracted directly from the micro vial by the addition of 1 mL of 50% methanol containing 20 mL-norvaline (procedural, internal control) and separated into polar (aqueous layer) and insoluble pellet (protein/DNA/RNA/glycogen) by centrifugation at 4 °C, 15,000 rpm for 10 min. The pellet was subsequently washed four times with 50% methanol and once with 100% methanol. The pellet was then hydrolyzed in 200 µL of 3N hydrochloric acid and then 200 µL of 100% methanol was added before drying. The polar and pellet fraction was dried at 10^{-3} mBar using a SpeedVac (ThermoFisher Scientific, Waltham, MA, USA) followed by derivatization. The insoluble pellet was hydrolyzed as described [67].

4.8. Sample Derivatization and Gas Chromatography-Mass Spectrometry (GC-MS) Quantification

Dried polar and insoluble samples were derivatized by the addition of 50 μL of 20 mg/mL methoxyamine hydrochloride in pyridine, vortexed thoroughly, and incubated for 1.5 h at 30 °C. Sequential addition of 80 μL of N-methyl-trimethylsilyl-trifluoroacetamide followed with an incubation time of 30 min at 37 °C with thorough vortexing between addition of solvents. The mixture was then transferred to an amber, v-shaped glass chromatography vial and analyzed by GC-MS.

An Agilent 7800B gas-chromatography coupled to a 5977B mass spectrometry detector was used for this study (Agilent, Santa Clara, CA, USA). GC-MS protocols were similar to those described previously [68,69] except a modified temperature gradient was used for GC: Initial temperature was 130 °C, held for 4 min, rising at 6 °C/min to 243 °C, rising at 60 °C/min to 280 °C, held for 2 min. The electron ionization energy was set to 70 eV. Scan (m/z: 50–800) and full scan mode were used for metabolomics analysis. Mass spectra were translated to relative metabolite abundance using MassHunter MS quantitative software matched to the FiehnLib metabolomics library (available through Agilent) for retention time and fragmentation pattern [69–71]. Relative abundance was corrected for recovery using the L-norvaline standard and adjusted to protein input represented by the sum of amino acids from the pellet fraction also analyzed by GC-MS.

4.9. Metabolite Analysis

Data were uploaded into MetaboAnalyst version 5.0 (Xia Lab, McGill University, Montreal, QC, Canada) for partial least-squares discriminant analysis (PLS-DA), variable importance in projection (VIP) analysis, and clustering heat map analysis. Data was uploaded as a CSV file and auto-scaled (mean-centered and divided by the standard deviation of each variable). The VIP score of a metabolite is calculated as a weighted sum of the squared correlations between the PLS-DA components. Heatmaps were organized using all metabolic features and distance measured by a Euclidean analysis.

4.10. Lipidomic Analysis

Lipid analysis of mouse adenoma tissues was performed by Lipidomics Shared Resources (Analytical Unit) at the Medical University of South Carolina according to their standard protocols (https://hollingscancercenter.musc.edu/research/shared-resources/lipidomics, accessed on 31 July 2021). The data for free fatty acids, diglycerides, and sphingolipids were processed and analyzed by Markey Cancer Center Biostatistics and Bioinformatics Shared Resource Facility based on the following procedure. Below the detection limit measurements of a metabolite were imputed by the minimum of detected values of the metabolite across samples divided by square root of 2. Data were then log2-transformed and used as input for the limma package to compare metabolic profile between Apc/Cre and FASN$^{+/\Delta}$/Apc/Cre groups [72]. p values and fold changes were calculated based on the moderated t-statistics. Multiple comparisons adjustment was performed by controlling the false discovery rate (FDR) based on the Benjamini and Hochberg method. An FDR < 0.05 was considered statistically significant.

4.11. Reverse Phase Protein Analysis

RPPA was performed on protein lyates of adenomas from Apc/Villin-Cre, FASN$^{+/\Delta}$/Apc/Villin-Cre and FASN$^{\Delta/\Delta}$/Apc/Villin-Cre mice by the Center for Environmental and Systems Biochemistry (Redox Metabolism Shared Resource Facility, University of Kentucky) as previously described [28]. The raw RPPA data obtained in 6 dilution steps were processed based on the following procedure. First, data quality control was performed by plotting the relationship between background-corrected intensity and dilution step. Proteins with low measurement quality as reflected by large variation across replicates or unreliable curve trend were excluded. Secondly, nonlinear curve fitting for the background corrected intensity vs. dilution step was applied based on the "serial dilution curve" algorithm [73] to infer the concentration of each protein in the original undiluted sample. Due to the low protein concentration measurement in the 6th dilution step, the nonlinear curve fitting was only based on data from the first 5 dilution steps. Third, protein concentrations were log2-transformed and normalized based on the median normalization method described in https://www.tcpaportal.org/tcpa/faq.html (accessed on 13 December 2020) and [74]. Finally, the Wilcoxon Rank Sum test was used for differential expression analysis comparing experimental groups. Multiple comparisons adjustment was performed by the Benjamini and Hochberg method. Differentially expressed proteins were identified by false discovery rate < 0.05. Heatmap and volcano plots were generated to demonstrate the relative expression changes of proteins of interest.

4.12. Analysis of Correlation between FASN and Citrate Synthase (CS)

Correlations between FASN and CS were determined based on RNA-Seq data of CRC patient tumor tissues from The Cancer Genome Atlas [75]. The RNA-Seq data (FPKM values) were downloaded from the Genomic Data Commons (https://www.cancer.gov/tcga, accessed on 4 August 2021) and converted to TPM values. Spearman's rank correlation coefficient was used to quantify the correlation between FASN and CS expressions.

Supplementary Materials: The following supporting information can be downloaded at: https://www.mdpi.com/article/10.3390/ijms23126510/s1.

Author Contributions: Conceptualization, J.D., M.S.G., R.S. and Y.Y.Z.; Data curation, L.E.A.Y., T.L.S., D.H., Y.W., C.W. and R.S.; Formal analysis, L.E.A.Y., T.L.S., C.O.K., D.H., J.L., Y.W., C.W. and R.S.; Funding acquisition, Y.Y.Z.; Investigation, J.D., L.E.A.Y., T.L.S. and Y.Y.Z.; Methodology, C.W., H.L.W., T.F., M.S.G. and R.S.; Project administration, Y.Y.Z.; Writing—original draft, J.D., R.S. and Y.Y.Z.; Writing—review & editing, M.S.G., R.S. and Y.Y.Z. All authors have read and agreed to the published version of the manuscript.

Funding: This research was supported by NCI R01 CA249734 (Y.Y.Z.), R03 CA262720 (Y.Y.Z.), NIGMS P20 GM121327 (Y.Y.Z.), R35 NS116824 (M.S.G.), R01 AG066653 (R.S.), St. Baldrick's Career Development Award (R.S.), V-Scholar Grant (R.S.), Rally Foundation Independent Investigator (R.S.).

Institutional Review Board Statement: The study was conducted in accordance with the Declaration of Helsinki, and human tissue specimens were collected under an Institutional Review Board of the University of Kentucky (#52094, continuous review approved 14 May 2022). The animal study protocol was approved by the University of Kentucky Animal Care and Use Committee (#2016-2521, approved 15 October 2019).

Informed Consent Statement: Patients/participants provided written informed consent to participate in this study.

Data Availability Statement: The raw data supporting the conclusions of this article will be made available by the authors, without undue reservation. The datasets presented in this study can be found in supplemental materials and freely available in Dryad repository.

Acknowledgments: University of Kentucky Markey Cancer Center's Biospecimen Procurement and Translational Pathology Shared Resource Facility (SRF) aided in preparation of tissue slides for immunohistochemistry, Center for Environmental and Systems Biochemistry (Redox Metabolism SRF) performed the RPPA, and the Biostatistics and Bioinformatics SRF provided statistical analysis of animal survival/tumor number, RPPA and lipid data. SRFs are supported by National Cancer Institute grant P30 CA177558. Special thank you to Li Xu in Zaytseva's laboratory, for assisting with animal work and Markey Cancer Center's Research Communications Office and Donna Gilbreath for assisting with preparation of this manuscript.

Conflicts of Interest: The authors declare no conflict of interest.

References

1. Menendez, J.A.; Lupu, R. Fatty acid synthase and the lipogenic phenotype in cancer pathogenesis. *Nat. Rev. Cancer* **2007**, *7*, 763–777. [CrossRef] [PubMed]
2. Cheng, C.; Geng, F.; Cheng, X.; Guo, D. Lipid metabolism reprogramming and its potential targets in cancer. *Cancer Commun.* **2018**, *38*, 27. [CrossRef] [PubMed]
3. Zaytseva, Y. Lipid Metabolism as a Targetable Metabolic Vulnerability in Colorectal Cancer. *Cancers* **2021**, *13*, 301. [CrossRef] [PubMed]
4. Buckley, D.; Duke, G.; Heuer, T.S.; O'Farrell, M.; Wagman, A.S.; McCulloch, W.; Kemble, G. Fatty acid synthase—Modern tumor cell biology insights into a classical oncology target. *Pharmacol. Ther.* **2017**, *177*, 23–31. [CrossRef] [PubMed]
5. Luo, X.; Cheng, C.; Tan, Z.; Li, N.; Tang, M.; Yang, L.; Cao, Y. Emerging roles of lipid metabolism in cancer metastasis. *Mol. Cancer* **2017**, *16*, 76. [CrossRef]
6. Zaytseva, Y.Y.; Rychahou, P.G.; Gulhati, P.; Elliott, V.A.; Mustain, W.C.; O'Connor, K.; Morris, A.J.; Sunkara, M.; Weiss, H.L.; Lee, E.Y.; et al. Inhibition of fatty acid synthase attenuates CD44-associated signaling and reduces metastasis in colorectal cancer. *Cancer Res.* **2012**, *72*, 1504–1517. [CrossRef]
7. Elliott, V.A.; Rychahou, P.; Zaytseva, Y.Y.; Evers, B.M. Activation of c-Met and upregulation of CD44 expression are associated with the metastatic phenotype in the colorectal cancer liver metastasis model. *PLoS ONE* **2014**, *9*, e97432. [CrossRef]
8. Zaytseva, Y.Y.; Harris, J.W.; Mitov, M.I.; Kim, J.T.; Butterfield, D.A.; Lee, E.Y.; Weiss, H.L.; Gao, T.; Evers, B.M. Increased expression of fatty acid synthase provides a survival advantage to colorectal cancer cells via upregulation of cellular respiration. *Oncotarget* **2015**, *6*, 18891–18904. [CrossRef]
9. Zaytseva, Y.Y.; Rychahou, P.G.; Le, A.T.; Scott, T.L.; Flight, R.M.; Kim, J.T.; Harris, J.; Liu, J.; Wang, C.; Morris, A.J.; et al. Preclinical evaluation of novel fatty acid synthase inhibitors in primary colorectal cancer cells and a patient-derived xenograft model of colorectal cancer. *Oncotarget* **2018**, *9*, 24787–24800. [CrossRef]
10. Ventura, R.; Mordec, K.; Waszczuk, J.; Wang, Z.; Lai, J.; Fridlib, M.; Buckley, D.; Kemble, G.; Heuer, T.S. Inhibition of de novo Palmitate Synthesis by Fatty Acid Synthase Induces Apoptosis in Tumor Cells by Remodeling Cell Membranes, Inhibiting Signaling Pathways, and Reprogramming Gene Expression. *eBioMedicine* **2015**, *2*, 806–822. [CrossRef]
11. Flavin, R.; Peluso, S.; Nguyen, P.L.; Loda, M. Fatty acid synthase as a potential therapeutic target in cancer. *Future Oncol.* **2010**, *6*, 551–562. [CrossRef] [PubMed]
12. National Cancer Institute. Clinical Trials Using FASN Inhibitor TVB-2640. Available online: www.cancer.gov/about-cancer/treatment/clinical-trials/intervention/C118285 (accessed on 31 January 2022).
13. Satoh, K.; Yachida, S.; Sugimoto, M.; Oshima, M.; Nakagawa, T.; Akamoto, S.; Tabata, S.; Saitoh, K.; Kato, K.; Sato, S.; et al. Global metabolic reprogramming of colorectal cancer occurs at adenoma stage and is induced by MYC. *Proc. Natl. Acad. Sci. USA* **2017**, *114*, E7697–E7706. [CrossRef] [PubMed]
14. Lau, D.S.; Archer, M.C. Fatty acid synthase is over-expressed in large aberrant crypt foci in rats treated with azoxymethane. *Int. J. Cancer* **2009**, *124*, 2750–2753. [CrossRef] [PubMed]
15. Visca, P.; Alo, P.L.; Del Nonno, F.; Botti, C.; Trombetta, G.; Marandino, F.; Filippi, S.; Di Tondo, U.; Donnorso, R.P. Immunohistochemical expression of fatty acid synthase, apoptotic-regulating genes, proliferating factors, and ras protein product in colorectal adenomas, carcinomas, and adjacent nonneoplastic mucosa. *Clin. Cancer Res.* **1999**, *5*, 4111–4118.
16. Kearney, K.E.; Pretlow, T.G.; Pretlow, T.P. Increased expression of fatty acid synthase in human aberrant crypt foci: Possible target for colorectal cancer prevention. *Int. J. Cancer* **2009**, *125*, 249–252. [CrossRef]
17. Cruz, M.D.; Wali, R.K.; Bianchi, L.K.; Radosevich, A.J.; Crawford, S.E.; Jepeal, L.; Goldberg, M.J.; Weinstein, J.; Momi, N.; Roy, P.; et al. Colonic mucosal fatty acid synthase as an early biomarker for colorectal neoplasia: Modulation by obesity and gender. *Cancer Epidemiol. Biomark. Prev.* **2014**, *23*, 2413–2421. [CrossRef]
18. Bueno, M.J.; Jimenez-Renard, V.; Samino, S.; Capellades, J.; Junza, A.; Lopez-Rodriguez, M.L.; Garcia-Carceles, J.; Lopez-Fabuel, I.; Bolanos, J.P.; Chandel, N.S.; et al. Essentiality of fatty acid synthase in the 2D to anchorage-independent growth transition in transforming cells. *Nat. Commun.* **2019**, *10*, 5011. [CrossRef]
19. Yang, J.; Yu, J. The association of diet, gut microbiota and colorectal cancer: What we eat may imply what we get. *Protein Cell* **2018**, *9*, 474–487. [CrossRef]
20. Tabung, F.K.; Liu, L.; Wang, W.; Fung, T.T.; Wu, K.; Smith-Warner, S.A.; Cao, Y.; Hu, F.B.; Ogino, S.; Fuchs, C.S.; et al. Association of Dietary Inflammatory Potential with Colorectal Cancer Risk in Men and Women. *JAMA Oncol.* **2018**, *4*, 366–373. [CrossRef]
21. Drury, J.; Rychahou, P.G.; He, D.; Jafari, N.; Wang, C.; Lee, E.Y.; Weiss, H.L.; Evers, B.M.; Zaytseva, Y.Y. Inhibition of Fatty Acid Synthase Upregulates Expression of CD36 to Sustain Proliferation of Colorectal Cancer Cells. *Front. Oncol.* **2020**, *10*, 1185. [CrossRef]
22. McCart, A.E.; Vickaryous, N.K.; Silver, A. Apc mice: Models, modifiers and mutants. *Pathol. Res. Pract.* **2008**, *204*, 479–490. [CrossRef] [PubMed]
23. Chakravarthy, M.V.; Pan, Z.; Zhu, Y.; Tordjman, K.; Schneider, J.G.; Coleman, T.; Turk, J.; Semenkovich, C.F. "New" hepatic fat activates PPARalpha to maintain glucose, lipid, and cholesterol homeostasis. *Cell Metab.* **2005**, *1*, 309–322. [CrossRef] [PubMed]
24. Wei, X.; Yang, Z.; Rey, F.E.; Ridaura, V.K.; Davidson, N.O.; Gordon, J.I.; Semenkovich, C.F. Fatty acid synthase modulates intestinal barrier function through palmitoylation of mucin 2. *Cell Host Microbe* **2012**, *11*, 140–152. [CrossRef] [PubMed]
25. Nagy, A. Cre recombinase: The universal reagent for genome tailoring. *Genesis* **2000**, *26*, 99–109. [CrossRef]

26. Li, L.T.; Jiang, G.; Chen, Q.; Zheng, J.N. Ki67 is a promising molecular target in the diagnosis of cancer (review). *Mol. Med. Rep.* **2015**, *11*, 1566–1572. [CrossRef]
27. Jafari, N.; Drury, J.; Morris, A.J.; Onono, F.O.; Stevens, P.D.; Gao, T.; Liu, J.; Wang, C.; Lee, E.Y.; Weiss, H.L.; et al. De Novo Fatty Acid Synthesis-Driven Sphingolipid Metabolism Promotes Metastatic Potential of Colorectal Cancer. *Mol. Cancer Res.* **2019**, *17*, 140–152. [CrossRef]
28. Fan, T.W.M.; Bruntz, R.C.; Yang, Y.; Song, H.; Chernyavskaya, Y.; Deng, P.; Zhang, Y.; Shah, P.P.; Beverly, L.J.; Qi, Z.; et al. De novo synthesis of serine and glycine fuels purine nucleotide biosynthesis in human lung cancer tissues. *J. Biol. Chem.* **2019**, *294*, 13464–13477. [CrossRef]
29. Liu, Z.; Yu, M.; Fei, B.; Fang, X.; Ma, T.; Wang, D. miR215p targets PDHA1 to regulate glycolysis and cancer progression in gastric cancer. *Oncol. Rep.* **2018**, *40*, 2955–2963. [CrossRef]
30. Murai, S.; Ando, A.; Ebara, S.; Hirayama, M.; Satomi, Y.; Hara, T. Inhibition of malic enzyme 1 disrupts cellular metabolism and leads to vulnerability in cancer cells in glucose-restricted conditions. *Oncogenesis* **2017**, *6*, e329. [CrossRef]
31. Martinez-Reyes, I.; Chandel, N.S. Mitochondrial TCA cycle metabolites control physiology and disease. *Nat. Commun.* **2020**, *11*, 102. [CrossRef]
32. Balsa-Martinez, E.; Puigserver, P. Cancer Cells Hijack Gluconeogenic Enzymes to Fuel Cell Growth. *Mol. Cell* **2015**, *60*, 509–511. [CrossRef] [PubMed]
33. Lieu, E.L.; Nguyen, T.; Rhyne, S.; Kim, J. Amino acids in cancer. *Exp. Mol. Med.* **2020**, *52*, 15–30. [CrossRef] [PubMed]
34. Wei, Z.; Liu, X.; Cheng, C.; Yu, W.; Yi, P. Metabolism of Amino Acids in Cancer. *Front. Cell Dev. Biol.* **2020**, *8*, 603837. [CrossRef]
35. Galim, E.B.; Hruska, K.; Bier, D.M.; Matthews, D.E.; Haymond, M.W. Branched-chain amino acid nitrogen transfer to alamine in vivo in dogs. Direct isotopic determination with [15N]leucine. *J. Clin. Investig.* **1980**, *66*, 1295–1304. [CrossRef] [PubMed]
36. Sousa, C.M.; Biancur, D.E.; Wang, X.; Halbrook, C.J.; Sherman, M.H.; Zhang, L.; Kremer, D.; Hwang, R.F.; Witkiewicz, A.K.; Ying, H.; et al. Pancreatic stellate cells support tumour metabolism through autophagic alanine secretion. *Nature* **2016**, *536*, 479–483. [CrossRef] [PubMed]
37. Schlichtholz, B.; Turyn, J.; Goyke, E.; Biernacki, M.; Jaskiewicz, K.; Sledzinski, Z.; Swierczynski, J. Enhanced citrate synthase activity in human pancreatic cancer. *Pancreas* **2005**, *30*, 99–104. [CrossRef]
38. Kridel, S.J.; Axelrod, F.; Rozenkrantz, N.; Smith, J.W. Orlistat is a novel inhibitor of fatty acid synthase with antitumor activity. *Cancer Res.* **2004**, *64*, 2070–2075. [CrossRef]
39. Dowling, S.; Cox, J.; Cenedella, R.J. Inhibition of fatty acid synthase by Orlistat accelerates gastric tumor cell apoptosis in culture and increases survival rates in gastric tumor bearing mice in vivo. *Lipids* **2009**, *44*, 489–498. [CrossRef]
40. Notarnicola, M.; Barone, M.; Francavilla, A.; Tutino, V.; Bianco, G.; Tafaro, A.; Minoia, M.; Polimeno, L.; Napoli, A.; Scavo, M.P.; et al. Lovastatin, but not orlistat, reduces intestinal polyp volume in an ApcMin/+ mouse model. *Oncol. Rep.* **2016**, *36*, 893–899. [CrossRef]
41. Migita, T.; Ruiz, S.; Fornari, A.; Fiorentino, M.; Priolo, C.; Zadra, G.; Inazuka, F.; Grisanzio, C.; Palescandolo, E.; Shin, E.; et al. Fatty acid synthase: A metabolic enzyme and candidate oncogene in prostate cancer. *J. Natl. Cancer Inst.* **2009**, *101*, 519–532. [CrossRef]
42. Alwarawrah, Y.; Hughes, P.; Loiselle, D.; Carlson, D.A.; Darr, D.B.; Jordan, J.L.; Xiong, J.; Hunter, L.M.; Dubois, L.G.; Thompson, J.W.; et al. Fasnall, a Selective FASN Inhibitor, Shows Potent Anti-tumor Activity in the MMTV-Neu Model of HER2(+) Breast Cancer. *Cell Chem. Biol.* **2016**, *23*, 678–688. [CrossRef] [PubMed]
43. Alli, P.M.; Pinn, M.L.; Jaffee, E.M.; McFadden, J.M.; Kuhajda, F.P. Fatty acid synthase inhibitors are chemopreventive for mammary cancer in neu-N transgenic mice. *Oncogene* **2005**, *24*, 39–46. [CrossRef]
44. Luo, Z.W.; Zhu, M.G.; Zhang, Z.Q.; Ye, F.J.; Huang, W.H.; Luo, X.Z. Increased expression of Ki-67 is a poor prognostic marker for colorectal cancer patients: A meta analysis. *BMC Cancer* **2019**, *19*, 123. [CrossRef] [PubMed]
45. Falchook, G.; Infante, J.; Arkenau, H.T.; Patel, M.R.; Dean, E.; Borazanci, E.; Brenner, A.; Cook, N.; Lopez, J.; Pant, S.; et al. First-in-human study of the safety, pharmacokinetics, and pharmacodynamics of first-in-class fatty acid synthase inhibitor TVB-2640 alone and with a taxane in advanced tumors. *eClinicalMedicine* **2021**, *34*, 100797. [CrossRef] [PubMed]
46. Jensen-Urstad, A.P.; Semenkovich, C.F. Fatty acid synthase and liver triglyceride metabolism: Housekeeper or messenger? *Biochim. Biophys. Acta* **2012**, *1821*, 747–753. [CrossRef] [PubMed]
47. Nenkov, M.; Ma, Y.; Gassler, N.; Chen, Y. Metabolic Reprogramming of Colorectal Cancer Cells and the Microenvironment: Implication for Therapy. *Int. J. Mol. Sci.* **2021**, *22*, 6262. [CrossRef] [PubMed]
48. Bruning, U.; Morales-Rodriguez, F.; Kalucka, J.; Goveia, J.; Taverna, F.; Queiroz, K.C.S.; Dubois, C.; Cantelmo, A.R.; Chen, R.; Loroch, S.; et al. Impairment of Angiogenesis by Fatty Acid Synthase Inhibition Involves mTOR Malonylation. *Cell Metab.* **2018**, *28*, 866–880.e15. [CrossRef] [PubMed]
49. Browne, C.D.; Hindmarsh, E.J.; Smith, J.W. Inhibition of endothelial cell proliferation and angiogenesis by orlistat, a fatty acid synthase inhibitor. *FASEB J.* **2006**, *20*, 2027–2035. [CrossRef]
50. Xiong, W.; Sun, K.Y.; Zhu, Y.; Zhang, X.; Zhou, Y.H.; Zou, X. Metformin alleviates inflammation through suppressing FASN-dependent palmitoylation of Akt. *Cell Death Dis.* **2021**, *12*, 934. [CrossRef]
51. Qian, X.; Yang, Z.; Mao, E.; Chen, E. Regulation of fatty acid synthesis in immune cells. *Scand. J. Immunol.* **2018**, *88*, e12713. [CrossRef]

52. Gong, J.; Lin, Y.; Zhang, H.; Liu, C.; Cheng, Z.; Yang, X.; Zhang, J.; Xiao, Y.; Sang, N.; Qian, X.; et al. Reprogramming of lipid metabolism in cancer-associated fibroblasts potentiates migration of colorectal cancer cells. *Cell Death Dis.* **2020**, *11*, 267. [CrossRef] [PubMed]
53. Icard, P.; Coquerel, A.; Wu, Z.; Gligorov, J.; Fuks, D.; Fournel, L.; Lincet, H.; Simula, L. Understanding the Central Role of Citrate in the Metabolism of Cancer Cells and Tumors: An Update. *Int. J. Mol. Sci.* **2021**, *22*, 6587. [CrossRef] [PubMed]
54. Chen, L.; Liu, T.; Zhou, J.; Wang, Y.; Wang, X.; Di, W.; Zhang, S. Citrate synthase expression affects tumor phenotype and drug resistance in human ovarian carcinoma. *PLoS ONE* **2014**, *9*, e115708. [CrossRef]
55. Haferkamp, S.; Drexler, K.; Federlin, M.; Schlitt, H.J.; Berneburg, M.; Adamski, J.; Gaumann, A.; Geissler, E.K.; Ganapathy, V.; Parkinson, E.K.; et al. Extracellular Citrate Fuels Cancer Cell Metabolism and Growth. *Front. Cell Dev. Biol.* **2020**, *8*, 602476. [CrossRef]
56. Drexler, K.; Schmidt, K.M.; Jordan, K.; Federlin, M.; Milenkovic, V.M.; Liebisch, G.; Artati, A.; Schmidl, C.; Madej, G.; Tokarz, J.; et al. Cancer-associated cells release citrate to support tumour metastatic progression. *Life Sci. Alliance* **2021**, *4*, e202000903. [CrossRef] [PubMed]
57. Salas, M.L.; Vinuela, E.; Salas, M.; Sols, A. Citrate Inhibition of Phosphofructokinase and the Pasteur Effect. *Biochem. Biophys. Res. Commun.* **1965**, *19*, 371–376. [CrossRef]
58. Ren, M.; Yang, X.; Bie, J.; Wang, Z.; Liu, M.; Li, Y.; Shao, G.; Luo, J. Citrate synthase desuccinylation by SIRT5 promotes colon cancer cell proliferation and migration. *Biol. Chem.* **2020**, *401*, 1031–1039. [CrossRef]
59. Denkert, C.; Budczies, J.; Weichert, W.; Wohlgemuth, G.; Scholz, M.; Kind, T.; Niesporek, S.; Noske, A.; Buckendahl, A.; Dietel, M.; et al. Metabolite profiling of human colon carcinoma–deregulation of TCA cycle and amino acid turnover. *Mol. Cancer* **2008**, *7*, 72. [CrossRef]
60. Vettore, L.; Westbrook, R.L.; Tennant, D.A. New aspects of amino acid metabolism in cancer. *Br. J. Cancer* **2020**, *122*, 150–156. [CrossRef]
61. Choi, B.H.; Coloff, J.L. The Diverse Functions of Non-Essential Amino Acids in Cancer. *Cancers* **2019**, *11*, 675. [CrossRef]
62. Parker, S.J.; Amendola, C.R.; Hollinshead, K.E.R.; Yu, Q.; Yamamoto, K.; Encarnacion-Rosado, J.; Rose, R.E.; LaRue, M.M.; Sohn, A.S.W.; Biancur, D.E.; et al. Selective Alanine Transporter Utilization Creates a Targetable Metabolic Niche in Pancreatic Cancer. *Cancer Discov.* **2020**, *10*, 1018–1037. [CrossRef] [PubMed]
63. Mika, A.; Duzowska, K.; Halinski, L.P.; Pakiet, A.; Czumaj, A.; Rostkowska, O.; Dobrzycka, M.; Kobiela, J.; Sledzinski, T. Rearrangements of Blood and Tissue Fatty Acid Profile in Colorectal Cancer-Molecular Mechanism and Diagnostic Potential. *Front. Oncol.* **2021**, *11*, 689701. [CrossRef] [PubMed]
64. Subramanian, A.; Tamayo, P.; Mootha, V.K.; Mukherjee, S.; Ebert, B.L.; Gillette, M.A.; Paulovich, A.; Pomeroy, S.L.; Golub, T.R.; Lander, E.S.; et al. Gene set enrichment analysis: A knowledge-based approach for interpreting genome-wide expression profiles. *Proc. Natl. Acad. Sci USA* **2005**, *102*, 15545–15550. [CrossRef] [PubMed]
65. Mootha, V.; Lindgren, C.; Eriksson, K.F.; Subramanian, A.; Sihag, S.; Lehar, J.; Puigserver, P.; Carlsson, E.; Ridderstråle, M.; Laurila, E.; et al. PGC-1α-responsive genes involved in oxidative phosphorylation are coordinately downregulated in human diabetes. *Nat. Genet.* **2003**, *34*, 267–273. [CrossRef] [PubMed]
66. Yu, G.; Wang, L.; Han, Y.; He, Q. clusterProfiler: An R package for comparing biological themes among gene clusters. *OMICS J. Integr. Biol.* **2012**, *16*, 284–287. [CrossRef]
67. Andres, D.A.; Young, L.E.A.; Veeranki, S.; Hawkinson, T.R.; Levitan, B.M.; He, D.; Wang, C.; Satin, J.; Sun, R.C. Improved workflow for mass spectrometry-based metabolomics analysis of the heart. *J. Biol. Chem.* **2020**, *295*, 2676–2686. [CrossRef]
68. Brewer, M.K.; Uittenbogaard, A.; Austin, G.L.; Segvich, D.M.; DePaoli-Roach, A.; Roach, P.J.; McCarthy, J.J.; Simmons, Z.R.; Brandon, J.A.; Zhou, Z. Targeting pathogenic Lafora bodies in Lafora disease using an antibody-enzyme fusion. *Cell Metab.* **2019**, *30*, 689–705.e6. [CrossRef]
69. Sun, R.C.; Dukhande, V.V.; Zhou, Z.; Young, L.E.; Emanuelle, S.; Brainson, C.F.; Gentry, M.S. Nuclear glycogenolysis modulates histone acetylation in human non-small cell lung cancers. *Cell Metab.* **2019**, *30*, 903–916.e7. [CrossRef]
70. Fiehn, O. Metabolomics by gas chromatography–mass spectrometry: Combined targeted and untargeted profiling. *Curr. Protoc. Mol. Biol.* **2016**, *114*, 30–34. [CrossRef]
71. Kind, T.; Wohlgemuth, G.; Lee, D.Y.; Lu, Y.; Palazoglu, M.; Shahbaz, S.; Fiehn, O. FiehnLib: Mass spectral and retention index libraries for metabolomics based on quadrupole and time-of-flight gas chromatography/mass spectrometry. *Anal. Chem.* **2009**, *81*, 10038–10048. [CrossRef]
72. Ritchie, M.E.; Phipson, B.; Wu, D.; Hu, Y.; Law, C.W.; Shi, W.; Smyth, G.K. limma powers differential expression analyses for RNA-sequencing and microarray studies. *Nucleic Acids Res.* **2015**, *43*, e47. [CrossRef] [PubMed]
73. Zhang, L.; Wei, Q.; Mao, L.; Liu, W.; Mills, G.B.; Coombes, K. Serial dilution curve: A new method for analysis of reverse phase protein array data. *Bioinformatics* **2009**, *25*, 650–654. [CrossRef] [PubMed]
74. Neeley, E.S.; Kornblau, S.M.; Coombes, K.R.; Baggerly, K.A. Variable slope normalization of reverse phase protein arrays. *Bioinformatics* **2009**, *25*, 1384–1389. [CrossRef]
75. Cancer Genome Atlas, N. Comprehensive molecular characterization of human colon and rectal cancer. *Nature* **2012**, *487*, 330–337. [CrossRef] [PubMed]

International Journal of *Molecular Sciences*

MDPI

Communication

Effect of Hypoxia-Induced Micro-RNAs Expression on Oncogenesis

Giorgia Moriondo [1,†], Giulia Scioscia [1,2,†], Piera Soccio [1], Pasquale Tondo [1,*], Cosimo Carlo De Pace [1], Roberto Sabato [2], Maria Pia Foschino Barbaro [1,2] and Donato Lacedonia [1,2]

1 Department of Medical and Surgical Sciences, University of Foggia, 71122 Foggia, Italy; giorgia.moriondo@unifg.it (G.M.); giulia.scioscia@unifg.it (G.S.); piera.soccio@unifg.it (P.S.); cosimo.depace@unifg.it (C.C.D.P.); mariapia.foschino@unifg.it (M.P.F.B.); donato.lacedonia@unifg.it (D.L.)

2 Institute of Respiratory Diseases, "Policlinico Riuniti" University Hospital of Foggia, 71122 Foggia, Italy; robsabato@libero.it

* Correspondence: pasquale.tondo@unifg.it

† These authors contributed equally to this work.

Abstract: MicroRNAs (miRNAs) are small non-coding RNAs that negatively regulate gene expression at the post-transcriptional level. An aberrant regulation of gene expression by miRNAs is associated with numerous diseases, including cancer. MiRNAs expression can be influenced by various stimuli, among which hypoxia; however, the effects of different types of continuous hypoxia (moderate or marked) on miRNAs are still poorly studied. Lately, some hypoxia-inducible miRNAs (HRMs, hypoxia-regulated miRNAs) have been identified. These HRMs are often activated in different types of cancers, suggesting their role in tumorigenesis. The aim of this study was to evaluate changes in miRNAs expression both in moderate continuous hypoxia and marked continuous hypoxia to better understand the possible relationship between hypoxia, miRNAs, and colorectal cancer. We used RT-PCR to detect the miRNAs expression in colorectal cancer cell lines in conditions of moderate and marked continuous hypoxia. The expression of miRNAs was analyzed using a two-way ANOVA test to compare the differential expression of miRNAs among groups. The levels of almost all analyzed miRNAs (miR-21, miR-23b, miR-26a, miR-27b, and miR-145) were greater in moderate hypoxia versus marked hypoxia, except for miR-23b and miR-21. This study identified a series of miRNAs involved in the response to different types of continuous hypoxia (moderate and marked), highlighting that they play a role in the development of cancer. To date, there are no other studies that demonstrate how these two types of continuous hypoxia could be able to activate different molecular pathways that lead to a different expression of specific miRNAs involved in tumorigenesis.

Keywords: microRNA; biomarkers; colorectal cancer; hypoxia; hypoxia-regulated microRNAs

Citation: Moriondo, G.; Scioscia, G.; Soccio, P.; Tondo, P.; De Pace, C.C.; Sabato, R.; Foschino Barbaro, M.P.; Lacedonia, D. Effect of Hypoxia-Induced Micro-RNAs Expression on Oncogenesis. *Int. J. Mol. Sci.* **2022**, *23*, 6294. https://doi.org/10.3390/ijms23116294

Academic Editors: Donatella Delle Cave and Alessandro Ottaiano

Received: 13 April 2022
Accepted: 2 June 2022
Published: 4 June 2022

1. Introduction

MicroRNAs (miRNAs) are small non-coding RNAs formed by about 18–22 nucleotides, whose main role is to negatively regulate gene expression at the post-transcriptional level. They act by recognizing specific mRNA targets in order to determine their translation degradation or repression [1]. To date, the function of many miRNAs is not known; however, their involvement in numerous physiological and pathological processes has been demonstrated as follows: in fact, they seem to play a role in cell proliferation, apoptosis, and differentiation [2]. An abnormal regulation of gene expression by microRNAs has been associated with the development and progression of numerous diseases, including cancer [3]. The expression of miRNAs can be influenced by various stimuli such as oxidative stress, inflammatory response, and hypoxia [4]. Assessment of miRNA expression changes during hypoxia is critical to understanding the role of miRNAs in many diseases and inflammatory processes. Although some guidelines have been established, there are no absolute partial pressure values of oxygen (pO_2) to define hypoxia. In general, pO_2 of 5% or less in cellular systems indicates hypoxia. Levels of pO_2 of between 5 and 2% correspond to a condition of moderate hypoxia, while levels <2% represent a condition of marked hypoxia [5].

Hypoxia can be continuous or intermittent; in the latter case, oxygen concentrations alternate between basal and low O_2 levels. In fact, intermittent hypoxia is caused by a series of repeated episodes of hypoxia and reoxygenation (in vivo, the classic model of intermittent hypoxemia occurs in the course of obstructive sleep apnea), while continuous hypoxia is characterized by constantly low oxygen levels. The pathogenetic mechanisms underlying these two types of hypoxia are completely different [4], and, in this study, we decided to focus our attention on continuous hypoxia.

There are the following two types of continuous hypoxia: moderate and marked. In recent years, some hypoxia-inducible microRNAs (HRMs, hypoxia-regulated microRNAs) have been identified. These HRMs are often activated in different types of cancers, such as breast and colon, suggesting their role in tumorigenesis [6]. Evaluating changes in miRNA expression during hypoxia is critical to understanding the role of miRNAs in many diseases, such as cancer and numerous inflammatory processes. To date, the mechanisms that regulate gene expression during hypoxia are not entirely clear; however, we know that many miRNAs are involved in the development of cancer. In general, it is known that miRNAs are directly involved in the formation of tumors [7], and we also know that about 6% of miRNAs have putative HRE (hypoxia response element) sites in their DNA, regions present in the promoters of hypoxia-inducible genes, indicating these miRNAs as possible targets of HIF-1 (Hypoxia Inducible Factor-1) and suggesting the possibility of their role associated with hypoxia [7]. As the cellular response to hypoxia involves the activation of several transcriptional regulators involved in inflammation, tumor invasion, angiogenesis, cell cycle block, and apoptosis, we believe that a better understanding of all these closely related mechanisms, as well as the identification of miRNAs sensitive to hypoxia, may prove fruitful in the search for new therapeutic targets and in the search for new and more effective anti-tumor therapies.

Based on the previous considerations, in this study we evaluated the expression of different miRNAs in conditions of continuous hypoxia, moderate (2% of oxygen) and marked (0.5% of oxygen), to study the possible relationship between these two types of continuous hypoxia, miRNA, and cancer and identify the miRNAs involved in carcinogenesis that are susceptible to hypoxia. In detail, we evaluated a group of miRNAs (miR-21, miR-23b, miR-26a, miR-27b, and miR-145), which are induced by a hypoxic environment, to better understand how continuous hypoxia could change their signature and to identify their eventual role in colorectal cancer (CRC), which is one of the most frequent tumors in women and men [8].

CRC is a very common malignant tumor, usually located between the junction of the rectum and sigmoid colon. According to a study conducted by Bray et al., CRC is the second cancer for mortality and the fourth for incidence [9]. The patient survival rate 5 years after diagnosis is approximately 65% [10].

Currently, the only prognostic indicator for CRC is represented by histological analysis, so it is very important to find useful biomarkers for the prognosis and diagnosis of this cancer.

To date, it is known that microRNAs can act as suppressors or promoters of CRC. The miRNAs play a role in CRC proliferation, metastasis, angiogenesis, and apoptosis, as well as play essential roles in various biological processes [11].

However, few studies have been carried out on their role in CRC in hypoxic conditions and, in particular, on the role of HMRs in CRC as potential biomarkers for the prognosis and diagnosis of CRC.

Under hypoxic conditions, cancer cells activate a series of molecular pathways driven by the transcription factor HIF-1 and, in response to these stimuli, modify their phenotype by implementing multiple survival strategies. It is, therefore, essential to identify molecular mediators through which HIF-1 controls tumor progression in order to identify new and specific molecular targets for the treatment of colorectal cancer [12].

2. Results

The MTT assay showed that at 2, 4, 8, and 24 h of exposure to the different experimental conditions, Caco-2 viability did not change as compared with the control. The survival rate was the same under all tested conditions (data not shown).

The expression of miRNAs was different in continuous hypoxia (moderate and marked) and normoxia (Figure 1).

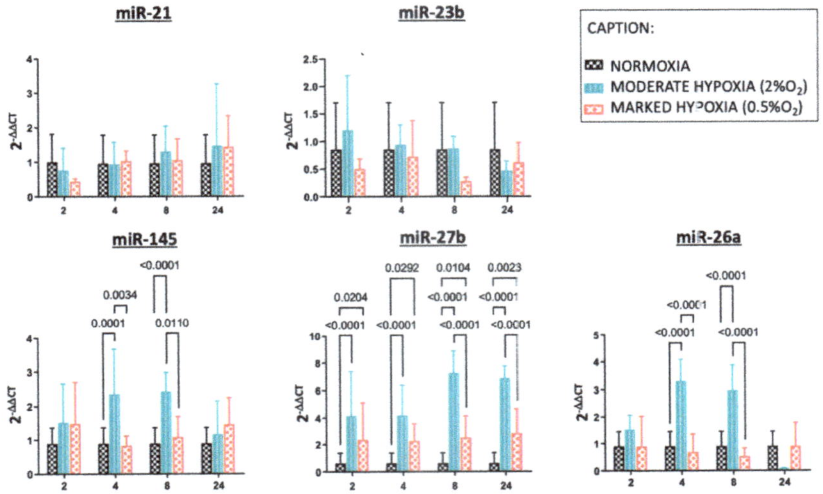

Figure 1. Expression of different miRNAs in normoxia, moderate continuous hypoxia (2% O_2) and marked continuous hypoxia (0.5% O_2). Quantitative real-time PCR analysis of differentially expressed microRNAs (miRNAs) in Caco-2 cells in condition of moderate (light blu) and marked (red) continuous hypoxia and normoxia (black). RNU-6B was used as endogenous control. The x axis shows the different times of exposure to hypoxia while the y axis shows the expression of each miRNA. Data represent the mean ± SD.

The levels of nearly all analyzed miRNAs were greater in moderate hypoxia versus marked hypoxia.

Notably, miR-145 and miR-26a showed higher levels of expression in moderate hypoxia than in marked one at 4 and 8 h, while for miR-27b, we proved this difference at 8 and 24 h.

We also detected an up-regulation among moderate hypoxia and normoxia at 4 and 8 h of miR-145 and miR-26a. Moreover, for miR-27b, there was the same up-regulation at all analyzed times (2, 4, 8, and 24 h).

In addition, miR-27b showed a higher expression at all times in marked hypoxia compared to normoxia.

On the other hand, miR-23b showed a lower expression at all times in marked hypoxia compared to normoxia, but for this result, there is no statistically significant difference, while only for miR-21, there seems to be no difference between moderate and marked hypoxia.

Therefore, the expression of nearly all analyzed miRNAs was greater in moderate hypoxia versus marked hypoxia, except for miR-23b and miR-21.

The mRNA expression of HIF-1α either in conditions of normoxia, moderate hypoxia, and marked hypoxia was assessed by qRT-PCR. As shown in Figure 2, the results showed, in agreement with the expression of nearly all analyzed miRNAs, a higher expression of HIF-1α in conditions of moderate hypoxia when compared with normoxia or marked hypoxia. This is true for 2, 4, and 8 h of hypoxia exposure; however, there is no significant expression difference at 24 h. This result could be explained by the fact that cells, under stressful conditions, are able to implement a series of strategies to react to external stimuli.

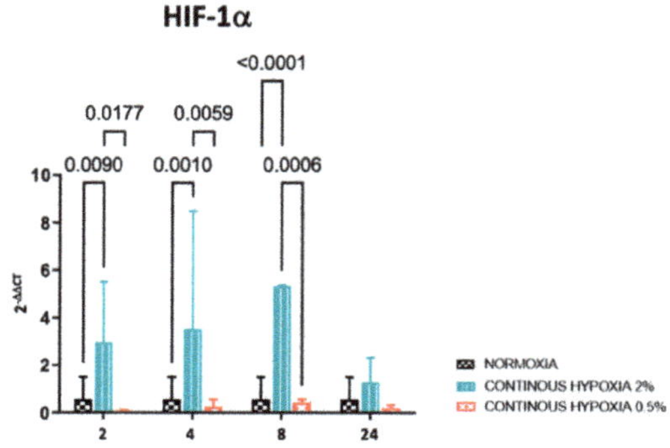

Figure 2. Relative mRNA expression of HIF-1α. Total RNA was extracted, and qRT-PCR was performed in order to quantify HIF-1α. β-actin was used as internal normalizer. Data represent the mean ± SD.

We also performed a cluster analysis on both moderate and marked continuous hypoxia miRNAs values. However, only for moderate continuous hypoxia do we obtain relevant data.

The differential miRNA expression in the condition of moderate continuous hypoxia (2% O_2) is shown in the heatmap in Figure 3.

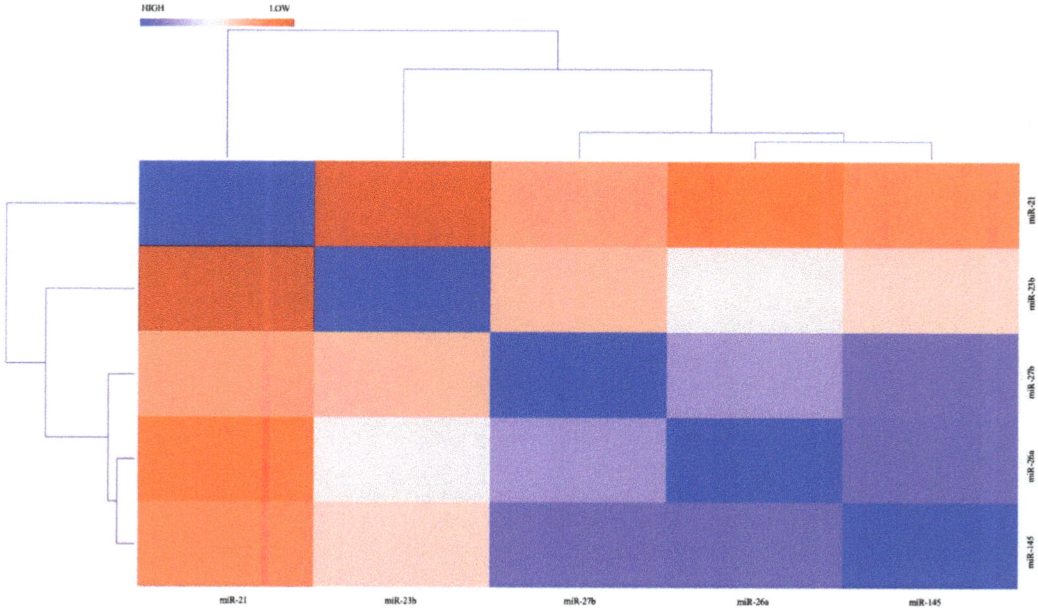

Figure 3. Heat map of miRNAs expression in condition of moderate continuous hypoxia (2% O_2). The heatmap shows the similarities between the expression profiles of the significantly changed miRNAs in condition of moderate continuous hypoxia. Blue color represents lower than mean intensity and red indicates higher than mean intensity. Each row and each column represent a miRNA.

The expression of miRNA was related to each other by means of the cluster analysis, resulting in a negative correlation between miR-21 and miR-23b (Figure 3). Moreover, we found a correlation between miR-26a, miR-27b, and miR-145, which form a cluster.

Generally, miRNAs with similar expression profiles during various experimental conditions are classified into clusters. Thus, this clustering allows the identification of miRNAs involved in the same cellular functions or the same regulatory pathway. Accordingly, the cluster analysis in our study shows a link between three miRNAs (miR-26a, miR-27b, and miR-145) because of their up-regulation during continuous moderate hypoxia (2% O_2).

3. Discussion

We currently know that miRNAs are involved in the development and progression of cancer and that they are able to regulate the expression of many oncogenes and tumor suppressor genes involved in the pathogenesis of cancer [13]. However, it is difficult to fully understand the role of miRNAs in carcinogenesis as their function can vary depending on the target tissue. To date, the molecular mechanisms by which miRNAs modulate cellular processes have yet to be fully elucidated, which is why the study of the specific functions of miRNAs in carcinogenesis could be useful to evaluate their therapeutic potential as diagnostic and prognostic markers of disease [14]. Recently, some studies have identified hypoxia-inducible miRNAs, HRMs, which are often activated in different types of tumors, suggesting their role in tumorigenesis [6]. The present study aimed to evaluate the expression of different miRNAs in conditions of moderate and marked hypoxia to study the possible relationship between these two types of hypoxia, miRNA, and cancer, and to identify the miRNAs involved in carcinogenesis that are susceptible to hypoxia. The main finding of this study is that nearly all of the miRNAs analyzed appear to have increased expression under conditions of moderate hypoxia; however, for miR-145 and miR-26a, this statement is not true when we consider their expression after 24 h of exposure to hypoxia.

In particular, three of them (miR-145, miR-27b, and miR-26a) are more expressed in moderate hypoxia than the marked one. For miR-145 and miR-26a, this is true for 4 and 8 h, whereas for miR-27b, we detected this up-regulation at 8 and 24 h. Vascular endothelial growth factor (VEGF) is a signal protein produced by cells that normally stimulate angiogenesis and that takes part in all those cellular mechanisms that restore normal oxygen supply to tissues following hypoxia. Tumors that over-express VEGF are able to grow and metastasize [15]. HIF-1α is a transcription factor that responds to hypoxia and stimulates the release of VEGF from parts of cells. VEGF binds to receptors on the endothelial cells and triggers a tyrosine kinase signaling pathway that leads to angiogenesis [16]. As previously mentioned, some miRNAs appear to be involved in the regulation of the HIF pathway by acting on specific signaling molecules that function as oncogenes or tumor suppressors. The miRNA most involved in this mechanism is certainly miR-26a [17].

All the miRNAs analyzed in this study are hypoxia-inducible miRNAs, however, there is evidence showing their altered expression in some types of tumors [18]. This, therefore, leads us to hypothesize that hypoxia (moderate or marked) can lead to an alteration of the expression of specific miRNAs involved in the formation of cancer.

MiR-23b is a hypoxia-regulated microRNA involved in apoptosis that appears to be up-regulated in some cancers such as those in the pancreas and colon [6]. In our study, miR-23b showed a lower expression at all times in the marked hypoxia compared to normoxia, but for this result, there is no statistically significant difference. Aberrant expression of miR-23b has been demonstrated in the development of several cancers. Chen L. et al., for example, investigated the oncogenic significance and function of miR-23b in glioma. However, they observed that miR-23b expression was elevated in glioma cells and that miR-23b acted through the HIF-1a/VEGF signaling pathway [19].

MiRNA-21 and miRNA-26a are released from endothelial cells. Previous work has shown that both miRNAs are expressed at the cellular level in response to hypoxic conditions. Therefore hypoxia seems to be a factor capable of inducing the activation of endothelial cells and, consequently, the release of these two miRNAs [20]. Our work shows results consistent with what has just been said. In fact, miR-26a appears higher in moderate hypoxia both compared to normoxia and to marked hypoxia at 4 and 8 h, while miR-21a seems to have a similar expression in marked and moderate hypoxia without any significant statistical difference. For miR-26a, this result is in line with what was previously demonstrated by Lacedonia et al. as far as it is concerned with continuous hypoxia [4].

To date, it is well known that miR-21 is overexpressed in most human tumors, and it promotes malignant growth and progression by acting on multiple targets [21]. In particular, miR-21 induces activation of PTEN (phosphatase and tensin homolog deleted on chromosome 10), AKT (protein kinase B), VEGF, and HIF-1 and consequently tumor progression [22]. Recent data suggest that miR-21 is also involved in promoting inflammation. Indeed, miR-21 appears to be able to reduce the expression of anti-inflammatory molecules such as TGF-β (transforming growth factor-β) [23]. MiR-21 thus constitutes a direct link between tumor-associated inflammatory state and cancer development or progression [24].

Moreover, for miR-26a, many studies have shown that it is dysregulated in various types of cancer [25,26]. Currently, many oncogenes, involved in multiple biological pathways such as proliferation, invasion, differentiation, and angiogenesis, appear to be targets of miR-26a. MiR-26a plays a role in tumorigenesis, acting both as a tumor suppressor and as an oncogene [27].

A recent study by Blick C. et al. showed that miR-145 plays an important role in hypoxia-dependent apoptosis in bladder cancer [28]. In this work, Blick C. and his collaborators demonstrated that miR-145 was significantly increased in response to hypoxia in bladder cancer cells and that this miRNA represents a target gene of HIF. Our work shows high levels of miR-145 in conditions of moderate hypoxia when compared to marked hypoxia or normoxia at 4 and 8 h.

Several studies in the literature suggest that miR-145 is a miRNA that acts as a tumor suppressor by inhibiting tumor growth and angiogenesis, and this miRNA appears to be down-regulated in various types of tumors [29]. Yu Yin et al. observed that miR-145 was significantly downregulated in plasma and tumor tissues of colorectal cancer patients and that miR-145 overexpression inhibited cell proliferation, migration, and invasion. In this previous work, they also demonstrated that miR-145 blocks the activation of the AKT and ERK1/2 pathways and the expression of HIF-1 and VEGF [30]. Based on our results, we, therefore, think that the low levels of miR-145 found in marked hypoxia may play a role in tumor development and progression, unlike when we observe moderate ones.

The same can be said for miR-27b. At present, many studies have reported that miR-27b plays an important role in cancer progression and have shown that miR-27b functions as a tumor suppressor in various types of cancers [31,32]. Chen Y et al. report that miR-27b-3p expression levels were lower than controls in both CRC patients and CRC cell lines. They also demonstrated that miR-27b-3p is capable of inhibiting the proliferation, migration, and invasion of colorectal cancer cells [33].

There are also evidences that miR-27 could facilitate the epithelial-mesenchymal transition (EMT) and the endothelial-mesenchymal transition (EndMT) in several types of cancer, including colorectal cancer through the activation of the transforming growth factor-β (TGF-β) [34,35]. It is well known that the EMT confers several traits to cancer cells that are required for malignant progression and TGF-β is considered to act as a primary inducer of this process [36].

So, based on these considerations and taking into account our results, we can therefore state that low levels of miR-27b in marked hypoxia may have a greater impact on cancer progression than in moderate hypoxia.

4. Materials and Methods

4.1. Cell Culture and Growth Conditions

Colorectal adenocarcinoma cell lines (CACO-2) were maintained in DMEM (Euroclone, Milan, Italy) supplemented with 10% fetal bovine serum (Euroclone, Milan, Italy), L-glutamine (Euroclone, Milan, Italy), and penicillin/streptomycin (Euroclone, Milan, Italy).

The cells were exposed to continuous hypoxia as follows:

- Moderate continuous hypoxia: we cultured the cells in an incubator (GALAXY 48 R, Eppendorf s.r.l., Milan, Italy) with oxygen maintained at 2% for 2 h, 4 h, 8 h, and 24 h;
- Marked continuous hypoxia: we cultured the cells in an incubator (GALAXY 48 R, Eppendorf s.r.l., Milan, Italy) with oxygen maintained at 0.5% for 2 h, 4 h, 8 h, and 24 h.

For both conditions of hypoxia, normoxic controls were maintained at 37 °C and 5% CO_2.

4.2. Viability Assay

Caco-2 cells (3×10^4 cells/well) were seeded in 96-well plates and exposed to complete medium. Both in normoxia, continuous hypoxia, and marked hypoxia for 2, 4, 8, and 24 h. Cell's viability was evaluated by MTT (1-(4,5-Dimethylthiazol-2-yl)-3,5-diphenylformazan) according to the manufacturer's protocol (Sigma-Aldrich, Milan, Italy). The cell's viability was calculated as follows: %viability = (optical density (OD) $_{560–655}$ of cell/OD $_{560–655}$ of control × 100).

4.3. Total RNA Purification and qRT-PCR Analysis

In order to perform miRNA analysis, total RNA was extracted by using TRIzol reagent (Thermo Fisher Scientific, Waltham, MA, USA), according to the manufacturer's protocol. Concentration and purity of RNA were measured using NanoDrop 1000 Spectrophotometer (Thermo Fisher Scientific). RNA purity was evaluated with the absorbance ratio OD260/OD280. RNA for miRNAs analysis (10 ng) was reverse transcribed into cDNA using TaqMan MicroRNA RT kit (Thermo Fisher Scientific), according to the manufacturer's protocol. The resulting cDNA transcript was used for detecting miRNA expression by quantitative real-time polymerase chain reaction (qRT-PCR) with Taqman miRNA assay (Thermo Fisher Scientific) according to the manufacturer's instructions. RNU-6B was used to normalize all RNA samples [37]. The miRNAs expression was calculated using the comparative $2^{-\Delta\Delta Ct}$ method.

Expression of HIF-1 was evaluated by qRT-PCR using SsoAdvanced ™ SYBR® Green Supermix (Bio-Rad, Hercules, CA, USA), as specified by the manufacturer.

Real-time reactions were set up in duplicate for each sample in 96-well plates in a reaction volume of 20 μL containing, respectively, 1X SsoAdvanced ™ SYBR® Green Supermix, 250 nM of specific primers, and 100 ng of cDNA.

The sequences of the primers used for amplification through qRT-PCR were the following:

HIF-1α forward 5′-AAAATCTCATCCAAGAAGCC-3′; HIF-1α reverse 5′-AATGTTCC AATTCCTACTGC-3′; β-actin forward 5′-GACGACATGGAGAAAATCTG-3′; β-actin reverse 5′-ATGATCTGGGTCATCTTCTC-3′.

The reaction was carried out on the ABI-PRISM 7300 instrument according to the manufacturer's instructions. Gene expression was analyzed according to $2^{-\Delta\Delta Ct}$ relative quantification method using β-actin as internal control.

4.4. Statistical Analysis

The results are expressed as mean ± SD. The ANOVA test was used to compare differences among groups. We used the two-way ANOVA test to compare the differential expression levels of miRNAs among moderate hypoxia (2% of oxygen), marked hypoxia (0.5% of oxygen), and normoxia and among different times (2, 4, 8, and 24 h) (GraphPad Software, 7825 Fay Avenue, Suite 230, La Jolla, CA 92037 USA). p-value < 0.05 was considered statistically significant.

Clustering analysis was performed on both moderate and marked continuous hypoxia values in order to (i) identify classes of miRNA based on their expression profiles and (ii) analyze the variations in the expression of miRNAs studied.

5. Conclusions

In conclusion, with this study, we have identified some miRNAs involved in different ways in the response to different types of continuous hypoxia (moderate or marked), highlighting that the expression of these small molecules can vary under hypoxic conditions and that they could play a role in the development of numerous diseases, including cancer.

Our work showed a change in the gene expression of some hypoxia-inducible miRNAs involved in tumor development and progression. Based on these preliminary results, we believe that moderate and marked hypoxia could activate different molecular pathways according to the severity of the hypoxia; however, further studies are needed in this regard.

To date, according to our knowledge, there are no similar studies in the literature that demonstrate a correlation between hypoxia (moderate and marked), miRNA, and cancer through the identification of a series of miRNAs involved in carcinogenesis that are susceptible to two different types of continuous hypoxia, moderate and marked.

In conclusion, this study demonstrates that continuous hypoxia induces the expression of several miRNAs, some of which appear to be directly involved in cancer formation and progression. Furthermore, it demonstrates how there is a different response between the condition of moderate continuous hypoxia (2%) and that of marked continuous hypoxia (0.5%), bringing to light that the former seems to be, in some cases, much more dangerous in terms of stimulation of the expression of some miRNAs.

Author Contributions: Conceptualization, G.M., G.S., P.S. and D.L.; methodology, G.M., P.S., G.S., R.S. and D.L.; software, G.M., P.S., P.T., C.C.D.P. and D.L.; validation, G.M., P.S. and D.L.; formal analysis, G.M., P.S. and G.S.; investigation, G.M., P.S. and G.S.; resources, G.M., P.S., P.T. and C.C.D.P.; data curation, G.M., P.S., G.S. and D.L.; writing—original draft preparation, G.M., P.S., G.S. and P.T.; writing—review and editing, G.M. and P.S.; visualization, D.L.; supervision, D.L., R.S. and M.P.F.B.; project administration, D.L. and M.P.F.B.; funding acquisition, D.L. All authors have read and agreed to the published version of the manuscript.

Funding: This study has been published with the financial support of the Department of Medical and Surgical Sciences of the University of Foggia (DR n. 285-2022, Prot. N. 0014908 - III/13, 16.03.2022).

Institutional Review Board Statement: Not applicable.

Informed Consent Statement: Not applicable.

Data Availability Statement: The data presented in this study are available on request from the corresponding author.

Conflicts of Interest: The authors declare no conflict of interest.

References

1. Bartel, D.P. MicroRNAs: Genomics, biogenesis, mechanism, and function. *Cell* **2004**, *116*, 281–297. [CrossRef]
2. Croce, C.M.; Calin, G.A. miRNAs, cancer, and stem cell division. *Cell* **2005**, *122*, 6–7. [CrossRef] [PubMed]
3. Harris, A.L. Hypoxia—A key regulatory factor in tumour growth. *Nat. Rev. Cancer* **2002**, *2*, 38–47. [CrossRef]
4. Lacedonia, D.; Scioscia, G.; Pia Palladino, G.; Gallo, C.; Carpagnano, G.E.; Sabato, R.; Foschino Barbaro, M.P. MicroRNA expression profile during different conditions of hypoxia. *Oncotarget* **2018**, *9*, 35114–35122. [CrossRef] [PubMed]
5. Bertout, J.A.; Patel, S.A.; Simon, M.C. The impact of O_2 availability on human cancer. *Nat. Rev. Cancer* **2008**, *8*, 967–975. [CrossRef] [PubMed]
6. Kulshreshtha, R.; Ferracin, M.; Wojcik, S.E.; Garzon, R.; Alder, H.; Agosto-Perez, F.J.; Davuluri, R.; Liu, C.G.; Croce, C.M.; Negrini, M.; et al. A microRNA signature of hypoxia. *Mol. Cell Biol.* **2007**, *27*, 1859–1867. [CrossRef] [PubMed]
7. Calin, G.A.; Sevignani, C.; Dumitru, C.D.; Hyslop, T.; Noch, E.; Yendamuri, S.; Shimizu, M.; Rattan, S.; Bullrich, F.; Negrini, M.; et al. Human microRNA genes are frequently located at fragile sites and genomic regions involved in cancers. *Proc. Natl. Acad. Sci. USA* **2004**, *101*, 2999–3004. [CrossRef]

8. Lacedonia, D.; Landriscina, M.; Scioscia, G.; Tondo, P.; Caccavo, I.; Bruno, G.; Giordano, G.; Piscazzi, A.; Foschino Barbaro, M.P. Obstructive Sleep Apnea Worsens Progression-Free and Overall Survival in Human Metastatic Colorectal Carcinoma. *J. Oncol.* **2021**, *2021*, 5528303. [CrossRef]
9. Bray, F.; Ferlay, J.; Soerjomataram, I.; Siegel, R.L.; Torre, L.A.; Jemal, A. Global cancer statistics 2018: GLOBOCAN estimates of incidence and mortality worldwide for 36 cancers in 185 countries. *CA Cancer J. Clin.* **2018**, *68*, 394–424. [CrossRef]
10. Siegel, R.L.; Miller, K.D.; Fedewa, S.A.; Ahnen, D.J.; Meester, R.G.S.; Barzi, A.; Jemal, A. Colorectal cancer statistics, 2017. *CA Cancer J. Clin.* **2017**, *67*, 177–193. [CrossRef]
11. Zhang, N.; Hu, X.; Du, Y.; Du, J. The role of miRNAs in colorectal cancer progression and chemoradiotherapy. *Biomed. Pharmacother.* **2021**, *134*, 111099. [CrossRef] [PubMed]
12. Vadde, R.; Vemula, S.; Jinka, R.; Merchant, N.; Bramhachari, P.V.; Nagaraju, G.P. Role of hypoxia-inducible factors (HIF) in the maintenance of stemness and malignancy of colorectal cancer. *Crit. Rev. Oncol. Hematol.* **2017**, *113*, 22–27. [CrossRef] [PubMed]
13. Fang, Z.; Tang, J.; Bai, Y.; Lin, H.; You, H.; Jin, H.; Lin, L.; You, P.; Li, J.; Dai, Z.; et al. Plasma levels of micro RNA-24, microRNA-320a and microRNA-423-5p are potential biomarkers for colorectal carcinoma. *J. Exp. Clin. Cancer Res.* **2015**, *34*, 86–96. [CrossRef] [PubMed]
14. Chi, Y.; Zhou, D. microRNAs in colorectal carcinoma- from pathogenesis to therapy. *J. Exp. Clin. Cancer Res.* **2016**, *35*, 43–54. [CrossRef]
15. Palmer, B.F.; Clegg, D.J. Oxygen sensing and metabolic homeostasis. *Mol. Cell Endocrinol.* **2014**, *397*, 51–57. [CrossRef]
16. Smith, T.G.; Robbins, P.A.; Ratcliffe, P.J. The human side of hypoxia-inducible factor. *Br. J. Haematol.* **2008**, *141*, 325–334. [CrossRef]
17. Li, C.; Li, Y.; Lu, Y.; Niu, Z.; Zhao, H.; Peng, Y.; Li, M. miR-26 family and its target genes in tumorigenesis and development. *Crit. Rev. Oncol. Hematol.* **2021**, *157*, 103124. [CrossRef]
18. Volinia, S.; Calin, G.A.; Liu, C.G.; Ambs, S.; Cimmino, A.; Petrocca, F.; Visone, R.; Iorio, M.; Roldo, C.; Ferracin, M.; et al. A microRNA expression signature of human solid tumors defines cancer gene targets. *Proc. Natl. Acad. Sci. USA* **2006**, *103*, 2257–2261. [CrossRef]
19. Chen, L.; Han, L.; Zhang, K.; Shi, Z.; Zhang, J.; Zhang, A.; Wang, Y.; Song, Y.; Li, Y.; Jiang, T.; et al. VHL regulates the effects of miR-23b on glioma survival and invasion via suppression of HIF-1α/VEGF and β-catenin/Tcf-4 signaling. *Neuro-Oncology* **2012**, *14*, 1026–1036. [CrossRef]
20. Eichhorn, L.; Dolscheid-Pommerich, R.; Erdfelder, F.; Ayub, M.A.; Schmitz, T.; Werner, N.; Jansen, F. Sustained apnea induces endothelial activation. *Clin. Cardiol.* **2017**, *40*, 704–709. [CrossRef]
21. Kumarswamy, R.; Volkmann, I.; Thum, T. Regulation and function of miRNA-21 in health and disease. *RNA Biol.* **2011**, *8*, 706–713. [CrossRef] [PubMed]
22. Cingarlini, S.; Bonomi, M.; Corbo, V.; Scarpa, A.; Tortora, G. Profiling mTOR pathway in neuroendocrine tumors. *Target Oncol.* **2012**, *7*, 183–188. [CrossRef] [PubMed]
23. Merline, R.; Moreth, K.; Beckmann, J.; Nastase, M.V.; Zeng-Brouwers, J.; Tralhão, J.G.; Lemarchand, P.; Pfeilschifter, J.; Schaefer, R.M.; Iozzo, R.V.; et al. Signaling by the matrix proteoglycan decorin controls inflammation and cancer through PDCD4 and MicroRNA-21. *Sci. Signal* **2011**, *4*, ra75. [CrossRef]
24. Olivieri, F.; Rippo, M.R.; Monsurrò, V.; Salvioli, S.; Capri, M.; Procopio, A.D.; Franceschi, C. MicroRNAs linking inflamm-aging, cellular senescence and cancer. *Ageing. Res. Rev.* **2013**, *12*, 1056–1068. [CrossRef]
25. Liu, P.; Tang, H.; Chen, B.; He, Z.; Deng, M.; Wu, M. miR-26a suppresses tumour proliferation and metastasis by targeting metadherin in triple negative breast cancer. *Cancer Lett.* **2015**, *357*, 384–392. [CrossRef]
26. Alajez, N.M.; Shi, W.; Hui, A.B.Y.; Bruce, J.; Lenarduzzi, M.; Ito, E. Enhancer of Zeste homolog 2 (EZH2) is overexpressed in recurrent nasopharyngeal carcinoma and is regulated by miR-26a, miR101, and miR-98. *Cell Death Dis.* **2010**, *1*, e85. [CrossRef] [PubMed]
27. Chen, J.; Zhang, K.; Xu, Y.; Gao, Y.; Li, C.; Wang, R.; Chen, L. The role of microRNA-26a in human cancer progression and clinical application. *Tumour. Biol.* **2016**, *37*, 7095–7108. [CrossRef]
28. Blick, C.; Ramachandran, A.; McCormick, R.; Wigfield, S.; Cranston, D.; Catto, J.; Harris, A.L. Identification of a hypoxia-regulated miRNA signature in bladder cancer and a role for miR-145 in hypoxia-dependent apoptosis. *Br. J. Cancer* **2015**, *113*, 634–644. [CrossRef]
29. Cui, S.Y.; Wang, R.; Chen, L.B. MicroRNA-145: A potent tumour suppressor that regulates multiple cellular pathways. *J. Cell Mol. Med.* **2014**, *18*, 1913–1926. [CrossRef]
30. Yin, Y.; Yan, Z.P.; Lu, N.N.; Xu, Q.; He, J.; Qian, X.; Yu, J.; Guan, X.; Jiang, B.H.; Liu, L.Z. Downregulation of miR-145 associated with cancer progression and VEGF transcriptional activation by targeting N-RAS and IRS1. *Biochim. Biophys. Acta* **2013**, *1829*, 239–247. [CrossRef]
31. Takahashi, R.U.; Miyazaki, H.; Takeshita, F.; Yamamoto, Y.; Minoura, K.; Ono, M.; Kodaira, M.; Tamura, K.; Mori, M.; Ochiya, T. Loss of microRNA-27b contributes to breast cancer stem cell genera- tion by activating ENPP1. *Nat. Commun.* **2015**, *6*, 7318. [CrossRef] [PubMed]
32. Wan, L.; Zhang, L.; Fan, K.; Wang, J.J. miR-27b targets LIMK1 to inhibit growth and invasion of NSCLC cells. *Mol. Cell Biochem.* **2014**, *390*, 85–91. [CrossRef] [PubMed]
33. Chen, Y.; Chen, G.; Zhang, B.; Liu, C.; Yu, Y.; Jin, Y. miR-27b-3p suppresses cell proliferation, migration and invasion by targeting LIMK1 in colorectal cancer. *Int. J. Clin. Exp. Pathol.* **2017**, *10*, 9251–9261. [PubMed]

34. Suzuki, H.I.; Katsura, A.; Mihira, H.; Horie, M.; Saito, A.; Miyazono, K. Regulation of TGF-β-mediated endothelial-mesenchymal transition by microRNA-27. *J. Biochem.* **2017**, *161*, 417–420. [CrossRef]
35. Haier, J.; Ströse, A.; Matuszcak, C.; Hummel, R. miR clusters target cellular functional complexes by defining their degree of regulatory freedom. *Cancer Metastasis Rev.* **2016**, *35*, 289–322. [CrossRef]
36. Puppo, M.; Bucci, G.; Rossi, M.; Giovarelli, M.; Bordo, D.; Moshiri, A.; Gorlero, F.; Gherzi, R.; Briata, P. miRNA-Mediated KHSRP Silencing Rewires Distinct Post-transcriptional Programs during TGF-β-Induced Epithelial-to-Mesenchymal Transition. *Cell Rep.* **2016**, *16*, 967–978. [CrossRef]
37. Furukawa, N.; Sakurai, F.; Katayama, K.; Seki, N.; Kawabata, K.; Mizuguchi, H. Optimization of a microRNA expression vector for function analysis of microRNA. *J. Control. Release* **2011**, *150*, 94–101. [CrossRef]

International Journal of *Molecular Sciences*

MDPI

Review

Proteomic Profiling and Biomarker Discovery in Colorectal Liver Metastases

Geoffrey Yuet Mun Wong [1,2,*], Connie Diakos [2,3], Thomas J. Hugh [1,2] and Mark P. Molloy [4]

1 Department of Upper Gastrointestinal Surgery, Royal North Shore Hospital, Sydney, NSW 2065, Australia; tom.hugh@sydney.edu.au
2 Northern Clinical School, The University of Sydney, Sydney, NSW 2065, Australia; connie.diakos@sydney.edu.au
3 Department of Medical Oncology, Royal North Shore Hospital, Sydney, NSW 2065, Australia
4 Bowel Cancer and Biomarker Research Laboratory, Faculty of Medicine and Health, School of Medical Sciences, The University of Sydney, Sydney, NSW 2006, Australia; m.molloy@sydney.edu.au
* Correspondence: gwon4318@uni.sydney.edu.au

Abstract: Colorectal liver metastases (CRLM) are the leading cause of death among patients with metastatic colorectal cancer (CRC). As part of multimodal therapy, liver resection is the mainstay of curative-intent treatment for select patients with CRLM. However, effective treatment of CRLM remains challenging as recurrence occurs in most patients after liver resection. Proposed clinicopathologic factors for predicting recurrence are inconsistent and lose prognostic significance over time. The rapid development of next-generation sequencing technologies and decreasing DNA sequencing costs have accelerated the genomic profiling of various cancers. The characterisation of genomic alterations in CRC has significantly improved our understanding of its carcinogenesis. However, the functional context at the protein level has not been established for most of this genomic information. Furthermore, genomic alterations do not always result in predicted changes in the corresponding proteins and cancer phenotype, while post-transcriptional and post-translational regulation may alter synthesised protein levels, affecting phenotypes. More recent advancements in mass spectrometry-based technology enable accurate protein quantitation and comprehensive proteomic profiling of cancers. Several studies have explored proteomic biomarkers for predicting CRLM after oncologic resection of primary CRC and recurrence after curative-intent resection of CRLM. The current review aims to rationalise the proteomic complexity of CRC and explore the potential applications of proteomic biomarkers in CRLM.

Keywords: colorectal cancer; colorectal liver metastases; proteomics; prognosis; biomarkers; mass spectrometry

Citation: Wong, G.Y.M.; Diakos, C.; Hugh, T.J.; Molloy, M.P. Proteomic Profiling and Biomarker Discovery in Colorectal Liver Metastases. *Int. J. Mol. Sci.* **2022**, *23*, 6091. https://doi.org/10.3390/ijms23116091

Academic Editors: Donatella Delle Cave and Alessandro Ottaiano

Received: 5 May 2022
Accepted: 27 May 2022
Published: 29 May 2022

1. Introduction

Globally, colorectal cancer (CRC) is the third most common cancer (10.0%) and the second leading cause of cancer death (9.4%) [1]. The liver is the most common site of CRC metastasis due to the portal venous drainage from the colon and rectum to the liver [2,3]. Colorectal liver metastases (CRLM) are detected in approximately 20% of patients at initial diagnosis and are the leading cause of death among patients with metastatic CRC [4–6]. Although CRLM portends a poor prognosis, liver resection is potentially curative in select patients, with actual 10-year recurrence-free survival reported in an estimated 20% of patients [7,8]. Multimodal treatment approaches have led to remarkable improvements in the prognosis of patients with CRLM over the past two decades. Although five-year overall survival rates after liver resection are as high as 50–60% in contemporary series, an estimated 75% of patients develop recurrence, and most occur within two years [9,10]. Whilst specific clinicopathologic variables are prognostic at baseline, conditional survival analysis in patients with resected CRLM demonstrates that these preoperative factors are inconsistent and lose prognostic significance over a relatively short time [11–13]. Early recurrence is the most useful single prognostic and clinical feature in estimating disease-specific survival, but the ability to predict this is currently limited [13–15]. Tumour genetics, location and treatment effect heterogeneity give rise to challenges in selecting treatment and predicting whether an individual might benefit from a particular treatment [16–18].

Patients with resectable CRLM require a nuanced approach given the expanding criteria of resectability and increasing treatment options [19,20]. This group presents a unique opportunity to understand the molecular underpinnings of metastatic CRC because they are free of detectable metastasis at a defined time point. The rapid development of next-generation sequencing technologies and the declining cost of human genome sequencing has accelerated genomic profiling of various cancers, including CRC [21,22]. Although the characterisation of genomic alterations in CRC has significantly improved our understanding of its carcinogenesis, the functional context at the protein level has not been established for most of this genomic information [23–25]. Starting from the genome, multiple biological regulatory and processing steps take place to arrive at the proteome, each step driving increasing complexity and diversity. Consequently, chemical modifications that affect protein function and protein–protein interactions that carry out critical biological activities cannot reliably be predicted from genomic and transcriptomic analyses. Moreover, because mutations do not always result in a predicted change in the corresponding proteins and phenotype, the gap between gene expression and the biological capability of cancer is not straightforward [26–29]. Therefore, precision oncology requires the examination of the co-expression of multiple genes and proteins under different disease states and the impact these have on clinically meaningful outcomes such as recurrence and survival.

Liquid chromatography–mass spectrometry (LC-MS) of digested proteins conducted with high-resolution instruments allows us to quantitate thousands of proteins from complex biological specimens in either data-dependent acquisition, or more recently, data-independent acquisition workflows [30,31]. Preclinical exploratory studies on proteomic profiling of cancer biospecimens have provided new insights into the molecular alterations in cancer and have identified leads for potentially useful clinical biomarkers. Proteomic mass spectrometry data have often accompanied landmark cancer genomic studies; for example, those reported on colorectal cancer, pancreatic cancer, ovarian cancer and lung cancer [32–36]. Publications have increased steadily in mass-spectrometry-driven proteomics analysis of differential protein expression and cancer-specific biomarkers derived from tissue and body fluids. Several studies have explored proteomic biomarkers in predicting CRLM after oncologic resection of primary CRC and recurrence after curative-intent resection of CRLM; however, there is no up-to-date overview of these findings. This review aims to rationalise the proteomic complexity of CRC and explore the potential applications of proteomic biomarkers in CRLM by critically appraising mass-spectrometry-based proteomic profiling of human CRLM biospecimens published over the last ten years.

2. Characteristics of Preclinical Exploratory Studies on Proteomic Biomarkers in Colorectal Liver Metastases

Seventeen exploratory studies on proteomic profiling and five studies on proteogenomic profiling of human CRLM were identified through a search of the literature using PubMed, Medline and ScienceDirect, with the main search terms including "colorectal", "cancer", "liver", metastasis" or "metastases", "proteomics" or "proteome", "proteogenomics", "biomarker", "mass spectrometry" and "prognosis". Relevant studies on human biospecimens from 2011 to 2021 were included and studies that focused on animal models, cell lines and patient-derived xenograft models were excluded. References contained in the included studies were reviewed for appropriate publications that the electronic search strategy may have missed. Proteomic and proteogenomic studies included are summarised separately in Tables 1 and 2, with the most recent publications listed first. The studies included samples from 301 patients with CRLM, with a range of 1–44 patients in each exploratory cohort. New proteomic signatures were revealed even in studies with small sample sizes, and therefore these were included. Sixteen studies used fresh frozen tissue and six used formalin-fixed paraffin-embedded (FFPE) tissue. Although only studies that reported MS-based proteomics were selected and the majority utilised LC-MS, there were variations and nuances in the techniques utilised across studies. Most of the included studies performed differential protein expression analysis to quantify protein abundance between two or more groups within the same experiment. The comparison groups included a combination of matched or unmatched primary CRC, normal colonic tissue, normal liver tissue, or prognostically different patient groups stratified by clinicopathological factors.

Table 1. Studies on the prognostic relevance of mass spectrometry-based proteomic biomarkers in human colorectal cancer liver metastases over the last 10 years in descending chronological order.

First Author/ Reference/Year/Journal	Biospecimen	Mass-Spectrometry-Based Technique	Discovery Cohort Characteristics (Sample Size and Comparator)	Key Biomarkers and Findings
Michal S et al. [37] 2021 *Journal of Personalized Medicine*	FFPE tissue	Label-free LC-MS/MS	$n = 29$ with recurrence < 6 months after resection of CLRM Comparison: $n = 29$ with recurrence 6–12 months after resection of CRLM	Upregulation of matrix metalloproteinase 7 (MMP7) and dehydropeptidase 1 (DPEP1) in poor-prognosis group. Downregulation of lysyl oxidase-like 1 (LOXL1) in poor-prognosis group. A third of differentially expressed proteins associated with extracellular matrix.
Fahrner M et al. [38] 2021 *Neoplasia*	FFPE tissue	Label-free LC-MS/MS	$n = 7$ synchronous CRLM Comparison: $n = 7$ matched primary CRC	Metabolic proteins: pyruvate carboxylase (PC) and fructose-bisphosphate aldolase B (ALDOB), and fructose-1,6-bisphosphatase 1 (FBP1) upregulated in CRLM. Immune system proteins: enrichment of complement components C1, C4, C5, C9 in CRLM. Structural proteins: depletion of desmin (DES), synemin (SYNM) and filamin-C (FLNC) in CRLM.
Liu X et al. [39] 2020 *Clinical and Translational Oncology*	Fresh frozen tissue	TMT-labelling, LC-MS/MS	$n = 8$ CRLM Comparison: $n = 8$ primary tumour	Upregulation of fibronectin (FN1), metalloproteinase inhibitor 1 (TIMP1), thrombospondin-1 (THBS1), periostin (POSTN) and in CRLM.
Voß H et al. [40] 2020 *Clinical and Experimental Metastasis*	Fresh frozen tissue	Label-free LC-MS/MS	$n = 1$ with 3 metachronous CRLM Comparison: N/A	Upregulation of 56 extracellular matrix-associated proteins including tenascin C (TNC), nidogen-1 (NID1), fibulin-1 (FBLN1), vitronectin (VTN).
van Huizen NA [41] 2020. *Frontiers in Oncology*	FFPE tissue	Label-free nano-LC-MS/MS	$n = 14$ CRLM Comparison: $n = 14$ matched liver tissue, matched primary CRC and normal colonic tissue	Overall degree of collagen hydroxylation was significantly lower in CRLM and primary CRC compared to normal colon Downregulation of 11 peptides with a specific number of hydroxylation in CRLM compared to normal liver tissue.
van Huizen et al. [42] 2019 *Journal of Proteome Research*	FFPE tissue	Nano-LC-ESI-ETD-HCD	$n = 2$ CLRM Comparison: $n = 2$ matched normal liver tissue	Lower ratio of 4xHyp at position 584 of collagen alpha-2(I) chain (COL1A2) in CRLM.
van Huizen NA [43] 2019 *Journal of Biological Chemistry*	FFPE tissue	Label-free nano-LC-MS/MS	$n = 30$ patients Comparison: $n = 30$ matched normal liver tissue, primary CRC and normal colon tissue	Upregulation of four collagen types in CRLM: COL10A1, COL12A1 (most abundant), COL14A1, COL15A1. Upregulation of six non-collagen colon-specific proteins in CRLM: cadherin-17 (CDH17), protein phosphatase 1 regulatory subunit 1B (PPP1R1B/DARP-32), keratin, type 1 cytoskeletal 20 (KRT20), carcinoembryonic antigen-related cell-adhesion molecule 5 (CEACAM5), cell-surface AA33 antigen (GPA33), mucin-13 (MUC13).

Table 1. *Cont.*

First Author/ Reference/Year/Journal	Biospecimen	Mass-Spectrometry-Based Technique	Discovery Cohort Characteristics (Sample Size and Comparator)	Key Biomarkers and Findings
Ku X et al. [44] 2019 *Analytical Cellular Pathology*	Fresh frozen tissue	TMT labelling, nano-LC-MS/MS	n = 9 CRLM Comparison: n = 9 matched primary tumour and normal colonic tissue	Upregulation of filamin A-interacting protein 1-like (FILIP1L) and plasminogen (PLG) in CRLM.
Yang W et al. [45] 2019 *Proteomics Clinical Applications*	Fresh frozen tissue	Label-free nano-LC-MS/MS	n = 17 CRLM Comparison: n = 20 Stage III CRC who did not develop CRLM	Nine key proteins identified in CRLM: heat shock protein family D member 1 (HSPD1), eukaryotic translation elongation factor 1 gamma, heterogeneous nuclear ribonucleoprotein A2/B1 (HNRNPA2B1), fibrinogen beta chain (FGB), Talin 1 (TLN 1), adaptor-related protein complex 2 subunit alpha-2 (AP2A2), serrated RNA effector molecule homolog (SRRT), apolipoprotein C3 (APOC3), and phosphoglucomutase 5 (PGM5). Fibrinogen beta chain is a key biomarker for CRLM.
Kim EK et al. [46] 2019 *Cancer Genomics Proteomics*	Fresh frozen tissue	2D-PAGE, MALDI-TOF MS	n = 5 CRLM Comparison: n = 5 synchronous primary CRC	Upregulation of serpin family A member 1 (SERPINA1), apolipoprotein AI (APOA1), intelectin 1 (ITLN1), desmin (DES), diazepam-binding inhibitor (DBI), succinate dehydrogenase complex flavoprotein subunit A (SDHA), and carbonic anhydrase 1 (CA1) in CRLM.
Kirana C et al. [47] 2019 *Clinical Proteomics*	Fresh frozen tissue	2D-DIGE, MALDI-TOF MS	n = 8 stage II CRC with CRLM within 5 years after surgery Comparison: n = 11 stage II CRC patients with no metastasis within 5 years after surgery	Upregulation of HLA class I histocompatibility antigen, B alpha chain (HLAB), A disintegrin and metalloproteinase with thrombospondin motifs 2 (ADAMTS2), latent-transforming growth factor beta-binding protein 3 (LTBP3), protein jagged-2 (JAG2) and nucleoside diphosphate kinase B (NME2) on tumour cells was associated with CRC progression and invasion, metastasis and CRC-specific survival.
Yuzhalin AE et al. [48] 2018 *Nature Communications*	Fresh frozen tissue	ECM enrichment, label-free, nano-LC-MS/MS	n = 5 CRLM Comparison: n = 5 matched normal liver, primary CRC and normal colon.	Increased amounts of citrullinated proteins in CRLM compared to normal liver. Primary CRC and normal colonic mucosa. Peptidylarginine deiminase 4 (PAD4)-driven citrullination of the extracellular matrix is essential for CRLM growth. Other upregulated proteins included versican (VCAN), metalloproteinase inhibitor 1 precursor (T1MP1), latent-transforming growth factor beta-binding protein (LTBP) 1–3, epithelial discoidin domain-containing receptor 1 (DDR1), and protein S100-A10 (S100A10).

Table 1. *Cont.*

First Author/ Reference/Year/Journal	Biospecimen	Mass-Spectrometry-Based Technique	Discovery Cohort Characteristics (Sample Size and Comparator)	Key Biomarkers and Findings
Yang Q et al. [49] 2017 *Journal of Proteomics*	Fresh frozen tissue	1D and 2D-PAGE, nano-LC-MS/MS	$n = 8$ CRLM Comparison: $n = 8$ matched primary, CRLM and adjacent normal colon and liver tissues.	Olfactomedin 4 (OLFM4), CD11b/integrin alpha m (ITGAM) and integrin alpha-2 (ITGA2) significantly overexpressed in primary CRC and CRLM
Shen Z et al. [50] 2016 *Journal of Proteomics*	Fresh frozen tissue	Acetylated peptide enrichment, TMT labelling, LC-MS/MS	$n = 3$ CRLM Comparison: $n = 3$ matched primary CRC	HIST2H3AK19Ac and H2BLK121Ac were the acetylated histones most changed. Tropomyosin beta chain (TPM2), K152Ac and alcohol dehydrogenase 1B (ADH1B), K331Ac were the acetylated non-histones most altered in CRLM.
Naba et al. [51] 2014 *BMC Cancer*	Fresh frozen tissue	ECM enrichment, off-gel electrophoresis, LC-MS/MS	$n = 3$ CRLM Comparison: $n = 3$ matched primary CRC and normal colonic tissue	Hemopexin (HPX), osteopontin/secreted phospho-protein 1 (SPP1), cartilage oligomeric matrix protein (COMP), insulin-like growth factor-binding protein complex acid labile subunit (IGFALS), fibronectin type III domain-containing protein1 (FNDC1), bone morphogenetic protein 1 (BMP1) and complement C1q tumour necrosis factor-related protein 5 (C1QTNF5). Extracellular matrix protein signatures are potential tissue or serological biomarkers.
Turtoi A et al. [52] 2014 *Hepatology*	FFPE tissue	MALDI-MS imaging, nano-UPLC-qTOF MS	$n = 8$ CRLM Comparison: $n = 8$ normal liver, $n = 3$ matched primary CRC	High expression of latent-transforming growth factor beta-binding protein 2 (LTBP2) and transforming growth factor-beta-induced protein ig-h3 (TGFBI) were consistent features of CRLM and are absent in normal tissues.
Kirana et al. [53] 2012 *International Journal of Proteomics*	Fresh frozen tissue	2D-DIGE, MALDI-TOF MS	$n = 8$ CRLM Comparison: $n = 8$ matched primary CRC	Overexpression of cathepsin D (CTSD) in cells from the main tumour body showed significant correlation with subsequent distant metastasis and shorter cancer-specific survival.

Table 2. Studies on proteogenomics of colorectal cancer liver metastases.

Authors	Biospecimen	MS Technique	Sample Number with CRLM	Key Findings
Li C et al. [54] 2020 *Cancer Cell*	Fresh frozen tissue	Phosphopeptide enrichment, nano-LC-MS/MS	$n = 43$ Comparator: $n = 146$ primary CRC, adjacent normal colon and normal liver	Three CRC subtypes with distinct molecular signatures and clinical prognosis were defined using proteomic profiling. Phosphoproteomic pattern distinguishes metastatic from non-metastatic colorectal cancer.
Blank-Landeshammer B et al. [55] 2019 *Cancers (Basel)*	Fresh frozen tissue	Phosphopeptide enrichment, stable heavy isotope peptide labelling, nano-LC-MS/MS	$n = 8$ Comparator: $n = 6$ paired normal liver tissue	Low expression of actionable somatic mutations including KRASG12V can be predicted by precise quantitation of altered proteins such as SRPX2, S6K-alpha-5, GTPase KRas, PTBP1, ARL2, PPP1R14C and HAUS7
Ma YS. [56] 2019 *Molecular Therapy Oncolytics*	Fresh frozen tissue	Label-free nano-LC-MS/MS	$n = 23$ Comparator: $n = 21$ paired normal colorectal cancer tissue with or without liver metastasis	UQCR5 and FDFT1 were frequently overexpressed in the CRLM cohort and shown to have potential prognostic value. High expression of UQCR5 and was associated with worse overall survival and progression-free survival. High expression of FDFT1 was associated with better overall survival and progression-free survival.
Ma YS et al. [57] 2018 *Molecular Cancer*	Fresh frozen tissue	Nano-LC-MS/MS-based shotgun proteomics profiling	$n = 23$ Comparator: $n = 21$ non-metastatic CRC	Four CNV-mRNA-protein correlated proteins were associated with worse overall survival: HSP90AB1, COL1A2, FABP5 and BGN. Two single amino acid variants were associated with shorter overall and disease-free survival: MYH9 and CCT6A
Snoeren N et al. [58] 2013. *British Journal of Cancer*	Fresh frozen tissue	SDS-PAGE gel electrophoresis and in-gel digestion, label-free nano-LC-MS/MS	$n = 10$ <6 months to recurrence (early), $n = 5$ >24 months to recurrence (prolonged), $n = 5$	SERPINB5 which encodes for Maspin was the most upregulated (~2.1 times higher, $p = 0.01$) in patients with early recurrence compared to prolonged (>24 months) time to recurrence. Maspin was the only overlapping factor among 14 genes and 46 genes that showed a significant association with recurrence.

3. Proteomic Profiling of Colorectal Liver Metastases Tissue Identifies Prognostically Distinct Groups

The clinical utility of prognostic prediction models in CRLM has been limited, as these scoring systems do not consistently stratify recurrence and survival after curative-intent surgery [13,59]. Clinical risk-scoring systems were marginally better than chance alone in predicting outcomes in some cohorts [12]. Prognostic biomarkers indicate the likelihood of a future clinical event, disease recurrence, or disease progression among patients with the same characteristics [60,61]. One existing strategy to overcome the current limitations of clinical risk scores is to identify prognostic biomarkers that indicate the likelihood of recurrence after resection of CRLM. Early recurrence is associated with poor prognosis and is a useful clinical feature in estimating conditional disease-specific survival [14,15,62].

Michal et al. characterised proteomic biomarkers in prognostically distinct clinical groups based on the time interval between the resection of CRLM and recurrence. A 12-month cut-off was used to divide patients into those with a 'good prognosis' ($n = 29$) and 'poor prognosis' ($n = 29$). Microdissection of FFPE tissue followed by label-free LC-MS identified 99 differentially expressed proteins, of which a third were associated with the extracellular matrix pathway. MMP7 and DPEP1 were upregulated, while LOXL1 was downregulated. MMP7 promotes invasion through proteolysis of the ECM proteins and proliferation of cancer cells through upregulation of MM2 and MMP9. In addition, MSH2 and MCM4, associated with DNA replication and repair pathways, were upregulated, and several components of the immune pathway—such as C5, C1RL, C8A, CD163, chymase 1, and HLA-B—were downregulated in the poor-prognosis group. This study indicated that components of the tumour microenvironment, especially the extracellular matrix pathway, may be critical drivers in early recurrence after resection of CRLM [37].

The study by Snoeren et al. used gene expression profiling and label-free nano-LC-MS/MS to identify genes and proteins that correlate with early (<6 months) and late (>24 months) recurrence after resection of CRLM. Upregulation of SERPINB5 and increased expression of Maspin were the only overlapping factor among 14 genes and 46 proteins that showed a significant association with recurrence. Immunohistochemical analysis of Maspin expression in stage III CRC correlated with early time to recurrence and disease-specific survival, but not in stage II CRC. Altogether, these findings point to Maspin as a potential biomarker for early recurrence in primary stage III and IV colorectal cancer [58].

4. Adjuvant Treatment Stratification for Stage II and Stage III Colorectal Cancer

Following oncologic resection of primary CRC, approximately 20% of patients with stage II and 30–40% of patients with stage III colorectal cancers will develop recurrence [63–65]. Follow-up after oncologic resection for primary colorectal cancer includes regular monitoring of carcinoembryonic antigen (CEA) and cross-sectional imaging to ultimately increase patient survival rates and quality of life through the early detection of recurrent disease. Adjuvant chemotherapy aims to eradicate cancer micrometastases. However, most patients with stage II CRC (i.e., those without regional lymph node metastasis) undergo clinical surveillance following oncologic resection of the primary tumour as the benefit of adjuvant chemotherapy has not been demonstrated in low-risk stage II CRC. Several high-risk clinicopathological features for recurrence have been identified but there is no clear evidence of patient selection and limited evidence on the benefit of adjuvant chemotherapy in this situation [66–68]. Therefore, the rationale for proteomic profiling in this patient group is to identify those at higher risk of recurrence, as these patients may benefit from adjuvant chemotherapy or more intensive surveillance. Kirana et al. used a combination of laser microdissection of primary CRC, two-dimensional differential gel electrophoresis (2D-DIGE) and matrix-assisted laser desorption ionisation time-of-flight mass spectrometry (MALDI-TOF MS) to identify protein biomarkers that stratify the risk of CRLM in patients with stage II disease. Cancer cells from patients who developed recurrence ($n = 11$) and those who did not develop recurrence ($n = 8$) within five years of surgery were isolated using laser microdissection to minimise protein contamination from non-tumour tissue. A total of 55 differentially expressed proteins were identified by 2D-DIGE and MALDI-TOF MS. The expression of HLAB, ADAMTS2, LTBP3, JAG2, and NME2 was among ten differentially expressed proteins significantly associated with vascular invasion and CRLM. These prognostic protein biomarkers may be useful in complementing current cancer staging systems and predicting the risk of CRLM in stage II CRC [47].

Adjuvant chemotherapy with fluoropyrimidine combined with oxaliplatin has been the standard of care for stage III CRC patients with good performance status and who can tolerate cytotoxic combination chemotherapy [69,70]. Although the therapeutic indication for this patient group is significantly less controversial than for patients with stage II CRC, the optimal duration of adjuvant chemotherapy is unclear [71]. The absolute difference of 0.9% in 3-year disease-free survival between patients receiving six versus three months of adjuvant chemotherapy is associated with increased toxicity and potential impairment of quality of life [72,73]. Yang et al. reported the protein expression profiles of tissue samples from patients with stage III and CRLM to identify key proteins related to progression in CRC. Protein expression profiles of patients with stage III CRC (n = 20) and CRLM (n = 17) were acquired using a label-free proteomics approach and nanoflow liquid chromatography coupled to an ultra-high-resolution mass spectrometer (nano-LC-MS/MS) [45]. Weighted correlation network analysis enabled clustering of co-expressed proteins into modules that correlated with traits [74]. Three modules were significantly correlated with CRC, from which nine proteins were identified through protein–protein interaction networks. Fibrinogen beta chain (FBG), Talin 1 (TLN1), and adaptor-related protein complex 2 subunit alpha 2 (AP2A2) all had a strong positive correlation with CRLM. HSPD1, EEF1G, and HNRNPA2B1 were positively correlated with primary CRC and CRLM. SRRT, APOC3, and PGM5 were key proteins associated with primary stage III CRC and provide insight into the progression from stage III CRC to CRLM [45].

5. Comparison of Colorectal Liver Metastases and Primary Colorectal Tumours

The molecular classification of colorectal cancers (CRCs) into intrinsic subtypes may be useful in refining prognosis and predicting patient outcomes [75,76]. Several studies have evaluated the proteome of primary CRC and matched colorectal liver metastases (CRLM) to establish the molecular basis of metastatic CRC [38,39,44,46,49]. Understanding the continuous evolution of CRC underscores the development of effective and targeted approaches across the spectrum of CRC [77].

A pilot study by Farhner et al. compared the proteome of seven matched FFPE specimens of primary CRC and CRLM using liquid chromatography–mass spectrometry (LC-MS/MS). Unsupervised clustering of over 2600 proteins demonstrated differences in the proteome of primary CRC and corresponding liver metastases. Many upregulated proteins in CRLM involve glucose metabolism, including pyruvate carboxylase, fructose-bisphosphate aldolase B and fructose-1,6-bisphosphatase 1. CRLM demonstrated an active immune response compared to primary CRC, as reflected by the upregulation of several complement system components, including C1, C4, C5 and C9. Multiple structural proteins associated with muscle contraction and cell junction assembly, such as desmin, synemin and filamin-C, were depleted in CRLM compared to primary CRC [38]. The molecular changes from primary CRC and CRLM highlight the distinct proteome of primary CRC and corresponding CRLM.

The proteomes of primary CRC and CRLM were compared by Liu et al. and Ku et al. using tandem mass tag (TMT)-labelling and LC-MS/MS [39,44]. TMTs are chemically reactive agents that impart isotope-based differences to peptide amines, enabling multiplexed LC-MS for peptide identification and simultaneous quantitation. TMT sample multiplexing facilitates high-throughput, large-scale quantitative proteomics data acquisition [78,79]. Liu et al. conducted a comparative analysis of proteomics between the primary CRC and CRLM in eight patients (n = 8). Several extracellular matrix components including FN1, TIMP1, THBS1, POSTN and VCAN, were upregulated in CRLM. Secondary analysis with immunohistochemistry revealed that increased THBS1 expression was significantly correlated with CRLM and poor prognosis. The role of THBS1 (Thrombospondin 1) in facilitating CRLM through enhancing epithelial–mesenchymal transition was supported by transwell cell migration and invasion assays, which in turn demonstrated that THBS1 depletion inhibited the migration and invasion of CRC cells [39]. Ku et al. used TMT labelling with LC-MS to compare of the proteomic profiles of fresh frozen tissue from nine patients ($n = 9$) and demonstrated protein signatures that distinguished CRLM and its primary and normal colon tissues. In total, 47 differentially expressed proteins were statistically significant between primary CRC and CRLM, of which Filamin A-interacting protein 1-like (FILIP1L) and plasminogen (PLG) were novel signature proteins described in CRLM. FILIP1L has been shown to suppress tumour progression by inhibiting cell proliferation and angiogenesis in CRC; hence, underexpression may contribute to CRC metastasis. Plasminogen, which showed significantly high expression, may allow for tumour attachment and invasion through the basement membrane and is associated with a worse CRC prognosis [44].

Synchronous CRLM are present in 15–25% at the index presentation with CRC [6]. The management of synchronous CRLM is more complex than metachronous CRLM and the prognosis for these patients is worse [80,81]. Kim et al. utilised an approach that combined 2D polyacrylamide gel electrophoresis and MALDI-TOF MS to identify metastasis-related factors differentially expressed in primary CRC and CRLM. The study identified 58 differentially expressed proteins between primary CRC and synchronous CRLM. Seven differentially abundant proteins were upregulated: SERPINA1, APOA1, ITLN1, DES, DBI, SDHA and CA1. Compared to primary CRC, pertinent biological processes altered in CRLM included increased energy metabolism and decreased immune-cell-related migration. The location of these differentially expressed proteins in the extracellular region and exosome or membrane-bound vesicles make these potentially useful circulating biomarkers [46].

Autoantibodies are produced by an immunological response to cancer cells [82]. Yang et al. used an immune-proteomic strategy to discover tumour tissue autoantigens from eight matched primary CRC, CRLM and adjacent normal liver tissue. Antigens from paired CRLM and normal liver tissue were identified using serum from patients with autoimmune disease. Furthermore, 1D and 2D gel electrophoresis and Western blotting were used to detect reactive protein bands, then these were analysed using mass spectrometry. Overall, 48 proteins were uniquely found in CRLM and absent in normal liver tissue. Olfactomedin 4 (OLFM4), CD11b, integrin $\alpha 2$ (ITGA2), periostin and thrombospondin-2 were reproducibly identified on Western blotting and mass spectrometry. These antigens were also overexpressed in primary CRC. OLFM4, CD11b and ITGA2 were validated in two cohorts [49]. OLFM4 is an anti-apoptotic factor and colon stem cell marker, whereas CD11b and ITGA2 are integrins that have a recognised role in promoting epithelial–mesenchymal transition and metastasis in CRC [83–85]. The concordant overexpression of these three biomarkers in both primary CRC and CRLM may be helpful in predicting the risk of CRLM and inform the development of immunotherapy for the treatment of CRC.

6. In-Depth Proteomic Characterisation of Colorectal Liver Metastases

Tumour heterogeneity describes differences between cancer cells within a tumour and leads to challenges in precision oncology. Genomic instability is a significant cause of genetic heterogeneity, a genetic feature of adenomatous tumours [86,87]. Although similar changes would be expected at the protein level, Turtoi et al. rationalised the proteome heterogeneity in CRLM by demonstrating a distinct and organised pattern of molecular alterations. Matrix-assisted laser desorption ionisation (MALDI)–mass spectrometry-based imaging and in-depth proteomic analysis of eight fresh CRLM samples and their corresponding normal tissue showed a reproducible, zonally delineated spatial distribution of over 1000 proteins. The centre of the lesion was characterised by elevated carbohydrate metabolism and DNA-repair activity, the rim of the metastasis displayed increased cellular growth movement and drug metabolism, and the peritumoral region featured elevated lipid metabolism and protein synthesis. LTBP2 and TGFB1 were two novel antigens consistently expressed in CRLM and were amenable to antibody-based tumour targeting in vivo, highlighting their therapeutic potential [52].

Progressive alterations in the proteome characterised by different extracellular matrix phenotypes have been reported in a single patient with metachronous CRLM and three curative-intent hepatic resections. Proteome analysis using LC-MS/MS identified 481 differentially regulated proteins, 81 of which were associated with the extracellular matrix and previously reported as negative prognostic markers, including tenascin C, nidogen 1, fibulin 1 and vitronectin. The clinical and proteomic findings correlate with increasing metastatic potential with each subsequent recurrence and support the rationale for comprehensive molecular analysis of metastases from different time points during disease progression [40].

7. Proteomic Profiling of the Extracellular Matrix in Colorectal Liver Metastases

The extracellular matrix (ECM) is a major component of the tumour microenvironment and comprises a complex network of macromolecules, such as proteins and polysaccharides secreted locally by cells [88]. In addition to its commonly recognised function of providing cells with structural support and mechanical integrity, the ECM is involved in biochemical signalling that modulates the hallmarks of cancer [89,90]. Cancer-associated ECM plays a significant role in sustaining proliferative signalling, evading growth suppressors, resisting cell death, enabling replicative immortality, inducing angiogenesis, as well as activating invasion and metastasis. ECM has also been implicated in emerging cancer hallmarks, including avoiding immune destruction, dysregulating cellular energetics, promoting genomic mutation and instability, and modulating immune cell behaviour and inflammation [89,90]. The abundance of protein in the ECM and its complex role in tumorigenesis have been the focus of several studies on proteomic profiling of the ECM in colorectal liver metastases (CRLM) [41–43,51].

A three-part series of mass spectrometry-based studies by van Huizen et al. on the role of collagen and its posttranslational modifications in CRLM have provided a deeper understanding of ECM's role in CRLM tumour biology [41–43]. The authors first demonstrated that specific collagen proteins were upregulated in CRLM compared to adjacent normal liver tissue in 30 FFPE CRLM samples. Out of 22 collagen-α chains, 19 were significantly ($p < 0.05$) upregulated in CRLM. The upregulation of 16 proteins required for collagen synthesis further supported increased collagen synthesis in metastatic CRC. Further, six non-collagen proteins (CDH17, KRT20, CEACAM5, GPA33, MUC13, and PPP1R1B/DARPP-32) were upregulated in CRLM, where CHD17 and PPP1R1B/DARPP-32 have not been described previously [43]. A subsequent study comparing CRLM and adjacent normal tissue ($n = 2$) identified posttranslational modification by enzymatic hydroxylation of proline at the Xaa position in collagen. Here, reduced 4-hydroxyproline in CRLM clearly distinguished it from control liver tissue [42]. Validation of these findings using a reference to a synthetic standard peptide in a larger sample ($n = 14$) showed consistent down-regulation of collagen hydroxylation in CRLM. Furthermore, the degree of hydroxylation of control liver and colonic tissue were similar, differentiating CRLM based on these posttranslational modifications [41]. Altogether, the differences in collagen types in CRLM may reflect altered collagen stability and could serve as potential prognostic biomarkers.

Naba et al. characterised the ECM composition of matched primary CRC, CRLM, and normal colonic tissue ($n = 3$) using ECM enrichment and liquid chromatography-tandem mass spectrometry [51]. The properties of ECM proteins, such as their large size, cross-linked and covalent bonds, and heavy glycosylation, render them challenging to analyse. The subcellular fractionation protocol described by Naba et al. takes advantage of the insolubility of ECM proteins to preferentially remove cytosolic proteins, nuclear proteins, membrane proteins, and cytoskeletal proteins, leaving a final insoluble fraction enriched for ECM. The ECM-enriched protein preparations are then digested into peptides for subsequent MS analysis [91,92]. Using these methods, robust signatures of ECM proteins that characterised each tissue were defined, amongst which COMP, FNDC1, IGFALS, SPP1, BMP1, CIQTNF5, and HPX were characteristic for CRLM. The ECM composition of CRLM and primary CRC showed a closer resemblance than normal liver tissue, with 23 proteins shared by primary CRC and CRLM. EGF-containing fibulin-like ECM protein 2 or Fibulin 4, thrombospondin 2, and tissue inhibitor of metalloproteinase-1, which were detected in both primary and secondary colon tumour tissue—but not in healthy tissue—have been shown to be detectable in serum of patients with CRC [93–95]. The studies to date point to a crucial contribution of the extracellular matrix. The dynamic range of ECM proteins may prove to be valuable indicators of progression or recurrence in CRC but warrant further validation.

8. Post-Translational Protein Modification in Colorectal Liver Metastases

Post-translational modification of cancer-associated extracellular matrix (ECM) alters the interaction of cancer cells with its microenvironment and influences malignancy and tumour growth [96]. Citrullination is produced through post-translational deamination of peptidyl-arginine and is catalysed by peptidylarginine deiminase (PAD). PAD and citrullination have been implicated in cancer development through several mechanisms such as activation of cancer cell signalling, alteration of epithelial-to-mesenchymal transition, formation of neutrophil extracellular traps and induction of antitumour activity [97]. Yuzhalin et al. identified tumour-derived peptidylarginine deiminase 4 (PAD4)-driven citrullination of ECM proteins to be essential for CRLM growth. ECM-enrichment and quantitative label-free analysis of proteomic data identified 287 proteins with statistically significant abundance between CRLM and normal liver, of which 69 proteins were upregulated or downregulated within the ECM by more than 3-fold. Among the upregulated proteins, versican, TIMP1, LTBP1–3, DDR1, and S100A10 have been previously linked to metastasis. ECM proteins are highly citrullinated in CRLM compared to normal liver, primary CRC and normal colonic mucosa. The upregulation of PAD4, which was 11 times more abundant in the ECM of CRLM than in normal liver, was also specific to CRLM and likely accounts for the increased citrullination of proteins in CRLM. CRC cell lines showed greater adhesion when cultured on citrullinated collagen type I than on non-citrullinated control and an increased expression of epithelial markers. These findings suggest that citrullination confers metastatic properties to CRC cells through enhanced epithelial–mesenchymal transition. Additionally, the functional significance of these post-translational modifications was demonstrated in murine models, where inhibition of PAD4 activity reduced citrullination and CRLM growth [48].

Enzymes and proteins involved in acetylation, typically on lysine residues, regulate many cellular physiological processes but are deregulated in cancer [98]. These alterations may have functional implications in cancer biology, such as metabolic reprogramming and adaptation to the tumour microenvironment [99,100]. Global-scale profiling of differentially expressed lysine-acetylated proteins in matched primary CRC and CRLM was first reported by Shen et al. This study characterised the acetylome paired primary CRC and CRLM samples ($n = 3$) using tandem mass tag protein labelling, high-affinity enrichment of acetylated peptides and LC-MS/MS analysis. A total of 603 acetylation sites from 316 proteins were identified, and 462 acetylation sites corresponding to 243 proteins were quantified. Further, 31 acetylated sites of 22 proteins were downregulated, while 40 acetylated sites of 32 proteins were upregulated in CRLM. Among differentially expressed acetylated histone proteins between primary CRC and CRLM, acetylated histone H3.2 at Lys 19 (HIST2H3AK19Ac) showed the most significant downregulation. In contrast, acetylated histone H2B type 1-L at Lys 121 (H2BLK121Ac) was the most overexpressed acetylated histone in CRLM. TPM2 K152Ac was the most downregulated acetylated non-histone protein and ADH1B K331Ac was the most upregulated non-histone protein in CRLM. Most of the identified acetylated proteins were localised within the cytoplasm, associated with binding, and involved in multiple biological processes such as metabolic pathways, carbon metabolism and biosynthesis of amino acids. The findings in this study demonstrate that protein acetylation may be pivotal in biological processes that drive the development and progression of CRLM [50].

9. Proteomics as a Principal Component of Multiomics in Colorectal Liver Metastases

Multiomics provides an integrated biological analysis approach and enables a more comprehensive understanding of molecular changes across multiple levels of biology [101,102]. The downstream signalling effects of genomic alterations can be characterised and evaluated by including proteomic and post-translational modification data [103,104]. Table 2 summarises five studies that have layered various omics in addition to proteomics to understand how cell processes in CRC are connected and communicate with each other [54–58]. The use of fresh frozen CRLM biospecimens and nanoscale liquid chromatography coupled to tandem mass spectrometry (nano-LC-MS/MS) characterises the few multiomic studies on CRLM to date.

Multiomic analysis has demonstrated that primary CRC and CRLM are highly similar at the genetic but not at the proteomic or phosphoproteomic levels. Furthermore, proteomic profiling has identified biologically and prognostically distinct CRC subtypes that may be useful in risk stratification [54]. Altered peptides can be precisely quantified and potentially utilised to predict distinct somatic mutations such as KRAS G12V, even when these genes are expressed at low levels and despite protein diversity [55]. Correlation of survival analysis and multiomic data have also identified new leads for prognostic biomarkers [54,56–58]. However, the multiomics space in CRC, as with many other cancers, is still in the preclinical discovery phase. Therefore, differentially expressed proteins in these studies are hypothesis-generating but require further validation and development to close the gap between biology and translation. The integration of proteomics and post-translational modification data nevertheless represents a substantial advance over prior genomic studies of CRC and directs the way to improve the molecular characterisation of clinical cohorts.

10. Limitations in Proteomic Biomarker Discovery in Colorectal Liver Metastases

Proteomic biomarker discovery and development presents many biological, technical, and technological challenges [105,106]. Consequently, proteomic biomarkers in CRLM have not progressed beyond the discovery phase after more than a decade of intensive research, and none of the identified leads for potentially useful biomarkers has translated to clinical practice. Recognising the general and context-specific challenges in proteomic biomarker development in CRLM is essential to increasing value and reducing waste in the future [107].

The fundamental challenge with proteomic profiling in CRLM to date is the small sample size of existing studies [107,108]. Multiple biological processes occur from the genome to the proteome and result in many levels of protein diversity [109,110]. Such tremendous protein diversity favours the chance to discover lead biomarkers, even in studies with small sample sizes. As a result, there may be a strong reporting bias of statistically significant proteomic biomarkers in CRLM. Small-sample-size studies also reduce the chance of detecting a true effect and result in the overfitting of identified lead biomarkers. Clinical and cancer heterogeneity, both within and between these exploratory studies, further complicates the differentiation of candidate biomarkers from spurious results due to random variation [111]. The dynamic range of the cellular proteome and lack of direct amplification mechanisms, unlike in DNA, create technological challenges for identifying proteins at low concentrations that may significantly impact tumour biology, particularly from specimens preserved with formalin [112,113]. Most potential and novel biomarkers in CRLM have subsequently not been validated beyond the initial discovery study to determine their reproducibility and generalisability. Part of this problem lies with the academic endeavour, which rewards the detection of new markers in favour of large-scale studies that can be tedious and high-risk. Although several studies have been published on proteomic biomarkers in CRLM, data cannot be directly combined or compared due to the complexity of overlaying proteomic data from different platforms.

To increase the discovery of accurate biomarkers, the sample selection needs to be based on clearly defined and clinically meaningful endpoints with distinct prognostic implications, e.g., recurrence within six months (i.e., early recurrence) of resection of CRLM and 10-year actual recurrence-free survival. Proteomic biomarkers are most needed at decision nodes where the performance of prognostic or predictive factors is uncertain or inconsistent. Cohorts from which samples are selected need to be well characterised and closely matched to the specific clinical context to which these proteomic biomarkers are intended to be applied. However, simply conducting more siloed work will not address the current challenges in translational proteomics in CRLM. The complexities of proteomic biomarker development require a new level of collaboration to enable large-scale, unbiased, deep, proteomic profiling and subsequent validation to capture the true value of proteomics in improving clinical outcomes for patients with CRLM.

11. Translational Proteomics in Colorectal Liver Metastases

Proteomic biomarker discovery and profiling reflect novel opportunities and approaches beyond genomics in managing patients with CRLM patients. The next step is to translate such proteomic discoveries into clinical applications (Figure 1). Practical applications include risk stratification, surveillance, detection of disappearing CRLM, monitoring cancer recurrence and treatment response through using blood/urine biomarkers, treatment stratification and development of targeted treatment.

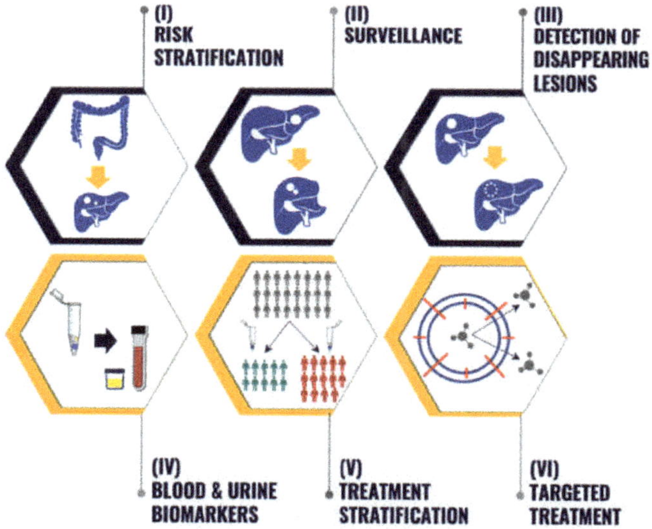

Figure 1. The potential clinical application of proteomic biomarkers in colorectal liver metastases (CRLM). **(I)** Risk stratification—assess the likelihood that CRLM will develop after oncologic resection of primary colorectal cancer (CRC). **(II)** Surveillance—early detection and treatment of recurrence after curative-intent resection of CRLM. **(III)** Detection of disappearing lesions—characterise the disease course and inform the surgical management of disappearing CRLM after preoperative systemic therapy. **(IV)** Blood and urine biomarkers—identify blood and urine biomarkers to monitor metastatic CRC progression and treatment. **(V)** Treatment stratification—predict response to multimodal therapy and select treatment that is most likely to yield a favourable response. **(VI)** Targeted treatment—patient selection for biomarker-driven clinical oncology trials.

Differing proteomic profiles among patients with stage II/III CRC who subsequently develop metastatic disease or remain disease-free demonstrate the prognostic value of the proteomics-based applications for predicting recurrence after oncologic resection of the primary tumour [45,47]. Similarly, prognosis-based stratification can be determined through differing proteomic expression profiles following curative-intent CRLM resection [37,58]. Consequently, proteomic biomarkers can possibly be applied as part of a tailored surveillance and treatment strategy based on a patient's molecular risk profile. Given the limitations of clinicopathologic predictors for CRLM, such potential CRLM proteomics-based applications represent additional analytical tools for molecular stratification.

The distinct CRLM and ECM proteome from normal hepatic tissue can serve as valuable proteomic targets for detecting post-chemotherapy disappearing CRLM [51,52]. The characterisation of differentially expressed proteins within tissue is also essential to identify candidate proteins for liquid biopsies. As most drugs target proteins, proteomic profiling of CRLM also has high value for drug discovery and development for metastatic CRC. The above-described studies offer a glance into the future of proteomics in CRLM, though further research efforts are required to translate these discoveries into clinical applications that improve individual CRC patient outcomes.

12. Conclusions

This article critically reviewed the current state of emerging proteomic biomarkers using human CRC tissue specimens and mass spectrometry-based techniques. Variations and nuances in mass spectrometry-based approaches and proteomic analysis highlight the breadth of customisable methods to study the proteome in CRC. The diversity of proteomic profiles identified to date reflects cancer heterogeneity at different stages in the disease course of CRC. While multimodal cancer therapy advancements have significantly expanded curative-intent therapy in CRLM, some patients only derive limited benefits from current clinical approaches. The comprehensive characterisation of proteins in both primary CRC and CRLM has provided insight into similarities and continuations in molecular alterations, which form the basis for developing new biomarkers, targeted treatments, and therapeutic strategies. The preclinical exploratory phase of proteomic biomarkers has identified promising directions for future research. Coordinated research efforts are needed to streamline and optimise critical workflow steps to enable reproducible and accurate protein quantitation in specific clinical contexts and to unravel the complex mechanistic biology of CRC.

Author Contributions: Conceptualization, G.Y.M.W.; data curation, G.Y.M.W.; writing—original draft preparation, G.Y.M.W. and M.P.M.; writing—review and editing, C.D. and T.J.H.; visualization, G.Y.M.W.; supervision, C.D., T.J.H. and M.P.M. All authors have read and agreed to the published version of the manuscript.

Funding: This first author is funded by the Sir Roy McCaughey Surgical Research Scholarship, Royal Australasian College of Surgeons.

Conflicts of Interest: The authors declare no conflict of interest.

References

1. Sung, H.; Ferlay, J.; Siegel, R.L.; Laversanne, M.; Soerjomataram, I.; Jemal, A.; Bray, F. Global Cancer Statistics 2020: GLOBOCAN Estimates of Incidence and Mortality Worldwide for 36 Cancers in 185 Countries. *CA Cancer J. Clin.* **2021**, *71*, 209–249. [CrossRef] [PubMed]
2. Siebenhüner, A.R.; Güller, U.; Warschkow, R. Population-based SEER analysis of survival in colorectal cancer patients with or without resection of lung and liver metastases. *BMC Cancer* **2020**, *20*, 246. [CrossRef] [PubMed]
3. Birnbaum, E. Surgical Anatomy of the Colon, Rectum and Anus. In *Coloproctology*; Ratto, C., Parrello, A., Dionisi, L., Litta, F., Eds.; Springer: Cham, Switzerland, 2015; pp. 9–19, ISBN 978-3-319-09807-4. [CrossRef]
4. Riihimäki, M.; Hemminki, A.; Sundquist, J.; Hemminki, K. Patterns of metastasis in colon and rectal cancer. *Sci. Rep.* **2016**, *6*, 29765. [CrossRef] [PubMed]

5. Van der Geest, L.G.; Lam-Bower, J.; Koopman, M.; Verhoef, C.; Elferink, M.A.; de Wilt, J.H. Nationwide trends in incidence, treatment, and survival of colorectal cancer patients with synchronous metastases. *Clin. Exp. Metastasis* **2015**, *32*, 457–465. [CrossRef] [PubMed]
6. Leporrier, J.; Maurel, J.; Chiche, L.; Bara, S.; Segol, P.; Launoy, G. A population-based study of the incidence, management and rognosis of hepatic metastases from colorectal cancer. *Br. J. Surg.* **2006**, *93*, 465–474. [CrossRef] [PubMed]
7. Tomlinson, J.S.; Jarnagin, W.R.; DeMatteo, R.P.; Fong, Y.; Kornprat, P.; Gonen, M.; Kemeny, N.; Brennan, M.F.; Blumgart, L.H.; D'Angelica, M. Actual 10-Year Survival After Resection of Colorectal Liver Metastases Defines Cure. *J. Clin. Oncol.* **2007**, *25*, 4575–4580. [CrossRef]
8. Pulitano, C.; Castillo, F.; Aldrighetti, L.; Bodingbauer, M.; Parks, R.W.; Ferla, G.; Wigmore, S.J.; Garden, O.J. What defines 'cure' after liver resection for colorectal metastases? Results after 10 years of follow-up. *HPB* **2010**, *12*, 244–249. [CrossRef]
9. D'Angelica, M.; Kornprat, P.; Gonen, M.; DeMatteo, R.P.; Fong, Y.; Blumgart, L.H.; Jarnagin, W.R. Effect on Outcome of Recurrence Patterns After Hepatectomy for Colorectal Metastases. *Ann. Surg. Oncol.* **2010**, *18*, 1096–1103. [CrossRef]
10. De Jong, M.C.; Pulitano, C.; Ribero, D.; Strub, J.; Mentha, G.; Schulick, R.D.; Choti, M.A.; Aldrighetti, L.; Capussotti, L.; Pawlik, T.M. Rates and Patterns of Recurrence Following Curative Intent Surgery for Colorectal Liver Metastasis: An international multi-institutional analysis of 1669 patients: An international multi-institutional analysis of 1669 patients. *Ann. Surg.* **2009**, *250*, 440–448. [CrossRef]
11. Fong, Y.; Fortner, J.; Sun, R.L.; Brennan, M.F.; Blumgart, L.H. Clinical Score for Predicting Recurrence After Hepatic Resection for Metastatic Colorectal Cancer: Analysis of 1001 consecutive cases. *Ann. Surg.* **1999**, *230*, 309–318. [CrossRef]
12. Zakaria, S.; Donohue, J.H.; Que, F.G.; Farnell, M.B.; Schleck, C.D.; Ilstrup, D.M.; Nagorney, D.M. Hepatic Resection for Colorectal Metastases. *Ann. Surg.* **2007**, *246*, 183–191. [CrossRef] [PubMed]
13. Tan, M.C.; Butte, J.M.; Gonen, M.; Kemeny, N.; Fong, Y.; Allen, P.J.; Kingham, T.P.; DeMatteo, R.P.; Jarnagin, W.R.; D'Angelica, M.I. Prognostic significance of early recurrence: A conditional survival analysis in patients with resected colorectal liver metastasis. *HPB* **2013**, *15*, 803–813. [CrossRef] [PubMed]
14. Inoue, Y.; Fujii, K.; Kagota, S.; Tomioka, A.; Yamaguchi, T.; Ohama, H.; Hamamoto, H.; Ishii, M.; Osumi, W.; Tsuchimoto, Y.; et al. The management of Recurrence within Six Months after Hepatic Resection for Colorectal Liver Metastases. *Dig. Surg.* **2020**, *37*, 282–291. [CrossRef] [PubMed]
15. Viganò, L.; Gentile, D.; Galvanin, J.; Corleone, P.; Costa, G.; Cimino, M.; Procopio, F.; Torzilli, G. Very Early Recurrence After Liver Resection for Colorectal Metastases: Incidence, Risk Factors, and Prognostic Impact. *J. Gastrointest. Surg.* **2021**, *26*, 570–582. [CrossRef] [PubMed]
16. Molinari, C.; Marisi, G.; Passardi, A.; Matteucci, L.; De Maio, G.; Ulivi, P. Heterogeneity in Colorectal Cancer: A Challenge for Personalized Medicine? *Int. J. Mol. Sci.* **2018**, *19*, 3733. [CrossRef]
17. Loupakis, F.; Yang, D.; Yau, L.; Feng, S.; Cremolini, C.; Zhang, W.; Maus, M.K.; Antoniotti, C.; Langer, C.; Scherer, S.J.; et al. Primary tumour location as a prognostic factor in metastatic colorectal cancer. *J. Natl. Cancer Inst.* **2015**, *107*, dju427. [CrossRef]
18. Punt, C.J.; Koopman, M.; Vermeulen, L. From tumour heterogeneity to advances in precision treatment of colorectal cancer. *Nat. Rev. Clin. Oncol.* **2017**, *14*, 234–246. [CrossRef]
19. Pawlik, T.M.; Schulick, R.D.; Chito, M.A. Expanding the criteria for Resectability of Colorectal Liver Metastases. *Oncologist* **2008**, *13*, 51–64. [CrossRef]
20. Van Custem, E.; Cervantes, A.; Adam, R.; Sobrero, A.; Van Krieken, J.H.; Aderka, D.; Aranda Aguilar, E.; Bardelli, A.; Benson, A.; Bodoky, G.; et al. ESMO concensus guidelines for the management of patients with metastatic colorectal cancer. *Ann. Oncol.* **2016**, *27*, 1386–1422. [CrossRef]
21. Mardis, E. A decade's perspective on DNA sequencing technology. *Nature* **2011**, *470*, 198–203. [CrossRef]
22. Drmanac, R. The advent of personal genome sequencing. *Genet. Med.* **2011**, *13*, 188–190. [CrossRef] [PubMed]
23. Stratton, M.R.; Campbell, P.J.; Futreal, P.A. The cancer genome. *Nature* **2009**, *458*, 719–724. [CrossRef] [PubMed]
24. Sidoli, S.; Kulej, K.; Garcia, B.A. Why proteomics is not the new genomics and the future of mass spectrometry in cell biology. *J. Cell Biol.* **2017**, *216*, 21–24. [CrossRef] [PubMed]
25. Vaisakar, S.; Huang, C.; Wang, X.; Petyuk, V.A.; Savage, S.R.; Wen, B.; Dou, Y.; Zhang, Y.; Shi, Z.; Arshad, O.A.; et al. Proteogenomic analysis of Human and Colon Cancer Reveals new Therapeutic Opportunities. *Cell* **2019**, *177*, 1035–1049. [CrossRef]
26. Black, D.L. Protein diversity from alternative splicing: A challenge for bioinformatics and post-genome biology. *Cell* **2000**, *103*, 367–370. [CrossRef]
27. Genuth, N.R.; Barna, M. Heterogeneity and specialized functions of translation machinery: From genes to organisms. *Nat. Rev. Genet.* **2018**, *19*, 431–452. [CrossRef]
28. Sonneveld, S.; Verhagen, B.M.; Tanenbaum, M.E. Heterogeneity in mRNA Translation. *Trends Cell Biol.* **2020**, *30*, 606–618. [CrossRef]
29. Rodriguez, H.; Zenklusen, J.C.; Staudt, L.M.; Doroshow, J.H.; Lowy, D.R. The next horizon in precision oncology: Proteogenomics to inform cancer diagnosis and treatment. *Cell* **2021**, *184*, 1661–1670. [CrossRef]
30. Wang, H.; Shi, T.; Qian, W.J.; Liu, T.; Kagan, J.; Srivastava, S.; Smith, R.D.; Rodland, K.D.; Camp, D.G. The clinical impact of recent advances in LC-MS for cancer biomarker discovery and verification. *Expert Rev. Proteom.* **2016**, *13*, 99–114. [CrossRef]
31. Islam Khan, M.Z.; Tam, S.Y.; Law, H.K.W. Advances in High Throughput Proteomics Profiling in Establishing Potential Biomarkers for Gastrointestinal Cancer. *Cells* **2022**, *11*, 973. [CrossRef]

32. Macklin, A.; Khan, S.; Kislinger, T. Recent advances in mass spectrometry based clinical proteomics: Applications to cancer research. *Clin. Proteom.* **2020**, *17*, 17. [CrossRef] [PubMed]
33. Zhang, B.; Wang, J.; Wang, X.; Zhu, J.; Shi, Z.; Chambers, M.C.; Zimmerman, L.J.; Shaddox, K.F.; Kim, S.; Davies, S.R.; et al. Proteogenomic characterization of human colon and rectal cancer. *Nature* **2014**, *513*, 382–387. [CrossRef] [PubMed]
34. Cao, L.; Huang, C.; Cui Zhou, D.; Hu, Y.; Lih, T.M.; Savage, S.R.; Krug, K.; Clark, D.J.; Schnaubelt, M.; Chen, L.; et al. Proteogenomic characterization of pancreatic ductal adenocarcinoma. *Cell* **2021**, *184*, 5031–5052. [CrossRef] [PubMed]
35. Zhang, H.; Liu, T.; Zhang, Z.; Payne, S.H.; Zhang, B.; McDermott, J.E.; Zhou, J.Y.; Petyuk, V.A.; Chen, L.; Ray, D.; et al. Integrated Proteogenomic Characterization of Human High-Grade Serous Ovarian Cancer. *Cell* **2016**, *166*, 755–765. [CrossRef]
36. Lehtiö, J.; Arslan, T.; Siavelis, I.; Pan, Y.; Socciarelli, F.; Berkovska, O.; Umer, J.M.; Mermelekas, G.; Pirmoradian, M.; Jönsson, M.; et al. Proteogenomics of non-small cell lung cancer reveals molecular subtypes associated with specific therapeutic targets and immune evasion mechanisms. *Nat. Cancer* **2021**, *2*, 1224–1242. [CrossRef]
37. Michal, S.; Tal, G.-L.; Gali, P.; Miki, G.; Elana, B.; Baroch, B.; Hanoch, K.; Irit, B.A.; Riad, H. Characterization of Biomarkers in Colorectal Cancer Liver Metastases as a Prognostic Tool. *J. Pers. Med.* **2021**, *11*, 1059. [CrossRef]
38. Fahrner, M.; Bronsert, P.; Fichtner-Feigl, S.; Jud, A.; Schilling, O. Proteome biology of primary colorectal carcinoma and corresponding liver metastases. *Neoplasia* **2021**, *23*, 1240–1251. [CrossRef]
39. Liu, X.; Xu, D.; Liu, Z.; Li, Y.; Zhang, C.; Gong, Y.; Jiang, Y.; Xing, B. THBS1 facilitates colorectal liver metastasis through enhancing epithelial–mesenchymal transition. *Clin. Transl. Oncol.* **2020**, *22*, 1730–1740. [CrossRef]
40. Voß, H.; Wurlitzer, M.; Smit, D.J.; Ewald, F.; Alawi, M.; Spohn, M.; Indenbirken, D.; Omidi, M.; David, K.; Juhl, H.; et al. Differential regulation of extracellular matrix proteins in three recurrent liver metastases of a single patient with colorectal cancer. *Clin. Exp. Metastasis* **2020**, *37*, 649–656. [CrossRef]
41. Van Huizen, N.A.; Burgers, P.C.; Van Rosmalen, J.; Doukas, M.; Ijzermans, J.N.M.; Luider, T.M. Down-Regulation of Collagen Hydroxylation in Colorectal Liver Metastasis. *Front. Oncol.* **2020**, *10*, 557737. [CrossRef]
42. Van Huizen, N.A.; Burgers, P.C.; Saintmont, F.; Brocorens, P.; Gerbaux, P.; Stingl, C.; Dekker, L.J.M.; Ijzermans, J.N.; Luider, T.M. Identification of 4-Hydroxyproline at the Xaa Position in Collagen by Mass Spectrometry. *J. Proteome Res.* **2019**, *18*, 2045–2051. [CrossRef] [PubMed]
43. Van Huizen, N.A.; Braak, R.R.C.V.D.; Doukas, M.; Dekker, L.J.; Ijzermans, J.N.; Luider, T.M. Up-regulation of collagen proteins in colorectal liver metastasis compared with normal liver tissue. *J. Biol. Chem.* **2019**, *294*, 281–289. [CrossRef] [PubMed]
44. Ku, X.; Xu, Y.; Cai, C.; Yang, Y.; Cui, L.; Yan, W. In-Depth Characterization of Mass Spectrometry-Based Proteomic Profiles Revealed Novel Signature Proteins Associated with Liver Metastatic Colorectal Cancers. *Anal. Cell. Pathol.* **2019**, *2019*, 7653230. [CrossRef] [PubMed]
45. Yang, W.; Shi, J.; Zhou, Y.; Liu, T.; Li, J.; Hong, F.; Zhang, K.; Liu, N. Co-expression Network Analysis Identified Key Proteins in Association with Hepatic Metastatic Colorectal Cancer. *Proteom. Clin. Appl.* **2019**, *13*, e1900017. [CrossRef]
46. Kim, E.-K.; Song, M.-J.; Jung, Y.; Lee, W.-S.; Jang, H.H. Proteomic Analysis of Primary Colon Cancer and Synchronous Solitary Liver Metastasis. *Cancer Genom. Proteom.* **2019**, *16*, 583–592. [CrossRef]
47. Kirana, C.; Peng, L.; Miller, R.; Keating, J.P.; Glenn, C.; Shi, H.; Jordan, T.W.; Maddern, G.; Stubbs, R.S. Combination of laser microdissection, 2D-DIGE and MALDI-TOF MS to identify protein biomarkers to predict colorectal cancer spread. *Clin. Proteom.* **2019**, *16*, 3. [CrossRef]
48. Yuzhalin, A.E.; Gordon-Weeks, A.N.; Tognoli, M.L.; Jones, K.; Markelc, B.; Konietzny, R.; Fischer, R.; Muth, A.; O'Neill, E.; Thompson, P.R.; et al. Colorectal cancer liver metastatic growth depends on PAD4-driven citrullination of the extracellular matrix. *Nat. Commun.* **2018**, *9*, 4783. [CrossRef]
49. Yang, Q.; Bavi, P.; Wang, J.Y.; Roehrl, M.H. Immuno-proteomic discovery of tumor tissue autoantigens identifies olfactomedin 4, CD11b, and integrin alpha-2 as markers of colorectal cancer with liver metastases. *J. Proteom.* **2017**, *168*, 53–65. [CrossRef]
50. Shen, Z.; Wang, B.; Luo, J.; Jiang, K.; Zhang, H.; Mustonen, H.K.; Puolakkainen, P.; Zhu, J.; Ye, Y.; Wang, S. Global-scale profiling of differential expressed lysine acetylated proteins in colorectal cancer tumors and paired liver metastases. *J. Proteom.* **2016**, *142*, 24–32. [CrossRef]
51. Naba, A.; Clauser, K.R.; Whittaker, C.A.; Carr, S.A.; Tanabe, K.K.; Hynes, R.O. Extracellular matrix signatures of human primary metastatic colon cancers and their metastases to liver. *BMC Cancer* **2014**, *14*, 518. [CrossRef]
52. Turtoi, A.; Blomme, A.; Debois, D.; Somja, J.; Delvaux, D.; Patsos, G.; Di Valentin, E.; Peulen, O.; Mutijima, E.N.; De Pauw, E.; et al. Organized proteomic heterogeneity in colorectal cancer liver metastases and implications for therapies. *Hepatology* **2013**, *59*, 924–934. [CrossRef] [PubMed]
53. Kirana, C.; Shi, H.; Laing, E.; Hood, K.; Miller, R.; Bethwaite, P.; Keating, J.; Jordan, T.W.; Hayes, M.; Stubbs, R. Cathepsin D Expression in Colorectal Cancer: From Proteomic Discovery through Validation Using Western Blotting, Immunohistochemistry, and Tissue Microarrays. *Int. J. Proteom.* **2012**, *2012*, 245819. [CrossRef] [PubMed]
54. Li, C.; Sun, Y.-D.; Yu, G.-Y.; Cui, J.-R.; Lou, Z.; Zhang, H.; Huang, Y.; Bai, C.-G.; Deng, L.-L.; Liu, P.; et al. Integrated Omics of Metastatic Colorectal Cancer. *Cancer Cell* **2020**, *38*, 734–747. [CrossRef] [PubMed]
55. Blank-Landeshammer, B.; Richard, V.R.; Mitsa, G.; Marques, M.; Leblanc, A.; Kollipara, L.; Feldmann, I.; Du Tertre, M.C.; Gambaro, K.; McNamara, S.; et al. Proteogenomics of Colorectal Cancer Liver Metastases: Complementing Precision Oncology with Phenotypic Data. *Cancers* **2019**, *11*, 1907. [CrossRef]

56. Ma, Y.-S.; Wu, Z.-J.; Zhang, H.-W.; Cai, B.; Huang, T.; Long, H.-D.; Xu, H.; Zhao, Y.-Z.; Yin, Y.-Z.; Xue, S.-B.; et al. Dual Regulatory Mechanisms of Expression and Mutation Involving Metabolism-Related Genes FDFT1 and UQCR5 during CLM. *Mol. Ther. Oncolytics* **2019**, *14*, 172–178. [CrossRef]
57. Ma, Y.-S.; Huang, T.; Zhong, X.-M.; Zhang, H.-W.; Cong, X.-L.; Xu, H.; Lu, G.-X.; Yu, F.; Xue, S.-B.; Lv, Z.-W.; et al. Proteogenomic characterization and comprehensive integrative genomic analysis of human colorectal cancer liver metastasis. *Mol. Cancer* **2018**, *17*, 139. [CrossRef]
58. Snoeren, N.; Emmink, B.L.; Koerkamp, M.J.G.; van Hooff, S.R.; Goos, J.A.C.M.; van Houdt, W.J.; de Wit, M.; Prins, A.M.; Piersma, S.R.; Pham, T.V.; et al. Maspin is a marker for early recurrence in primary stage III and IV colorectal cancer. *Br. J. Cancer* **2013**, *109*, 1636–1647. [CrossRef]
59. Gomez, D.; Cameron, I.C. Prognostic scores for colorectal liver metastasis: Clinically important or an academic exercise? *HPB (Oxford)* **2010**, *12*, 227–238. [CrossRef]
60. Strimbu, K.; Tavel, J.A. What Are Biomarkers? *Curr. Opin. HIV AIDS* **2010**, *5*, 463–466. [CrossRef]
61. Søreide, K. Receiver-operating characteristic curve analysis in diagnostic, prognostic and predictive biomarker research. *J. Clin. Pathol.* **2008**, *62*, 1–5. [CrossRef]
62. Takahashi, S.; Konishi, M.; Nakagohri, T.; Gotohda, N.; Saito, N.; Kinoshita, T. Short Time to Recurrence After Hepatic Resection Correlates with Poor Prognosis in Colorectal Hepatic Metastasis. *Jpn. J. Clin. Oncol.* **2006**, *36*, 368–375. [CrossRef] [PubMed]
63. Manfredi, S.; Lepage, C.; Hatem, C.; Coatmeur, O.; Faivre, J.; Bouvier, A.-M. Epidemiology and Management of Liver Metastases from Colorectal Cancer. *Ann. Surg.* **2006**, *244*, 254–259. [CrossRef] [PubMed]
64. Böckelman, C.; Engelmann, B.E.; Kaprio, T.; Hansen, T.; Glimelius, B. Risk of recurrence in patients with colon cancer stage II and III: A systematic review and meta-analysis of recent literature. *Acta Oncol.* **2014**, *54*, 5–16. [CrossRef] [PubMed]
65. O'Connell, M.J.; Campbell, M.E.; Goldberg, R.M.; Grothey, A.; Seitz, J.-F.; Benedetti, J.K.; André, T.; Haller, D.G.; Sargent, D. Survival Following Recurrence in Stage II and III Colon Cancer: Findings from the ACCENT Data Set. *J. Clin. Oncol.* **2008**, *26*, 2336–2341. [CrossRef]
66. Quasar Collaborative Group; Gray, R.; Barnwell, J.; Hills, R.K.; Williams, R.S.; Kerr, D.J. Adjuvant chemotherapy versus observation in patients with colorectal cancer: A randomised study. *Lancet* **2007**, *370*, 2020–2029. [CrossRef]
67. André, T.; Boni, C.; Navarro, M.; Tabernero, J.; Hickish, T.; Topham, C.; Bonetti, A.; Clingan, P.; Bridgewater, J.; Rivera, F.; et al. Improved overall survival with oxaliplatin, fluorouracil, and leucovorin as adjuvant treatment in stage II or III colon cancer in the MOSAIC trial. *J. Clin. Oncol.* **2009**, *27*, 3109–3116. [CrossRef]
68. Figueredo, A.; E Coombes, M.E.; Mukherjee, S. Adjuvant Therapy for completely resected Stage II Colon Cancer. *Cochrane Database Syst. Rev.* **2008**, *2010*, CD005390. [CrossRef]
69. Lieu, C.; Kennedy, E.B.; Bergsland, E.; Berlin, J.; George, T.J.; Gill, S.; Gold, P.J.; Hantel, A.; Jones, L.; Mahmoud, N.; et al. Duration of Oxaliplatin-Containing Adjuvant Therapy for Stage III Colon Cancer: ASO Clinical Practice Guideline. *J. Clin. Oncol.* **2019**, *37*, 1436–1447. [CrossRef]
70. Argilés, G.; Tabernero, J.; Labianca, R.; Hochhauser, D.; Salazar, R.; Iveson, T.; Laurent-Puig, P.; Quirke, P.; Yoshino, T.; Taieb, J.; et al. Localised colon cancer: ESMO Clinical Practice Guidelines for diagnosis, treatment, and follow-up. *Ann. Oncol.* **2020**, *31*, 1291–1305. [CrossRef]
71. Iveson, T.J.; Kerr, R.; Saunders, M.P.; Cassidy, J.; Hollander, N.H.; Tabernero, J.; Haydon, A.; Glimelius, B.; Harkin, A.; Allan, K.; et al. 3 versus 6 months of adjuvant oxaliplatin-fluoropyrimidine combination therapy for colorectal cancer (SCOT): An international, randomised, phase 3, non-inferiority trial. *Lancet Oncol.* **2018**, *19*, 562–578. [CrossRef]
72. André, T.; Meyerhardt, J.; Iveson, T.; Sobrero, A.; Yoshino, T.; Souglakos, I.; Grothey, A.; Niedzwiecki, D.; Saunders, M.; Labianca, R.; et al. Effect of duration of adjuvant chemotherapy for patients with stage III colon cancer (IDEA collaboration): Final results from a prospective, pooled analysis of six randomised, phase 3 trials. *Lancet Oncol.* **2020**, *21*, 1620–1629. [CrossRef]
73. Taieb, J.; Gallois, C. Adjuvant Chemotherapy for Stage III Colon Cancer. *Cancers* **2020**, *12*, 2697. [CrossRef] [PubMed]
74. Langfelder, P.; Horvath, S. WGCNA: An R package for weighted correlation network analysis. *BMC Bioinform.* **2008**, *9*, 559. [CrossRef] [PubMed]
75. Guinney, J.; Dienstmann, R.; Wang, X.; De Reyniès, A.; Schlicker, A.; Soneson, C.; Marisa, L.; Roepman, P.; Nyamundanda, G.; Angelino, P.; et al. The consensus molecular subtypes of colorectal cancer. *Nat. Med.* **2015**, *21*, 1350–1356. [CrossRef]
76. Pitroda, S.P.; Khodarev, N.N.; Huang, L.; Uppal, A.; Wightman, S.C.; Ganai, S.; Joseph, N.; Pitt, J.; Brown, M.; Forde, M.; et al. Integrated molecular subtyping defines a curable oligometastatic state in colorectal liver metastasis. *Nat. Commun.* **2018**, *9*, 1793. [CrossRef]
77. Fares, J.; Fares, M.Y.; Khachfe, H.H.; Salhab, H.A.; Fares, Y. Molecular principles of metastasis: A hallmark of cancer revisited. *Signal Transduct. Target. Ther.* **2020**, *5*, 28. [CrossRef]
78. Erickson, B.K.; Rose, C.M.; Braun, C.R.; Erickson, A.R.; Knott, J.; McAlister, G.C.; Wühr, M.; Paulo, J.A.; Everley, R.A.; Gygi, S.P. A Strategy to Combine Sample Multiplexing with Targeted Proteomics Assays for High-Throughput Protein Signature Characterization. *Mol. Cell* **2017**, *65*, 361–370. [CrossRef]
79. Zecha, J.; Satpathy, S.; Kanashova, T.; Avanessian, S.C.; Kane, H.; Clauser, K.; Mertins, P.; Carr, S.A.; Kuster, B. TMT Labeling for the Masses: A Robust and Cost-efficient, In-solution Labeling Approach. *Mol. Cell. Proteom.* **2019**, *18*, 1468–1478. [CrossRef]
80. Jegatheeswaran, S.; Mason, J.; Hancock, H.C.; Siriwardena, A.K. The Liver-First Approach to the Management of Colorectal Cancer with Synchronous Hepatic Metastases. *JAMA Surg.* **2013**, *148*, 385–391. [CrossRef]

81. Adam, R.; de Gramont, A.; Figueras, J.; Kokudo, N.; Kunstlinger, F.; Loyer, E.; Poston, G.; Rougier, P.; Rubbia-Brandt, L.; Sobrero, A.; et al. Managing synchronous liver metastases from colorectal cancer: A multidisciplinary international consensus. *Cancer Treat. Rev.* **2015**, *41*, 729–741. [CrossRef]
82. De Jonge, H.; Iamele, L.; Maggi, M.; Pessino, G.; Scotti, C. Anti-Cancer Auto-Antibodies: Roles, Applications and Open Issues. *Cancers* **2021**, *13*, 813. [CrossRef] [PubMed]
83. Yu, L.; Wang, L.; Chen, S. Olfactomedin 4, a novel marker for the differentiation and progression of gastrointestinal cancers. *Neoplasma* **2011**, *58*, 9–13. [CrossRef] [PubMed]
84. Van der Flier, L.G.; Haegebarth, A.; Stange, D.; van de Wetering, M.; Clevers, H. OLFM4 Is a Robust Marker for Stem Cells in Human Intestine and Marks a Subset of Colorectal Cancer Cells. *Gastroenterology* **2009**, *137*, 15–17. [CrossRef] [PubMed]
85. Xiong, J.; Balcioglu, H.E.; Danen, E.H. Integrin signaling in control of tumor growth and progression. *Int. J. Biochem. Cell Biol.* **2013**, *45*, 1012–1015. [CrossRef]
86. Burrell, R.A.; McGranahan, N.; Bartek, J.; Swanton, C. The causes and consequences of genetic heterogeneity in cancer evolution. *Nature* **2013**, *501*, 338–345. [CrossRef]
87. Grady, V.M.; Carethers, J.M. Genomic and epigenetic instability in colorectal cancer pathogenesis. *Gastroenterology* **2008**, *135*, 1079–1099. [CrossRef]
88. Frantz, C.; Stewart, K.M.; Weaver, V.M. The extracellular matrix at a glance. *J. Cell Sci.* **2010**, *123*, 4195–4200. [CrossRef]
89. Pickup, M.W.; Mouw, J.K.; Weaver, V.M. The extracellular matrix modulates the hallmarks of cancer. *EMBO Rep.* **2014**, *15*, 1243–1253. [CrossRef]
90. Hanahan, D. Hallmarks of Cancer: New Dimensions. *Cancer Discov.* **2022**, *12*, 31–46. [CrossRef]
91. Naba, A.; Clauser, K.; Hoersch, S.; Liu, H.; Carr, S.A.; Hynes, R.O. The Matrisome: In Silico Definition and In Vivo Characterization by Proteomics of Normal and Tumor Extracellular Matrices. *Mol. Cell. Proteom.* **2012**, *11*, M111.014647. [CrossRef]
92. Naba, A.; Clauser, K.; Hynes, R.O. Enrichment of Extracellular Matrix Proteins from Tissues and Digestion into Peptides for Mass Spectrometry Analysis. *J. Vis. Exp.* **2015**, *101*, e53057. [CrossRef] [PubMed]
93. Yao, L.; Lao, W.; Zhang, Y.; Tang, X.; Hu, X.; He, C.; Hu, X.; Xu, L.X. Identification of EFEMP2 as a Serum Biomarker for the Early Detection of Colorectal Cancer with Lectin Affinity Capture Assisted Secretome Analysis of Cultured Fresh Tissues. *J. Proteome Res.* **2012**, *11*, 3281–3294. [CrossRef] [PubMed]
94. Holten-Andersen, M.N.; Christensen, I.J.; Nielsen, H.J.; Stephens, R.W.; Jensen, V.; Nielsen, O.H.; Sørensen, S.; Overgaard, J.; Lilja, H.; Harris, A.; et al. Total levels of tissue inhibitor of metalloproteinases 1 in plasma yield high diagnostic sensitivity and specificity in patients with colon cancer. *Clin. Cancer Res.* **2002**, *8*, 156–164. [PubMed]
95. Sørensen, N.M.; Byström, P.; Christensen, I.J.; Berglund, Å.; Nielsen, H.J.; Brünner, N.; Glimelius, B. TIMP-1 Is Significantly Associated with Objective Response and Survival in Metastatic Colorectal Cancer Patients Receiving Combination of Irinotecan, 5-Fluorouracil, and Folinic Acid. *Clin. Cancer Res.* **2007**, *13*, 4117–4122. [CrossRef]
96. Holstein, E.; Dittmann, A.; Kääriäinen, A.; Pesola, V.; Koivunen, J.; Pihlajaniemi, T.; Naba, A.; Izzi, V. The Burden of Post-Translational Modification (PTM)—Disrupting Mutations in the Tumor Matrisome. *Cancers* **2021**, *13*, 1081. [CrossRef]
97. Yuzhalin, A.E. Citrullination in Cancer. *Cancer Res.* **2019**, *79*, 1274–1284. [CrossRef]
98. Gil, J.; Ramírez-Torres, A.; Encarnación-Guevara, S. Lysine acetylation and cancer: A proteomics perspective. *J. Proteom.* **2017**, *150*, 297–309. [CrossRef]
99. Harachi, M.; Masui, K.; Cavenee, W.K.; Mischel, P.S.; Shibata, N. Protein Acetylation at the Interface of Genetics, Epigenetics and Environment in Cancer. *Metabolites* **2021**, *11*, 216. [CrossRef]
100. Li, W.; Li, F.; Zhang, X.; Lin, H.K.; Xu, C. Insights into the post-translational modification and its emerging role in shaping the tumor microenvironment. *Signal Transduct. Target. Ther.* **2021**, *6*, 422. [CrossRef]
101. Tong, M.; Zheng, W.; Li, H.; Li, X.; Ao, L.; Shen, Y.; Liang, Q.; Li, J.; Hong, G.; Yan, H.; et al. Multi-omics landscapes of colorectal cancer subtypes discriminated by an individualized prognostic signature for 5-fluorouracil-based chemotherapy. *Oncogenesis* **2016**, *5*, e242. [CrossRef]
102. Pettini, F.; Visibelli, A.; Cicaloni, V.; Iovinelli, D.; Spiga, O. Multi-Omics Model Applied to Cancer Genetics. *Int. J. Mol. Sci.* **2021**, *22*, 5751. [CrossRef] [PubMed]
103. Wang, H.; Yang, L.; Liu, M.; Luo, J. Protein post-translational modifications in the regulation of cancer hallmarks. *Cancer Gene Ther.* **2022**, 1–19. [CrossRef] [PubMed]
104. Deribe, Y.L.; Pawson, T.; Dikic, I. Post-translational modifications in signal integration. *Nat. Struct. Mol. Biol.* **2010**, *17*, 666–672. [CrossRef] [PubMed]
105. Sallam, R.M. Proteomics in Cancer Biomarkers Discovery: Challenges and Applications. *Dis. Markers* **2015**, *2015*, 321370. [CrossRef]
106. Kern, S.E. Why Your New Cancer Biomarker May Never Work: Recurrent Patterns and Remarkable Diversity in Biomarker Failures. *Cancer Res.* **2012**, *72*, 6097–6101. [CrossRef]
107. Ioannidis, J.P.A.; Bossuyt, P.M.M. Waste, Leaks, and Failures in the Biomarker Pipeline. *Clin. Chem.* **2017**, *63*, 963–972. [CrossRef]
108. Hernández, B.; Parnell, A.; Pennington, S.R. Why have so few proteomic biomarkers "survived" validation? (Sample size and independent validation considerations). *Proteomics* **2014**, *14*, 1587–1592. [CrossRef]
109. Harper, J.; Bennett, E.J. Proteome complexity and the forces that drive proteome imbalance. *Nature* **2016**, *537*, 328–338. [CrossRef]

110. Bludau, I.; Aebersold, R. Proteomic and interactomic insights into the molecular basis of cell functional diversity. *Nat. Rev. Mol. Cell Biol.* **2020**, *21*, 327–340. [CrossRef]
111. Pepe, M.S.; Etzioni, R.; Feng, Z.; Potter, J.; Thompson, M.L.; Thornquist, M.D.; Winget, M.; Yasui, Y. Phases of Biomarker Development for Early Detection of Cancer. *JNCI J. Natl. Cancer Inst.* **2001**, *93*, 1054–1061. [CrossRef]
112. Chen, E.I.; Yates, J.R. Cancer proteomics by quantitative shotgun proteomics. *Mol. Oncol.* **2007**, *1*, 144–159. [CrossRef] [PubMed]
113. Cho, W.C. Proteomics Technologies and Challenges. *Genom. Proteom. Bioinform.* **2007**, *5*, 77–85. [CrossRef]

International Journal of *Molecular Sciences*

MDPI

Article

Captopril, a Renin–Angiotensin System Inhibitor, Attenuates Tumour Progression in the Regenerating Liver Following Partial Hepatectomy

Georgina E. Riddiough [1,2], Katrina A. Walsh [1], Theodora Fifis [1], Georgios Kastrappis [1], Bang M. Tran [2], Elizabeth Vincan [2,3,4], Vijayaragavan Muralidharan [1], Christopher Christophi [1], Claire L. Gordon [5,6,7,†] and Marcos V. Perini [1,*,†]

1 Department of Surgery, The University of Melbourne, Austin Health, Lance Townsend Building, Level 8, 145 Studley Road, Heidelberg, VIC 3084, Australia; georgina.riddiough@unimelb.edu.au (G.E.R.); kawalsh@unimelb.edu.au (K.A.W.); tfifis@unimelb.edu.au (T.F.); glk@student.unimelb.edu.au (G.K.); v.muralidharan@unimelb.edu.au (V.M.); c.christophi@unimelb.edu.au (C.C.)
2 Department of Infectious Diseases, The Peter Doherty Institute for Infection and Immunity, The University of Melbourne, Melbourne, VIC 3000, Australia; manht@unimelb.edu.au (B.M.T.); evincan@unimelb.edu.au (E.V.)
3 Victorian Infectious Disease Reference Laboratory, The Peter Doherty Institute for Infection and Immunity, Melbourne, VIC 3000, Australia
4 Curtin Medical School, Curtin University, Perth, WA 6102, Australia
5 Department of Infectious Diseases, Austin Health, 145 Studley Road, Heidelberg, VIC 3084, Australia; claire.gordon@austin.org.au
6 Department of Microbiology & Immunology, The Peter Doherty Institute for Infection and Immunity, The University of Melbourne, Melbourne, VIC 3000, Australia
7 North Eastern Public Health Unit, Austin Health, 145 Studley Road, Heidelberg, VIC 3084, Australia
* Correspondence: marcos.perini@unimelb.edu.au; Tel.: +61-(3)-9496-3670
† These authors contributed equally to this work.

Citation: Riddiough, G.E.; Walsh, K.A.; Fifis, T.; Kastrappis, G.; Tran, B.M.; Vincan, E.; Muralidharan, V.; Christophi, C.; Gordon, C.L.; Perini, M.V. Captopril, a Renin–Angiotensin System Inhibitor, Attenuates Tumour Progression in the Regenerating Liver Following Partial Hepatectomy. *Int. J. Mol. Sci.* **2022**, *23*, 5281. https://doi.org/10.3390/ijms23095281

Academic Editors: Donatella Delle Cave and Alessandro Ottaiano

Received: 6 April 2022
Accepted: 6 May 2022
Published: 9 May 2022

Abstract: (1) Liver regeneration following partial hepatectomy for colorectal liver metastasis (CRLM) has been linked to tumour recurrence. Inhibition of the renin–angiotensin system (RASi) attenuates CRLM growth in the non-regenerating liver. This study investigates whether RASi exerts an anti-tumour effect within the regenerating liver following partial hepatectomy for CRLM and examines RASi-induced changes in the tumour immune microenvironment; (2) CRLM in mice was induced via intrasplenic injection of mouse colorectal tumour cells, followed by splenectomy on Day 0. Mice were treated with RASi captopril (250 mg/kg/day), or saline (control) from Day 4 to Day 16 (endpoint) and underwent 70% partial hepatectomy on Day 7. Liver and tumour samples were characterised by flow cytometry and immunofluorescence; (3) captopril treatment reduced tumour burden in mice following partial hepatectomy ($p < 0.01$). Captopril treatment reduced populations of myeloid-derived suppressor cells (MDSCs) ($CD11b^+Ly6C^{Hi}$ $p < 0.05$, $CD11b^+Ly6C^{Lo}$ $p < 0.01$) and increased PD-1 expression on infiltrating hepatic tissue-resident memory (T_{RM})-like $CD8^+$ ($p < 0.001$) and double-negative ($CD4^-CD8^-$; $p < 0.001$) T cells; (4) RASi reduced CRLM growth in the regenerating liver and altered immune cell composition by reducing populations of immunosuppressive MDSCs and boosting populations of PD-1$^+$ hepatic T_{RM}s. Thus, RASi should be explored as an adjunct therapy for patients undergoing partial hepatectomy for CRLM.

Keywords: surgical oncology; liver regeneration; immunology; hepatic tissue-resident memory T cells; liver neoplasms; neoplasm metastasis

1. Introduction

Liver resection offers the best chance of cure for patients with colorectal liver metastasis (CRLM); however, tumour recurrence in the future liver remnant (FLR) is not infrequent and affects 40% of patients [1,2]. The biological processes and signalling pathways underlying physiological liver regeneration are implicated in tumour recurrence following liver resection [3]. Most liver regeneration occurs within two weeks of surgery and complete restoration of hepatic mass is achieved within 3 months in humans and 10 days in mice [4]. Neoadjuvant and adjuvant chemotherapy has been shown to improve patient survival after CRLM resection [5,6]; however, chemotherapeutic agents are avoided during the immediate perioperative period due to hepatotoxicity and impairment of early postoperative liver regeneration. Novel therapies which safely promote antitumour responses during the immediate postoperative period are, therefore, required.

Renin–angiotensin system inhibitors (RASis), commonly used to treat cardiac failure and hypertension, have been shown to prolong survival, improve response to neoadjuvant therapy, and reduce recurrence in a range of solid cancers [7–9]. Studies have reported that RASis exhibit a range of antitumour activities, including immunomodulatory, antiproliferative, and antiangiogenic effects [3,10,11]. Additionally, we recently reported that one type of RASi, captopril, exhibits its antitumour effects in the regenerating liver via *c-myc* and *cyclin D1* downregulation [12]. In the non-regenerating liver, it has been shown that RASi therapy reprograms the adaptive antitumour immune response by enhancing populations of $CD8^+$ cytotoxic and $CD4^-CD8^-$ (double-negative (DN)) T cells [13]; however, it is not known if these changes are also observed in the regenerating liver. To address this question, this study investigates the effect of captopril in the regenerating liver, on both innate and adaptive immune responses, using a model of CRLM and partial hepatectomy, which closely mimics the clinical scenario in which patients undergo liver resection for CRLM when clinically undetectable micrometastases may be present in the FLR.

2. Results

2.1. Captopril Treatment Reduces CRLM Tumour Burden in the Regenerating Liver

Captopril treatment significantly reduced CRLM tumour burden in the FLR following partial hepatectomy, compared with the control group at Day 16 (Figure 1a,b). This demonstrates that captopril is capable of inhibiting tumour growth within the milieu of the regenerating liver following partial hepatectomy.

2.2. Captopril Treatment Significantly Increases PD-1 Expression in T Cells

We recently reported that the RASi captopril induces changes in T-lymphocyte subpopulations and their activation status in a non-regenerating model of CRLM [13]. The present study was specifically focused on the immunomodulatory changes observed in the regenerating liver, the situation most relevant following liver resection.

Changes observed in the regenerating liver were complimentary to those previously observed in the non-regenerating liver [13]. Captopril treatment significantly reduced the ratio of $CD4^+$:$CD8^+$ (control 1.57 vs. captopril 0.91, $p = 0.01$) (Table 1) and $CD4^+$:DN T (control 1.63 vs. captopril 0.87, $p = 0.001$) cells in liver (Table S1).

To gain insight into the activation status of T lymphocytes, we examined the immune regulatory marker, PD-1. Treatment with captopril significantly increased the proportion of $CD8^+$ and double-negative (DN; $CD4^-CD8^-$) T-cell populations expressing PD-1, and the intensity of PD-1 expression, in both regenerating liver and tumour specimens (Figure 1e,f). Immunofluorescence staining demonstrated that $PD-1^+$ $CD8^+$ T cells are located predominantly within tumour tissue (Figure 1g). These data suggest that captopril induces a phenotypic shift towards the expression of the activation marker and immune regulator, PD-1, on $CD8^+$ and DN T cells.

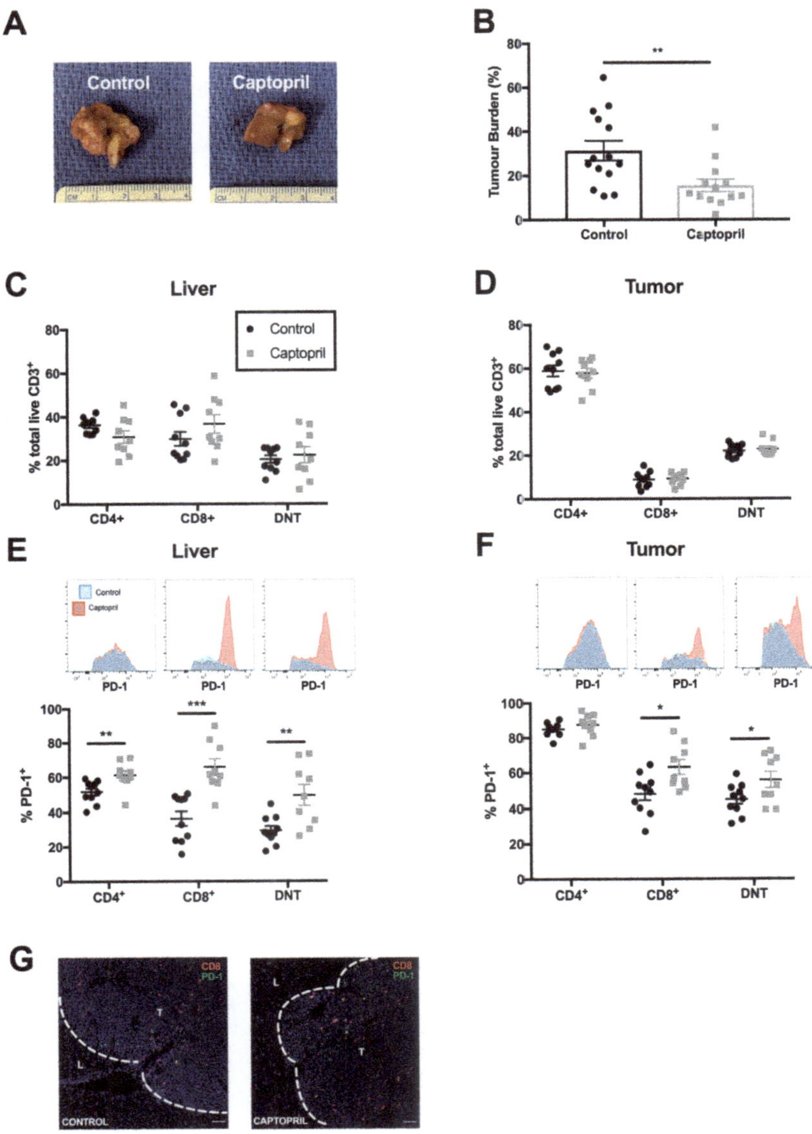

Figure 1. Captopril reduces tumour burden in regenerating liver and alters T-cell composition within liver and tumour tissues. Mice underwent surgical induction of colorectal liver metastasis via splenic injection, followed by splenectomy (Day 0). From Day 4 to Day 16, intraperitoneal injections of captopril 250 mg/kg/day or control (saline) were administered. On Day 7, mice underwent a 70% partial hepatectomy. Mice were culled on Day 16, and liver and tumour samples were retrieved for analysis: (**A**) representative photographs of liver and tumour appearance in the control and captopril treatment groups showed significantly greater tumour burden in the control group, compared with treatment; (**B**) tumour burden in the regenerating liver following partial hepatectomy in control and captopril-treated mice. Tumour burden = (tumour volume/total liver volume) × 100. Data compiled from two independently performed experiments are shown (control n = 14, captopril n = 13); (**C**,**D**) proportion of total $CD3^+$ T lymphocytes that were $CD4^+$, $CD8^+$ and DNT ($CD4^-$/$CD8^-$) in liver (**C**) and tumour (**D**); (**E**,**F**) percentage of $PD1^+$ T lymphocyte populations and insets of concatenated histograms of PD-1

fluorescence intensity on T lymphocyte populations in liver (**E**) and tumour (**F**). Data are compiled from two independently performed murine experiments (control n = 10 mice, captopril n = 9 mice). Significant differences were determined by an unpaired, two-tailed Student's t-test. $p < 0.05$ (*), $p < 0.01$ (**), $p < 0.001$ (***); (**G**) representative immunofluorescence staining of CD8 (red) and PD-1 (green) in control and captopril-treated liver. Shown are tumour (T) and surrounding liver parenchyma (L).

Table 1. Ratios of total $CD4^+$:$CD8^+$ lymphocytes with their respective ratios for $CD44^+CD69^+$ and $CD44^+CD69^+PD\text{-}1^+$ populations in liver.

CD4:CD8 Ratios	Total	$CD44^+CD69^+$	$CD44^+CD69^+PD\text{-}1^+$
Treatment—Captopril (mean ± SD)	0.91 ± 0.36)	0.45 ± 0.25	0.44 ± 0.27
Control—Saline (mean ± SD)	1.57 ± 0.35	1.00 ± 0.33	1.33 ± 0.31
p value	**0.01**	**0.01**	**0.0007**

Significant differences were determined by the unpaired, two-tailed student's t-test. Significant p values are in bold. SD—standard deviation.

2.3. Captopril Treatment Enhances Populations of T_{RM}-like Cells and Increases PD-1 Expression on $CD8^+$ T_{RM}-Like Cells

Non-circulating T cells, termed tissue-resident memory T cells (T_{RM}) [14], are increasingly implicated in mediating effective and sustained antitumour responses in solid cancers such as lung [15] and breast cancer [16]. We investigated whether the T-cell populations observed in regenerating liver and tumour tissue had features of T_{RM}, defined by the expression of murine hepatic T_{RM} markers, CD44 and CD69 (termed "T_{RM}-like cells"), and assessed the effect of captopril on these populations and their PD-1 expression. We found that, although the proportion of T_{RM}-like $CD8^+$ T cells did not increase with captopril treatment in the liver (Figure 2a), the ratio of $CD4^+$ T_{RM}-like:$CD8^+$ T_{RM}-like cells significantly reduced (control 1.00 vs. captopril 0.45, $p = 0.01$, Table 1). PD-1 expression on $CD8^+$ T_{RM}-like T cells significantly increased (Figure 2d–f). Conversely, the proportion of $CD4^+$ T_{RM}-like cells decreased in the regenerating liver following captopril treatment (Figure 2a), while their PD-1 expression was unchanged (Figure 2c,e,f).

2.4. Captopril Treatment Significantly Enhances Populations of T_{RM}-Like DN T Lymphocytes and Increases PD-1 Expression on T_{RM}-Like DN T Cells

The proportion of DN T cells co-expressing CD44 and CD69 (Figure 3a, "DN T_{RM}-like") in captopril-treated regenerating liver tissue significantly increased, and this corresponded with an increase in PD-1 expression on DN T_{RM}-like cells in captopril-treated regenerating liver and tumour (Figure 3c,d).

2.5. In the Absence of Tumour, Captopril Treatment Enhances Populations of PD-1 Expressing T_{RM}-Like $CD8^+$ and DN T Cells

As tumour induces many changes in immune cell infiltrates and responses, we sought to extract the effect of captopril on tumour-related changes by comparing them with those changes that occurred in response to liver regeneration alone. In the absence of a tumour, captopril significantly increased the proportion of $CD8^+$ T lymphocytes (Figure 4a; 37.0% vs. 59.0%, control vs. treatment) while reducing the proportion of $CD4^+$ and DN T lymphocytes (Figure 4a). Additionally, $PD\text{-}1^+$ expression significantly increased in $CD8^+$ T lymphocytes (Figure 4b). Furthermore, captopril treatment increased the proportions of both $CD8^+$ T_{RM}-like cells (Figure 4c) and $PD1^+CD8^+$ T_{RM}-like cells (Figure 4d) in regenerating liver without a tumour. The increase in $CD8^+$ T_{RM}-like cells with captopril treatment was more substantial in the absence of tumour (Figure 4c; 15.8% vs. 41.9%, control vs. treatment, $p < 0.001$), compared with that observed in the CLRM model (Figure 2a; 10.3% vs. 18.3%, $p = 0.2$) and this was also accompanied by a large reduction in the $CD4^+$ T_{RM}-like:$CD8^+$

T_{RM}-like ratio (control 0.47 vs. captopril 0.13, $p < 0.001$, Table S2). Additionally, we found that captopril reduced the population of T_{RM}-like DN T cells (Figure 4c; 9.5% vs. 5.3%, control vs. treatment, $p < 0.05$), compared with the result seen in the presence of a tumour in which case this population was significantly increased (Figure 3a; 7.1% vs. 10.7%, control vs. treatment, $p < 0.05$). Like the CRLM model, we observed that captopril treatment increased PD-1 expression on both $CD8^+$ and DN T T_{RM}-like cells (Figure 4b). These data suggest that captopril boosts the proportion of T_{RM}-like $CD8^+$ T cells (but not DN T cells) and increases the expression of PD-1 in $CD8^+$ and DN T_{RM}-like populations in physiological liver regeneration. This suggests that the expansion of DN and PD-1$^+$ DN T_{RM}-like populations is a tumour-specific response.

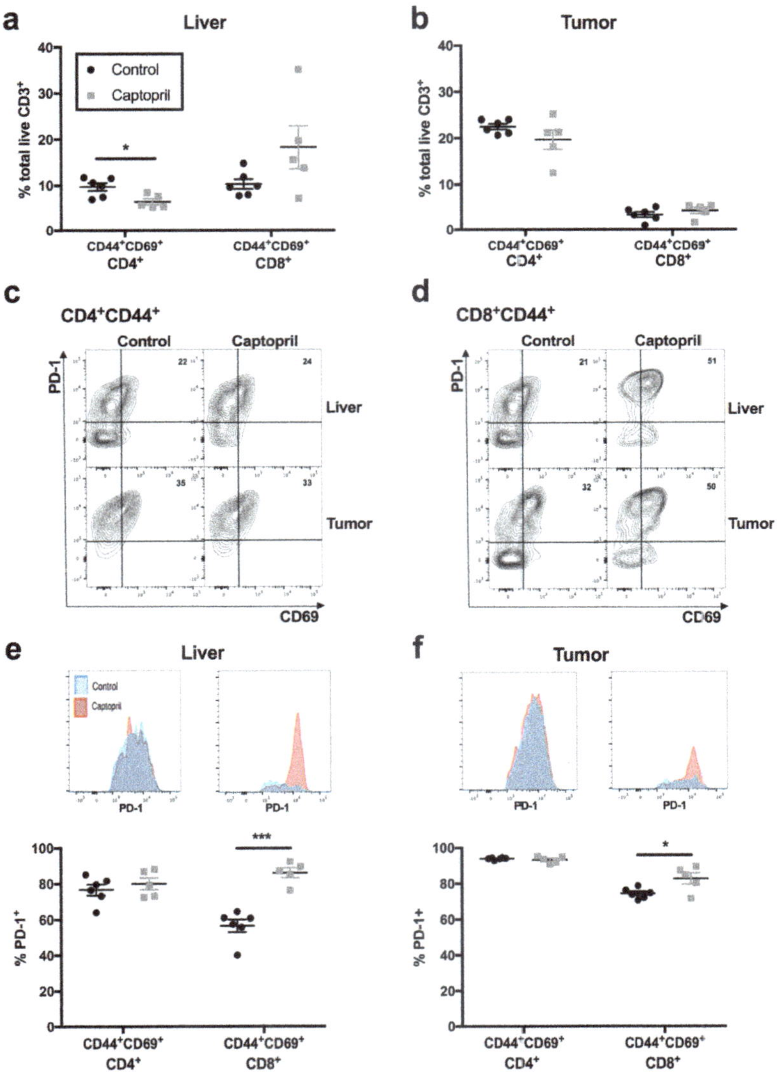

Figure 2. Captopril increases expression of the immune regulator PD-1 on tissue-resident memory (T_{RM})-like T cells in regenerating liver and tumour. Mice underwent surgical induction of colorectal liver metastasis via splenic injection (Day 0). From Day 4 to Day 16, intraperitoneal injections of captopril 250 mg/kg/day or saline (control) were administered. On Day 7, mice underwent a 70% partial hepatectomy. Mice were

culled on Day 16, and liver and tumour specimens were retrieved for flow cytometry: (**a**,**b**) proportion of $CD4^+$ or $CD8^+$ T lymphocytes that were T_{RM}-like (co-expressing $CD44^+/CD69^+$) in liver (**a**) and tumour (**b**); (**c**) concatenated contour plots demonstrating expression of CD69 and PD-1 on $CD4^+CD44^+$ T lymphocytes in liver (top panels) and tumour (bottom panels); (**d**) concatenated contour plots demonstrating expression of CD69 and PD-1 on $CD8^+CD44^+$ T lymphocytes in liver (top panels) and tumour (bottom panels); (**e**,**f**) percentage of PD-1^+ T_{RM}-like T lymphocyte populations and insets of concatenated histograms of PD-1 fluorescence intensity on T_{RM}-like T lymphocyte populations in liver (**e**) and tumour (**f**). Data are representative of a single experiment (control n = 6 mice, captopril n = 5 mice). Significant differences were determined by an unpaired, two-tailed Student's *t*-test. $p < 0.05$ (*), $p < 0.001$ (***).

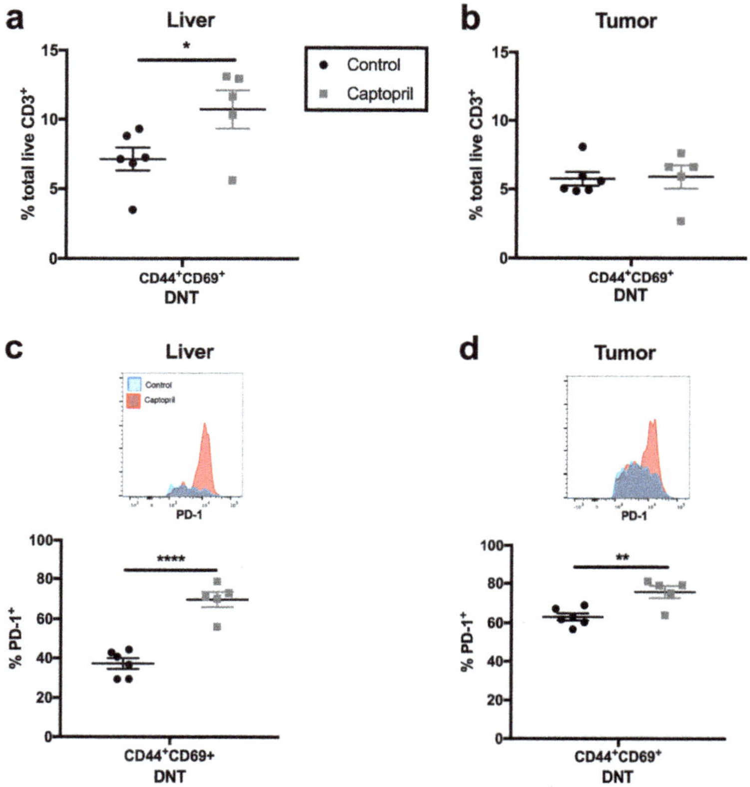

Figure 3. The renin–angiotensin system inhibitor (RASi) captopril increases the expression of the immune regulator PD-1 on tissue-resident memory (T_{RM})-like DNT cells in regenerating liver and tumour. Mice underwent colorectal liver metastasis tumour induction and were treated with either control (saline) or RASi captopril. From Day 4 to Day 16, intraperitoneal injections of captopril 250 mg/kg/day or saline (control) were administered. On Day 7, mice underwent a 70% partial hepatectomy. Mice were culled on Day 16, and liver and tumour specimens were retrieved for flow cytometry: (**a**,**b**) proportion of total $CD3^+$ T lymphocytes that were DN T T_{RM}-like (co-expressing $CD44^+/CD69^+$) in liver (**a**) and tumour (**b**); (**c**,**d**) percentage of PD-1^+ T_{RM}-like DN T lymphocytes and insets of concatenated histograms of PD-1 fluorescence intensity on T_{RM}-like DN T lymphocytes in liver (**c**) and tumour (**d**). Data are representative of a single murine experiment (control n = 6 mice, captopril, n = 5 mice). Significant differences were determined by the unpaired, two-tailed Student's *t*-test. $p < 0.05$ (*), $p < 0.01$ (**), $p < 0.0001$ (****).

2.6. Captopril Treatment Modulates Myeloid-Derived Suppressor Cell Populations

Myeloid-derived suppressor cells (MDSCs) have been recognised to have immunosuppressive effects and are associated with cancer progression and poorer prognosis [17].

In the mouse model with CRLM, captopril treatment was associated with an overall reduction in MDSC populations, demonstrated by a reduction in $CD11b^+$ leukocytes in the liver (Figure 5a). MDSC populations were distinguished into three ontological phenotypes based on their expression of Ly6C; CD11b+ $Ly6C^{Hi}$ (monocytic MDSCs), $Ly6C^{Int}$ (intermediate), and $Ly6C^{Lo}$ (granulocytic MDSCs) [18,19]. Captopril treatment was associated with a reduction in both monocytic ($Ly6C^{Hi}$) and granulocytic ($Ly6C^{Lo}$) MDSC populations (Figure 5a).

Conversely, in the mouse model without CRLM, captopril did not modulate MDSC populations; no difference was found in either the total $CD11b^+$ MDSCs or the MDSC subtypes (Figure 5b). Taken together, these data suggest that the RASi captopril may improve the overall antitumour immune response in the regenerating liver by reducing populations of MDSCs.

We have previously shown that captopril treatment modulates macrophage levels in CRLM, within the non-regenerating liver [20]. To investigate whether captopril modulated macrophage populations within the regenerating liver, we examined the expression of F4/80 on $CD11b^+$Ly6C populations. Captopril treatment did not alter the expression of macrophage marker F4/80 on $CD11b^+$Ly6C populations in either the regenerating model with or without CRLM (Figure S1a,b).

PDL1 is an inhibitory ligand most commonly expressed on MDSCs including $F4/80^+$ macrophages and tumour cells. Captopril treatment was associated with an increase in PDL1 expression on $Ly6C^{Hi}F4/80^+$ and $Ly6C^{Int}F4/80^+$ leukocytes in the regenerating livers in the model with CRLM (Figure S1c). However, this response was not observed in the model without CRLM (Figure S1d).

Notably, populations of MDSCs were higher in the presence of CRLM, compared with those in the absence of CRLM (control mean with tumour, total $CD11b^+$ = 46% vs. control mean without tumour, total $CD11b^+$ = 33%) (Figure 5a,b). Interestingly, captopril treatment reduced these populations in the presence of tumour (captopril mean total $CD11b^+$ = 34%, Figure 5a), back down to the baseline level observed in the regenerating liver in the absence of CRLM (Figure 5b). Taken together, these findings highlight that, although the MDSCs present in the tissue respond by upregulating PDL1 expression, the sum effect of captopril is to reduce the presence of MDSCs.

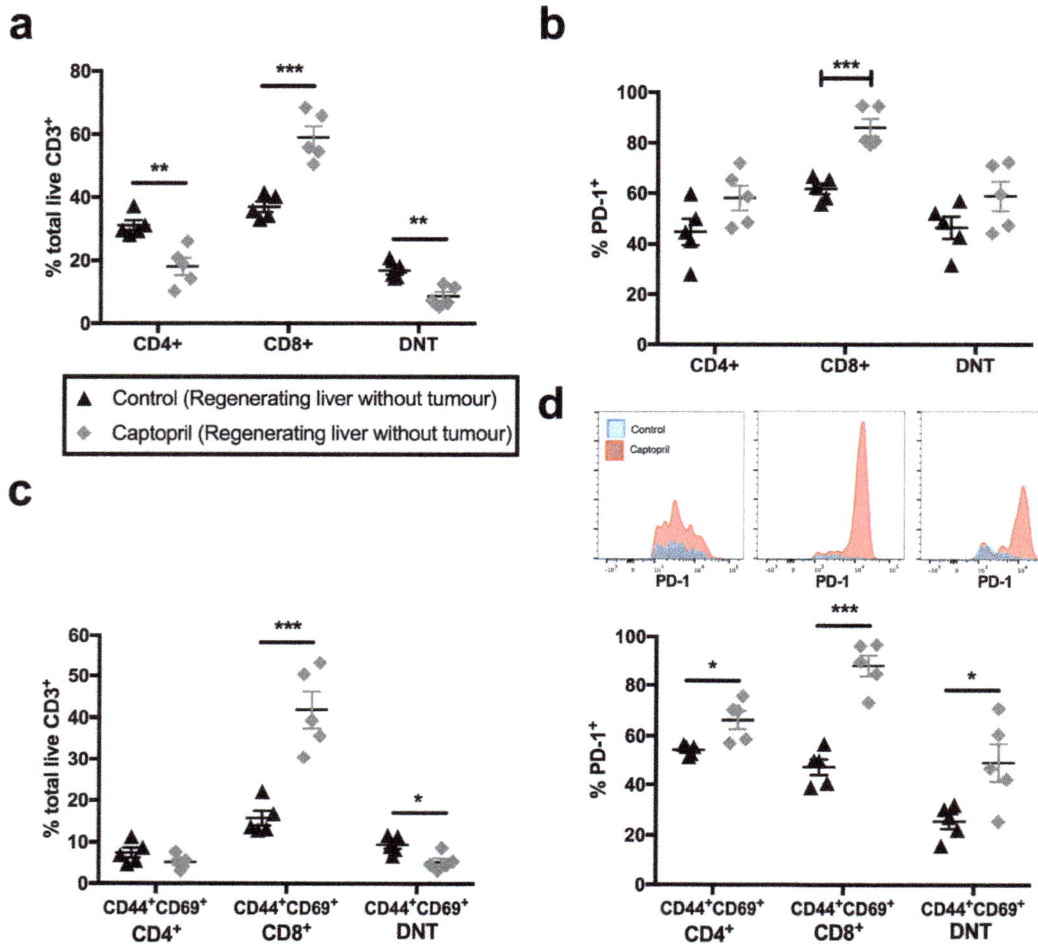

Figure 4. Captopril modulates T-lymphocyte populations in the regenerating liver, in the absence of tumour. Mice commenced treatment with either saline (control) or captopril 250 mg/kg/day and did not undergo tumour induction. Briefly, 70% hepatectomy was performed 3 days after the commencement of treatment, and mice were culled nine days following partial hepatectomy in keeping with prior experiments. Liver specimens were retrieved for flow cytometry: (**a**) proportion of total $CD3^+$ T lymphocytes that were $CD4^+$, $CD8^+$ and DN T ($CD4^-/CD8^-$); (**b**) percentage of PD-1^+ T lymphocyte subpopulations; (**c**) proportion of $CD4^+$, $CD8^+$ and DN T lymphocytes co-expressing T_{RM} markers, $CD44^+$ and $CD69^+$; (**d**) percentage of PD-1^+ T_{RM}-like T lymphocyte populations and insets of concatenated histograms demonstrating fluorescence intensity of PD-1 on T_{RM}-like populations. Data are representative of a single experiment (control, n = 5 mice, captopril, n = 5 mice). Significant differences were determined by an unpaired, two-tailed Student's *t*-test. $p < 0.05$ (*), $p < 0.01$ (**), $p < 0.001$ (***).

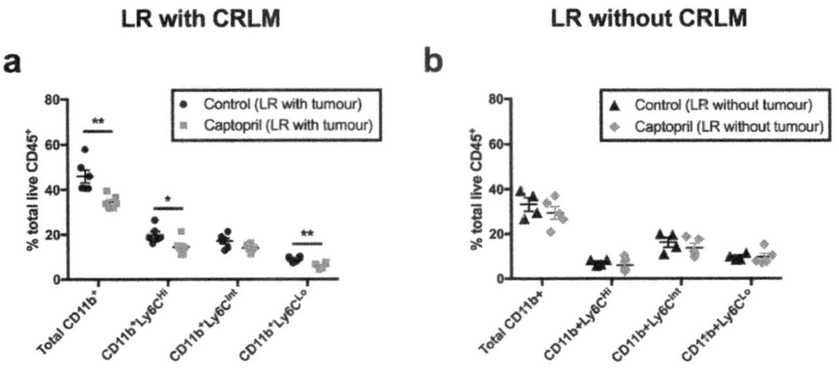

Figure 5. The renin–angiotensin system inhibitor (RASi) captopril reduces CD11b$^+$ myeloid-derived suppressor cell populations within the regenerating liver (LR) following tumour induction, down to baseline levels observed in regenerating liver without tumour. One group of mice underwent surgical induction of colorectal liver metastasis via splenic injection (Day 0) and one group of mice did not undergo tumour induction. From Day 4 to Day 16, intraperitoneal injections of captopril 250 mg/kg/day or saline (control) were administered to all mice in both groups. On Day 7, all mice underwent a 70% partial hepatectomy. All mice were culled on Day 16, and liver and tumour specimens were retrieved for flow cytometry: (**a**,**b**) proportion of total CD45$^+$ myeloid cells that were CD11b$^+$, CD11b$^+$Lyc6C^{Hi}, CD11b$^+$Ly6C^{Int}, and CD11b$^+$Ly6C^{Lo} in livers of mice following tumour induction and partial hepatectomy (**a**) and in livers of mice following partial hepatectomy alone (no tumour) (**b**). Data are representative of two independently performed murine experiments (control n = 6 mice, captopril n = 6 mice). Significant differences were determined by an unpaired, two-tailed Student's *t*-test. $p < 0.05$ (*), $p < 0.01$ (**).

3. Discussion

Tumour recurrence stimulated by the process of liver regeneration, following liver resection, is a major barrier to the long-term survival of patients with primary or secondary liver tumours. Presently, 40% of patients undergoing liver resection will experience intrahepatic recurrence. In this study, using a homogenous tumour mouse model that closely mimics the clinical scenario, we showed that RASi therapy with captopril inhibits the growth of CRLM in the regenerating FLR and is associated with phenotypic changes in immune cells in the tumour microenvironment. These changes included a reduction in populations of MDSCs and enhanced populations of PD-1$^+$ hepatic-T_{RM}-like T lymphocytes. Further research is required to determine whether the changes in the immune microenvironment induced by captopril therapy directly reduce tumour burden in the regenerating liver and how these immune changes relate to other known antitumour effects of captopril [3,10,11]. Nevertheless, in the future, perioperative administration of a RASi could be an effective therapy for reducing micrometastatic tumour growth in the FLR and improving clinical outcomes following liver resection for CRLM.

Studies have demonstrated that T_{RM}s are involved in maintaining effective antitumour immune responses and correlate positively with survival for a range of solid cancers [15,16]. In contrast to other cancers, for example, melanoma and renal cell carcinoma, where inhibition of the PD-1-PDL1 pathway using checkpoint inhibitors has improved clinical outcomes by reinvigorating antitumour T cell responses [21,22], T_{RM} expressing PD-1 have been shown to maintain cytotoxicity in the tumour microenvironment (TME) [15,16,23,24]. This challenges the assumption that cytotoxic T cells expressing checkpoint markers such as PD-1 are tolerogenic or exhausted, which may explain why responsiveness to checkpoint inhibitors has not been universally observed in all cancer types, particularly if cancers rely on T_{RM} for antitumour activity, or there is an abundance of MDSCs. These findings correlate with those revealed in our study, in which a significant reduction in tumour burden was observed alongside the expansion of PD-1$^+$ T_{RM} populations. Our findings support a role for PD1$^+$ CD8$^+$ T_{RM}-like lymphocytes in maintaining cytotoxicity against CRLM. This may explain why some cancers, including mismatch repair proficient colorectal cancers, have exhibited poor responses to checkpoint inhibitor therapy and suggests that RASi could play a role in the treatment of CRC by modulating the immune infiltrate.

MDSCs are crucial components of TME that facilitate cancer growth by secreting immunosuppressive cytokines such as TNF and TGFβ, inadvertently allowing cancer cells to evade the patient's immune system [25,26]. Clinical studies have shown that an abundance of tumour-infiltrating MDSCs is associated with a poorer prognosis [17]. Both granulocytic and monocytic MDSCs possess immunosuppressive properties such as inhibiting effector T cells and promoting the inhibitory effects of regulatory T cells [27]. These findings support the data presented here, which show that tumour burden is reduced alongside a significant reduction in granulocytic (Ly6C^{Lo}) and monocytic (Ly6C^{Hi}) MDSCs.

In acute inflammation, the adaptive and innate arms of the immune system work in concert with one another to restore homeostasis. In contrast, during chronic inflammation, such as within the TME, the immune response is dysregulated and characterised by the accumulation of immunoinhibitory cells, including MDSCs. Therefore, the finding we report here—namely, that certain T_{RM}-cell populations are enhanced, while MDSC populations are reduced—is not likely to be coincidental. Studies in mice have demonstrated that both granulocytic and monocytic MDSCs suppress cytotoxic T cells [28]. However, further experiments are required to confirm that the MDSC changes observed here are directly responsible for the expansion of T_{RM} populations. To date, immunotherapy has largely focused on improving the cytotoxic function of effector T cells, rather than reducing populations of MDSCs, but the latter is also emerging as a viable approach to improve clinical cancer outcomes and trials examining this are underway [29]. Clinical trials examining the outcomes of treatment with anti-PD-1 therapies in CRC have only demonstrated benefits for patients with mismatch repair deficient (dMMR) CRC [30]. However, only 15% of CRCs are dMMR; therefore, the majority of CRC patients have tumours that are unresponsive to PD-1 inhibitor therapy. Immune therapies for mismatch repair proficient (pMMR) CRC could benefit from the addition of anti-MDSC therapies and our findings support exploring RASi as an immunomodulatory agent in this setting.

To our knowledge, this is the first study to demonstrate that the RASi captopril modulates the immune responses in the regenerating liver, in the absence of a tumour. We found that captopril significantly increased populations of CD8$^+$ T cells, including T_{RM}-like CD8$^+$ T cells, in the regenerating liver in the absence of a tumour. This is in line with other studies that have also demonstrated that CD8$^+$ T cells are key for hepatocyte proliferation and successful liver regeneration [31]. Conversely, the MDSC levels and subtype composition were not altered by captopril at this time point of liver regeneration, in the absence of a tumour. MDSC levels in the untreated controls were significantly higher in the combined tumour induction and liver regeneration group, compared with that seen in the absence of tumour. Furthermore, captopril treatment reduced MDSC levels in the combined tumour induction and liver regeneration group back to the baseline levels observed in the group with the absence of the tumour and liver regeneration.

Our study also demonstrated a crucial role for DN T cells in attaining effective antitumour immunity. The expansion of PD-1$^+$ T_{RM}-like DN T cells was unique to the CRLM model, suggestive of tumour specificity. This is in agreement with other studies that have demonstrated DN T cells possess important antitumour functions [32], and adoptive transfer of DN T cells is presently being explored as a novel cancer therapy [33].

In conclusion, this study showed that RASis reduced CRLM growth in the regenerating liver and altered immune cell composition by reducing populations of immunosuppressive MDSCs and enhancing populations of PD-1+ hepatic T_{RM}-like lymphocytes. These results contribute to a growing body of literature suggesting that perioperative RASis may improve treatment outcomes for patients undergoing CRLM resection and should be explored further.

4. Materials and Methods

4.1. Animal Experiments

Animal experiments were approved by Austin Health Animal Ethics Committee (AEC-05435). Male CBA mice aged 10 weeks old and weighing 20–25 g were obtained from the Animal Resources Centre, Western Australia.

4.2. In Vivo Model and Cell Line

A mouse colorectal cancer cell line (MoCR), developed via dimethyl-hydrazine induction of colon carcinoma in CBA mice [34], was used for CRLM induction. Animals were divided into groups: with tumour induction and without tumour induction. The groups with tumour induction underwent surgical inoculation of CRLM via intrasplenic injection of 50,000 MoCR cells in 50 μL Ringer's solution, followed by splenectomy (performed as one surgical procedure) (Figure 6) [35]. From Day 4, captopril (S-3 mercapto-2-methylpropionyl-L-proline, Sigma-Aldrich, St. Louis, Missouri, USA) 250 mg/kg/day or saline (control) was administered via intraperitoneal injection (Figure 6). On Day 7 following tumour induction, mice underwent a 70% partial hepatectomy, as previously described [35]. Carprofen (5 mg/kg) (Pfizer, New York, NY, USA) was administered for analgesia. Mice were culled on Day 16, and liver and tumour samples were retrieved for analysis.

The groups of mice not undergoing tumour induction followed the same treatment and hepatectomy timeline as outlined above but did not undergo tumour inoculation or splenectomy.

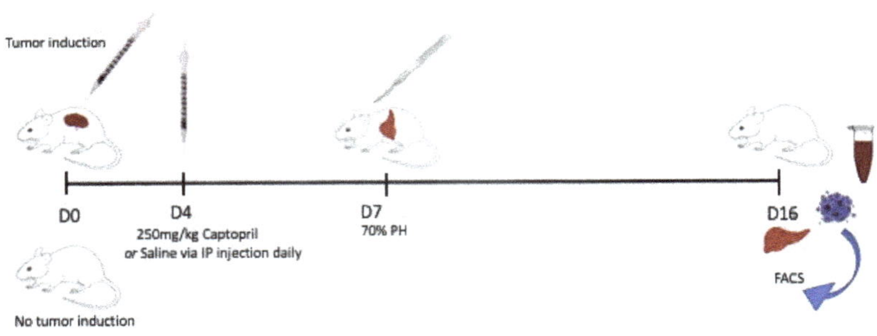

Figure 6. Experimental methodology. Mice were allocated to either tumour induction or no tumour induction groups, and this was performed on Day 0. From Day 4 to Day 16, half the mice in each group received treatment with intraperitoneal injections of captopril 250 mg/kg/day and the other half received intraperitoneal injections of saline (control). On Day 7, all mice underwent a 70% partial hepatectomy. Mice were culled on Day 16, and liver and tumour (if applicable) samples were retrieved for analysis.

4.3. Stereometric Tumour Burden Analysis

Livers were fixed in 10% formalin for 24 h and stereometric tumour burden analysis was calculated, as previously described [13]. Briefly, livers were transected using a tissue fractionator. Liver slices were photographed and analysed using Image-Pro Plus. Tumour burden was calculated as a percentage of the whole liver volume.

4.4. Flow Cytometry

Tumour and liver samples were enzymatically digested (0.25 mg/mL liberase (Sigma) and 1 mg/mL DNAse (Sigma) in DMEM) at 37 °C for 40 min. Cell suspensions were filtered (100 μm cell filters, Corning), treated with red blood cell lysis buffer for 2 min at 37 °C, washed, and resuspended at 1×10^7 cells/mL in cold FACS wash buffer (10% bovine serum albumin/5 mM EDTA/0.01% sodium azide in PBS pH 7). Briefly, 1×10^6 cell aliquots were stained with either Cocktail 1: CD45-BV510, CD4-FITC, CD3-PE, CD44-PerCP, CD8-PECy7, PD-1-APC, CD69-APC-Cy7 (BD, Biosciences) and viability dye DAPI, Cocktail 2: CD45-

BV510, CD103-FITC, CD3-PE, CD8-PECy7, PD-1-APC, CD4-APC-Cy7 (BD, Biosciences) and DAPI, or Cocktail 3: CD45-BV510, F4/80-FITC, Ly6C-PE, PD-L1-PECy7, CD11b-APC-Cy7 (BD, Biosciences), and DAPI. Samples were run on the Canto II (BD Biosciences) and analysed using FlowJo™ software.

4.5. Immunofluorescence

Following formalin fixation, livers were processed into paraffin blocks using standard techniques (Austin Pathology, Austin Health) and 4 μm sections were slide-mounted. Slides were dewaxed, and antigen retrieval (Tris buffer, pH 9) at 99 °C for 30 min was performed. Goat serum (10%) was used for blocking. Primary antibodies to PD-1 (rabbit monoclonal, Abcam, Cambridge, UK, 214421) and CD8 (rat monoclonal, Invitrogen, Waltham, MA, USA, 14-0808-82) were incubated overnight at 4 °C. Corresponding secondary antibodies were applied at 1:250 (Life Technologies AlexaFluor 488, Carlsbad, CA, USA, goat anti-rabbit IgG, A11034 and Life Technologies AlexaFluor 568, goat anti-rat IgG, A11077). DAPI 1:5000 was used for nuclear staining. Images were processed using confocal microscopy (Zeiss LSM780, Jena, Germany).

4.6. Statistical Analysis

GraphPad Prism was used for statistical analysis. Data were analysed using a two-tailed Student's *t*-test. Graphical data are expressed as the mean +/− standard error of the mean.

Supplementary Materials: The following supporting information can be downloaded at: https://www.mdpi.com/article/10.3390/ijms23095281/s1.

Author Contributions: Conceptualisation, M.V.P., C.C., V.M. and G.E.R.; methodology, G.E.R., G.K., T.F., K.A.W., M.V.P. and C.L.G.; validation, T.F., K.A.W., C.L.G., B.M.T. and E.V.; formal analysis, G.E.R., T.F., K.A.W., C.L.G. and M.V.P.; resources, M.V.P., C.C. and V.M.; data curation, G.E.R.; writing—original draft preparation, M.V.P., C.L.G., T.F., K.A.W., B.M.T. and E.V.; writing—review and editing.; supervision, M.V.P., C.C., V.M. and E.V.; project administration, M.V.P. and C.C.; funding acquisition, G.E.R. All authors have read and agreed to the published version of the manuscript.

Funding: This research was funded by the Royal Australasian College of Surgeons, Reg Worcester Scholarship 2019, and Tour de Cure Scholarship 2020 awarded to G.E.R., C.L.G (GNT 1160963) is supported by an NHMRC Early Career Fellowship.

Institutional Review Board Statement: The animal study protocol was approved by Austin Health Animal Ethics Committee (AEC-05435 v12, March 2020).

Informed Consent Statement: Not applicable.

Data Availability Statement: The data presented in this study are available on request from the corresponding author.

Acknowledgments: We would like to express our appreciation for the support from Austin Health BioResources Facility; Ashleigh Poh, Cancer and Inflammation Laboratory; Olivia Newton-John Cancer Research Institute; Biological Optical Microscopy Platform (BOMP), University of Melbourne.

Conflicts of Interest: The authors declare no conflict of interest. The funders had no role in the design of the study; in the collection, analyses, or interpretation of data; in the writing of the manuscript; or in the decision to publish the results.

References

1. Oldhafer, K.J.; Donati, M.; Jenner, R.M.; Stang, A.; Stavrou, G.A. ALPPS for Patients with Colorectal Liver Metastases: Effective Liver Hypertrophy, but Early Tumor Recurrence. *World J. Surg.* **2014**, *38*, 1504–1509. [CrossRef] [PubMed]
2. Mao, R.; Zhao, J.-J.; Bi, X.-Y.; Zhang, Y.-F.; Li, Z.-Y.; Zhou, J.-G.; Wu, X.-L.; Xiao, C.; Zhao, H.; Cai, J.-Q. A Postoperative Scoring System for Post-Hepatectomy Early Recurrence of Colorectal Liver Metastases. *Oncotarget* **2017**, *8*, 102531–102539. [CrossRef]
3. Riddiough, G.E.; Fifis, T.; Muralidharan, V.; Perini, M.V.; Christophi, C. Searching for the Link; Mechanisms Underlying Liver Regeneration and Recurrence of Colorectal Liver Metastasis Post Partial Hepatectomy. *J. Gastroen. Hepatol.* **2019**, *34*, 1276–1286. [CrossRef] [PubMed]

4. Yagi, S.; Hirata, M.; Miyachi, Y.; Uemoto, S. Liver Regeneration after Hepatectomy and Partial Liver Transplantation. *Int. J. Mol. Sci.* **2020**, *21*, 8414. [CrossRef] [PubMed]
5. Nordlinger, B.; Sorbye, H.; Glimelius, B.; Poston, G.J.; Schlag, P.M.; Rougier, P.; Bechstein, W.O.; Primrose, J.N.; Walpole, E.T.; Finch-Jones, M.; et al. Perioperative FOLFOX4 Chemotherapy and Surgery versus Surgery Alone for Resectable Liver Metastases from Colorectal Cancer (EORTC 40983): Long-Term Results of a Randomised, Controlled, Phase 3 Trial. *Lancet Oncol.* **2013**, *14*, 1208–1215. [CrossRef]
6. Khoo, E.; O'Neill, S.; Brown, E.; Wigmore, S.J.; Harrison, E.M. Systematic Review of Systemic Adjuvant, Neoadjuvant and Perioperative Chemotherapy for Resectable Colorectal-Liver Metastases. *Hpb* **2016**, *18*, 485–493. [CrossRef] [PubMed]
7. Morris, Z.S.; Saha, S.; Magnuson, W.J.; Morris, B.A.; Borkenhagen, J.F.; Ching, A.; Hirose, G.; McMurry, V.; Francis, D.M.; Harari, P.M.; et al. Increased Tumor Response to Neoadjuvant Therapy among Rectal Cancer Patients Taking Angiotensin-converting Enzyme Inhibitors or Angiotensin Receptor Blockers. *Cancer* **2016**, *122*, 2487–2495. [CrossRef]
8. Nakai, Y.; Isayama, H.; Ijichi, H.; Sasaki, T.; Sasahira, N.; Hirano, K.; Kogure, H.; Kawakubo, K.; Yagioka, H.; Yashima, Y.; et al. Inhibition of Renin–Angiotensin System Affects Prognosis of Advanced Pancreatic Cancer Receiving Gemcitabine. *Brit. J. Cancer* **2010**, *103*, 1644–1648. [CrossRef]
9. Pinter, M.; Weinmann, A.; Wörns, M.-A.; Hucke, F.; Bota, S.; Marquardt, J.U.; Duda, D.G.; Jain, R.K.; Galle, P.R.; Trauner, M.; et al. Use of Inhibitors of the Renin–Angiotensin System Is Associated with Longer Survival in Patients with Hepatocellular Carci-noma. *United Eur. Gastroent.* **2017**, *5*, 987–996. [CrossRef]
10. Perini, M.V.; Dmello, R.S.; Nero, T.L.; Chand, A.L. Evaluating the Benefits of Renin-Angiotensin System Inhibitors as Cancer Treatments. *Pharmacol. Therapeut.* **2020**, *211*, 107527. [CrossRef]
11. Volpert, O.V.; Ward, W.F.; Lingen, M.W.; Chesler, L.; Solt, D.B.; Johnson, M.D.; Molteni, A.; Polverini, P.J.; Bouck, N.P. Cap-topril Inhibits Angiogenesis and Slows the Growth of Experimental Tumors in Rats. *J. Clin. Investig.* **1996**, *98*, 671–679. [CrossRef]
12. Riddiough, G.E.; Fifis, T.; Walsh, K.A.; Muralidharan, V.; Christophi, C.; Tran, B.M.; Vincan, E.; Perini, M.V. Captopril, a Renin-Angiotensin System Inhibitor, Attenuates Features of Tumor Invasion and Down-Regulates C-Myc Expression in a Mouse Model of Colorectal Cancer Liver Metastasis. *Cancers* **2021**, *13*, 2734. [CrossRef] [PubMed]
13. Ardila, D.L.V.; Walsh, K.A.; Fifis, T.; Paolini, R.; Kastrappis, G.; Christophi, C.; Perini, M.V. Immunomodulatory Effects of Renin–Angiotensin System Inhibitors on T Lymphocytes in Mice with Colorectal Liver Metastases. *J. Immunother. Cancer* **2020**, *8*, e000487. [CrossRef] [PubMed]
14. Sasson, S.C.; Gordon, C.L.; Christo, S.N.; Klenerman, P.; Mackay, L.K. Local Heroes or Villains: Tissue-Resident Memory T Cells in Human Health and Disease. *Cell Mol. Immunol.* **2020**, *17*, 113–122. [CrossRef] [PubMed]
15. Ganesan, A.-P.; Clarke, J.; Wood, O.; Garrido-Martin, E.M.; Chee, S.J.; Mellows, T.; Samaniego-Castruita, D.; Singh, D.; Seumois, G.; Alzetani, A.; et al. Tissue-Resident Memory Features Are Linked to the Magnitude of Cytotoxic T Cell Responses in Human Lung Cancer. *Nat. Immunol.* **2017**, *18*, 940–950. [CrossRef]
16. Kathleen Cuningham Foundation Consortium for Research into Familial Breast Cancer (kConFab); Savas, P.; Virassamy, B.; Ye, C.; Salim, A.; Mintoff, C.P.; Caramia, F.; Salgado, R.; Byrne, D.J.; Teo, Z.L.; et al. Single-Cell Profiling of Breast Cancer T Cells Reveals a Tissue-Resident Memory Subset Associated with Im-proved Prognosis. *Nat. Med.* **2018**, *24*, 986–993. [CrossRef] [PubMed]
17. Zhang, S.; Ma, X.; Zhu, C.; Liu, L.; Wang, G.; Yuan, X. The Role of Myeloid-Derived Suppressor Cells in Patients with Solid Tumors: A Meta-Analysis. *PLoS ONE* **2016**, *11*, e0164514. [CrossRef]
18. Wolf, A.A.; Yáñez, A.; Barman, P.K.; Goodridge, H.S. The Ontogeny of Monocyte Subsets. *Front. Immunol.* **2019**, *10*, 1642. [CrossRef]
19. Mildner, A.; Schönheit, J.; Giladi, A.; David, E.; Lara-Astiaso, D.; Lorenzo-Vivas, E.; Paul, F.; Chappell-Maor, L.; Priller, J.; Leutz, A.; et al. Genomic Characterization of Murine Monocytes Reveals C/EBPβ Transcription Factor Dependence of Ly6C− Cells. *Immunity* **2017**, *46*, 849–862.e7. [CrossRef]
20. Wen, S.W.; Ager, E.I.; Neo, J.; Christophi, C. The Renin Angiotensin System Regulates Kupffer Cells in Colorectal Liver Me-tastases. *Cancer Biol. Ther.* **2014**, *14*, 720–727. [CrossRef]
21. Flynn, M.; Pickering, L.; Larkin, J.; Turajlic, S. Immune-Checkpoint Inhibitors in Melanoma and Kidney Cancer: From Se-quencing to Rational Selection. *Ther. Adv. Med. Oncol.* **2018**, *10*, 1758835918777427. [CrossRef] [PubMed]
22. Topalian, S.L.; Hodi, F.S.; Brahmer, J.R.; Gettinger, S.N.; Smith, D.C.; McDermott, D.F.; Powderly, J.D.; Carvajal, R.D.; Sosman, J.A.; Atkins, M.B.; et al. Safety, Activity, and Immune Correlates of Anti–PD-1 Antibody in Cancer. *N. Engl. J. Med.* **2012**, *366*, 2443–2454. [CrossRef] [PubMed]
23. Clarke, J.; Panwar, B.; Madrigal, A.; Singh, D.; Gujar, R.; Wood, O.; Chee, S.J.; Eschweiler, S.; King, E.V.; Awad, A.S.; et al. Single-Cell Transcriptomic Analysis of Tissue-Resident Memory T Cells in Human Lung Cancer. *J. Exp. Med.* **2019**, *216*, 2128–2149. [CrossRef] [PubMed]
24. Komdeur, F.L.; Prins, T.M.; van de Wall, S.; Plat, A.; Wisman, G.B.A.; Hollema, H.; Daemen, T.; Church, D.N.; de Bruyn, M.; Nijman, H.W. CD103+ Tumor-Infiltrating Lymphocytes Are Tumor-Reactive Intraepithelial CD8+ T Cells Associated with Prog-nostic Benefit and Therapy Response in Cervical Cancer. *Oncoimmunology* **2017**, *6*, e1338230. [CrossRef]
25. Tesi, R.J. MDSC; the Most Important Cell You Have Never Heard Of. *Trends Pharm. Sci.* **2018**, *40*, 4–7. [CrossRef]
26. Nagaraj, S.; Schrum, A.G.; Cho, H.-I.; Celis, E.; Gabrilovich, D.I. Mechanism of T Cell Tolerance Induced by Myeloid-Derived Suppressor Cells. *J. Immunol.* **2010**, *184*, 3106–3116. [CrossRef]

27. Yang, Y.; Li, C.; Liu, T.; Dai, X.; Bazhin, A.V. Myeloid-Derived Suppressor Cells in Tumors: From Mechanisms to Antigen Specificity and Microenvironmental Regulation. *Front. Immunol.* **2020**, *11*, 1371. [CrossRef]
28. Youn, J.-I.; Nagaraj, S.; Collazo, M.; Gabrilovich, D.I. Subsets of Myeloid-Derived Suppressor Cells in Tumor-Bearing Mice. *J. Immunol.* **2008**, *181*, 5791–5802. [CrossRef]
29. Law, A.M.K.; Valdes-Mora, F.; Gallego-Ortega, D. Myeloid-Derived Suppressor Cells as a Therapeutic Target for Cancer. *Cells* **2020**, *9*, 561. [CrossRef]
30. Le, D.T.; Uram, J.N.; Wang, H.; Bartlett, B.R.; Kemberling, H.; Eyring, A.D.; Skora, A.D.; Luber, B.S.; Azad, N.S.; Laheru, D.; et al. PD-1 Blockade in Tumors with Mismatch-Repair Deficiency. *N. Engl. J. Med.* **2015**, *372*, 2509–2520. [CrossRef]
31. Tumanov, A.V.; Koroleva, E.P.; Christiansen, P.A.; Khan, M.A.; Ruddy, M.J.; Burnette, B.; Papa, S.; Franzoso, G.; Nedospasov, S.A.; Fu, Y.; et al. T Cell-Derived Lymphotoxin Regulates Liver Regeneration. *Gastroenterology* **2009**, *136*, 694–704.e4. [CrossRef] [PubMed]
32. Young, K.J.; Kay, L.S.; Phillips, M.J.; Zhang, L. Antitumor Activity Mediated by Double-Negative T Cells. *Cancer Res.* **2003**, *63*, 8014–8021. [PubMed]
33. Li, Y.; Dong, K.; Fan, X.; Xie, J.; Wang, M.; Fu, S.; Li, Q. DNT Cell-Based Immunotherapy: Progress and Applications. *J. Cancer* **2020**, *11*, 3717–3724. [CrossRef] [PubMed]
34. Kuruppu, D.; Christophi, C.; Bertram, J.F.; Brien, P.E.O. Characterization of an Animal Model of Hepatic Metastasis. *J. Gastroen. Hepatol.* **1996**, *11*, 26–32. [CrossRef] [PubMed]
35. Koh, S.L.; Ager, E.I.; Costa, P.L.N.; Malcontenti-Wilson, C.; Muralidharan, V.; Christophi, C. Blockade of the Renin–Angiotensin System Inhibits Growth of Colorectal Cancer Liver Metastases in the Regenerating Liver. *Clin. Exp. Metastas* **2014**, *31*, 395–405. [CrossRef]

International Journal of *Molecular Sciences*

MDPI

Article

Lost by Transcription: Fork Failures, Elevated Expression, and Clinical Consequences Related to Deletions in Metastatic Colorectal Cancer

Marcel Smid , Saskia M. Wilting and John W. M. Martens *

Erasmus MC Cancer Institute, Department of Medical Oncology, University Medical Center Rotterdam, 3015 GD Rotterdam, The Netherlands; m.smid@erasmusmc.nl (M.S.); s.wilting@erasmusmc.nl (S.M.W.)
* Correspondence: j.martens@erasmusmc.nl

Abstract: Among the structural variants observed in metastatic colorectal cancer (mCRC), deletions (DELs) show a size preference of ~10 kb–1 Mb and are often found in common fragile sites (CFSs). To gain more insight into the biology behind the occurrence of these specific DELs in mCRC, and their possible association with outcome, we here studied them in detail in metastatic lesions of 429 CRC patients using available whole-genome sequencing and corresponding RNA-seq data. Breakpoints of DELs within CFSs are significantly more often located between two consecutive replication origins compared to DELs outside CFSs. DELs are more frequently located at the midpoint of genes inside CFSs with duplications (DUPs) at the flanks of the genes. The median expression of genes inside CFSs was significantly higher than those of similarly-sized genes outside CFSs. Patients with high numbers of these specific DELs showed a shorter progression-free survival time on platinum-containing therapy. Taken together, we propose that the observed DEL/DUP patterns in expressed genes located in CFSs are consistent with a model of transcription-dependent double-fork failure, and, importantly, that the ability to overcome the resulting stalled replication forks decreases sensitivity to platinum-containing treatment, known to induce stalled replication forks as well. Therefore, we propose that our DEL score can be used as predictive biomarker for decreased sensitivity to platinum-containing treatment, which, upon validation, may augment future therapeutic choices.

Keywords: metastatic colorectal cancer; structural variants; common fragile sites; platinum therapy; stalled replication fork

Citation: Smid, M.; Wilting, S.M.; Martens, J.W.M. Lost by Transcription: Fork Failures, Elevated Expression, and Clinical Consequences Related to Deletions in Metastatic Colorectal Cancer. *Int. J. Mol. Sci.* **2022**, *23*, 5080. https://doi.org/10.3390/ijms23095080

Academic Editors: Alessandro Ottaiano and Donatella Delle Cave

Received: 31 March 2022
Accepted: 28 April 2022
Published: 3 May 2022

1. Introduction

Large cohorts of primary cancers analyzed by whole-genome sequencing (WGS) were recently described [1], yielding a deluge of defects observed in cancer genomes. Most emphasis in these studies was on driver genes and mutational signatures [2–4], their disease-specific patterns, and their possible etiology. Building upon these shoulders, metastatic lesions of cancer patients have also been investigated in depth using WGS [5–8], describing enriched genetic aberrations and mutational signatures related to progression or to prior treatment. In metastatic colorectal cancer (mCRC), we previously detailed various somatic events and compared these to those found in primary CRC [9].

One result was that deletions in genes present in common fragile sites (CFS) were associated with response to therapy, and this current manuscript further investigates this. Large deletions, along with tandem duplications (DUPs), inversions, and translocations, are a class of structural variants (SVs) observed in most cancer types. In a primary pan-cancer study, these SVs have recently been extensively described, reporting high numbers of complex SVs, deletions, and tandem duplications in over 20 cancer types, including colorectal cancer. Other types of SVs are reported in virtually all cancer types, but at lower frequency. However, CFSs were only briefly discussed [10]. Among all cases, CFSs showed a peak DEL size of around 100 kb, and among the CRC cases (n = 52), the well-known

CFS regions containing genes *FHIT* and *MACROD2* were most often identified (>40% of CRC cases).

CFSs are a decades-old observation [11], quickly gaining interest from the cancer field after reports of recurrent CFSs, with structural variation most often in the form of deletions [12–14]. Generally, CFSs are genomic regions that are late-replicating, AT-rich, often contain large genes, and showing tissue specific patterns (reviewed by [13]). In cancer, focal deletions are often associated with CFS; in a pan-cancer analysis [15] of 4934 cases, 70 recurrent focal deletions were reported, of which 22 were found in large genes and several genes known from CFS regions (*FHIT, WWOX, PDE4D, PARK2*). In terms of the mechanism driving the genome fragility of these regions, Wilson et al. distilled their observations in cell lines to a possible model involving a transcription-dependent double-fork failure (TrDoFF) [14]. In short, transcription of large genes possibly persists into the S-phase. This prevents firing of late (or dormant) replication origins that would otherwise be used to rescue stalled replication forks and thereby complete replication. In CFS, the stalled converging replication forks (i.e., a double-fork failure) create large, non-replicated regions which ultimately are resolved by deletion of the non-replicated DNA and duplications arising on the flanks, suggested to occur via template switching and microhomology-mediated break-induced replication (MMBIR) or alternative end joining [14,16].

We here aimed to have a detailed look into the role of the DELs in metastatic lesions of 429 mCRC patients, using WGS and corresponding RNA-sequencing data. We evaluated whether the instability at CFSs described in primary CRC persists in metastatic disease. We specifically studied the localization and size of DELs and DUPs, and included expression of the genes affected by these SVs. Lastly, we investigated the potential clinical consequences of these characteristics.

2. Results

2.1. Position and Size of Chromosomal Deletions in Metastatic mCRC

We previously described [9] density profiles of SVs (Figure 1a) in a large cohort of 429 metastases of colorectal cancer (mCRC) and noticed that the deletions (DELs) profile showed a peak at ~10 kb–1 Mb, which is not observed in metastatic prostate or breast cancer (Figure 1b). Furthermore, these DELs were not distributed randomly across the genome (Figure 1c), but instead showed a clear preference for specific chromosomal locations. Using a threshold of at least 150 events, we established 13 hotspot locations (see Supplementary Figure S1a and Table 1) which were covering, or, for all but one, located nearby known common fragile sites (CFSs) [17]. For all hotspot locations, many samples had more than a single DEL in the region, often overlapping (Table 1 and an example in Supplementary Figure S1b). We evaluated the sequence around the breakpoints of the DELs but were unable to find enriched motifs, besides the sequence logos showing AT-rich sequences (Supplementary Figure S1c), which is a well-known feature of CFSs (reviewed by [13]). In summary, the mCRC genome contains DELs of a specific size at specific chromosomal locations, linked to well-known CFSs.

We next used previously reported locations of several potentially relevant genomic features to investigate the reason behind the preference for the DEL size, including topologically associated domains (TADs) [18] and sites where initiation of replication was found to occur (Ini-seq peaks) [19]. TADs are structural features of genomic organization, where regions bordering TAD boundaries are thought to regulate gene expression. Both TADs and Ini-seq peaks often flanked the regions were the DELs were found. See for example *RBFOX1* in Figure 2a. Next, we systematically evaluated whether both breakpoints of a DEL (size 10 kb–1 Mb) were located between consecutive Ini-seq peaks or TADs (Figure 2b). Outside the 13 hotspot regions, 68% of DELs (6488 out of 9553) were located between consecutive Ini-seq peaks, whereas inside the hotspot regions, this was the case for 92% of DELs (4836 out of 5237, chi-sq $p = 1.29 \times 10^{-246}$, odds ratio 5.697). For TADs, 84% vs. 90% of DELs were in between TADs outside and inside hotspot regions, respectively (chi-square

$p = 5.96 \times 10^{-25}$), and with an odds ratio of 1.738, having a much smaller effect size compared to Ini-seq peaks. A multivariable logistic regression of DELs within Ini-seq peaks and TADs with being inside or outside hotspot region as outcome variable showed odds ratios of 5.43 and 1.14 (p-values 1.33×10^{-189} and 0.025), respectively, indicating that Ini-seq peaks have an exceedingly larger role in bordering DELs in the hotspot regions.

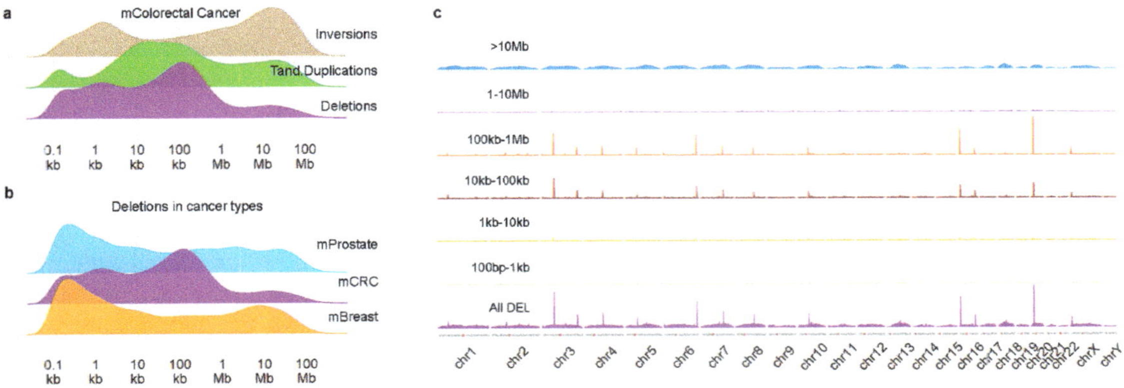

Figure 1. Deletions in mCRC show spatial and size preference. Density plots of structural variant in mCRC (**a**) and of deletions in 3 different metastatic cancers (**b**); (**c**) shows density plots by the indicated sizes across the genome. Figure 1a reproduced from Mendelaar et al., Nat Commun 2021, 12:574 (CC BY 4.0, https://creativecommons.org/licenses/by/4.0/, accessed on 1 December 2021) with modified colors.

Table 1. Regions with at least 150 DELs in mCRC.

Chromosome	Start Position of First DEL	End Position of Last DEL	Gene	Fragile Site (HumCFS)	No. of Samples with >1 DEL
1	48435860	50548351	*AGBL4*	~0.6 Mb 5′ of FRA1B	34
3	59572890	61536465	*FHIT*	FRA3B	168
3	174058837	175755903	*NAALADL2*	~6.7 Mb 5′ of FRA3C	80
4	91027502	92737250	*CCSER1*	FRA4E	85
5	58229103	59931219	*PDE4D*	FRA5H	55
6	161666892	163416771	*PRKN*	FRA6E	125
7	109283013	111357245	*IMMP2L*	FRA7K	76
8	89804552	91152321	*NBN* *	~3.2 Mb 5′ of FRA8B	36
10	52569645	54085262	*PRKG1*	~0.6 Mb 5′ of FRA10C	40
16	5690370	7815010	*RBFOX1*	~7 Mb 5′ of FRA16A	173
16	78099748	79260329	*WWOX*	FRA16D	61
20	13921064	16054649	*MACROD2*	~1.8 Mb 5′ of FRA20A	207
X	6511238	7782627	*PUDP STS*	~21.9 Mb 5′ of FRAXB	59

* Nearest gene. HumCFS is data from [17]. Coordinates by hg19 reference.

Since it is well known that CFS regions are late-replicating, it seems reasonable that replication origins (identified by the Ini-seq peaks) are of importance for SVs therein, and it is likely that the observed preference of DEL sizes in mCRC are restricted by the location of the bordering Ini-seq peaks. A histogram of the distances between all consecutive Ini-seq peaks (Figure 2c) showed a median of 24.9 kb (95% confidence interval (CI) 24.2–25.7 kb) with 95% of observations below 425 kb. However, the median distance between Ini-seq peaks at the hotspot regions is 1647 kb, which would be the upper bound if bordering replication origins determine the DELs size in CFS. Furthermore, assuming that converging replication forks starting from the bordering Ini-seq peaks are arrested at some point followed by deletion of the intervening region, DELs are expected to concentrate around the midpoint of the region. This assumption matched with the observations in our cohort; DELs within the 13 hotspot regions were much more localized toward the midpoint of the bordering Ini-seq peaks, compared to the more random localization of DELs outside the hotspot regions (Figure 2d).

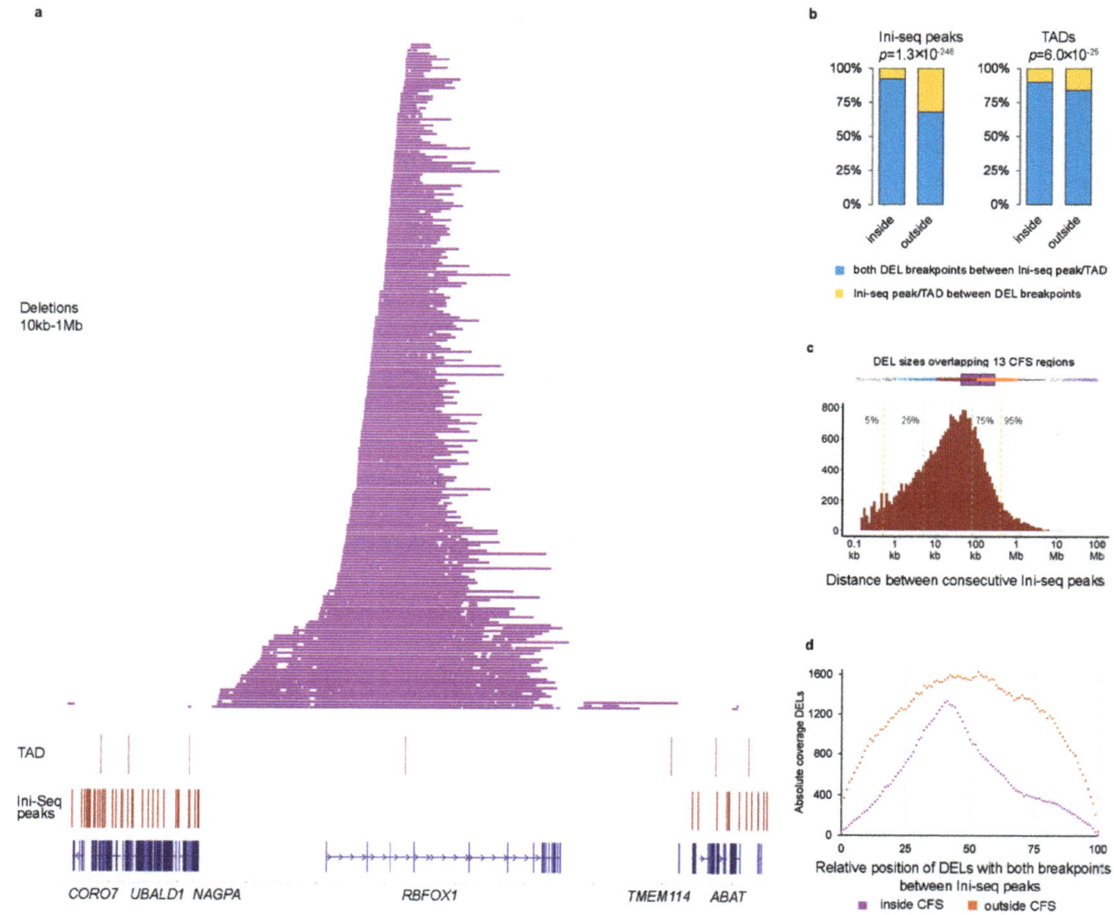

Figure 2. Origins of replication determine size of deletions in CFS. (**a**) Chromosome 16 region containing *RBFOX1*. Indicated are (bottom to top) the locations of genes, Ini-seq peaks, TADs, and DELs between 10 kb–1 Mb. (**b**) Bar graph showing enrichment of Ini-seq peaks/TADs inside CFS regions. *p*-values via Mann–Whitney *U*-test. (**c**) Top horizontal boxplot shows the size of all DELs within or overlapping the 13 CFS regions. DELs are colored by size according to the groups in (**c**). Bottom histogram shows the distribution of distance between 2 consecutive Ini-seq peaks. Vertical lines show the indicated percentiles. (**d**) Relative position of DELs between consecutive Ini-seq peaks.

2.2. Deletion and Duplication Patterns Concur with the Transcription-Dependent Double-Fork Failure Model

These observations suggest that the DELs may be a consequence of stalled replication forks, the reason for their stalling likely linked to the particular genomic milieu. A potential solution for these stalled replication forks was proposed by the transcription-dependent double-fork failure (TrDoFF) model [13,14], in which resolution of these stalled replication forks leads to DELs centering around the midpoint of the gene and duplication (DUP) at the flanks of the region. In support of this model, we find that DUPs (Figure 3a,b) are indeed also enriched at the same loci as DELs in mCRC, although with a much lower prevalence and of smaller size (Supplementary Figure S2). The exact location of DUPs and DELs is exemplified in Figure 3c, showing the events at the FHIT locus. Systematic evaluation of the genes located in the hotspot CFS regions, taking the transcription direction into account, showed that DELs are indeed centered around the midpoint of the genes, with DUPs more present at the flanks of the region (Figure 3d).

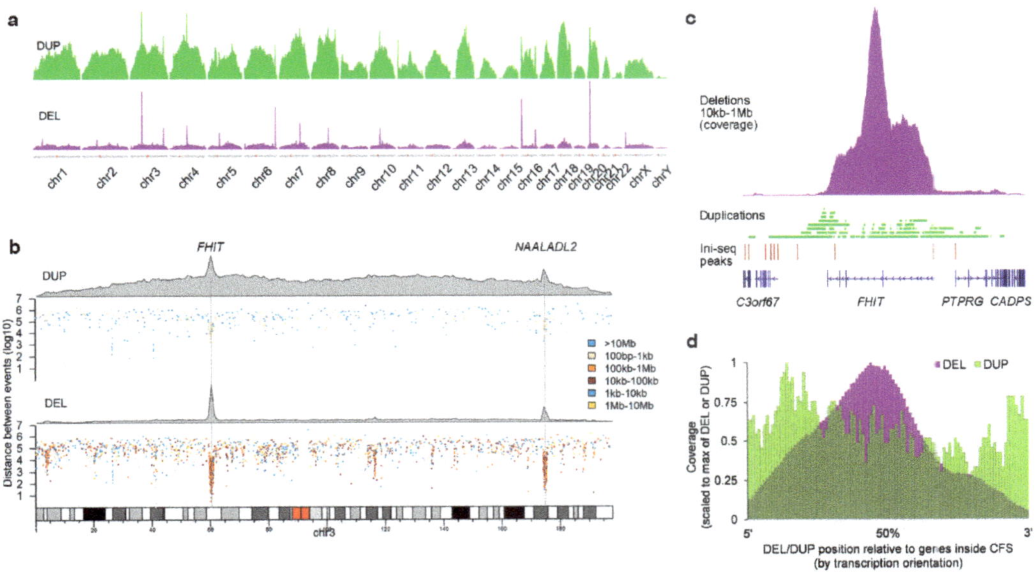

Figure 3. Spatial and size preference for duplications in mCRC. (**a**) Density plots of DELs and DUPs in mCRC. Each track is scaled to the maximum density within that track. (**b**) DELs and DUPs in chromosome 3, each showing a density plot and a rainfall plot to indicate sizes. (**c**) More detailed DEL/DUP locations for the *FHIT* gene. (**d**) Relative position of DELs/DUPs in hotspot CFS regions, corrected for transcription orientation of the genes.

Additional support for an actual role of TrDoFF in mCRC is provided, employing available gene expression data (Figure 4). For example, *NAALADL2* (located in FRA3C, genome size 951 kb), one of the genes with many DELs, was expressed and located between Ini-seq peaks, while the neighboring *NLGN1* gene, also large (size 898 kb) and located between Ini-seq peaks, but having no accompanying DELs, was not expressed. We verified this observation genome-wide by selecting large genes (>500 kb) within and outside the 13 CFS regions but positioned between Ini-seq peaks (n = 147 genes). We showed that the median gene expression level of the genes inside the hotspot regions was significantly higher (Mann–Whitney p = 0.0013) than that of the genes outside hotspot regions (Figure 4b).

Of note, not all 13 hotspot regions contained a single large gene at the location where DELs were observed (Table 1 and Supplementary Figure S3). For the hotspot located on chromosome (chr) X, multiple genes are located within consecutive Ini-seq peaks, with DELs and DUPs, respectively, centering and flanking two genes (*PUDP* and *STS*). The hotspot region on chr 8 has no gene annotated to the region where DELs/DUPs are located. An uncharacterized noncoding RNA LOC105375631 is located nearby, and visual inspection of RNA-seq bam files shows some mapped reads there (data not shown).

In summary, the specific DEL size and DEL/DUP position seen in mCRC appeared to be guided by two observations: (1) the origins of replication flanking the CFS regions and (2) active transcription. The TrDoFF model proposes that the combination of late replication and transcription of large genes gives rise to double-fork failure. The disentanglement of that complex yields DELs and DUPs at specific locations which precisely fit with the observations presented here.

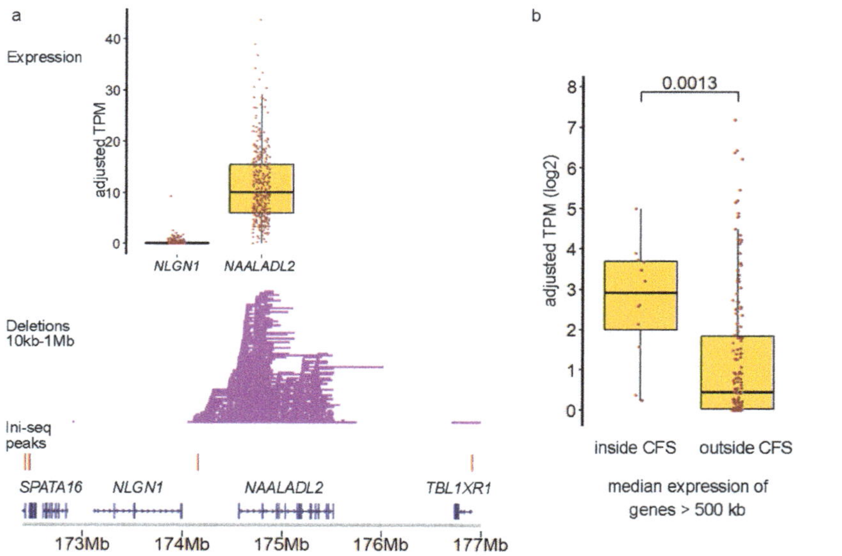

Figure 4. Transcription drives deletions in CFS regions. (**a**) Detailed plot of the *NAALADL2* and *NLGN1* genes, both large genes that locate between consecutive Ini-seq peaks, showing DELs occurring only in the expressed *NAALADL2* gene. (**b**) Boxplot of median expression levels of genes inside CFS vs. large (>500 kb) genes outside CFS. *p*-Value via Mann–Whitney *U*-test.

2.3. Cells with DELs in CFS Retain Transcription of the Locus

For samples with available RNA-seq data, we searched for sequence reads that corroborate the DEL reported at DNA level in that sample, by identifying reads that cross the junction created by the DEL. In total, 305 sequence reads were identified in 44 samples (Figure 5). Though often only a single read showed the presence of the DEL, in all 13 CFS regions multiple samples showed evidence of such reads. Each of these sequence-reads were uniquely aligned to the genome, with part of the read mapping immediately before the 5′ position of a DEL with the remaining part of the sequence mapping directly after the 3′ position of the DEL. Supplementary Figure S4 shows an example of these in *WWOX* in a single sample.

2.4. Clinical Implications

We considered the number of DELs (10 kb–1 Mb) within the 13 hotspot regions as a biomarker for the ability to circumvent stalled replication forks. We hypothesized that such a mechanism would act on any double-fork failure, not necessarily transcription-dependent or just those in CFS regions. Since we previously observed that loss of some of the genes in hotspot regions (e.g., *WWOX*, *PARK2*) were associated with response to platinum-containing therapy [9], we reasoned that the mechanism that resolves stalled replication forks may also be able to resolve platinum-caused cross-links in DNA that potentially lead to stalled forks as well. We considered samples >20 DELs in the 13 hotspot regions (see Supplementary Figure S5) as resolved stalled fork positive (RSF+).

In total, 118 patients received platinum-containing therapy after biopsy (14 of whom are RSF+, see Table 2), and associating RSF groups with progression-free survival (PFS) showed that this was significantly shorter in RSF+ patients (Figure 6, left panel, log rank

$p = 0.0001$). A subset of patients that were given platinum-containing therapy after biopsy were also previously treated with this agent (i.e., prior to biopsy of the metastasis that was sequenced), a clinical decision often taken when the patient had a relatively good response (defined as the interval between last platinum-containing therapy and time of progression >6 months). A multivariable analysis to test the association of RSF status, prior platinum therapy, and the number of prior-treatment lines with PFS on subsequent platinum-containing therapy showed a relative hazard ratio (HR) for RSF+ of 3.01 ($p = 0.0029$, 95% CI 1.46–6.21), HR = 0.83 ($p = 0.10$, 95% CI 0.20–1.15) for prior platinum therapy and an HR of 1.58 ($p = 0.038$, 95% CI 1.03–2.42) for the number of prior-treatment lines.

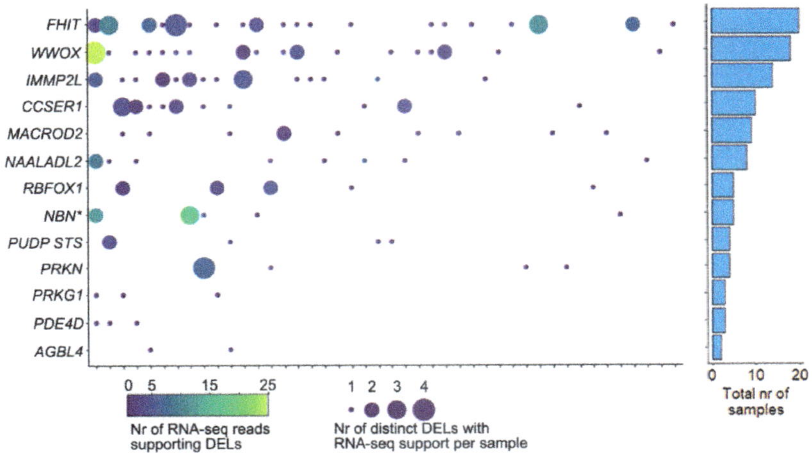

Figure 5. RNA-seq reads supporting DELs. The left panel shows the number of RNA-seq reads that support a DEL in one of the CFS regions. X-axis shows 44 samples, y-axis the 13 CFS regions (* indicates that *NBN* is the nearest gene). The color of the bubble indicates the total number of reads observed for all DELs in that sample/CFS region, and the size of the bubble indicates the total number of DELs with supporting RNA-seq reads in that sample/CFS region. The right panel shows the total number of samples that have RNA-seq support of DELs in the CFS region.

Table 2. Patients treated with platinum-containing therapy.

Prior Treatment	Total	RSF+	RSF-
All	118	14	104
Platinum-containing	18	2	16
Other	19	2	17
None	81	10	71

Lastly, patients who did not receive any prior systemic treatment before biopsy also showed a significantly shorter PFS on platinum-containing therapy in RSF+ patients (Figure 6, right panel, log rank $p = 0.0007$). The group of patients with any prior treatment (platinum or other) was too small to perform a statistically meaningful analysis, with only two RSF+ patients in either prior containing platinum or prior other. These four patients did show a similar trend with a PFS < 250 days.

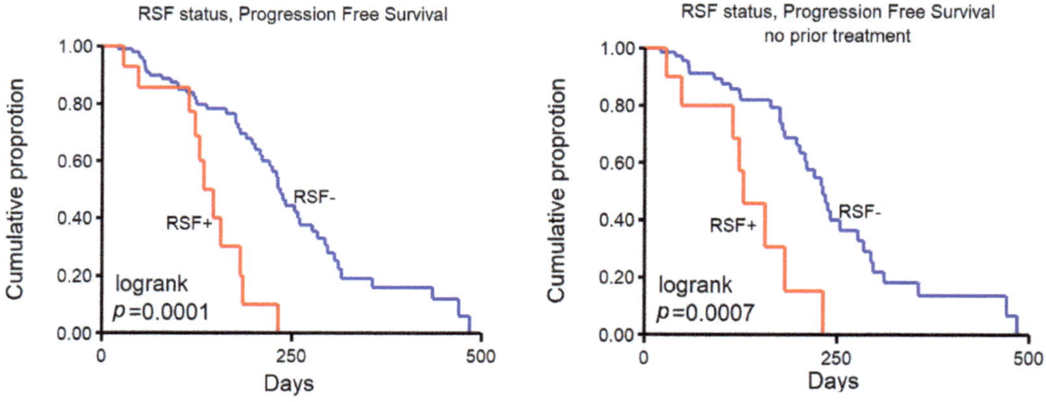

Figure 6. Number of DELs and progression-free survival. Kaplan–Meier progression-free survival curves of patients treated with platinum-containing therapy. Samples were grouped in those >20 DELs, labeled as resolved stalled fork positive (RSF+) vs. <20 (RSF-). Left panel shows all patients, and right panel shows patients with no prior therapy.

3. Discussion

We here describe an in-depth analysis of DELs in metastatic lesions of 429 CRC patients. The density profile of DELs in mCRC is distinct from those in metastatic breast- and prostate-cancer but bears resemblance to a profile seen in, e.g., primary esophagus and pancreas cancer [10], with the top affected CFS loci and their connected genes (*FHIT*, *MACROD2* and *WWOX*) also among the reported regions here. Using their pan-cancer set of primary tumors, Li et al. reported 18 CFS regions [10], of which 10 out of our 13 regions overlapped. The observed DEL and DUP patterns fully support the TrDoFF mechanism, whereby these SVs are generated in CFS [14]. In short, the model, which is based on experimental data in in vitro model systems, suggests that in large genes during replication stress, transcription is still active in the S-phase, precluding activation of dormant replication origins to resolve stalled replication forks. As a result, in late-replicating regions that are actively transcribed, converging stalled replication forks cannot be rescued. To resolve these, template switching across both stalled forks gives rise to DELs while occasional re-initiation of replication results in DUPs near the origins of replication. Another frequently observed aspect of the phenotype is that within a single tumor lesion, multiple, often overlapping, DELs are present within the same gene/CFS. This is likely due to clonal heterogeneity where subpopulations of clonal cells each have their own specific deletion. The RNA-seq data agree with this notion, since sequence reads can positively identify multiple and overlapping DELs within the same sample. We hypothesize that in different cells of a tumor, the transcription complex is probably stochastically located at different positions in the gene, stalling the converging replication forks at different locations, resulting in distinct DELs. The fact that CFS regions are not all entirely lost in CRC during progression into these metastatic lesions implies that the TrDoFF mechanism is not active in all cancer cells or in every cell division. The events we do observe are those derived from expansion of individual cells in which the initial, and specific, event occurred.

To the best of our knowledge, we are the first to match observations in a large cohort of metastatic CRC patients with four essential features of the TrDoFF model: (i) sparsity of replication origins, (ii) transcription, and the (iii) presence and (iv) position of DELs and DUPs. For the latter three features, we were able to use in-house data only; for the sparsity of origins we additionally required public data (Ini-seq peaks) which were based on a replication origin mapping technique whereby newly replicated DNA is labeled and sequenced [19]. The position of these Ini-seq peaks show that DELs are enriched between two consecutive Ini-seq peaks in CFS regions, which would be in line with CFS regions having no active origin of replication to rescue any stalled replication forks converging from the flanks of the CFS regions. Of note, the locations of the Ini-seq peaks in CFS regions thus provide an upper bound for the size of the DELs observed in mCRC. Lastly, it is important to realize, as also addressed by the original authors [13,14], that a key event in the formation of DELs in CFS is that transcription appears as the reason for the absence of an active replication origin that could rescue stalled replication forks. There may be underlying causes other than transcription for the inactivation of origins, or other mechanisms responsible for double-fork failure or incomplete replication in general.

Indirect clues for other contributing factors come from the observation that patients that received platinum-containing treatment prior to removal of the metastatic biopsy showed a significantly higher number of DELs in CFS. Platinum compounds induce crosslinks between two bases, either GG or AG, in DNA [20,21] while interstrand crosslinks potentially block transcription and replication [22]. We argue that in mCRC patients the mechanism that is used to resolve the transcription-dependent double-fork failure may also act on stalled forks caused by platinum crosslinks. The observation that PFS on platinum-containing therapy is significantly shorter in patients with a high number of DELs (RSF+) is in line with this hypothesis. The clinical decision to administer platinum-containing therapy may be augmented by only including RSF- patients, since the benefit of platinum-containing therapy for RSF+ patients seems limited. Prospective trials where patients are properly stratified during inclusion are required to evaluate the true clinical applicability of the RSF status.

In conclusion, we here present data that show that CFSs are still present in mCRC, providing an explanation for why DELs observed in mCRC are of a certain size, and that the observations agree with a model of transcription-dependent double-fork failure as modus operandi for DELs (and DUPs) in CFS. This ability to resolve stalled replication forks is associated with resistance to platinum-based therapy, which may provide new prospects for clinical decision making.

4. Materials and Methods

4.1. Study Cohort

The cohort consists of 429 metastatic biopsies taken from colorectal cancer patients gathered as part of the Center for Personalized Cancer Treatment (CPCT) consortium (CPCT-02 Biopsy Protocol, ClinicalTrial.gov no. NCT01855477, 16 May 2013), which was approved by the medical ethics committee of the University Medical Center Utrecht, the Netherlands [6]. All patients gave explicit consent for whole-genome sequencing and data sharing for cancer research purposes. Additional characteristics were previously described [9], but in short, whole-genome sequencing (WGS) of paired tumor/normal was performed in all cases. Raw sequencing data were processed using bcl2fastq (versions 2.17 to 2.20) and mapped to the human reference genome GRCh37 using BWA-mem v0.7.5a (https://github.com/lh3/bwa, accessed on 1 December 2021). Of all tumor biopsies, 98% had a coverage of at least 30 × (95% with >60 × coverage), whereas for the normal blood, 98% had >10 × coverage and 94% >20 × coverage. Structural variants were called using GRIDSS v1.8.0 [23]. RNA-sequencing data were available for 343 samples and RNA-seq was performed as described previously [9] with gene expression data (adjusted TPM) obtained using Isofox (https://github.com/hartwigmedical/hmftools/tree/master/isofox, accessed on 1 December 2021). Metastatic patients starting any new systemic treatment were included, and the particular treatment given to participating patients was decided by the local clinicians. The types and lines of systemic treatment given before the biopsy was taken were recorded; in total, 241 patients received a platinum-containing therapy before biopsy, mostly Oxaliplatin in combination with Capecitabine (n = 231), while 124 patients were treatment-naïve at the time of biopsy. Furthermore, response to treatment after biopsy was evaluated; 118 patients received a platinum-containing therapy after the biopsy was taken with available progression-free survival data (again, mostly (n = 85) in the form of a combination of Oxaliplatin + Capecitabine with/without Bevacizumab).

4.2. Statistics

For categorical data, a Pearson's chi-squared test was used while continuous variables were evaluated using a Mann–Whitney U-test (MWU). All statistical tests were two-sided and considered statistically significant when $p < 0.05$. Stata 13.0 (StataCorp, College Station, TX, USA) and R (v3.6.0) were used for the statistical analyses. R package "mcp" (https://github.com/lindeloev/mcp/, accessed on 1 December 2021) was used to perform piecewise linear regression in order to find change points in the number of deletions per sample. The midpoint between the first two changepoints was selected to indicate samples with a high nr of DELs (>20). KaryoploteR [24] was used for visualizations. The sequence logo was generated via WebLogo [25].

Supplementary Materials: The Supplementary Materials can be downloaded at: https://www.mdpi.com/article/10.3390/ijms23095080/s1.

Author Contributions: Conceptualization, M.S. and J.W.M.M.; methodology, M.S. and S.M.W.; formal analysis, M.S. and S.M.W.; resources, J.W.M.M.; writing—original draft preparation, M.S.; writing—review and editing, M.S., S.M.W. and J.W.M.M.; visualization, M.S.; supervision, J.W.M.M.; All authors have read and agreed to the published version of the manuscript.

Funding: This research was funded by the Cancer Genomics Netherlands (CGC.nl) through a grant from the Netherlands Organization of Scientific Research (NWO).

Institutional Review Board Statement: The study was conducted according to the guidelines of the Declaration of Helsinki, and approved by the Medical Ethics Committee of the University Medical Center Utrecht, the Netherlands (ClinicalTrial.gov no. NCT01855477, 16 May 2013).

Informed Consent Statement: Informed consent was obtained from all subjects involved in the study.

Data Availability Statement: All data (WGS, RNA-seq, and clinical data) used in this study are made available by the Hartwig Medical Foundation (Dutch non-profit biobank organization) after signing a license agreement stating that data cannot be made publicly available via third party organizations. Therefore, the data are available under data request code DR-058 and can be requested by contacting the Hartwig Medical Foundation (https://www.hartwigmedicalfoundation.nl/applying-for-data/).

Acknowledgments: This publication uses data made available through the Hartwig Medical Foundation and the Center for Personalized Cancer Treatment (CPCT), The Netherlands. We especially appreciate all participating patients and would like to furthermore thank all local principal investigators, medical specialists, and nurses of all contributing centers for their help with patient accrual.

Conflicts of Interest: The authors declare no conflict of interest.

References

1. The ICGC/TCGA Pan-Cancer Analysis of Whole Genomes Consortium. Pan-cancer analysis of whole genomes. *Nature* **2020**, *578*, 82–93. [CrossRef] [PubMed]
2. Alexandrov, L.B.; Kim, J.; Haradhvala, N.J.; Huang, M.N.; Tian Ng, A.W.; Wu, Y.; Boot, A.; Covington, K.R.; Gordenin, D.A.; Bergstrom, E.N.; et al. The repertoire of mutational signatures in human cancer. *Nature* **2020**, *578*, 94–101. [CrossRef] [PubMed]
3. Alexandrov, L.B.; Nik-Zainal, S.; Wedge, D.C.; Aparicio, S.A.; Behjati, S.; Biankin, A.V.; Bignell, G.R.; Bolli, N.; Borg, A.; Borresen-Dale, A.L.; et al. Signatures of mutational processes in human cancer. *Nature* **2013**, *500*, 415–421. [CrossRef] [PubMed]
4. Nik-Zainal, S.; Alexandrov, L.B.; Wedge, D.C.; Van Loo, P.; Greenman, C.D.; Raine, K.; Jones, D.; Hinton, J.; Marshall, J.; Stebbings, L.A.; et al. Mutational processes molding the genomes of 21 breast cancers. *Cell* **2012**, *149*, 979–993. [CrossRef] [PubMed]
5. Angus, L.; Smid, M.; Wilting, S.M.; van Riet, J.; Van Hoeck, A.; Nguyen, L.; Nik-Zainal, S.; Steenbruggen, T.G.; Tjan-Heijnen, V.C.G.; Labots, M.; et al. The genomic landscape of metastatic breast cancer highlights changes in mutation and signature frequencies. *Nat. Genet.* **2019**, *51*, 1450–1458. [CrossRef]
6. Priestley, P.; Baber, J.; Lolkema, M.P.; Steeghs, N.; de Bruijn, E.; Shale, C.; Duyvesteyn, K.; Haidari, S.; van Hoeck, A.; Onstenk, W.; et al. Pan-cancer whole-genome analyses of metastatic solid tumours. *Nature* **2019**, *575*, 210–216. [CrossRef]
7. van Dessel, L.F.; van Riet, J.; Smits, M.; Zhu, Y.; Hamberg, P.; van der Heijden, M.S.; Bergman, A.M.; van Oort, I.M.; de Wit, R.; Voest, E.E.; et al. The genomic landscape of metastatic castration-resistant prostate cancers reveals multiple distinct genotypes with potential clinical impact. *Nat. Commun.* **2019**, *10*, 5251. [CrossRef]
8. Christensen, S.; Van der Roest, B.; Besselink, N.; Janssen, R.; Boymans, S.; Martens, J.W.M.; Yaspo, M.L.; Priestley, P.; Kuijk, E.; Cuppen, E.; et al. 5-Fluorouracil treatment induces characteristic T > G mutations in human cancer. *Nat. Commun.* **2019**, *10*, 4571. [CrossRef]

9. Mendelaar, P.A.J.; Smid, M.; van Riet, J.; Angus, L.; Labots, M.; Steeghs, N.; Hendriks, M.P.; Cirkel, G.A.; van Rooijen, J.M.; Ten Tije, A.J.; et al. Whole genome sequencing of metastatic colorectal cancer reveals prior treatment effects and specific metastasis features. *Nat. Commun.* **2021**, *12*, 574. [CrossRef]
10. Li, Y.; Roberts, N.D.; Wala, J.A.; Shapira, O.; Schumacher, S.E.; Kumar, K.; Khurana, E.; Waszak, S.; Korbel, J.O.; Haber, J.E.; et al. Patterns of somatic structural variation in human cancer genomes. *Nature* **2020**, *578*, 112–121. [CrossRef]
11. Glover, T.W.; Berger, C.; Coyle, J.; Echo, B. DNA polymerase alpha inhibition by aphidicolin induces gaps and breaks at common fragile sites in human chromosomes. *Hum. Genet.* **1984**, *67*, 136–142. [CrossRef] [PubMed]
12. Arlt, M.F.; Durkin, S.G.; Ragland, R.L.; Glover, T.W. Common fragile sites as targets for chromosome rearrangements. *DNA Repair* **2006**, *5*, 1126–1135. [CrossRef] [PubMed]
13. Glover, T.W.; Wilson, T.E.; Arlt, M.F. Fragile sites in cancer: More than meets the eye. *Nat. Rev. Cancer* **2017**, *17*, 489–501. [CrossRef]
14. Wilson, T.E.; Arlt, M.F.; Park, S.H.; Rajendran, S.; Paulsen, M.; Ljungman, M.; Glover, T.W. Large transcription units unify copy number variants and common fragile sites arising under replication stress. *Genome Res.* **2015**, *25*, 189–200. [CrossRef]
15. Zack, T.I.; Schumacher, S.E.; Carter, S.L.; Cherniack, A.D.; Saksena, G.; Tabak, B.; Lawrence, M.S.; Zhsng, C.Z.; Wala, J.; Mermel, C.H.; et al. Pan-cancer patterns of somatic copy number alteration. *Nat. Genet.* **2013**, *45*, 1134–1140. [CrossRef]
16. Hastings, P.J.; Ira, G.; Lupski, J.R. A microhomology-mediated break-induced replication model for the origin of human copy number variation. *PLoS Genet.* **2009**, *5*, e1000327. [CrossRef] [PubMed]
17. Kumar, R.; Nagpal, G.; Kumar, V.; Usmani, S.S.; Agrawal, P.; Raghava, G.P.S. HumCFS: A database of fragile sites in human chromosomes. *BMC Genom.* **2019**, *19*, 985. [CrossRef]
18. Dixon, J.R.; Selvaraj, S.; Yue, F.; Kim, A.; Li, Y.; Shen, Y.; Hu, M.; Liu, J.S.; Ren, B. Topological domains in mammalian genomes identified by analysis of chromatin interactions. *Nature* **2012**, *485*, 376–380. [CrossRef]
19. Langley, A.R.; Graf, S.; Smith, J.C.; Krude, T. Genome-wide identification and characterisation of human DNA replication origins by initiation site sequencing (ini-seq). *Nucleic Acids Res.* **2016**, *44*, 10230–10247. [CrossRef]
20. Boot, A.; Huang, M.N.; Ng, A.W.T.; Ho, S.C.; Lim, J.Q.; Kawakami, Y.; Chayama, K.; Teh, B.T.; Nakagawa, H.; Rozen, S.G. In-depth characterization of the cisplatin mutational signature in human cell lines and in esophageal and liver tumors. *Genome Res.* **2018**, *28*, 654–665. [CrossRef]
21. Harrington, C.F.; Le Pla, R.C.; Jones, G.D.; Thomas, A.L.; Farmer, P.B. Determination of cisplatin 1,2-intrastrand guanine-guanine DNA adducts in human leukocytes by high-performance liquid chromatography coupled to inductively coupled plasma mass spectrometry. *Chem. Res. Toxicol.* **2010**, *23*, 1313–1321. [CrossRef] [PubMed]
22. Roy, U.; Scharer, O.D. Involvement of translesion synthesis DNA polymerases in DNA interstrand crosslink repair. *DNA Repair* **2016**, *44*, 33–41. [CrossRef] [PubMed]
23. Cameron, D.L.; Schroder, J.; Penington, J.S.; Do, H.; Molania, R.; Dobrovic, A.; Speed, T.P.; Papenfuss, A.T. GRIDSS: Sensitive and specific genomic rearrangement detection using positional de Bruijn graph assembly. *Genome Res.* **2017**, *27*, 2050–2060. [CrossRef] [PubMed]
24. Gel, B.; Serra, E. karyoploteR: An R/Bioconductor package to plot customizable genomes displaying arbitrary data. *Bioinformatics* **2017**, *33*, 3088–3090. [CrossRef]
25. Crooks, G.E.; Hon, G.; Chandonia, J.M.; Brenner, S.E. WebLogo: A sequence logo generator. *Genome Res.* **2004**, *14*, 1188–1190. [CrossRef]

International Journal of
Molecular Sciences

MDPI

Article

HOXB9 Overexpression Promotes Colorectal Cancer Progression and Is Associated with Worse Survival in Liver Resection Patients for Colorectal Liver Metastases

Eirini Martinou [1,2,*], Carla Moller-Levet [3], Dimitrios Karamanis [4,5], Izhar Bagwan [6] and Angeliki M. Angelidi [7,*]

1 Department of Hepatobiliary and Pancreatic Surgery, Royal Surrey County Hospital, Guildford GU2 7XX, UK
2 School of Biosciences and Medicine, Faculty of Health and Medical Sciences, University of Surrey, Guildford GU2 7HX, UK
3 Department of Bioinformatics, Faculty of Health and Medical Sciences, University of Surrey, Guildford GU2 7HX, UK; c.moller-levet@surrey.ac.uk
4 Department of Economics, University of Piraeus, 185 34 Piraeus, Greece; karamanis@unipi.gr
5 Department of Health Informatics, Rutgers School of Health Professions, Newark, NJ 07107, USA
6 Department of Histopathology, Royal Surrey County Hospital, Guildford GU2 7XX, UK; izhar.bagwan@nhs.net
7 Department of Medicine, Beth Israel Deaconess Medical Centre, Harvard Medical School, Boston, MA 02215, USA
* Correspondence: eirini.martinou@nhs.net (E.M.); aangelid@broadinstitute.org (A.M.A.)

Abstract: As is known, HOXB9 is an important factor affecting disease progression and overall survival (OS) in cancer. However, its role in colorectal cancer (CRC) remains unclear. We aimed to explore the role of HOXB9 in CRC progression and its association with OS in colorectal liver metastases (CRLM). We analysed differential *HOXB9* expression in CRC using the Tissue Cancer Genome Atlas database (TCGA). We modulated *HOXB9* expression in vitro to assess its impact on cell proliferation and epithelial-mesenchymal transition (EMT). Lastly, we explored the association of HOXB9 protein expression with OS, using an institutional patient cohort ($n = 110$) who underwent liver resection for CRLM. Furthermore, *HOXB9* was upregulated in TCGA-CRC ($n = 644$) vs. normal tissue ($n = 51$) and its expression levels were elevated in *KRAS* mutations ($p < 0.0001$). In vitro, HOXB9 overexpression increased cell proliferation ($p < 0.001$) and upregulated the mRNA expression of EMT markers (*VIM, CDH2, ZEB1, ZEB2, SNAI1* and *SNAI2*) while downregulated *CDH1*, ($p < 0.05$ for all comparisons). Conversely, *HOXB9* silencing disrupted cell growth ($p < 0.0001$). High HOXB9 expression (HR = 3.82, 95% CI: 1.59–9.2, $p = 0.003$) was independently associated with worse OS in CRLM-HOXB9-expressing patients after liver resection. In conclusion, HOXB9 may be associated with worse OS in CRLM and may promote CRC progression, whereas HOXB9 silencing may inhibit CRC growth.

Keywords: HOX; *HOXB9*; colorectal cancer; colorectal liver metastases

Citation: Martinou, E.; Moller-Levet, C.; Karamanis, D.; Bagwan, I.; Angelidi, A.M. *HOXB9* Overexpression Promotes Colorectal Cancer Progression and Is Associated with Worse Survival in Liver Resection Patients for Colorectal Liver Metastases. *Int. J. Mol. Sci.* **2022**, *23*, 2281. https://doi.org/10.3390/ijms23042281

Academic Editor: Sun-Hee Leem

Received: 28 December 2021
Accepted: 16 February 2022
Published: 18 February 2022

1. Introduction

Colorectal cancer (CRC) is the most common gastrointestinal malignancy and the third leading cause of cancer-related deaths worldwide [1]. Alarming evidence shows that its incidence is rising, especially in the younger population [1]. Despite significant advances in diagnostic and therapeutic strategies, the prognosis remains poor because most patients develop synchronous or metachronous colorectal liver metastases (CRLM) [2]. The development of metastatic disease indicates that cancer cells are not entirely eradicated by current therapies and are the primary cause of cancer-related mortality [2]. CRC is a highly heterogeneous disease which led to the formation of an international consortium in 2015, proposing the molecular classification of CRC into four categories based on transcriptomic characteristics (consensus of molecular subtypes) [3]. In the era of precision medicine, recognising that transcriptomics represents molecular data that are ultimately linked to tumour biology and clinical behaviour, has led to a paradigm shift in the research toward identifying novel transcription factors (TFs) which are linked to the aggressive behaviour of CRC [3]. TFs are important not only in the pathogenesis of CRC but also in the progression and formation of metastases [4]. They also seem to have a prognostic role in overall survival (OS). Thus, they may serve as useful biomarkers and therapeutic targets for the treatment of primary and metastatic CRC [4].

Homeobox containing (HOX) factors are a family of TFs characterised as master regulators of embryonic development that play a pivotal role in regulating cellular functions such as proliferation, invasion, and migration [5]. Humans have 39 HOX genes in their genome, which are organised into four chromosomal clusters (A, B, C, and D), and their importance in cancer has been reported in many studies as alterations in their expression have been found to affect cancer progression [6,7]. The *HOXB9* gene belongs to the HOX family and has been identified as a critical TF involved in numerous human solid tumours as its aberrant expression contributes to tumour growth, progression, and metastases [8]. Several studies have reported that HOXB9 overexpression increases the metastatic potential of cancer cells by activating an important process called epithelial-mesenchymal transition (EMT) [8]. In CRC, EMT is characterised by the loss of epithelial markers (E-cadherin) with the subsequent upregulation of mesenchymal markers (N-cadherin and vimentin) which allows cancer cells to obtain invasive and metastatic potential [9]. Additionally, high HOXB9 protein levels have also been reported by many studies to be associated with a poor prognosis in patients with lung, breast, hepatocellular, and pancreatic carcinoma [8]. In CRC, few studies have reported contradictory findings regarding the HOXB9 prognostic role and function in CRC progression [10]. However, no studies have examined colorectal liver metastases (CRLM) [11–13]. Therefore, this study aimed to investigate the impact of HOXB9 on CRC progression and its prognostic importance in CRLM.

2. Materials and Methods

2.1. Gene Expression Bioinformatics Analysis

To investigate the difference in *HOXB9* expression between cancer and normal tissues in CRC, gene expression data from the Cancer Genome Atlas (TCGA) for colon (TCGA-COAD) and rectal adenocarcinoma (TCGA-READ) were downloaded from the Genomic Data Commons Data Portal (https://portal.gdc.cancer.gov/, (accessed on 23 May 2019)). The edgeR Bioconductor package (v. 3.24.3) was used for data pre-processing and differential expression analyses [14]. A negative binomial generalised log-linear model was fitted to the read counts for *HOXB9*, and likelihood ratio tests for tumour vs. normal tissue differences were conducted using the R package edgeR [14,15]. *p*-values were adjusted for multiple comparisons using the Benjamini-Hochberg (BH) approach [16]. The UALCAN online platform (http://ualcan.path.uab.edu/, (accessed on 23 December 2021) was used to compare the transcriptional levels of *HOXB9* in CRC compared to other types of cancers and the GEPIA tool (http://gepia.cancer-pku.cn/index.html, (accessed on 27 December 2021)) was used to compare the transcriptional levels of *HOXB9* in CRC in comparison with the rest of the *HOX* genes [17–19]. Lastly, the OmicSoft Suite with the integrated OncoLand database (Qiagen, Manchester, UK) was used to assess the *HOXB9* gene expression levels in CRC mutant versus wild-type for the top three somatic mutations in CRC which were identified using the COSMIC database (https://cancer.sanger.ac.uk/cosmic, (accessed on 22 June 2020)) [20].

2.2. Gene Expression Editing Mechanistic Studies

We initially used the STRING server (https://string-db.org/, (accessed on 23 December 2021)) to define HOXB9 functional partners and further explore its potential action. For gene-expression-editing studies, the human HCT116 colon adenocarcinoma cell line was obtained from the American Type Culture Collection (ATCC, Manassas, VA,

USA) [21]. The plasmid vectors (OriGene, Köln, Germany), pCMV6-AC-HOXB9-GFP (RG213735), and pCMV6-AC-GFP (PS100010) were used as *HOXB9*, thereby overexpressing negative control vectors, respectively. For *HOXB9* gene silencing, the Silencer® Select small interfering RNA (siRNA) (Life Technologies, Loughborough, UK) against *HOXB9* was used. The non-targeting Silencer® Select siRNA#1 was used as a negative control. The overexpression and knockdown of HOXB9 efficiency were evaluated at the mRNA and protein levels using RT-qPCR and Western blotting, respectively. The outcomes were cell proliferation and fold-change in the RNA expression of EMT markers between the HOXB9 overexpressing cell group and the control group (*VIM, CDH1, CDH2, ZEB1, ZEB2, SNAI1, SNAI2, TWIST*) [22]. Cell proliferation was assessed using the Alamar Blue proliferation assay for up to 120 h post-transfection [22,23]. A detailed methodology of the in vitro studies is provided in Supplementary Digital Content 1.doc (Supplementary Material S1) and the flow chart is shown in Figure 1. The normality of the data was evaluated using the Shapiro–Wilk test. Unpaired two-tailed Student's *t*-test (for normally distributed values) or Mann–Whitney test (for non-normally distributed values) was used to compare differences between control and treated groups using GraphPad Prism 8 and SPSS v27.

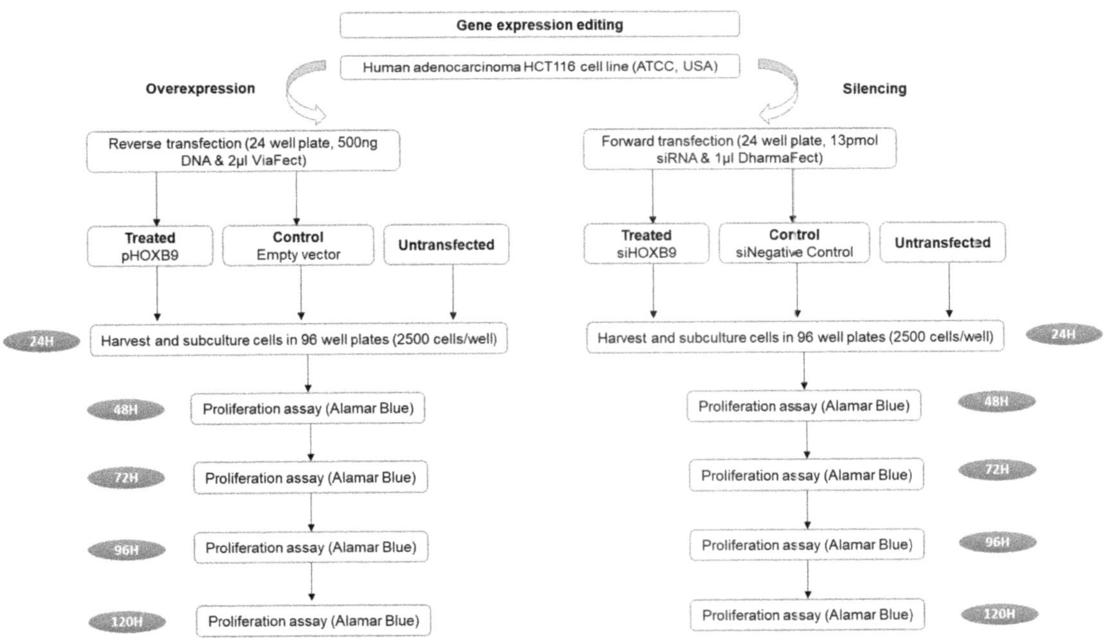

Figure 1. Flow chart of the in vitro experimental studies. (ATCC: American Type Culture Collection, H: hours).

2.3. In Silico Transcriptional Regulation Prediction of HOXB9 and Gene Set Enrichment Analysis

We used the Cistrome Data Browser (http://dbtoolkit.cistrome.org/, (accessed on 20 December 2021)) to identify *HOXB9* putative regulators to further dissect its action with regard to CRC proliferation. A gene set enrichment analysis (GSEA) of the gene list identified through the Cistrome DB was performed on the Enrichr server (https://maayanlab.cloud/Enrichr/, (accessed on 2 December 2021)) to identify potentially related biological processes.

2.4. Patient Tissue Samples, Clinicopathological Variables and Immunohistochemistry

Approval from the National Research Ethics Committee (Brighton and Sussex REC, Southcoast, 09/H1103/50/AM05) was obtained for the retrospective use of archived formalin-fixed paraffin-embedded (FFPE) human tissue. Available FFPE specimens from patients (n = 211) who underwent liver resection for CRLM between 2007 and 2014 were obtained from the institutional archive-management service (http://www.cellnass.com, (accessed on 15 February 2019)). Demographic, clinicopathological, and treatment-related variables were collected from institutional electronic records. Survival data were obtained using the NHS Summary Care Record (SCR) electronic system (NHS Digital, https://digital.nhs.uk/spine, (accessed on 28 October 2020)). Clinicopathological variables were defined based on the Tumour-Node-Metastases staging system (Table 1) [24,25].

Table 1. Definition of clinicopathological variables in patients with CRLM.

Variable	Definition
Age (years)	[Date of Operation–Date of Birth]
T	T1–T4, Tumour depth as per American Joint Committee on Cancer (AJCC) 8th edition
N	N0, N1, N2, Lymph nodal invasion as per AJCC 8th edition
M	M0: No metastatic disease at the time of diagnosis of CRC, (liver metastases were developed later: metachronous) M1: Liver metastatic disease present at the time of diagnosis of colorectal cancer (synchronous)
Stage	I–IV, as per AJCC 8th edition
Grade	1: Low differentiation of CRC cells 2: Moderate differentiation of CRC cells 3: High differentiation of CRC cells
Primary Tumour Location	Right site: CRC located from the caecum to the transverse colon up to the splenic flexure Left site: CRC located from the splenic flexure to the rectum
CRLM location	Unilobar: metastases present at either the left or right liver lobe Bilobar: metastases present at both liver lobes
Size of CLRM	Size of largest metastatic deposit measured at histopathological examination (measured in cm)
Number of CRLM	Number of metastatic deposits mentioned at histopathology report
CEA	CEA level measured at the time of the diagnosis of metastatic liver disease (ng/mL)
Response to neoadjuvant chemotherapy	Yes: Patient demonstrating either complete or partial response to chemo on CT according to Response evaluation criteria in solid tumours (RECIST) criteria No: Patient demonstrating either stable disease or disease progression on CT according to RECIST criteria
Resection	R0: resection margin $\geq$1 mm R1: resection margin <1 mm
Local Recurrence	Patient demonstrating new intrahepatic disease after first liver resection
Overall Survival	Date of death or the date of status checked in the NHS Spine (28 October 2020) minus the date of discharge.

CRLM: colorectal liver metastases, CRC: colorectal cancer, CEA: carcinoembryonic antigen, NHS: National health system.

Eligible FFPE blocks (n = 110) containing viable tumours >40% of the surrounding tissue were selected as donor blocks for tissue microarray (TMA) construction. TMA blocks consisted of 1.5 mm core biopsies taken from the donor blocks and contained CRLM and normal liver tissue [26]. Immunohistochemical staining of TMA slides for HOXB9 was performed with the BenchMark automated Ventana system (Roche Tissue Diagnostics, Dundee, UK), supplementary digital content 1.doc (Supplementary Material S1). Furthermore, HOXB9 expression was semi-quantified by a consultant pathologist blinded to the clinical data, in duplicate, with a cooling period of 4 weeks [27]. Staining intensity was graded as follows, 0: no staining; 1+, weak; 2+, intermediate/strong. The percentage of stained cells was also estimated and the H-score was calculated by multiplying the staining intensity by the percentage of stained cells [28]. To analyse the association of HOXB9 expression with OS in the TMA-CRLM patient cohort, the Reporting Recommendations for Tumour Marker Prognostic Studies (REMARK) were followed and compliance is reported in supplementary digital content 2.doc [27] (Supplementary Material S2).

Sample-size calculation requiring a minimum sample of 43 patients was performed based on previous studies with 85% power and a p-value of 0.05 [11,12,29]. Patients were categorised based on their H-Score by selecting the median value of the observed H-Score as a threshold to characterise tumours as H-negative (<10) or H-positive ($\geq$10) [15]. Additionally, among the H-positive patient group, the 30th percentile of the observed H-score range was used to categorise tumours with high expression ($\geq$50) or low expression (<50) [27]. Patients corresponding to core biopsies that were lost during TMA slide cutting ($n = 11$) as well as patients with 90-day postoperative mortality ($n = 3$) were excluded from the final survival analysis. The Kaplan–Meier curves were produced and log-rank test was conducted to compare OS between different groups based on their HOXB9 expression (intensity, cell percentage and H-score). Univariable Cox regression was performed to identify variables that were associated with OS. Multivariable Cox regression analysis was conducted to adjust for competing prognostic factors. Various multivariable models were built containing HOXB9 expression as well as statistically and/or clinically significant variables, which were identified from the univariable analysis [27,30]. Each multivariable model was assessed for "goodness of fit", with the Omnibus test of model coefficients producing the model's p-value. Models with a p-value ≤ 0.001 have been reported [31]. Analysis was performed using the SPSS package v27.

The association between HOXB9 expression and clinicopathological characteristics was also explored. Three groups were compared based on their HOXB9 expression: (1) negative: H-score < 10, low: H-score (10–50) and high: H-score $\geq$ 50. Differences in continuous variables were compared using one-way ANOVA, whereas in categorical variables with 2×3 Fisher's exact test using GraphPad Prism 8.

3. Results

3.1. HOXB9 Differential Expression in CRC

There were 644 primary solid tumours and 51 normal samples available in the combined TCGA COAD and READ datasets. Bioinformatics analysis showed that HOXB9 expression was significantly increased in CRC vs. normal colon ($p < 0.0001$), Figure 2a. Additionally the UALCAN platform showed that *HOXB9* demonstrated the highest expression levels in CRC among all types of cancers as shown in Figure 2b whilst the GEPIA tool showed that *HOXB9* demonstrated the highest expression levels in CRC in comparison with the rest of the *HOX* gene family, Figure 2c. The COSMIC database identified *APC*, *TP53*, and *KRAS* as the top three somatic CRC mutations with frequencies of 51%, 46%, and 34%, respectively. The OmicSoft analysis revealed that *HOXB9* expression was higher in mutant CRC versus wild type with highly significant upregulation in *KRAS*-mutated samples ($p < 0.0001$) (Figure 2c–e).

3.2. Impact of HOXB9 Dysregulation in CRC Progression In Vitro

The STRING server showed that the anti-proliferative proteins BTG1 and BTG2 are among the top ten predicted partners that interact with HOXB9. The gain-of-function experiments in HCT116 cells included 194 treated versus 196 control samples. Additionally, HOXB9 overexpression significantly increased cell proliferation in the overexpressing group compared to that in the control group (Figure 3a,b). Additionally, HOXB9 overexpression significantly upregulated the mRNA expression of mesenchymal markers *VIM* ($p < 0.0001$) and *CDH2* ($p < 0.0001$), while downregulating the epithelial marker *CDH1* ($p < 0.0001$). Additionally, the upregulation of important EMT activators *ZEB1* ($p < 0.0001$), *ZEB2* ($p < 0.0001$), *SNAI1* ($p < 0.01$), and *SNAI2* ($p = 0.018$) were also observed (Figure 3c). Loss-of-function siRNA interference experiments consisting of 189 treated versus 179 control

HCT116 samples showed that the silencing of HOXB9 markedly suppressed CRC cell proliferation over five days post gene expression modulation ($p < 0.001$) (Figure 3d,e).

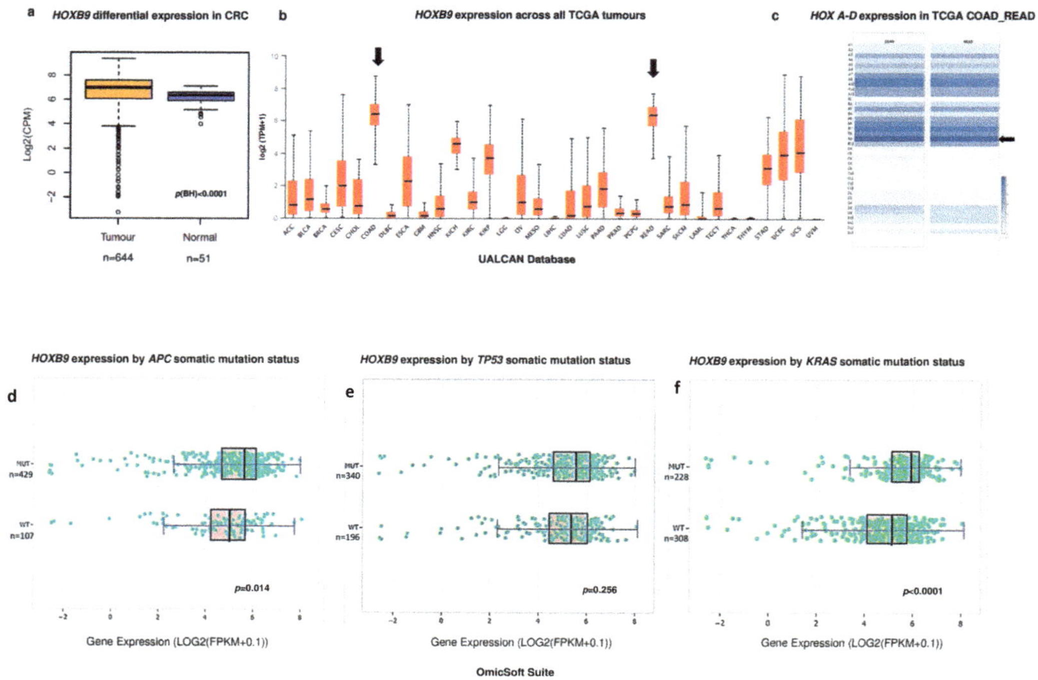

Figure 2. Differential *HOXB9* expression in CRC. (**a**) Box plot of bioinformatics differential HOXB9 expression in CRC TCGA samples vs. normal tissue samples. Values are expressed in Log2 counts per million (Log2CPM). (**b**) Box plot graph produced by the UALCAN web computational server showing the *HOXB9* gene expression levels across all types of cancers in the TCGA datasets. Black arrows represent the expression levels of *HOXB9* in the COAD (colonic adenocarcinoma) and READ (rectal adenocarcinoma) datasets. Values are shown as Log2transcripts per million (log2TPM). (**c**) Expression intensity of 39 *HOX* genes in CRC from COAD (left column) and READ (right column) datasets. Colour intensity corresponds to the value of z-score automatically produced by GEPIA server, black arrow indicates the *HOXB9* gene (**d**) Box plot of *HOXB9* differential expression in *APC* mutant CRC samples vs. wild type CRC. (**e**) Box plot of *HOXB9* differential expression in *TP53* mutant CRC samples vs. wild type CRC. (**f**) Box plot of *HOXB9* differential expression in *KRAS* mutant CRC samples vs. wild type CRC, values expressed as Log2Fragments Per Kilobase of transcript per Million mapped reads (Log2(FPKM + 0.1)). Box plots of figures (**c**–**e**), as well as the *p*-values, were automatically generated by the OmicSoft Suite/OncoLand platform (Qiagen, UK) by selecting the TCGA COADREAD dataset group and the gene-expression command.

3.3. *Predicted HOXB9 Regulators and Related Biological Processes*

Thirteen TFs were found to potentially regulate the transcription of *HOXB9* in CRC (*CDK9, SP1, HEXIM1, CNOT3, TCF7L1/2, TRIM28, TFAP4, MYC, ZBTB17, CDX2*, and *POLR2A*). Enrichment analysis of the predicted HOXB9 regulators with the Enrichr server revealed that biological processes related to the regulation of cell proliferation and cell cycle were among the significantly enriched ones. An interactive illustration of the GSEA results is provided in the link (https://maayanlab.cloud/Enrichr/enrich?dataset=10db559 14af0d55c6d4a8ee83c2b3936, (accessed on 27 December 2021)).

3.4. Association of HOXB9 with OS in Patients with CRLM

We investigated the clinical significance of HOXB9 dysregulation by exploring the association between HOXB9 protein expression levels and OS in a cohort of patients who underwent liver resection for CRLM. After excluding TMA procedural tissue loss and 90-day mortality, 96 of the initial 110 patients with a mean age of 66 ± 11 years were included in the final survival analysis. Patient demographics, clinicopathological characteristics, and treatment characteristics along with HOXB9 expression are shown in Table 2.

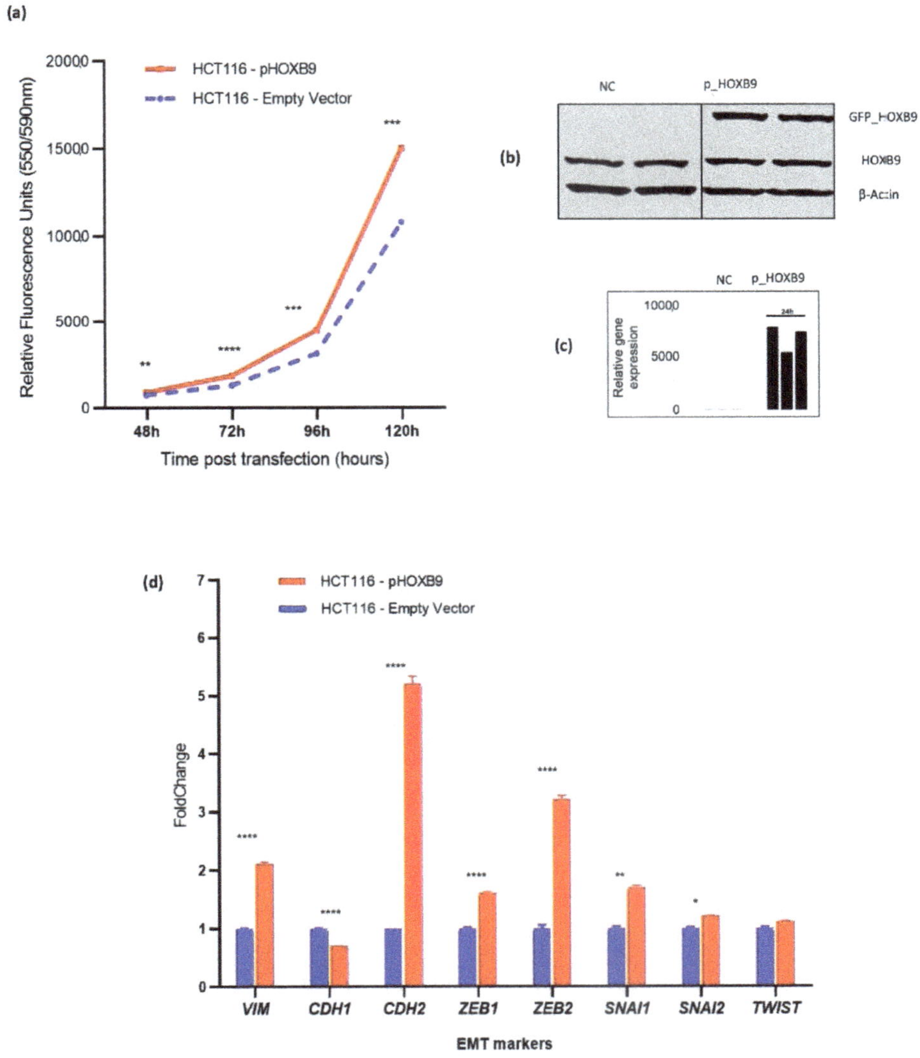

Figure 3. *Cont.*

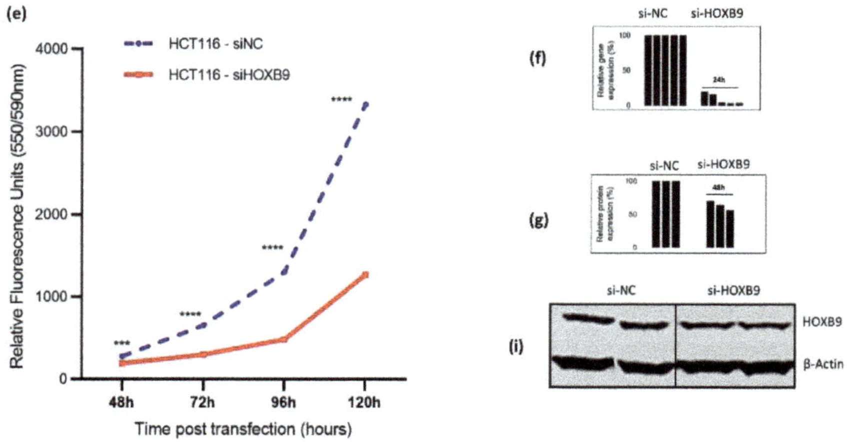

Figure 3. Impact of *HOXB9* gene expression modulation on HCT116 cell proliferation and EMT markers expression in vitro. (**a**) HCT116 cell proliferation measured as relative fluorescence (RFU) after HOXB9 overexpression. Comparison groups were pHOXB9 (overexpressing) and the control group which was transfected with an empty vector. Y-axis represents time points post-transfection, (data derived from 3 biological replicates with 8–12 technical replicates). (**b**) Western blot evaluation of HOXB9 overexpression in HCT116 cells. In the pHOXB9 group (right) the top band shows the GFP_HOXB9 fusion overexpressed protein, the middle band shows the endogenous HOXB9 protein expression and the bottom band shows the expression of β-actin which was used as a loading control. (**c**) Histogram showing relative *HOXB9* gene expression assessed by RT-qPCR in HCT116 cells from 3 biological replicates in triplicates (*ACTB* was used as endogenous control gene). (**d**) RNA fold change expression of EMT-related transcription factors in HCT116 overexpressing HOXB9 vs. control group, (data derived from 3 biological replicates assessed in triplicates). (**e**) HCT116 cell proliferation measured as relative fluorescence (RFU) after HOXB9 silencing. Comparison groups were siHOXB9 (silenced) and the negative control group (siNC). Y-axis represents time points post-transfection, (data derived from 5 biological replicates with 8–12 technical replicates). (**f**) Histogram showing the evaluation of *HOXB9* % knockdown at mRNA level 24 h post transfection with RT-qPCR using the ΔΔCq method. Y-axis represents the % of the relative gene expression normalised to si-NC samples, the difference between si-NC and si-HOXB9 columns represents the % knockdown efficiency, (data are derived from 5 biological experiments assessed in triplicates). (**g**) Histogram showing relative % protein expression in the si-HOXB9 samples in relation to si-NC, the difference between si-NC and si-HOXB9 columns represents the % reduction in HOXB9 protein expression 48 h post transfection, β-actin expression was used as a loading control, data derived from 3 biological experiments. (**h**) Western blot evaluation of HOXB9 silencing in HCT116 cells. The top band shows the HOXB9 protein expression intensity, and the bottom band shows the expression of β-actin which was used as a loading control. Values in (**a**,**d**,**e**) are presented as mean ± standard error of mean (SEM) * $p < 0.05$, ** $p < 0.01$, *** $p < 0.001$, **** $p < 0.0001$.

Table 2. TMA CRLM patient cohort demographics, clinicopathological and treatment-related characteristics categorised per HOXB9 expression (H-Score).

	Total (n = 96)	Neg: <10 (n = 46)	Low: [10–50] (n = 39)	High: ≥50 (n = 11)	p-Value *
Age (mean, SD), [range]	66 (33), [32–81]	68 (11), [32–89]	64 (11), [35–81]	66 (10), [52–82]	0.187
Gender, n (%)					
Male	63 (67%)	28 (61%)	25 (64%)	10 (91%)	0.164
Female	33 (33%)	18 (39%)	14 (36%)	1 (9%)	
Deceased	74 (77%)	40 (87%)	25 (64%)	9 (82%)	0.195
Primary CRC characteristics					
Tumour Location, n (%)					
Right colon	15 (16%)	9 (20%)	5 (13%)	1 (9%)	0.402
Left colon	81 (84%)	37 (80%)	34 (87%)	10 (91%)	
Tumour Depth, n (%)					
T1/2	18 (19%)	8 (17%)	9 (23%)	1 (9%)	0.546
T3/4	78 (81%)	38 (83%)	30 (77%)	10 (91%)	
Lymph node status, n (%)					
Negative	40 (42%)	19 (41%)	20 (51%)	1 (9%)	0.035
Positive	56 (58%)	28 (59%)	18 (49%)	10 (91%)	
Metastases, n (%)					
M0	60 (63%)	28 (61%)	23 (59%)	9 (82%)	0.366
M1	36 (37%)	18 (39%)	16 (41%)	2 (18%)	
Stage, n (%)					
I/II	17 (18%)	8 (17%)	8 (21%)	1 (9%)	0.680
III/IV	79 (82%)	38 (83%)	31 (79%)	10 (91%)	
Grade, n (%)					
Well/Moderate	70 (73%)	36 (84%)	27 (82%)	7 (88%)	0.923
Poor	14 (15%)	7 (16%)	6 (18%)	1 (12%)	
CRLM characteristics					
CRLM Location, n (%)					
Unilobar	65 (68%)	31 (67%)	25 (64%)	9 (82%)	0.537
Bilobar	31 (35%)	15 (33%)	14 (36%)	2 (18%)	
Number of CRLM, n (%)					
<4	77 (80%)	37 (80%)	31 (80%)	9 (82%)	0.985
≥4	19 (20%)	9 (20%)	8 (20%)	2 (8%)	
Size of CRLM (cm), n (%)					
<5	77 (80%)	37 (80%)	30 (77%)	10 (91%)	0.589
≥5	19 (20%)	9 (20%)	9 (23%)	1 (9%)	
CEA (ng/mL), n (%)					
<20	33 (34%)	20 (77%)	12 (100%)	1 (50%)	0.387
≥20	7 (7%)	6 (23%)	0 (0%)	1 (50%)	
Neoadjuvant Chemo, n (%)					
Yes	74 (77%)	35 (76%)	30 (77%)	9 (82%)	0.919
No	22 (23%)	11 (24%)	9 (23%)	2 (18%)	
Local Recurrence, n (%)					
Yes	31 (32%)	14 (30%)	14 (36%)	3 (27%)	0.865
No	61 (64%)	28 (70%)	25 (44%)	8 (73%)	

TMA: tissue microarray, **SD**: standard deviation, **CRLM**: colorectal liver metastases, **CRC**: colorectal cancer, **CEA**: carcinoembryonic antigen, *: 2 × 3 fisher's exact test.

Regarding the HOXB9 protein, no statistical difference was observed in survival Kaplan–Meier curves between patients with a 0, 1+, and 2+ HOXB9 staining intensity (Figure 4a–e). Among patients who expressed HOXB9, those who had a high percentage of stained cells had worse survival compared to patients with a low percentage of stained cells (Figure 4f). Patients were also compared based on their H-scores, as shown in Figure 4g. Among the patients who demonstrated positive endogenous HOXB9 (H-positive) expression, those who had high HOXB9 levels demonstrated significantly worse OS than those in the low-level group ($p = 0.036$). To further evaluate whether HOXB9 expression levels could be a potential independent prognostic factor, a Cox regression analysis was conducted in the patient group who demonstrated endogenous HOXB9 expression (H-positive) ($n = 50$). Univariable Cox regression identified factors that have a prognostic role in OS after liver resection for CRLM and are shown in Table 3. In the univariable analysis, an inverse tendency between HOXB9 levels and OS was found (HR: 2.1; 95% CI: 0.98–4.63, $p = 0.056$) (Table 3). In terms of a multivariable assessment, all three multivariable models showed that among patients who expressed HOXB9, those with high expression seemed to demonstrate an increased risk for worse OS with an HR between 3.8, 95% CI 1.2–12, $p = 0.023$ to 4.2, 95% CI 1.7–10.1, $p = 0.002$ (Table 3). Local recurrence was another factor that demonstrated a significant adverse prognostic role in all three models ($p = 0.001$). Lastly, the size of CRLM ≥5 cm also seemed to increase the likelihood of worse OS (multivariable models 2 and 3) (Table 3).

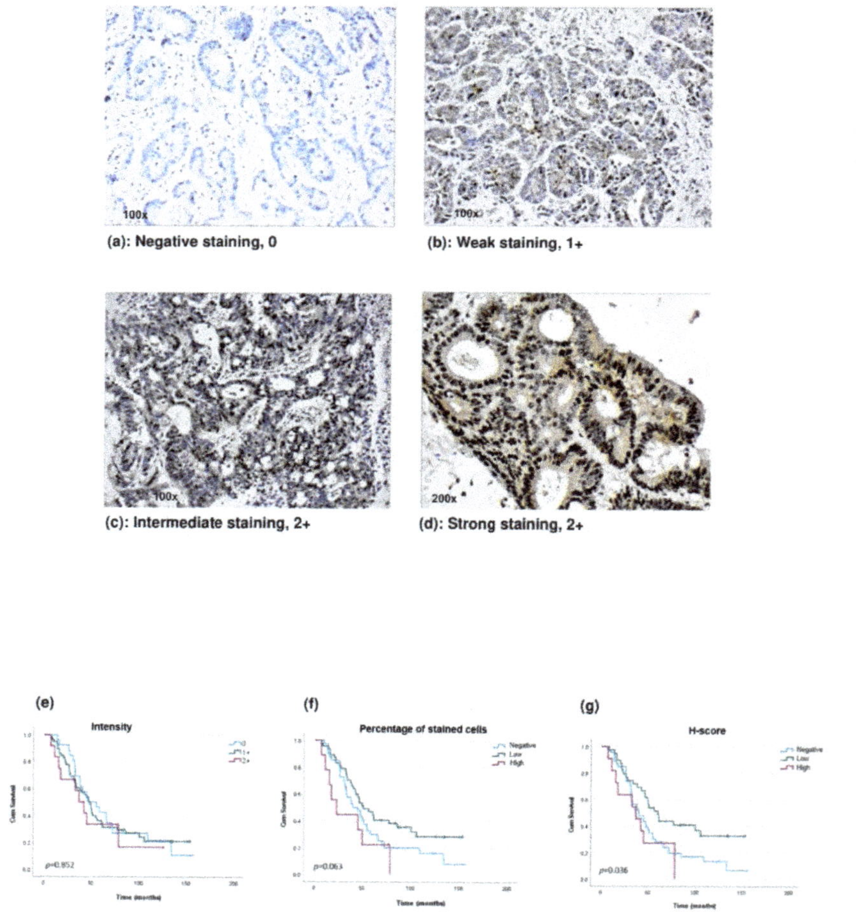

Figure 4. HOXB9 protein expression and OS in patients with CRLM after liver resection. (**a–d**) Intensity of HOXB9 protein expression assessed by immunohistochemistry in CRLM tissues. (**e**) Kaplan–Meier curve of OS in CRLM patients based on staining intensity: 0 (blue line), 1+ (green line) and 2+ (red line). (**f**) Kaplan–Meier curve of OS in CRLM patients based on the percentage of stained cells: <10% (negative, blue line), 10–50% (low, green line) and ≥50% (high, red line). (**g**) Kaplan–Meier curve of OS in CRLM patients based on H-score: <10 (negative, blue line), 10–50 (low, green line) and ≥50 (high, red line).

Table 3. Univariable and Multivariable Cox hazards analyses of factors associated with OS after liver resection in CRLM patients who demonstrated endogenous HOXB9 expression (n = 50).

Variables	Univariable	Multivariable (1)	Multivariable (2)	Multivariable (3)
	HR (95% CI) p-Value	HR (95% CI) p-Value	HR (95% CI) p-Value	HR (95% CI) p-Value
Age	1.02 (0.10–1.04) $p = 0.121$		1.04 (1.00–1.08) $p = 0.048$	1.02 (0.98–1.07) $p = 0.333$
Gender (Male)	1.29 (0.79–2.09) $p = 0.303$			
Local Recurrence *	2.29 (1.40–3.56) $p = 0.001$	4.28 (1.88–9.72) $p = 0.001$	5.73 (2.33–14.08) $p < 0.001$	5.83 (2.11–16.11) $p = 0.001$
HOXB9 staining (2+)	1.18 (0.58–2.43) $p = 0.648$			
HOXB9 H-Score (High)	2.13 (0.98–4.63) $p = 0.056$	3.82 (1.59–9.19) $p = 0.003$	4.15 (1.71–10.06) $p = 0.002$	3.79 (1.20–11.98) $p = 0.023$
Tumour Location * (left)	0.48 (0.26–0.87) $p = 0.017$	0.39 (0.13–1.13) $p = 0.083$	0.38 (0.13–1.10) $p = 0.074$	
Number of CRLM * (≥4)	1.78 (1.03–3.08) $p = 0.040$	1.25 (0.45–3.45) $p = 0.665$	1.41 (0.54–3.71) $p = 0.489$	1.83 (0.58–5.74) $p = 0.302$
Size of CRLM *(≥5 cm)	1.87 (1.08–3.25) $p = 0.027$	2.27 (0.88–5.88) $p = 0.091$	2.76 (1.06–7.20) $p = 0.038$	4.44 (1.11–17.75) $p = 0.035$
T3/4	1.34 (0.64–2.81) $p = 0.438$			
N1/2	1.41 (0.87–2.29) $p = 0.168$			1.04 (0.33–3.28) $p = 0.946$
M1	0.99 (0.51–1.90) $p = 0.970$			
Stage (III/IV)	1.23 (0.64–1.97) $p = 0.535$			
Grade 2/3	1.18 (0.71–1.97) $p = 0.518$			
CRLM Location (bilobar)	1.26 (0.78–2.02) $p = 0.342$			0.42 (0.12–1.46) $p = 0.170$
CEA(≥20 ng/mL)	1.54 (0.79–3.01) $p = 0.207$			
R1 resection	1.09 (0.51–2.35) $p = 0.827$			
Neoadjuvant Chemotherapy	1.26 (0.72–2.23) $p = 0.422$			
Response to Chemotherapy	0.83 (0.42–1.66) $p = 0.598$			

CRLM: colorectal liver metastasis, **HR**: hazard ratio, **CI**: confidence interval, T: tumour depth, N: lymph node status, M: metastatic disease, **CEA**: carcinoembryonic antigen.

4. Discussion

In this study, we observed that *HOXB9* gene was not only significantly upregulated in cancer vs. normal colon, but its levels were significantly increased when *KRAS* mutations were present. *KRAS* mutant CRC is a molecular subtype of CRC, which demonstrates resistance to standard chemotherapy and immunotherapy [32]. Additionally, KRAS is an established marker of a negative prognosis in patients with primary and metastatic CRC, and the upregulation of *HOXB9* in *KRAS* mutant samples indicates its potential association with aggressive tumour biology [33]. Indeed, studies by Hoshino et al. and Huang et al. have reported a positive association between high HOXB9 protein levels and lymph node invasion, presence of distant metastases, and poor differentiation in patients with CRC [11,12]. Additionally, in our systematic review, we found by conducting a post hoc meta-analysis that high HOXB9 expression levels were associated with a significantly increased risk for metastases (OR 4.14, 95% CI: 1.64–10.43, $p = 0.003$) [10]. In our CRLM patient group, although we did not find a significant association with the adverse CRLM characteristics, we noticed that high HOXB9 levels were positively correlated with the presence of metastatic disease in the regional lymph nodes at the time of the primary cancer resection, indicating that HOXB9 may promote CRC progression and affect survival.

Four studies have demonstrated that HOXB9 significantly affects OS in patients with CRC. Interestingly, studies have shown contradictory results with those of Song et al. [34] and Zhan et al. [13] supporting a favourable prognosis, whereas Hoshino et al. [12] and Huang et al. [11] indicated a negative HOXB9 prognostic role in patients with high HOXB9 levels and with CRC after bowel resection. In our study, we included patients with CRLM after liver resection and our findings are more consistent with studies by Hoshino et al. and Huan et al., as Kaplan–Meier showed that among patients who express HOXB9, those with high staining intensity had worse OS than patients with low levels. Interestingly, we found no difference when patients were categorised based on their staining intensity. The HOXB9 expression level, as an independent risk factor for OS in CRC, has not been previously assessed in multivariable models. Carbone et al. explored the prognostic role of HOXB9 in disease-free survival (DFS) and reported that HOXB9 expression was an independent adverse risk factor for worse DFS in stage IV CRC and possibly more important compared to KRAS and BRAF mutations, which are well-known negative prognostic markers in CRC/CRLM [29,33]. From a bioinformatics analysis that we performed, we also found that patients with high *HOXB9* mRNA levels demonstrated lower DFS survival rates, whereas we observed no difference in OS rates between the high and low *HOXB9* mRNA expressing group (HR: 1 (0.92–1.1), $p = 0.620$, data not shown herein). In our study, in all three multivariable models, a high HOXB9 H-Score and intrahepatic recurrence were the two factors that retained significance as adverse independent prognostic factors in CRLM. The size and number of CRLMs, as well as the development of local recurrence after first liver resection, are well-established prognostic factors in CRLM, indicating that tumour biology plays a vital role in determining prognosis [33,35]. In our study, high HOXB9 levels appear to potentially increase the likelihood of worse OS, similar to the presence of intrahepatic recurrence, which highlights the importance of HOXB9 as a potential prognostic marker in CRLM and suggests that HOXB9 may play an oncopromoting role in CRC. Nevertheless, it has to be acknowledged that to date no definite conclusion can be made regarding the exact association of HOXB9 with OS in patients with CRC indicating the need for further research to elucidate the prognostic role of HOXB9. Additionally, given the fact that stage plays an important role as a selection criterion during a biomarker study, it is suggested that a larger biomarker study restricted to certain stages is needed to further explore the association of HOXB9 with OS in CRC [27].

HOXB9 protein appears to be the most frequently investigated protein among all other HOX proteins in CRC. However, it is interesting that studies report contradictory findings in terms of its clinicopathological significance as well as its mechanistic role in CRC progression. Studies including our own, report opposing findings regarding the association of HOXB9 in OS [10–13]. This could be attributed to the different methodological approaches implemented by the studies with regard to the categorization of high and low HOXB9 expression patient groups. For instance, despite the fact that studies used IHC as an evaluation method of HOXB9 protein expression, the categorization based on staining intensity varied between studies [10–13]. Additionally, in our study, we accounted for both the intensity as well as the percentage of stained cells to ensure a more robust classification method of HOXB9 protein expression. Likewise, the experimental observations also differed between studies with regard to the role of HOXB9 in CRC progression. Our study, Huang et al. [11] and Hoshino et al. [12] reported a potential tumour promoting role of HOXB9 whereas Zhan et al. [13] observed a potential tumour suppressive function of HOXB9 in CRC [10]. Variability in the selection of downstream functional assays could be one reason for the contradictory findings. Additionally, HOX proteins undergo significant post-translational modifications which can cause changes in their functions highlighting their potential dual role in cancer [36]. Acetylation has been found by Wan et al. to be an important post-translational modification of HOXB9, resulting in the downregulation of its target gene jumonji domain-containing protein 6 (JMJD6), and subsequently causing a suppression in tumour growth and the migration of in lung adenocarcinoma in vitro [37].

In our gain-of-function experiments, we found that HOXB9 overexpression significantly increased in vitro cell proliferation, indicating a tumour-promoting role; however, the mechanism by which HOXB9 affects cell proliferation in CRC is still unknown. Our protein–protein network analysis showed that important proteins related to cell proliferation may interact with HOXB9. Additionally, TFs that are predicted to regulate the transcription of HOXB9 were enriched in processes related to cell proliferation and the cell cycle, leading to the hypothesis that HOXB9 may play an important role in the cell cycle. This hypothesis is supported by findings from studies conducted in other types of cancer, showing that HOXB9 knockdown results in cell-cycle arrest, indicating that it may be an important molecular component of the cell cycle and may be a promising target for novel personalised gene therapy [38]. Nevertheless, further research in the area of CRC is needed to obtain more evidence on the role of HOXB9 in the cell cycle and cell proliferation. Our study also showed that the RNA expression of important EMT molecular markers and activators was significantly altered. We showed that the mesenchymal markers *VIM* and *CDH2*, which encode for vimentin and N-cadherin, respectively, were significantly upregulated. In contrast, *CDH1*, which encodes the epithelial marker E-cadherin was downregulated. These findings indicate that HOXB9 may contribute to the so-called "cadherin switch", which is a hallmark of EMT, enabling cancer cells to obtain metastatic potential [39]. Additionally, our experiments showed that the RNA expression of EMT activators such as *ZEB1*, *ZEB2*, *SNAI1*, and *SNAI2* was significantly upregulated after HOXB9 overexpression, supporting the hypothesis that HOXB9 may promote CRC progression. Interestingly, HOXB9 has recently been recognised as an important TF that plays a vital role in cancer progression by activating EMT through important signalling pathways, including the transforming growth factor beta (TGF-β) and wingless-related integration site (WNT) signalling pathways [6–8]. Furthermore, HOXB9 high expression has been attributed to the promotion of angiogenesis and resistance to anti-angiogenic treatment with bevacizumab in CRC, indicating that silencing HOXB9 could be a promising approach to modulate this resistance [8,29].

To assess whether HOXB9 could be a potential therapeutic target, we transiently silenced its expression, and we observed that the exponential logarithmic growth of HCT116 cells was significantly disrupted in the intervention group. Our in vitro findings are similar to the in vivo findings reported by Hoshino et al. and Huang et al., who also showed that HOXB9 overexpression increased tumour growth, whereas silencing caused the development of fewer lung and liver metastases in nude mice compared to their control group [11,12].

Our study has limitations which should be considered when interpreting its findings. First, this translational prognostic-marker study was based on a small retrospective cohort study. Challenges in optimal biological tissue collection were recognised as FFPE specimens were based on their availability. However, according to our a priori sample calculation based on published studies, our sample size was sufficient to allow for an accurate analysis of our data [11,12]. Additionally, in contrast with the currently published studies, we used various multivariable models to obtain more evidence on the effect of HOXB9 on OS in CRLM, in compliance with the REMARK criteria. Second, in our study, we used the TMA approach to analyze HOXB9 protein expression in CRLM tissues, which potentially introduces selection bias as it consists of core biopsies instead of a larger section and limits the tumour-heterogeneity inspection. In our initial optimization IHC experiments, we noticed that HOXB9 showed heterogeneous staining where some areas were negative, whereas in others, positive staining was observed. Considering this observation, the possible misclassification of a patient as a false negative for HOXB9 expression could not be excluded. Despite the limitations of this approach, TMA is a well-established and widely used technique for biomarker studies and biobanks. To overcome this limitation, we chose the maximum available TMA diameter of 1.5 mm instead of 0.6 mm. Finally, another limitation is that there are no gold-standard classification criteria for immunohistochemical evaluation of HOXB9 expression. To strengthen our study, we used two different categorisation approaches based on staining intensity and H-score. Considering that HOXB9 is emerging as a crucial prognostic factor in various cancers, a consensus to standardise HOXB9 grading in cancers is urgently needed and the above limitations could be potentially minimised by the design of a larger-scale HOXB9 biomarker study. In addition, to validate the hypothesis generated by the survival analysis, we conducted in vitro experiments in addition to our initial bioinformatics analysis.

Our study has several implications which should be explored in future research. CRC/CRLM patients, especially those with KRAS mutations, represent a major treatment challenge and have a worse prognosis [29,33]. Our findings showed that in the HCT116 cell line which harbours KRAS mutation according to the ATCC records, silencing of HOXB9 markedly suppressed cell growth, indicating that HOXB9 may be a novel target for the development of new anticancer agents for resistant CRC/CRLM. The possibility of achieving response and disease control with precision medicine by targeting HOXB9 in a selected group of patients may potentially improve the respectability rates for liver resection and may eventually improve outcomes.

5. Conclusions

In conclusion, our study found that HOXB9 may exert an oncopromoting role in CRC by accelerating cell growth and activating EMT. Additionally, our study demonstrates that HOXB9 may play an important role in the OS of patients with CRLM after liver resection. Lastly, we showed that HOXB9 knockdown disrupts CRC cell growth in vitro, indicating that silencing this gene might be a novel approach for the development of personalised gene-directed therapy in primary and metastatic CRC.

Supplementary Materials: The following supporting information can be downloaded at: https://www.mdpi.com/article/10.3390/ijms23042281/s1.

Author Contributions: Conceptualization, E.M.; methodology, E.M., C.M.-L., I.B. and A.M.A.; software, C.M.-L.; validation, E.M., C.M.-L., D.K., I.B. and A.M.A.; formal analysis, E.M., C.M.-L., D.K., I.B. and A.M.A.; investigation, E.M., C.M.-L. and I.B.; resources, E.M., C.M.-L. and I.B.; data curation, E.M., C.M.-L. and A.M.A.; writing—original draft preparation, E.M. and C.M.-L.; writing—review and editing, E.M., C.M.-L., D.K., I.B. and A.M.A.; visualization, E.M., C.M.-L., D.K., I.B. and A.M.A.; supervision, E.M. and A.M.A.; project administration, E.M.; funding acquisition, E.M. The international committee of medical journal editors (ICMJE) guidelines were followed. All authors have read and agreed to the published version of the manuscript.

Funding: This research was partially funded by BRIGHT Cancer Care charity.

Institutional Review Board Statement: The study was approved by the National Research Ethics Committee at Brighton and Sussex REC, Southcoast (09/H1103/50/AM05, 30 March 2018).

Informed Consent Statement: This research was conducted using retrospective archival de-identified FFPE samples, therefore informed consent was waived for HOXB9 analysis on CLRM tissues.

Data Availability Statement: Publicly available TCGA COADREAD datasets that were analysed in this study are available through https://portal.gdc.cancer.gov/, (accessed on 23 May 2019). Some datasets generated during the current study are not publicly available but are available from the corresponding author on reasonable request.

Acknowledgments: We thank Aikaterini Chatzipli, Medical Informatics, Harvard Medical School, USA, for scientific advice on mechanistic studies. We thank Giulia Falgari, Clinical and Experimental Medicine, University of Surrey for contribution on research topics presented herein. We thank Emma Clarke and Elaine Smith, Surrey Pathology Services, Royal Surrey County Hospital, UK, for her scientific contribution to immunohistochemistry. We thank the Healthcare Tissue Bank, Imperial College, UK, for providing the facilities, equipment, and technical help in TMA construction. We thank the scientific personnel of the AY and AX Laboratory, University of Surrey, UK, for the technical and material help. We thank the QIAGEN company, UK, for granting access to the OmicSoft suite software and OncoLand database. We thank the BRIGHT Cancer Care Charity trustees for funding this research. We thank the Association of Surgeons of Great Britain and Ireland for awarding this research with the Moynihan Prize.

Conflicts of Interest: The authors declare no conflict of interest.

References

1. Siegel, R.L.; Miller, K.D.; Jemal, A. Cancer statistics. *CA Cancer J. Clin.* **2020**, *70*, 7–30. [CrossRef] [PubMed]
2. Chow, F.C.-L.; Chok, K.S.-H. Colorectal liver metastases: An update on multidisciplinary approach. *World J. Hepatol.* **2019**, *11*, 150–172. [CrossRef]
3. Singh, M.P.; Rai, S.; Pandey, A.; Singh, N.K.; Srivastava, S. Molecular subtypes of colorectal cancer: An emerging therapeutic opportunity for personalized medicine. *Genes Dis.* **2021**, *8*, 133–145. [CrossRef] [PubMed]
4. Xu, H.; Liu, L.; Li, W.; Zou, D.; Yu, J.; Wang, L.; Wong, C.C. Transcription factors in colorectal cancer: Molecular mechanism and therapeutic implications. *Oncogene* **2021**, *40*, 1555–1569. [CrossRef]
5. Castelli-Gair Hombría, J.; Lovegrove, B. Beyond homeosis-HOX function in morphogenesis and organogenesis. *Differentiation* **2003**, *71*, 461–476. [CrossRef]
6. Li, B.; Huang, Q.; Wei, G.-H. The Role of HOX Transcription Factors in Cancer Predisposition and Progression. *Cancers* **2019**, *11*, 528. [CrossRef]
7. Paço, A.; de Bessa Garcia, S.A.; Castro, J.L.; Costa-Pinto, A.; Freitas, R. Roles of the HOX Proteins in Cancer Invasion and Metastasis. *Cancers* **2020**, *13*, 10. [CrossRef] [PubMed]
8. Contarelli, S.; Fedele, V.; Melisi, D. HOX Genes Family and Cancer: A Novel Role for Homeobox B9 in the Resistance to Anti-Angiogenic Therapies. *Cancers* **2020**, *12*, 3299. [CrossRef]
9. Paschos, K.A.; Majeed, A.W.; Bird, N.C. Natural history of hepatic metastases from colorectal cancer - Pathobiological pathways with clinical significance. *World J. Gastroenterol.* **2014**, *20*, 3719–3737. [CrossRef] [PubMed]
10. Martinou, E.; Falgari, G.; Bagwan, I.; Angelidi, A.M. A Systematic Review on *HOX* Genes as Potential Biomarkers in Colorectal Cancer: An Emerging Role of *HOXBInt. J. Mol. Sci.* **2021**, *22*, 13429. [CrossRef]
11. Huang, K.; Yuan, R.; Wang, K.; Hu, J.; Huang, Z.; Yan, C.; Shen, W.; Shao, J. Overexpression of HOXB9 promotes metastasis and indicates poor prognosis in colon cancer. *Chin. J. Cancer Res.* **2014**, *26*, 72–80. [CrossRef] [PubMed]
12. Hoshino, Y.; Hayashida, T.; Hirata, A.; Takahashi, H.; Chiba, N.; Ohmura, M.; Wakui, M.; Jinno, H.; Hasegawa, H.; Maheswaran, S.; et al. Bevacizumab terminates homeobox B9-induced tumor proliferation by silencing microenvironmental communication. *Mol. Cancer* **2014**, *13*, 102. [CrossRef] [PubMed]
13. Zhan, J.; Niu, M.; Wang, P.; Zhu, X.; Li, S.; Song, J.; He, H.; Wang, Y.; Xue, L.; Fang, W.; et al. Elevated HOXB9 expression promotes differentiation and predicts a favourable outcome in colon adenocarcinoma patients. *Br. J. Cancer* **2014**, *111*, 883–893. [CrossRef]
14. Robinson, M.D.; McCarthy, D.J.; Smyth, G.K. EdgeR: A Bioconductor package for differential expression analysis of digital gene expression data. *Bioinformatics* **2009**, *26*, 139–140. [CrossRef] [PubMed]
15. Robinson, M.D.; Oshlack, A. A scaling normalization method for differential expression analysis of RNA-seq data. *Genome Biol.* **2010**, *11*, R25. [CrossRef] [PubMed]
16. Benjamini, Y.; Hochberg, Y. Controlling the False Discovery Rate: A Practical and Powerful Approach to Multiple Testing. *J. R. Stat. Soc. Ser. B* **1995**, *57*, 289–300. [CrossRef]
17. Tang, Z.; Kang, B.; Li, C.; Chen, T.; Zhang, Z. GEPIA2: An enhanced web server for large-scale expression profiling and interactive analysis. *Nucleic Acids Res.* **2019**, *47*, W556–W560. [CrossRef]
18. Rhodes, D.R.; Kalyana-Sundaram, S.; Mahavisno, V.; Varambally, R.; Yu, J.; Briggs, B.B.; Barrette, T.R.; Anstet, M.J.; Kincead-Beal, C.; Kulkarni, P.; et al. Oncomine 3.0: Genes, Pathways, and Networks in a Collection of 18,000 Cancer Gene Expression Profiles. *Neoplasia* **2007**, *9*, 166–180. [CrossRef]

19. Chandrashekar, D.S.; Bashel, B.; Balasubramanya, S.A.H.; Creighton, C.J.; Ponce-Rodriguez, I.; Chakravarthi, B.V.S.K.; Varambally, S. UALCAN: A portal for facilitating tumor subgroup gene expression and survival analyses. *Neoplasia* **2017**, *19*, 649–658. [CrossRef]
20. Wu, C.; Huang, B.E.; Chen, G.; Lovenberg, T.W.; Pocalyko, D.J.; Yao, X. Integrative Analysis of DiseaseLand Omics Database for Disease Signatures and Treatments: A Bipolar Case Study. *Front. Genet.* **2019**, *10*, 10–3389. [CrossRef]
21. Ahmed, D.; Eide, P.W.; Eilertsen, I.A.; Danielsen, S.A.; Eknaes, M.; Hektoen, M.; Lind, G.E.; Lothe, R.A. Epigenetic and genetic features of 24 colon cancer cell lines. *Oncogenesis* **2013**, *2*, e71. [CrossRef]
22. Menyhárt, O.; Harami-Papp, H.; Sukumar, S.; Schäfer, R.; Magnani, L.; de Barrios, O.; Győrffy, B. Guidelines for the selection of functional assays to evaluate the hallmarks of cancer. *Biochim. Biophys. Acta Bioenerg.* **2016**, *1866*, 300–319. [CrossRef] [PubMed]
23. Zachari, M.; Chondrou, P.S.; Pouliliou, S.; Mitrakas, A.G.; Abatzoglou, I.; Zois, C.E.; Koukourakis, M.I. Evaluation of The Alamarblue Assay for Adherent Cell Irradiation Experiments. *Dose-Response* **2013**, *12*, 246–258. [CrossRef]
24. Weiser, M.R. AJCC 8th ed.; Colorectal Cancer. *Ann. Surg. Oncol.* **2018**, *25*, 1454–1455. [CrossRef] [PubMed]
25. Eisenhauer, E.A.; Therasse, P.; Bogaerts, J.; Schwartz, L.H.; Sargent, D.; Ford, R.; Dancey, J.; Arbuck, S.; Gwyther, S.; Mooney, M.; et al. New response evaluation criteria in solid tumours: Revised RECIST guideline (version 1.1). *Eur. J. Cancer* **2009**, *45*, 228–247. [CrossRef]
26. Leighton, X.; Bera, A.; Eidelman, O.; Bubendorf, L.; Zellweger, T.; Banerjee, J.; Gelmann, E.P.; Pollard, H.B.; Srivastava, M. Tissue microarray analysis delineate potential prognostic role of Annexin A7 in prostate cancer progression. *PLoS ONE* **2018**, *13*, e0205837. [CrossRef]
27. Altman, D.G.; McShane, L.M.; Sauerbrei, W.; Taube, S.E. Reporting Recommendations for Tumor Marker Prognostic Studies (REMARK): Explanation and Elaboration. *PLoS Med.* **2012**, *9*, e1001216. [CrossRef]
28. Numata, M.; Morinaga, S.; Watanabe, T.; Tamagawa, H.; Yamamoto, N.; Shiozawa, M.; Nakamura, Y.; Kameda, Y.; Okawa, S.; Rino, Y.; et al. The clinical significance of SWI/SNF complex in pancreatic cancer. *Int. J. Oncol.* **2012**, *42*, 403–410. [CrossRef]
29. Carbone, C.; Piro, G.; Simionato, F.; Ligorio, F.; Cremolini, C.; Loupakis, F.; Alì, G.; Rossini, D.; Merz, V.; Santoro, R.; et al. Homeobox B9 Mediates Resistance to Anti-VEGF Therapy in Colorectal Cancer Patients. *Clin. Cancer Res.* **2017**, *23*, 4312–4322. [CrossRef]
30. Pentheroudakis, G.; Kalogeras, K.T.; Wirtz, R.M.; Grimani, I.; Zografos, G.; Gogas, H.; Stropp, U.; Pectasides, D.; Skarlos, D.; Hennig, G.; et al. Gene expression of estrogen receptor, progesterone receptor and microtubule-associated protein Tau in high-risk early breast cancer: A quest for molecular predictors of treatment benefit in the context of a Hellenic Cooperative Oncology Group trial. *Breast Cancer Res. Treat.* **2008**, *116*, 131–143. [CrossRef] [PubMed]
31. Pallant, J.; Pallant, J. Logistic Regression. In *SPSS Survival Manual*; Open University Press: Berkshire, UK, 2020; p. 176.
32. Liao, W.; Overman, M.J.; Boutin, A.T.; Shang, X.; Zhao, D.; Dey, P.; Li, J.; Wang, G.; Lan, Z.; Li, J.; et al. KRAS-IRF2 Axis Drives Immune Suppression and Immune Therapy Resistance in Colorectal Cancer. *Cancer Cell* **2019**, *35*, 559–572.e7. [CrossRef] [PubMed]
33. Margonis, G.A.; Buettner, S.; Andreatos, N.; Kim, Y.; Wagner, D.; Sasaki, K.; Beer, A.; Schwarz, C.; Løes, I.M.; Smolle, M.; et al. Association of BRAF Mutations With Survival and Recurrence in Surgically Treated Patients With Metastatic Colorectal Liver Cancer. *JAMA Surg.* **2018**, *153*, e180996. [CrossRef] [PubMed]
34. Song, J.; Wang, T.; Xu, W.; Wang, P.; Wan, J.; Wang, Y.; Zhan, J.; Zhang, H. HOXB9 acetylation at K27 is responsible for its suppression of colon cancer progression. *Cancer Lett.* **2018**, *426*, 63–72. [CrossRef] [PubMed]
35. Margonis, G.A.; Sergentanis, T.N.; Ntanasis-Stathopoulos, I.; Andreatos, N.; Tzanninis, I.G.; Sasaki, K.; Psaltopoulou, T.; Wang, J.; Buettner, S.; He, J.; et al. Impact of Surgical Margin Width on Recurrence and Overall Survival Following R0 Hepatic Resection of Colorectal Metastases: A Systematic Review and Meta-analysis. *Ann. Surg.* **2018**, *267*, 1047–1055. [CrossRef]
36. Yu, M.; Zhan, J.; Zhang, H. HOX family transcription factors: Related signaling pathways and post-translational modifications in cancer. *Cell. Signal.* **2020**, *66*, 109469. [CrossRef]
37. Wan, J.; Xu, W.; Zhan, J.; Ma, J.; Li, X.; Xie, Y.; Wang, J.; Zhu, W.-G.; Luo, J.; Zhang, H. PCAF-mediated acetylation of transcriptional factor HOXB9 suppresses lung adenocarcinoma progression by targeting oncogenic protein JMJD. *Nucleic Acids Res.* **2016**, *44*, 10662–10675. [CrossRef]
38. Brotto, D.B.; Siena, Á.D.D.; de Barros, I.I.; Carvalho, S.; Muys, B.; Goedert, L.; Cardoso, C.; Plaça, J.R.; Ramão, A.; Squire, J.A.; et al. Contributions of HOX genes to cancer hallmarks: Enrichment pathway analysis and review. *Tumor Biol.* **2020**, *42*, 1010428320918050. [CrossRef]
39. Loh, C.-Y.; Chai, J.Y.; Tang, T.F.; Wong, W.F.; Sethi, G.; Shanmugam, M.K.; Chong, P.P.; Looi, C.Y. The E-Cadherin and N-Cadherin Switch in Epithelial-to-Mesenchymal Transition: Signaling, Therapeutic Implications, and Challenges. *Cells* **2019**, *8*, 1118. [CrossRef]

International Journal of *Molecular Sciences*

MDPI

Article

MicroRNA-199b Deregulation Shows Oncogenic Properties and Promising Clinical Value as Circulating Marker in Locally Advanced Rectal Cancer Patients

Andrea Santos [1,2,†], Ion Cristóbal [1,2,*,†], Jaime Rubio [1,2,3,†], Cristina Caramés [1,2,3], Melani Luque [4], Marta Sanz-Alvarez [4], Miriam Morales-Gallego [4], Juan Madoz-Gúrpide [4], Federico Rojo [4] and Jesús García-Foncillas [2,3,*]

1 Cancer Unit for Research on Novel Therapeutic Targets, Oncohealth Institute, IIS-Fundación Jiménez Díaz-UAM, 28040 Madrid, Spain; andrea.santos@quironsalud.es (A.S.); jaime.rubiop@quironsalud.es (J.R.); ccarames@fjd.es (C.C.)
2 Translational Oncology Division, Oncohealth Institute, IIS-Fundación Jiménez Díaz-UAM, 28040 Madrid, Spain
3 Medical Oncology Department, University Hospital "Fundación Jiménez Díaz", UAM, 28040 Madrid, Spain
4 Pathology Department, IIS-Fundación Jiménez Díaz-UAM, 28040 Madrid, Spain; melani.luque@quironsalud.es (M.L.); marta.sanza@quironsalud.es (M.S.-A.); miriam.moralesg@quironsalud.es (M.M.-G.); jmadoz@fjd.es (J.M.-G.); frojo@fjd.es (F.R.)
* Correspondence: ion.cristobal@idcsalud.es (I.C.); jesus.garciafoncillas@oncohealth.eu (J.G.-F.); Tel.: +34-915504800 (I.C. & J.G.-F.)
† These authors contributed equally to this work.

Citation: Santos, A.; Cristóbal, I.; Rubio, J.; Caramés, C.; Luque, M.; Sanz-Alvarez, M.; Morales-Gallego, M.; Madoz-Gúrpide, J.; Rojo, F.; García-Foncillas, J. MicroRNA-199b Deregulation Shows Oncogenic Properties and Promising Clinical Value as Circulating Marker in Locally Advanced Rectal Cancer Patients. *Int. J. Mol. Sci.* **2022**, *23*, 2203. https://doi.org/10.3390/ijms23042203

Academic Editors: Donatella Delle Cave and Alessandro Ottaiano

Received: 29 December 2021
Accepted: 14 February 2022
Published: 17 February 2022

Publisher's Note: MDPI stays neutral with regard to jurisdictional claims in published maps and institutional affiliations.

Abstract: The identification of robust prognostic markers still represents a need in locally advanced rectal cancer (LARC). MicroRNAs (miRs) have progressively emerged as promising circulating markers, overcoming some limitations that traditional biopsy comprises. Tissue miR-199b deregulation has been reported to predict outcome and response to neoadjuvant chemoradiotherapy (nCRT) in LARC, and was also found to be associated with disease progression in colorectal cancer. However, its biological and clinical relevance remains to be fully clarified. Thus, we observed here that miR-199b regulates cell migration, aggressiveness, and cell growth, and inhibits colonosphere formation and induces caspase-dependent apoptosis. Moreover, miR-199b expression was quantified by real-time PCR in plasma samples from LARC patients and its downregulation was observed in 22.7% of cases. This alteration was found to be associated with higher tumor size ($p = 0.002$) and pathological stage ($p = 0.020$) after nCRT. Notably, we observed substantially lower global miR-199b expression associated with patient downstaging ($p = 0.009$), as well as in non-responders compared to those cases who responded to nCRT in both pre- ($p = 0.003$) and post-treatment samples ($p = 0.038$). In concordance, we found that miR-199b served as a predictor marker of response to neoadjuvant therapy in our cohort ($p = 0.011$). Altogether, our findings here demonstrate the functional relevance of miR-199b in this disease and its potential value as a novel circulating marker in LARC.

Keywords: MiR-199b; tumor suppressor; prognosis; LARC

1. Introduction

Colorectal cancer (CRC) is already the third leading cause of cancer death in the world, with rectal cancer (RC) representing almost 30% of total CRC cases, and its incidence is steadily rising in developing countries and in younger patients [1]. According to GLOBOCAN data, CRC is also the third most commonly diagnosed form of cancer globally, comprising 10% of all cancer diagnoses [2]. In Spain, around 43,600 new cases of CRC have been diagnosed in 2021, from which more than 14,200 were RC patients [3]. Locally advanced rectal cancer (LARC), defined as clinical stage II (T3-4, lymph node negative) or stage III (lymph node positive) disease has experienced a great paradigm shift in management over the past few decades given the anatomical constrictions of the pelvis and the risk for local recurrence and distant metastasis. Despite advances in preoperative treatment and surgical techniques, response to neoadjuvant chemoradiotherapy (nCRT) varies among patients and, while nearly 40% of LARC patients achieve partial responses, about 20% have a pathological complete response (pCR). Thus, health-related quality of life and survival still remain suboptimal in patients with this disease [4,5]. Moreover, given the achievement of pCR in a wide range of patients undergoing nCRT and the major effects and complications of total mesorectal excision (TME) surgery, there is a growing interest in the identification of patients that could benefit from a watch and wait (W&W) clinical approach. This novel strategy could avoid surgery in around 25% of cases and supports organ preservation, avoiding unnecessary postoperative morbidity with good long-term oncological outcomes and improving quality of life in highly selected patients [6,7]. Thus, the RAPIDO and PRODIGE 23 phase III randomized clinical trials have recently shown that the addition of neoadjuvant chemotherapy to a standard short- or long-course radiation significantly decreases the risk of metastatic progression and associates with a better disease-free survival in LARC patients [8,9].

In the last decades, advances in medical oncology have improved the treatment of patients and their outcomes, but the current use of traditional biopsies based on obtaining tumor tissue has several limitations in the developing era of precision medicine, mainly due to the progression of cancer and the onset of therapy resistance [5]. In this regard, research is constantly moving forward to find more accurate and personalized biomarkers [10]. In fact, there is a critical need to identify novel biomarkers with high specificity and sensitivity in LARC patients [11,12]. Liquid biopsy represents a promising tool for biomarker detection due to its potential to overcome many of the limitations of classical biopsy. The obtention of liquid biopsies is minimally invasive and can be repeated several times. Moreover, it is a cost-effective method that can be used to screen and monitor both disease evolution and treatment response as well as to identify cellular subclones involved in relapse, metastasis or treatment resistance [13,14]. Liquid biopsy refers to the isolation of cancer-derived components, such as circulating tumor cells (CTC), circulating tumor DNA (ctDNA), microRNAs (miRs), long non-coding RNAs (lncRNAs) and proteins, from peripheral blood or other body fluids, and their genomic or proteomic assessment [15]. In this context, miRs appear to be good candidates as liquid biopsy-based cancer biomarkers. They are gaining more attention than other potential biomarkers detectable in liquid biopsies, and seem to be potential candidates in predicting LARC and CRC prognosis and therapy response, since they are easily detectable, highly stable in biological fluids, and show higher sensitivity and specificity compared to other circulating tumor components [14,16,17].

MiRs are highly conserved endogenous non-coding and single-stranded RNAs of 18–25 nucleotides in length. Generally, miRs negatively regulate gene expression via binding to the 3′-untranslated region (3′-UTR) of their target mRNAs that result in transcriptional repression with or without mRNA degradation [17–20]. Previous studies have indicated that altered expression levels of circulating miRs are related to cellular transformation and carcinogenesis progression in different tumor types [21–29]. MiRs have been implicated in development and progression of CRC by functioning as oncogenes or tumor suppressors [30]. MiR-199b downregulation has been associated with tumorigenesis and metastasis in various human cancers through the alteration of different signaling pathways. In fact, miR-199b overexpression has been reported to decrease proliferation, migration and invasion in hepatocellular carcinoma cells by directly binding and negatively regulating JAG1, thereby influencing Notch signaling [31,32]. The Notch signaling pathway has been described to be indirectly regulated in medulloblastoma by miR-199b. This miR negatively correlates with HES1, a key Notch effector, impairing CD133+ stem cell-like subpopulation of cancer cells, demonstrating the key role of miR-199b in directly targeting CD133 expression [33,34]. Moreover, this miR has a significant tumor suppressive function in the invasion and metastasis of prostate and triple negative breast cancers [35,36] regulating epithelial to mesenchymal transition (EMT) through the inhibition of the DDR1-ERK signaling axis. It has been described that miR-199b is repressed in breast cancer cells, and its overexpression reduces tumor growth and angiogenesis by directly targeting ALK1 [37]. All of these targets have also been described to be deregulated in CRC, which would confirm the possible contributing role of miR-199b in tumor progression and metastasis through the regulation of multiple signaling pathways. CircNSD2 was found to target miR-199b in CRC cells leading to DDR1/JAG1 activation and promoting the development of metastatic disease [38]. Our group has previously reported that miR-199b could be involved in regulating tumor stemness through CD133 modulation, since a negative correlation between this miR and CD133 in CRC cell lines has been shown and further confirmed in metastatic CRC patients [39]. Moreover, it has been described that the deregulation of the ALK1/TGF-β signaling pathway enhanced both EMT of tumor cells and self-renewal ability of cancer stem cells in CRC, thereby contributing to disease progression, suggesting a role of miR-199b in this issue [40,41], as it occurs in other tumor types described above. Notably, miR-199b downregulation has been described to be associated with increased cell invasion and migration in CRC, promoting cancer progression and metastasis by SIRT1/CREB/KISS1 signaling pathway regulation, SIRT1 being a direct target of this miR [42]. The contribution of the miR-199b target SET to these functional effects with a particular interest in LARC should also be considered. Thus, patients with LARC are treated with a 5-fluorouracil (5-FU) based preoperative CRT, and it has been reported that SET determines CRC cell sensitivity to this chemotherapeutic agent [43,44].

In summary, miR-199b has emerged as a promising clinical biomarker for CRC and LARC patients [45,46], but some aspects regarding miR-199b-induced effects at the functional level in this disease still remain to be fully clarified, as well as its potential clinical usefulness not only as a tissue tumor marker but also as a circulating marker.

In this report, we aimed to further explore the biological role of miR-199b expression in disease progression and analyze its potential value as a circulating marker in LARC. Thus, we observed that miR-199b regulates cell migration and aggressiveness, inhibits colonosphere formation and induces caspase-dependent apoptosis in CRC cells. Moreover, we studied miR-199b expression in plasma samples of LARC patients, observing that its downregulation is a common event strongly associated with relapse and lack of response to nCRT. Clinically, our results here are concordant with our previous findings in tumor biopsies and provide a preliminary basis for the use of miR-199b as a circulating marker in LARC patients.

2. Results

2.1. MiR-199b Regulates Cell Migration and Agressiveness

We aimed to investigate the functional relevance of miR-199b deregulation as an alteration that could be contributing to disease progression. We started analyzing potential changes in cell migration of the CRC cell lines SW480 and HT-29 after an ectopic modulation of this miR. Quantification of miR-199b was performed to confirm the efficacy of the transfections (Table S1). We performed wound-healing assays and observed that miR-199b overexpression substantially reduced migration in SW480 cells, whereas the opposite effect was found in the same cell line after miR-199b silencing. However, results did not achieve statistical significance in SW480 cells transfected with anti-miR-199b. These observations were confirmed in the HT-29 cell line, which showed similar regulation of its migration ability when miR-199b was overexpressed or silenced, respectively, and differences were statistically significant in both cases with this cell line (Figure 1).

Given that SET has been reported in previous works from our group as a key direct target of miR-199b, we also analyzed the role of the miR-199b/SET axis in cell migration. We first confirmed the previously published role of miR-199b as a direct negative SET regulator (Figure S1). Next, we performed transwell migration assays showing that SET overexpression totally restored the miR-199b-derived inhibition of cell migration in both SW480 and HT-29 cells (Figure S2).

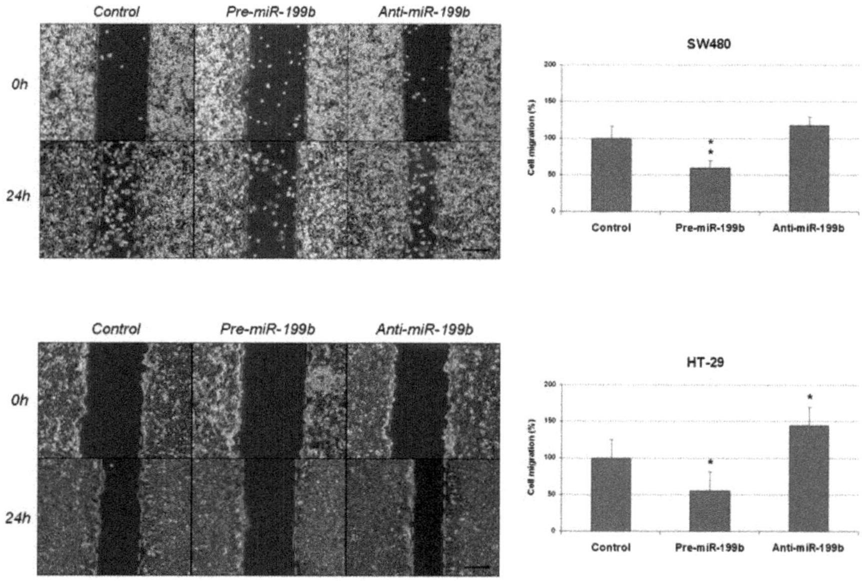

Figure 1. MiR-199b modulates cell migration in CRC cells. Wound-healing assay showing migration of SW480 and HT-29 cell lines after transfection with pre-miR-199b and anti-miR-199b. Dashed lines represent the migration border; * $p < 0.05$; ** $p < 0.01$. Scale bar: 200 µm.

To confirm the antitumor effects of miR-199b, we next performed colony-formation assays in soft agar to analyze whether this miR could also affect the aggressiveness of CRC cells by altering its anchorage-independent growth ability. We observed that colony formation was markedly inhibited in both SW480 and HT-29 cells ectopically overexpressing miR-199b. Of note, we did not find changes in these cell lines after miR-199b silencing compared with the negative controls (Figure 2). As expected, we found that the ectopic expression of SET was able to totally restore the miR-199b-induced effects in both the SW480 and HT-29 cell lines (Figure S3A). We also observed marked differences in the size of the colonies formed with the SW480 cell line (Figure S3B). Based on these findings, we also evaluated the role of this signaling axis in CRC cell growth. In concordance with soft agar assays, we observed that ectopic expression of SET significantly restored the antiproliferative effects of miR-199b in both SW480 and HT-29 cells (Figure S4). Altogether, our observations indicate that miR-199b plays a role in disease progression and aggressiveness by regulating cell migration and the colony-forming ability of CRC cells.

2.2. MiR-199b Induces Apoptosis and Inhibts Colonosphere Formation

The markedly reduced size of the colonies obtained in colony-forming assays prompted us to analyze whether miR-199b could be affecting apoptosis in CRC cells. Thus, we observed that miR-199b overexpression led to enhanced caspase-dependent apoptosis in both SW480 and HT-29 cells (Figure 3). However, we did not find differences in the apoptotic levels of CRC cell lines after miR-199b silencing compared to negative controls, which was in concordance with the results obtained in soft agar experiments (Figure 2).

We further explored the biologic effects of miR-199b deregulation in CRC by analyzing the colonosphere formation ability of SW480 and HT-29 cells. Notably, miR-199b overexpression led to significant decreased colonosphere formation ability in both number of colonospheres formed (Figure 4) and cells per colonosphere (Figure S5), suggesting that miR-199b would be playing an important role in colonosphere formation and self-renewal of CRC cells. Conversely, colonosphere formation was enhanced by the transfection with anti-miR-199b in both cell lines, but significance was only achieved in SW480 cells.

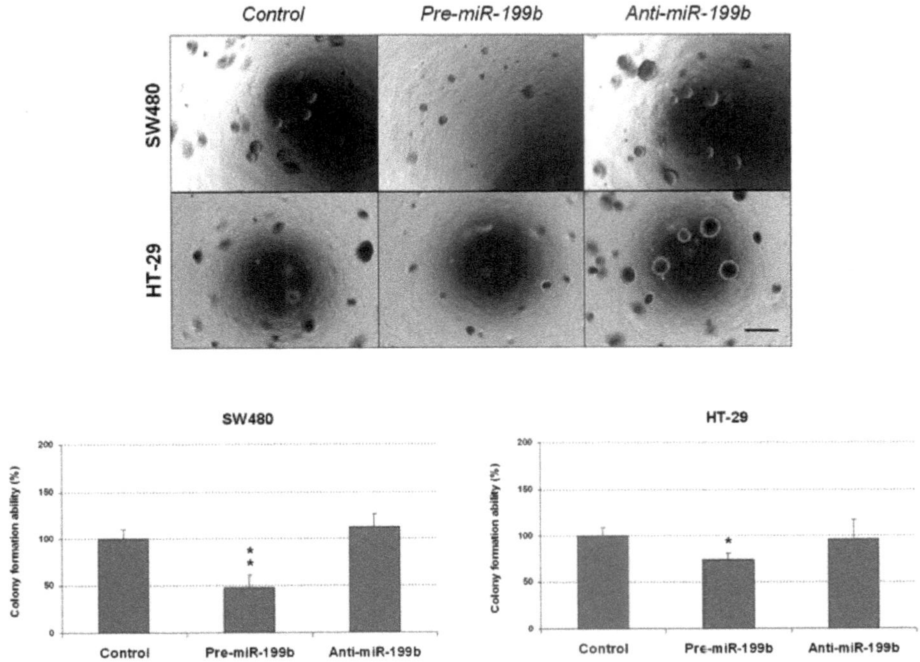

Figure 2. MiR-199b deregulation affects the colony-forming ability of CRC cells. Colony-forming assays showing the effect on the anchorage-independent cell growth of SW480 and HT-29 cell lines after ectopic miR-199b silencing or overexpression; * $p < 0.05$; ** $p < 0.01$. Scale bar: 200 μm.

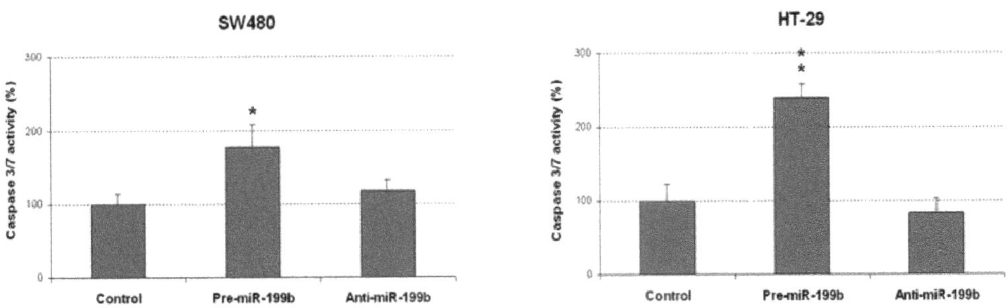

Figure 3. MiR-199b induces apoptosis in CRC cells. Caspase-3/7 activity assay showing levels of apoptotic SW480 and HT-29 cells after transfection with pre-miR-199b and anti-miR-199b; * $p < 0.05$; ** $p < 0.01$.

2.3. *Analysis of miR-199b Expression Levels as Circulating Markers in Larc Patients and Its Association with Molecular and Clinical Parameters*

To evaluate the potential usefulness of miR-199b as a circulating marker in LARC, we obtained liquid biopsies from a series of 22 patients that were diagnosed by LARC, underwent 5-FU-based nCRT in Fundación Jiménez Díaz Hospital and had clinical follow-up data available. A summary of the molecular and clinical characteristics of the cohort studied is shown in Table S2, and the same information of each case included in the study is provided in Table S3. We observed miR-199b downregulation in 22.7% of cases (5 out of 22). Notably, we found a significant correlation between miR-199b downregulation and higher tumor size after nCRT ($p = 0.002$), as well as higher pathological stage ($p = 0.020$). Interestingly, patients with low miR-199b levels also showed correlation with lymph node positivity after nCRT, but statistical significance was not achieved in this case ($p = 0.091$). The association between miR-199b expression and clinical and molecular characteristics is included in Table 1.

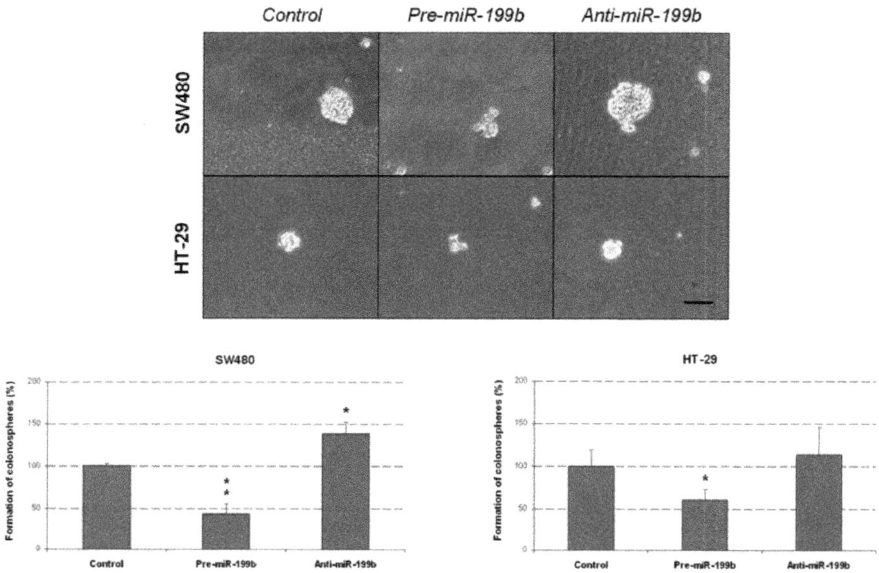

Figure 4. MiR-199b negatively regulates colonosphere formation from CRC cells. Optical microscope images showing the colonospheres formed in the different conditions. The graphs show the number of SW480 and HT-29-derived colonospheres obtained after transfection with pre-miR-199b and anti-miR-199b; * $p < 0.05$; ** $p < 0.01$. Scale bar: 200 µm.

Table 1. Association of miR-199b expression with clinical and molecular characteristics in a cohort of 22 liquid biopsies from LARC patients.

Parameter	No. Cases	No. miR-199b High (%)		No. miR-199 Low (%)		p
MiR-199b	22	17	(77.3)	5	(22.7)	
Gender	22	17		5		0.078
Male	12	11	(91.7)	1	(8.3)	
Female	10	6	(60)	4	(40)	
Age	22	17		5		0.962
<70	13	10	(76.9)	3	(23.1)	
≥70	9	7	(77.8)	2	(22.2)	
Grade pre-CRT [1]	22	17		5		0.211
Low	14	12	(85.7)	2	(14.3)	
Moderate-High	8	5	(62.5)	3	(37.5)	
Clinical stage pre-CRT	22	17		5		0.312
II	3	3	(100)	0	(0)	
III	19	14	(73.7)	5	(26.3)	
ECOG [2]	22	17		5		0.962
0	13	10	(76.9)	3	(23.1)	
1	9	7	(77.8)	2	(22.2)	

Table 1. *Cont.*

Parameter	No. Cases	No. miR-199b High (%)		No. miR-199 Low (%)		p
ypT [3]	20	15		5		0.002
0–2	12	12	(100)	0	(0)	
3–4	8	3	(37.5)	5	(62.5)	
ypN [4]	20	15		5		0.091
0	14	12	(85.7)	2	(14.3)	
1–2	6	3	(50)	3	(50)	
Pathological stage	20	15		5		0.020
yp0-I	9	19	(100)	0	(0)	
ypII-III	11	6	(54.5)	5	(45.5)	

[1] CRT = Chemoradiotherapy; [2] ECOG = Eastern Cooperative Oncology Group; [3] ypT = tumor size after CRT; [4] ypN = pathological lymph node after CRT.

2.4. Circulating miR-199b Expression Levels Predict Pathological Response to nCRT and Recurrence in LARC Patients

We next evaluated the clinical impact of miR-199b as a predictor of response to preoperative CRT in LARC. Our patient cohort was stratified into responders and non-responders to neoadjuvant therapy. As a preliminary analysis we compared global miR-199b expression between both subgroups, observing that the subgroup of non-responders had substantially lower miR-199b levels than responder patients ($p = 0.003$) (Figure 5).

In concordance with these observations, we found that miR-199b downregulation served as a circulating marker predictive of lack of response to nCRT in our patient cohort ($p = 0.011$) (Table 2).

To further evaluate the clinical relevance of miR-199b as a novel circulating marker in LARC patients, we also studied its prognostic value to predict both patient recurrence and downstaging. MiR-199b was found downregulated in two out the three cases with a reported recurrence in our cohort, and the association was very close but failed to achieve statistical significance in this case ($p = 0.051$) (Table S4). However, we observed that the subgroup of cases with low miR-199b levels was significantly associated with a lack of downstaging when we compared the clinical stage before and after nCRT in our patient cohort (Table 3).

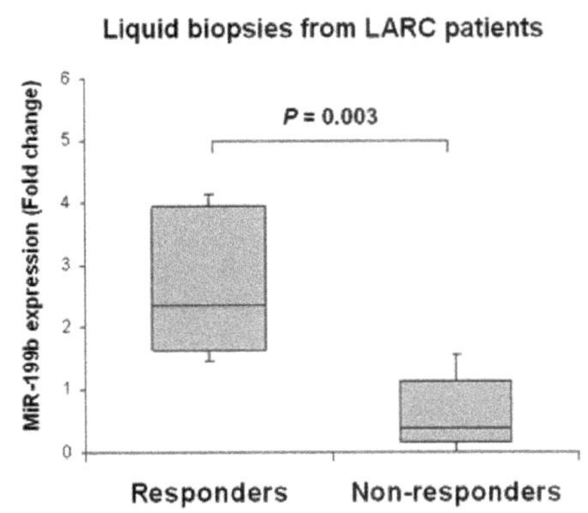

Figure 5. MiR-199b downregulation associates with lack of response to nCRT in pre-treatment samples of LARC patients. The box-plot shows miR-199b expression levels in LARC patients stratified by response or non-response to nCRT treatment. The responder group corresponds to cases with moderate or complete response (RYAN 0 and 1), including cases following the W&W protocol. The non-responder group includes those with minimal or none complete response (RYAN 2 and 3).

Table 2. Association between response to nCRT and miR-199b expression in liquid biopsies from LARC patients.

Response to Neoadjuvant CRT [1]						
Response	**No. Cases**	**Responders [2] (%)**		**Non-Responders [3] (%)**		**p**
MiR-199b expression	22	11		11		0.011
Low	5	0	(0)	5	(100)	
High	17	11	(64.7)	6	(35.3)	

[1] CRT = chemoradiotherapy; [2] Responders = cases with moderate or complete pathological response, and cases following W&W strategy; [3] Non-Responders = cases with poor or minimal pathological response.

Table 3. Association between downstaging and miR-199b expression in liquid biopsies from LARC patients.

Downstaging	No. Cases	Yes	(%)	No	(%)	p
MiR-199b expression	22	15		7		0.009
Low	5	1	(20)	4	(80)	
High	17	14	(82.4)	3	(17.6)	

Finally, we investigated potential changes in the miR-199b expression profile analyzing paired liquid biopsies from our initial cohort obtained after the neoadjuvant treatment. We could include post-treatment samples from 16 out of the 22 cases with enough material available. Interestingly, we found no differences in miR-199b expression levels between the subgroups of pre- and post-treatment samples ($p = 0.649$) (data not shown). Moreover, we observed decreased miR-199b levels with a fold change of 7.20 in post-treatment samples when we compared responders vs. non-responders, which is concordant with the results obtained in the pre-treatment specimens. Differences were statistically significant between subgroups ($p = 0.038$) (Figure S6).

3. Discussion

In recent years, many studies in the literature have evaluated the usefulness of miRs as liquid biopsy biomarkers. MiRs function as key regulators of diverse biological processes with high relevance in cancer, acting as oncogenes or tumor suppressors depending on their target genes, and have been involved in the pathogenesis of many tumor types [47–50]. Great insights into the implication of miRs in tumorigenesis and metastasis have allowed us to better understand the molecular mechanisms of disease development and progression, their potential clinical impact as biomarkers, and the investigation of miR-based therapies.

It has been reported that miR-199b is downregulated in several cancers and exerts tumor suppressive functions in tumor cell growth, invasion and metastasis [35,36,49,50]. In previous works from our group we have already studied the association of miR-199b with CRC progression, but it is still necessary to fully evaluate the biological role of miR-199b deregulation and its oncogenic properties. In the present work, we observed that miR-199b significantly reduces cell migration (Figure 1) and colony-forming ability (Figure 2). These results were in line with our initial hypothesis since miR-199b has been identified to negatively modulate SET [39], a key regulator of cell migration in CRC [51,52]. In fact, the potential role of the miR-199b/SET signaling axis in cell migration was also evaluated by transwell migration assays, observing that SET overexpression reversed the migration inhibitory effects of miR-199b (Figure S2). In fact, we show here that the ectopic expression of SET totally restores the miR-199b induced effects in colony formation ability (Figure S3), and the relevance of the miR-199b/SET axis was also evaluated and confirmed in cell proliferation (Figure S4). Furthermore, the observed effects of this miR in cell migration could also be explained by the fact that miR-199b also regulates additional targets such as JAG1 [31,32,38], DDR1 [35,36,38] or SIRT1 [42], which have been reported to be involved in the regulation of cell migration and invasion in CRC, as well as in other tumor types. Moreover, we analyzed the effect of miR-199b modulation on caspase-dependent apoptosis and we found a markedly enhanced number of apoptotic CRC cells after miR-199b overexpression. Thus, our results in Figure 3 are in concordance with previous observations reporting changes in CRC cell viability after ectopic miR-199b modulation [39]. These miR-199b-derived antitumor effects could probably be explained by its role as a negative regulator of several of those targets indicated above. We also showed that colonosphere formation ability was altered after miR-199b overexpression, substantially decreasing not only the number of colonosphere-derived cells formed (Figure 4), but also the number of cells per colonosphere (Figure S5). These observations could be due to the effect of miR-199b regulating targets such as SET or CD133 that have been described to regulate CRC cell stemness [34,37]. In fact, colonosphere derived CRC cells have been found to be enriched in CD133 expression, and to show a concomitant miR-199b downregulation. Of note, it can be noticed that in vitro experiments using anti-miR-199b showed that it had a significantly lower effect than pre-miR-199b. This issue can be explained by the fact that CRC cell lines including both SW480 and HT-29 have been described to express very low basal levels of this miR [39]. In fact, in concordance with the in vitro results showed in the present work, it has been previously reported that the same slight effects in both proliferation and modulating of PP2A activity exist with the anti-miR-199b due to this issue [39,44]. Notably, the miR-199b basal expression is even lower in SW480 cells than in HT-29 cells (as we also confirmed here, Table S1), which could explain that the effects of miR-199b silencing in cell migration that we found here only achieved significance in the HT-29 cell line (Figure 1). Altogether, the functional relevance of this miR is probably due to the contribution of several regulated targets of different signaling pathways, as explained above.

Regarding the clinical impact of miR-199b in LARC, we have previously studied its relevance as a prognostic tissue marker in LARC [44], results that were recently validated in a larger and independent cohort [45]. Moreover, miR-199b was also identified by Baek et al. [46] as one of the miRs with a potential clinical value as circulating preoperative marker in serum and exosomes-derived samples, determining its overexpression in higher survival rates. In concordance with the previous data and our findings with miR-199b as a tumor tissue marker in LARC patients, we found miR-199b downregulated in 22.7% of cases, compared with the prevalence of 26.4% and 22.2% obtained in our recent works [44,45]. Interestingly, we described that miR-199b downregulation associates with higher tumor size and higher pathological stage after nCRT (Table 1), as previously described [44]. We next stratified our cohort by the response to nCRT and observed lower miR-199b levels in the subgroup of non-responder cases (Figure 5). Notably, we also found that plasma expression levels of this miR had a predictive value of response to nCRT (Table 2), results that are similar to those obtained for the same miR as tumor tissue markers [44,45], and significantly associated with patient downstaging (Table 3). Moreover, we analyzed paired liquid biopsies after nCRT and also observed a marked reduction of miR-199b expression in the subgroup of non-responders (Figure S6), further supporting the findings described in pre-treatment samples. However, our findings here need to be confirmed in larger independent cohorts since the low number of cases analyzed in this work together with the lack of a training and a validation set of cases represents a strong limitation of the study, and the conclusions should be taken with caution.

The involvement of miR-199b deregulation in CRC progression has been previously reported in the literature. Thus, it has been described that miR-199b is downregulated in hepatic metastasis tissues from CRC patients [42], which was also described by our group in metastatic CRC [39] and LARC patients [45], observing that the subgroup of patients with low miR-199b expression showed markedly higher recurrence rates. These observations prompted us to explore the potential involvement of this miR in disease progression in liquid biopsies as well. We found a strong association between low miR-199b levels and relapse (Table S4), even though we did not reach statistical significance mainly due to the short cohort of patients that we could include in this study. Altogether, these findings strengthen the fact that miR-199b is hardly involved in tumor malignant features and progression, and could be used as a predictive circulating marker of response to nCRT.

Of note, there are some limitations of our study that have to be taken into consideration. These limitations include the need to validate the functional role of miR-199b observed using in vivo models and primary cultures from CRC patients. Moreover, the limited number of patients included in our study supposes a great limitation, and a future validation of these results in a larger, multicentric and independent cohort of LARC patients that would reinforce the clinical relevance of miR-199b as a circulating marker of response to nCRT in this disease is required. In addition, it would be interesting to lengthen the time of the study to analyze miR-199b as a predictor of patient outcome and relapse. This also could allow for the obtaining of periodic blood samples of each patient, thereby making an exhaustive follow-up of both responders and non-responders to nCRT groups. This would increase the number of patients that could avoid postoperative complications due to TME surgery and could be included in a W&W protocol, improving the quality of life of a larger number of selected patients. Moreover, an independent study with a larger W&W patient cohort could represent an opportunity to validate miR-199b as a circulating marker of recurrence in patients following this approach.

4. Materials and Methods

4.1. Cell Cultures and Transfection

The human CRC cell lines SW480 (ATCC CCL-228) and HT-29 (ATCC HTB-38) were purchased from American Type Culture Collection (ATCC, Manassas, VA, USA). Authentication was done by the authors in all cases (LGC Standards, Wesel, Germany). Cell lines were maintained in RPMI-1640 (Invitrogen, Carlsbad, CA, USA) with 10% fetal bovine serum (FBS) and were grown at 37 °C in a 5% CO_2 atmosphere. Media were supplemented with penicillin G (100 U/mL) and streptomycin (0.1 mg/mL). For transfection experiments, CRC cells were seeded in 6-well plates and transfected with 10 μL of Lipofectamine 2000 (Invitrogen, Carlsbad, CA, USA) and 20 nM of a miR-199b specific mirVana™ miRNA Mimic and Inhibitor (Ambion, Cambridge, UK).

4.2. Patient Samples

The study was carried out on plasma samples obtained from a total of 22 LARC patients who were selected retrospectively and treated between 2018 and 2021 in University Hospital Fundación Jiménez Díaz (Madrid, Spain). Plasma samples were taken from each patient before and after preoperative nCRT. Baseline, clinical and pathological characteristics of each patient at the time of inclusion are presented in Table S3. All patients were treated with nCRT and TME, except for two of them who were not subjected to surgery and followed a therapeutic strategy based on the W&W protocol. All of them were treated by the European Guidelines Recommendations with correct preoperative locoregional staging based on a magnetic resonance (MR), a transrectal ultrasound (TRUS), and a full body computed tomography (CT). The selection criteria included the presence of adenocarcinoma with operable disease, enough material collected before and after nCRT, clinical follow-up data available and absence of metastasis. TNM (tumor, node, metastases) staging was performed based on the 7th American Joint Committee on Cancer (AJCC) staging system established for CRC. Plasma samples were collected into EDTA Vacutainer®-tubes and immediately sent to the laboratory. Samples were subjected to centrifugation at 1500× g for 15 min at 4 °C for plasma separation. The supernatant was carefully transferred to a sterile tube and centrifuged again at 3000× g for 5 min at 4 °C in order to pellet any debris and insoluble components. Plasma samples were aliquoted and immediately frozen at −80 °C until the experiments were carried out. All patients gave written informed consent for sample storage and analysis at Fundación Jiménez Díaz Hospital biobank with the approval of the ethical committee and institutional review board of Fundación Jiménez Díaz (ref. PIC202-20).

4.3. Evaluation of Pathological Response

All tumor samples that resulted from the initial biopsies derived from colonoscopy and the surgical resection were classified according to the College of American Pathologist guidelines for invasive carcinomas (TNM, 7th ed.). Two independent pathologists who were blinded to patient outcome evaluated tumor regression grade according to the modified Ryan classification that categorizes tumors into four levels of response: complete response, moderate response, minimal response and poor response. Complete response indicates no viable cancer cells (RYAN 0), moderate response indicates single cells or little groups of cancer cells (RYAN 1), minimal response denotes residual cancer outgrown by fibrosis (RYAN 2), and poor response is associated with minimal or no tumor kill with extensive residual cancer (RYAN 3). According to clinical guidelines, every regression grade was compared with the primary tumor [53].

4.4. Wound-Healing Assay

A total of 8×10^5 cells per well were seeded in 6-well plates and allowed to adhere for 24 h in complete medium. The monolayer was artificially injured by scratching across the plate with a 10 µL pipette tip. Wells were then washed twice with phosphate-buffered saline (PBS) to remove detached cells and wound healing was monitored using a Leica DMi1 (Leica, Wetzlar, Germany) microscope and the image acquisition software Leica Application Suite version 4.5. Images were captured at the beginning and at regular intervals during cell migration to close the wound. Comparisons were performed to quantify the migration rate of the cells between the different experimental conditions. Relative cell migration is represented in the histograms considering the percentage of healed area after ectopic miR-199b silencing or overexpression and compared to control conditions.

4.5. Transwell Migration Assay

Migration assays were performed using 24-well plates with transwell permeable supports of 6.5 mm insert and a polycarbonate membrane with an 8 µm pore size (Costar 3422, Corning Inc., Corning, NY, USA). Cells were seeded in the upper chamber at 1×10^4 cells/mL in 0.1 mL of serum-free RPMI-1640 media. A volume of 0.8 mL of media supplemented with 10% FBS was placed in the bottom well as a chemo-attractant. After incubation for 24 h at 37 °C in an atmosphere containing 5% CO_2, migrated cells on the lower surface were stained using crystal violet and counted under a light microscope.

4.6. Colony-Forming Assay

Experiments were performed in 6-well plates coated with 3 mL of 0.6% soft agarose (Sigma, St. Louis, MO, USA) in RPMI medium. A total of 5×10^3 cells were suspended in 0.3% agarose in RPMI medium and plated in triplicates over the pre-coated wells. Fresh medium was supplied twice a week. After 10–15 days, colonies were stained with Thiazolyl Blue Tetrazolium Bromiden MTT (M-5655, Sigma, St. Louis, MO, USA) for 4 h at 37 °C. Then, colonies were fixed by adding dimethyl sulfoxide (DMSO) overnight at 37 °C. Colony numbers were determined from triplicates and three independent experiments were carried out for each condition and cell line.

4.7. Cell Viability Assay

Cell proliferation was measured in triplicate wells by the MTS assay in 96-well plates using the CellTiter 96 Aqueous One Solution Cell Proliferation Assay (Promega Corp., Madison, WI, USA), according to the manufacturer's instructions.

4.8. Analysis of Caspase Activation

Quantification of caspase-3/7 activity was carried out using the caspase Glo-3/7 assay kit (Promega Corp., Madison, WI, USA). First, 5×10^3 cells were seeded in a white-walled 96-well plate, and the Z-DEVD reagent, the luminogenic caspase-3/7 substrate containing a tetrapeptide Asp–Glu–Val–Asp, was added with a 1:1 ratio of reagent to sample. After incubation at room temperature for 90 min, the substrate cleavage by activated caspase-3 and -7 and the intensity of a luminescent signal was measured by a FLUOstar OPTIMA luminometer (BMG Labtech, Cary, NC, USA). Differences in caspase-3/7 activity are expressed as fold-change in luminescence.

4.9. Colonosphere Formation Assay

We generated colonosphere-derived cells from SW480 and HT-29 using 6-well ultra-low attachment plates (Corning Inc., Corning, NY, USA), where 10,000 cells per well were plated. Cells were grown in serum-free medium DMEM/F12 supplemented with GlutMAX^{TM}-I (ThermoFisher Scientific, Waltham, MA, USA), 1% N2 (ThermoFisher Scientific), 2% B27 (ThermoFisher Scientific), 20 ng/mL human FGF (Sigma, St. Louis, MO, USA) and 50 ng/mL EGF (Sigma). After seven days, plates were analyzed for colonosphere formation. For quantification of the number of cells per colonosphere, colonospheres were collected and dissociated with trypsin to give single-cell suspensions. Viable cells were counted in a Neubauer chamber using a Trypan Blue exclusion test.

4.10. Western Blot Analysis

Protein extracts were isolated using TRIzol Reagent (Invitrogen, Carlsbad, CA, USA) following manufacturer's indications, clarified (12,000× *g*, 15 min, 4 °C), denatured and subjected to SDS-PAGE and Western-blot. Antibodies used were goat polyclonal anti-SET (E-15) (Santa Cruz Biotechnology, Dallas, TX, USA), and mouse monoclonal anti-βactin (Sigma, St. Louis, MO, USA). Proteins were detected with the appropriate secondary antibodies conjugated to alkaline phosphatase (Sigma, St. Louis, MO, USA) by chemiluminescence using Tropix CSPD and Tropix Nitro Block II (Applied Biosystems, Foster City, CA, USA).

4.11. RNA Isolation

RNA extraction from all plasma samples was performed using the Qiagen miRNeasy Serum/Plasma Kit (QIAGEN, Hilden, Germany, catalog number: 217184). Plasma was thawed on ice and centrifuged at 3000 g for 5 min at 4 °C. An aliquot of 200 μL of plasma per sample was transferred to a new microcentrifuge tube and 1 mL of Qiazol was added. Total RNA, including small RNAs, were enriched and purified following the manufacturer's instructions. RNA obtained was quantified with a Nanodrop Spectrophotometer (Thermo Scientific, Waltham, MA, USA).

4.12. Quantification of miR Expression Levels

The RNA samples were reverse transcribed using the TaqManHMicroRNA Reverse Transcription Kit (Applied Biosystems, Foster City, CA, USA), and mature miRs were quantified by quantitative real-time reverse transcription polymerase chain reaction (RT-PCR) using TaqMan MicroRNA Assays (Applied Biosystems) specific for the miR-199b (reference number: 000500), and miR-1228 (reference number: 002919), which was used as internal control. Reactions were carried out using an Applied Biosystems 7500 Sequence Detection System under the following conditions: 95 °C for 10 min, followed by 40 cycles of 95 °C for 15 s and 60 °C for 1 min. Analysis of relative gene expression data was performed using the ΔΔCT method [54], where ΔΔCT = (CT, Target Gene-CT, control) Tumor-(CT, Target Gene-CT, control) Normal Control. MiR-199b downregulation was considered when the expression in a sample was lower than the mean minus standard deviation (SD) of the patient cohort, as previously described [39].

4.13. Statistical Analysis

All statistical analyses were carried out using the software tool SPSS v20 for Windows (SPSS Inc., Chicago, IL, USA). The association between miR-199b expression and clinical and molecular parameters were analyzed by applying the chi-square test (Fisher's exact test) based on bimodal distribution of data. Comparisons between miR-199b expression levels of each patient before and after preoperative CRT were performed using Mann–Whitney and paired t-tests. Data represented for transfection experiments are mean of three independent experiments ± SD. Statistical comparisons were obtained by two-sided t-test analyses. Statistical significance was considered when p-value was lower than 0.05. This study has been performed following the Reporting Recommendations for Tumor Marker Prognostic Studies (REMARK) guidelines [55,56].

5. Conclusions

In conclusion, our findings here provide novel important data regarding the relevance of miR-199b as a regulator of cellular processes crucial for malignant transformation such as cell growth, migration, aggressiveness, colonosphere formation and apoptosis, with the particular importance of the miR-199bSET axis in some of these phenotypes. At the clinical level, we found that miR-199b expression could serve as a marker in plasma samples from LARC patients, showing predictive value of response to nCRT and patient recurrence. Altogether, our results highlight the functional roles and clinical usefulness of miR-199b as a potential circulating biomarker in LARC patients.

Supplementary Materials: The following are available online at https://www.mdpi.com/article/10.3390/ijms23042203/s1.

Author Contributions: Conceptualization, I.C., A.S. and J.G.-F.; methodology, I.C., A.S. and J.R.; software, J.R., A.S. and C.C.; formal analysis, I.C. and A.S.; investigation, I.C., A.S., M.L., M.S.-A. and M.M.-G.; writing—original draft preparation, I.C. and A.S.; writing—review and editing, J.M.-G. and F.R.; funding acquisition, F.R. and J.G.-F. All authors have read and agreed to the published version of the manuscript.

Funding: This research was funded by PI18/00382 and PI16/01468 grants from "Instituto de Salud Carlos III FEDER". M.S.-A. is supported by "Fundación Conchita Rábago de Jiménez Díaz".

Institutional Review Board Statement: The study was conducted according to the guidelines of the Declaration of Helsinki, and approved by the Institutional Review Board of University Hospital Fundación Jiménez Díaz (PIC202-20).

Informed Consent Statement: Informed consent was obtained from all subjects involved in the study.

Data Availability Statement: Data sharing is not applicable for this article.

Conflicts of Interest: The authors declare that they have no conflict of interest.

References

1. Rawla, P.; Sunkara, T.; Barsouk, A. Epidemiology of colorectal cancer: Incidence, mortality, survival, and risk factors. *Prz. Gastroenterol.* **2019**, *14*, 89–103. [CrossRef]
2. Global Cancer Observatory. Colorectal Cancer. Available online: https://gco.iarc.fr/today/data/factsheets/cancers/10_8_9-Colorectum-fact-sheet.pdf (accessed on 2 November 2021).
3. Spanish Society of Medical Oncology. Cancer Data in Spain. Available online: https://seom.org/images/Cifras_del_cancer_en_Espnaha_2021.pdf (accessed on 31 October 2021).
4. Sanghera, P.; Wong, D.W.; McConkey, C.C.; Geh, J.I.; Hartley, A. Chemoradiotherapy for rectal cancer: An updated analysis of factors affecting pathological response. *Clin. Oncol. (R. Coll. Radiol.)* **2008**, *20*, 176–183. [CrossRef]
5. Dayde, D.; Gunther, J.; Hirayama, Y.; Weksberg, D.C.; Boutin, A.; Parhy, G.; Aguilar-Bonavides, C.; Wang, H.; Katayama, H.; Abe, Y.; et al. Identification of blood-based biomarkers for the prediction of the response to neoadjuvant chemoradiation in rectal cancer. *Cancers* **2021**, *13*, 3642. [CrossRef]
6. São Julião, G.P.; Habr-Gama, A.; Vailati, B.B.; Araujo, S.E.A.; Fernandez, L.M.; Perez, R.O. New strategies in rectal cancer. *Surg. Clin. N. Am.* **2017**, *97*, 587–604. [CrossRef]
7. Papaccio, F.; Roselló, S.; Huerta, M.; Gambardella, V.; Tarazona, N.; Fleitas, T.; Roda, D.; Cervantes, A. Neoadjuvant chemotherapy in locally advanced rectal cancer. *Cancers* **2020**, *12*, 3611. [CrossRef]
8. Bahadoer, R.R.; Dijkstra, E.A.; van Etten, B.; Marijnen, C.A.M.; Putter, H.; Kranenbarg, E.M.-K.; Roodvoets, A.G.H.; Nagtegaal, I.D.; Beets-Tan, R.G.H.; Blomqvist, L.K.; et al. Short-course radiotherapy followed by chemotherapy before total mesorectal excision (TME) versus preoperative chemoradiotherapy, TME, and optional adjuvant chemotherapy in locally advanced rectal cancer (RAPIDO): A randomised, open-label, phase 3 trial. *Lancet Oncol.* **2021**, *22*, 29–42. [CrossRef]

9. Conroy, T.; Bosset, J.-F.; Etienne, P.-L.; Rio, E.; François, É.; Mesgouez-Nebout, N.; Vendrely, V.; Artignan, X.; Bouché, O.; Gargot, D.; et al. Neoadjuvant chemotherapy with FOLFIRINOX and preoperative chemoradiotherapy for patients with locally advanced rectal cancer (UNICANCER-PRODIGE 23): A multicentre, randomised, open-label, phase 3 trial. *Lancet Oncol.* **2021**, *22*, 702–715. [CrossRef]
10. Norcic, G. Liquid biopsy in colorectal cancer-current status and potential clinical applications. *Micromachines* **2018**, *9*, 300. [CrossRef] [PubMed]
11. Vacante, M.; Borzì, A.M.; Basile, F.; Biondi, A. Biomarkers in colorectal cancer: Current clinical utility and future perspectives. *World J. Clin. Cases* **2018**, *6*, 869–881. [CrossRef] [PubMed]
12. Palmirotta, R.; Lovero, D.; Cafforio, P.; Felici, C.; Mannavola, F.; Pellè, E.; Quaresmini, D.; Tucci, M.; Silvestris, F. Liquid biopsy of cancer: A multimodal diagnostic tool in clinical oncology. *Ther. Adv. Med. Oncol.* **2018**, *10*, 1758835918794630. [CrossRef] [PubMed]
13. Malentacchi, F.; Sgromo, C.; Antonuzzo, L.; Pillozzi, S. Liquid biopsy in endometrial cancer. *J. Cancer Metastasis Treat.* **2020**, *6*, 34. [CrossRef]
14. Quirico, L.; Orso, F. The power of microRNAs as diagnostic and prognostic biomarkers in liquid biopsies. *Cancer Drug Resist.* **2020**, *3*, 117–139. [CrossRef]
15. Vacante, M.; Ciuni, R.; Basile, F.; Biondi, A. The liquid biopsy in the management of colorectal cancer: An overview. *Biomedicines* **2020**, *8*, 308. [CrossRef]
16. Shigeyasu, K.; Toden, S.; Zumwalt, T.J.; Okugawa, Y.; Goel, A. Emerging role of MicroRNAs as liquid biopsy biomarkers in gastrointestinal cancers. *Clin. Cancer Res.* **2017**, *23*, 2391–2399. [CrossRef]
17. De Palma, F.D.E.; Luglio, G.; Tropeano, F.P.; Pagano, G.; D'Armiento, M.; Kroemer, G.; Maiuri, M.C.; De Palma, G.D. The role of micro-RNAs and circulating tumor markers as predictors of response to neoadjuvant therapy in locally advanced rectal cancer. *Int. J. Mol. Sci.* **2020**, *21*, 7040. [CrossRef]
18. Peng, Y.; Croce, C.M. The role of MicroRNAs in human cancer. *Signal Transduct. Target Ther.* **2016**, *1*, 15004. [CrossRef]
19. Kong, Y.W.; Ferland-McCollough, D.; Jackson, T.J.; Bushell, M. MicroRNAs in cancer management. *Lancet Oncol.* **2012**, *13*, e249–e258. [CrossRef]
20. Imedio, L.; Cristóbal, I.; Rubio, J.; Santos, A.; Rojo, F.; García-Foncillas, J. MicroRNAs in rectal cancer: Functional significance and promising therapeutic value. *Cancers* **2020**, *12*, 2040. [CrossRef]
21. Iorio, M.V.; Ferracin, M.; Liu, C.G.; Veronese, A.; Spizzo, R.; Sabbioni, S.; Magri, E.; Pedriali, M.; Fabbri, M.; Campiglio, M.; et al. MicroRNA gene expression deregulation in human breast cancer. *Cancer Res.* **2005**, *65*, 7065–7070. [CrossRef]
22. Calin, G.A.; Ferracin, M.; Cimmino, A.; Di Leva, G.; Shimizu, M.; Wojcik, S.E.; Iorio, M.V.; Visone, R.; Sever, N.I.; Fabbri, M.; et al. A microRNA signature associated with prognosis and progression in chronic lymphocytic leukemia. *N. Engl. J. Med.* **2005**, *353*, 1793–1801. [CrossRef]
23. Qi, J.; Wang, J.; Katayama, H.; Sen, S.; Liu, S.M. Circulating microRNAs (cmiRNAs) as novel potential biomarkers for hepatocellular carcinoma. *Neoplasma* **2013**, *60*, 135–142. [CrossRef] [PubMed]
24. Roldo, C.; Missiaglia, E.; Hagan, J.P.; Falconi, M.; Capelli, P.; Bersani, S.; Calin, G.A.; Volinia, S.; Liu, C.G.; Scarpa, A.; et al. MicroRNA expression abnormalities in pancreatic endocrine and acinar tumors are associated with distinctive pathological features and clinical behavior. *J. Clin. Oncol.* **2006**, *24*, 4677–4684. [CrossRef] [PubMed]
25. Porkka, K.P.; Pfeiffer, M.J.; Waltering, K.K.; Vessella, R.L.; Tammela, T.L.J.; Visakorpi, T. MicroRNA expression profiling in prostate cancer. *Cancer Res.* **2007**, *67*, 6130–6135. [CrossRef]
26. Ciafrè, S.A.; Galardi, S.; Mangiola, A.; Ferracin, M.; Liu, C.G.; Sabatino, G.; Negrini, M.; Maira, G.; Croce, C.M.; Farace, M.G. Extensive modulation of a set of microRNAs in primary glioblastoma. *Biochem. Biophys. Res. Commun.* **2005**, *334*, 1351–1358. [CrossRef]
27. Caramés, C.; Cristóbal, I.; Moreno, V.; Del Puerto, L.; Moreno, I.; Rodriguez, M.; Marín, J.P.; Correa, A.V.; Hernández, R.; Zenzola, V.; et al. MicroRNA-21 predicts response to preoperative chemoradiotherapy in locally advanced rectal cancer. *Int. J. Colorectal Dis.* **2015**, *30*, 899–906. [CrossRef] [PubMed]
28. Caramés, C.; Cristobal, I.; Moreno, V.; Marín, J.P.; González-Alonso, P.; Torrejón, B.; Minguez, P.; Leon, A.; Martín, J.I.; Hernández, R.; et al. MicroRNA-31 emerges as a predictive biomarker of pathological response and outcome in locally advanced rectal cancer. *Int. J. Mol. Sci.* **2016**, *17*, 878. [CrossRef] [PubMed]
29. Zhu, Y.; Peng, Q.; Lin, Y.; Zou, L.; Shen, P.; Chen, F.; Min, M.; Shen, L.; Chen, J.; Shen, B. Identification of biomarker microRNAs for predicting the response of colorectal cancer to neoadjuvant chemoradiotherapy based on microRNA regulatory network. *Oncotarget* **2016**, *8*, 2233–2248. [CrossRef]
30. Roman-Canal, B.; Tarragona, J.; Moiola, C.P.; Gatius, S.; Bonnin, S.; Ruiz-Miró, M.; Sierra, J.E.; Rufas, M.; González, E.; Porcel, J.M.; et al. EV-associated miRNAs from peritoneal lavage as potential diagnostic biomarkers in colorectal cancer. *J. Transl. Med.* **2019**, *17*, 208. [CrossRef]
31. Li, G.L.; Yuan, J.H.; Zhuang, G.D.; Wu, D.Q. MiR-199b exerts tumor suppressive functions in hepatocellular carcinoma by directly targeting JAG1. *Eur. Rev. Med. Pharmacol. Sci.* **2018**, *22*, 7679–7687. [CrossRef]
32. Qu, X.; Chen, Z.; Fan, D.; Sun, C.; Zeng, Y.; Guo, Z.; Qi, Q.; Li, W. MiR-199b-5p inhibits osteogenic differentiation in ligamentum flavum cells by targeting JAG1 and modulating the Notch signalling pathway. *J. Cell Mol. Med.* **2017**, *21*, 1159–1170. [CrossRef]

33. Fan, X.; Matsui, W.; Khaki, L.; Stearns, D.; Chun, J.; Li, Y.M.; Eberhart, C.G. Notch pathway inhibition depletes stem-like cells and blocks engraftment in embryonal brain tumors. *Cancer Res.* **2006**, *66*, 7445–7452. [CrossRef]
34. Garzia, L.; Andolfo, I.; Cusanelli, E.; Marino, N.; Petrosino, G.; De Martino, D.; Esposito, V.; Galeone, A.; Navas, L.; Esposito, S.; et al. MicroRNA-199b-5p impairs cancer stem cells through negative regulation of HES1 in medulloblastoma. *PLoS ONE* **2009**, *4*, e4998. [CrossRef]
35. Zhao, Z.; Zhao, S.; Luo, L.; Xiang, Q.; Zhu, Z.; Wang, J.; Liu, Y.; Luo, J. MiR-199b-5p-DDR1-ERK signalling axis suppresses prostate cancer metastasis via inhibiting epithelial-mesenchymal transition. *Br. J. Cancer* **2021**, *124*, 982–994. [CrossRef]
36. Wu, A.; Chen, Y.; Liu, Y.; Lai, Y.; Liu, D. MiR-199b-5p inhibits triple negative breast cancer cell proliferation, migration and invasion by targeting DDR1. *Oncol. Lett.* **2018**, *16*, 4889–4896. [CrossRef]
37. Lin, X.; Qiu, W.; Xiao, Y.; Ma, J.; Xu, F.; Zhang, K.; Gao, Y.; Chen, Q.; Li, Y.; Li, H.; et al. MiR-199b-5p Suppresses tumor angiogenesis mediated by vascular endothelial cells in breast cancer by targeting ALK1. *Front. Genet.* **2020**, *10*, 1397. [CrossRef]
38. Chen, L.Y.; Zhi, Z.; Wang, L.; Zhao, Y.Y.; Deng, M.; Liu, Y.H.; Qin, Y.; Tian, M.M.; Liu, Y.; Shen, T.; et al. NSD2 circular RNA promotes metastasis of colorectal cancer by targeting miR-199b-5p-mediated DDR1 and JAG1 signalling. *J. Pathol.* **2019**, *248*, 103–115. [CrossRef]
39. Cristóbal, I.; Caramés, C.; Rincón, R.; Manso, R.; Madoz-Gúrpide, J.; Torrejón, B.; González-Alonso, P.; Rojo, F.; García-Foncillas, J. Downregulation of microRNA-199b predicts unfavorable prognosis and emerges as a novel therapeutic target which contributes to PP2A inhibition in metastatic colorectal cancer. *Oncotarget* **2017**, *8*, 40169–40180. [CrossRef]
40. Liu, R.; Wang, J.H.; Xu, C.; Sun, B.; Kang, S.O. Activin pathway enhances colorectal cancer stem cell self-renew and tumor progression. *Biochem. Biophys. Res. Commun.* **2016**, *479*, 715–720. [CrossRef]
41. Jung, B.; Staudacher, J.J.; Beauchamp, D. Transforming growth factor β superfamily signaling in development of colorectal cancer. *Gastroenterology* **2017**, *152*, 36–52. [CrossRef]
42. Shen, Z.L.; Wang, B.; Jiang, K.W.; Ye, C.X.; Cheng, C.; Yan, Y.C.; Zhang, J.Z.; Yang, Y.; Gao, Z.D.; Ye, Y.J.; et al. Downregulation of miR-199b is associated with distant metastasis in colorectal cancer via activation of SIRT1 and inhibition of CREB/KISS1 signaling. *Oncotarget* **2016**, *7*, 35092–35105. [CrossRef]
43. Cristóbal, I.; Rubio, J.; Torrejón, B.; Santos, A.; Caramés, C.; Luque, M.; Sanz-Álvarez, M.; Alonso, R.; Zazo, S.; Madoz-Gúrpide, J.; et al. MicroRNA-199b deregulation shows a strong SET-independent prognostic value in early-stage colorectal cancer. *J. Clin. Med.* **2020**, *9*, 2419. [CrossRef] [PubMed]
44. Cristóbal, I.; Rubio, J.; Santos, A.; Torrejón, B.; Caramés, C.; Imedio, L.; Mariblanca, S.; Luque, M.; Sanz-Alvarez, M.; Zazo, S.; et al. MicroRNA-199b Downregulation confers resistance to 5-fluorouracil treatment and predicts poor outcome and response to neoadjuvant chemoradiotherapy in locally advanced rectal cancer patients. *Cancers* **2020**, *12*, 1655. [CrossRef] [PubMed]
45. Cristóbal, I.; Santos, A.; Rubio, J.; Caramés, C.; Zazo, S.; Sanz-Álvarez, M.; Luque, M.; Madoz-Gúrpide, J.; Rojo, F.; García-Foncillas, J. Validation of microRNA-199b as A promising predictor of outcome and response to neoadjuvant treatment in locally advanced rectal cancer patients. *Cancers* **2021**, *13*, 5003. [CrossRef] [PubMed]
46. Baek, D.W.; Kim, G.; Kang, B.W.; Kim, H.J.; Park, S.Y.; Park, J.S.; Choi, G.S.; Kang, M.K.; Hur, K.; Kim, J.G. High expression of microRNA-199a-5p is associated with superior clinical outcomes in patients with locally advanced rectal cancer. *J. Cancer Res. Clin. Oncol.* **2020**, *146*, 105–115. [CrossRef]
47. Calin, G.A.; Croce, C.M. MicroRNA signatures in human cancers. *Nat. Rev. Cancer* **2006**, *6*, 857–866. [CrossRef]
48. Suer, I.; Guzel, E.; Karatas, O.F.; Creighton, C.J.; Ittmann, M.; Ozen, M. MicroRNAs as prognostic markers in prostate cancer. *Prostate* **2019**, *79*, 265–271. [CrossRef]
49. Lai, Y.; Quan, J.; Lin, C.; Li, H.; Hu, J.; Chen, P.; Xu, J.; Guan, X.; Xu, W.; Lai, Y.; et al. miR-199b-5p serves as a tumor suppressor in renal cell carcinoma. *Exp. Ther. Med.* **2018**, *16*, 436–444. [CrossRef]
50. Favreau, A.J.; McGlauflin, R.E.; Duarte, C.W.; Sathyanarayana, P. MiR-199b, a novel tumor suppressor miRNA in acute myeloid leukemia with prognostic implications. *Exp. Hematol. Oncol.* **2016**, *5*, 4. [CrossRef]
51. Ten Klooster, J.P.; Leeuwen, I.V.; Scheres, N.; Anthony, E.C.; Hordijk, P.L. Rac-1induced cell migration requires membranes recruitment of the nuclear oncogene SET. *EMBO J.* **2007**, *26*, 336–345. [CrossRef]
52. Cristóbal, I.; Torrejón, B.; Rubio, J.; Santos, A.; Pedregal, M.; Caramés, C.; Zazo, S.; Luque, M.; Sanz-Alvarez, M.; Madoz-Gúrpide, J.; et al. Deregulation of SET is associated with tumor progression and predicts adverse outcome in patients with early-stage colorectal cancer. *J. Clin. Med.* **2019**, *8*, 346. [CrossRef]
53. Greenson, J.K.; Huang, S.-C.; Herron, C.; Moreno, V.; Bonner, J.D.; Tomsho, L.P.; Ben-Izhak, O.; Cohen, H.I.; Trougouboff, P.; Bejhar, J.; et al. Pathologic predictors of microsatellite instability in colorectal cancer. *Am. J. Surg. Pathol.* **2009**, *33*, 126–133. [CrossRef]
54. Livak, K.J.; Schmittgen, T.D. Analysis of relative gene expression data using real-Time quantitative PCR and the 2(-Delta Delta C(T)) Method. *Methods* **2001**, *25*, 402–408. [CrossRef]
55. McShane, L.M.; Altman, D.G.; Sauerbrei, W.; Taube, S.E.; Gion, M.; Clark, G.M. Statistics Subcommittee of the NCI-EORTC Working Group on Cancer Diagnostics. Reporting recommendations for tumor marker prognostic studies. *J. Clin. Oncol.* **2005**, *23*, 9067–9072. [CrossRef]
56. Sauerbrei, W.; Taube, S.E.; McShane, L.M.; Cavenagh, M.M.; Altman, D.G. Reporting Recommendations for Tumor Marker Prognostic Studies (REMARK): An abridged explanation and elaboration. *J. Natl. Cancer Inst.* **2018**, *110*, 803–811. [CrossRef]

International Journal of
Molecular Sciences

MDPI

Review

MicroRNAs and 'Sponging' Competitive Endogenous RNAs Dysregulated in Colorectal Cancer: Potential as Noninvasive Biomarkers and Therapeutic Targets

Brian G. Jorgensen and Seungil Ro *

Department of Physiology & Cell Biology, University of Nevada, Reno School of Medicine, Reno, NV 89557, USA; brianjorgensen@med.unr.edu
* Correspondence: sro@med.unr.edu; Tel.: +1-775-784-1462

Abstract: The gastrointestinal (GI) tract in mammals is comprised of dozens of cell types with varied functions, structures, and histological locations that respond in a myriad of ways to epigenetic and genetic factors, environmental cues, diet, and microbiota. The homeostatic functioning of these cells contained within this complex organ system has been shown to be highly regulated by the effect of microRNAs (miRNA). Multiple efforts have uncovered that these miRNAs are often tightly influential in either the suppression or overexpression of inflammatory, apoptotic, and differentiation-related genes and proteins in a variety of cell types in colorectal cancer (CRC). The early detection of CRC and other GI cancers can be difficult, attributable to the invasive nature of prophylactic colonoscopies. Additionally, the levels of miRNAs associated with CRC in biofluids can be contradictory and, therefore, must be considered in the context of other inhibiting competitive endogenous RNAs (ceRNA) such as lncRNAs and circRNAs. There is now a high demand for disease treatments and noninvasive screenings such as testing for bloodborne or fecal miRNAs and their inhibitors/targets. The breadth of this review encompasses current literature on well-established CRC-related miRNAs and the possibilities for their use as biomarkers in the diagnoses of this potentially fatal GI cancer.

Keywords: colorectal cancer; miRNA; ceRNA; lncRNA; circRNA; sponging; noninvasive biomarkers

Citation: Jorgensen, B.G.; Ro, S. MicroRNAs and 'Sponging' Competitive Endogenous RNAs Dysregulated in Colorectal Cancer: Potential as Noninvasive Biomarkers and Therapeutic Targets. *Int. J. Mol. Sci.* **2022**, *23*, 2166. https://doi.org/10.3390/ijms23042166

Academic Editors: Alessandro Ottaiano and Donatella Delle Cave

Received: 20 January 2022
Accepted: 2 February 2022
Published: 16 February 2022

1. Introduction

As a generalized class of biomolecules, RNA species occupy a unique niche that can alternate between being transient and/or enduring, while being consequential or seemingly neutral to cellular homeostasis, growth, and survival across a myriad of taxonomic levels of life. While RNA has been known as a unique biomolecule since the 19th century, it came to the forefront of biological study in the mid-20th century upon the discovery that various RNA molecules were, collectively, the transcript, delivery, and production scaffolding for translating the genetic information contained within DNA into functional proteins via ribosomes. However, in the decades since those foundational studies, it has been found that a vast majority of the biological activity of RNA species is not as transcript templates (messenger RNA (mRNA)) for protein synthesis, but rather regulators of gene expression. These non-mRNA RNA transcripts do not primarily function as templates for protein translation and, thus, are called noncoding RNAs (ncRNA). One important class of endogenous regulatory ncRNAs are microRNAs (miRNA), which are 21–25 nucleotides in length when mature following enzymatic processing. Following their initial discoveries as endogenously produced transcripts, it was found that the dysregulation of many specific ncRNAs are tied to malignancies in humans [1–4]. Subsequent studies into each individual/category of ncRNA built upon these discoveries of dysregulated ncRNA expression to reveal the modes of action on cellular pathways known to be causative in carcinogenesis. Certain miRNAs were found to be key regulators of well-established carcinogenic pathways such as miR-143 and miR-145, acting as tumor suppressors on the p53/c-Myc pathway [5,6], the circular RNA (circRNA) Cdr1as sequestering miR-7, reducing its ability to regulate the proto-oncogenic PI3K/AKT pathway [7,8], and the long noncoding RNA (lncRNA) GAS5-controlling mTOR-mediated proliferation via the competitive binding of glucocorticoid receptors and miR-21 [9,10]. While these miRNAs can affect many distinct cellular pathways, there is oftentimes an overlap and/or interplay between unique ncRNAs classes that results in the pathological alterations of many signaling pathways at the cellular level. This can manifest, for example, when lncRNAs or circRNAs competitively sequester miRNAs from their targeted mRNAs (collectively known as competitive endogenous RNAs, ceRNA) through complementary base pairing, known as "sponging", as seen between the long noncoding RNA (lncRNA) HOTAIRM1 or the circular RNA (circRNA) circITCH and miR-17 [11,12]. Colorectal cancer (CRC) most commonly manifests as neoplasias consisting mainly of mutated intestinal epithelial cells that are unable to maintain a proper differentiation status and/or connections to neighboring cells, leading to unchecked cellular division and the dysregulation of oncogenic and/or tumor-suppressive genes [13,14]. Most CRC cases do not yet have a prescribed causative provenance and are, thus, considered sporadic (70%) [15]. For decades, invasive colonoscopies have been the most common form of early detection for CRC and they luckily reduce the death rate by about 50% [16,17]. Unfortunately, this reduction in mortality appears to be largely associated with carcinomas found on the left side of the colon as they are easier to observe and remove [18], leaving around 10% of all CRCs undetectable via colonoscopy [19]. This inability to robustly detect CRC regardless of the location provides the impetus to find biomarkers, such as miRNAs, in excreted bodily fluids (stool/serum/plasma) that could potentially diagnose CRC in various colonic regions. These strategies of measuring levels of known oncogenes in bodily fluids (mainly serum, plasma, or stool) have already begun to be investigated and employed for miRNA [20–22], lncRNA [23,24], and circRNA [25,26]. Unfortunately, due to miRNAs primarily exerting direct influence on other RNA molecules or proteins, which may or may not coexist together with the miRNA of interest, the basic presence or absence of any given miRNA is not necessarily enough to diagnose CRC. This conundrum adds a layer of difficulty in attempting to understand the relationship of these miRNAs to GI pathologies resulting in many unclear, or even contradictory, connections between the disease state and miRNA across samples and studies. Thus, to avoid a cavalcade listing of any miRNAs statistically associated with CRC, this review focuses on known miRNA signatures found within the biofluids of CRC patients, that also have accompanying evidence-based mechanistic hypotheses which represent ideal candidates for the use of biomarkers as diagnostics and/or therapeutics.

2. miRNAs and Colorectal Cancer

miRNAs are short RNA molecules, only 21–25 nucleotides in length upon the completion of the processing of the stem-loop pre-miRNA via the endoribonucleases Drosha and Dicer into single-stranded mature miRNAs, which are bound by Argonaute proteins for delivery to sequence-specific sites [27]. The most well-established method of the action of mature miRNAs is that of the binding targeted mRNAs through a nucleotide base pair complementarity between the miRNAs seed sequence (nucleotides 2–8 from the 5′ end) and the targeted mRNA sequence, most commonly at the 3′ UTR or coding sequences leading to varied changes in the translation efficiency of the bound mRNA into a protein product [28–31]. A reduction in translation in the associated protein is most frequently observed, but increases in the associated protein expression do occur [32–34]. Additionally, strong evidence suggests the noncanonical binding of miRNAs occurs with functional effects as seen with miR-21 binding/activating Toll-like receptor proteins (TLR8) [35], as well as binding the lncRNA GAS5 [36], with both leading to pro-inflammatory signaling cascades, which are also found in CRC. Cataloging and correlating miRNAs associated with malignant tissue in order to uncover biomarkers has been an intense area of research for decades after it was found that miRNAs are better candidates for biomarker investigation than mRNAs [37–39]. Current literature suggests there are well over 230 miRNAs, and likely many more, associated with CRC, with some frequently appearing in various tissues/fluids/cells and others only rarely appearing [20,40–53]. Due to the seemingly boundless nature of connecting miRNAs to CRC tissue/biofluids/cells, the focus of this review is on some of the most well-established miRNAs, their pathways that lead to CRC, and observed levels in biofluids (Table 1, Figure 1) instead of an exhaustive inventory of all potential candidates. Many miRNAs are possible contenders for biomarking CRC in bodily fluids, as miRNAs affect numerous known pathways to malignancy as found below.

Table 1. Dysregulated miRNAs detected in noninvasive biofluids in CRC patient samples.

miRNA	Biofluid Source	Levels in CRC	Reference
miR-21	Stool	Upregulated	[54]
	Stool	Upregulated	[55]
	Stool	Upregulated	[56]
	Serum	Upregulated	[57]
	Serum	Upregulated	[58]
	Serum	Upregulated	[59]
	Serum	Upregulated	[60]
	Serum	Upregulated	[61]
	Serum	Upregulated	[62]
	Serum	Upregulated	[63]
	Serum	Upregulated	[64]
miR-17	Serum	Upregulated	[65]
	Serum	Upregulated	[66]
	Serum	Upregulated	[67]
	Serum	Upregulated	[62]
miR-18a	Plasma	Downregulated	[68]
	Serum	Upregulated	[69]
	Serum	Upregulated	[70]
miR-19a	Serum	Upregulated	[65]
	Serum	Upregulated	[71]
	Serum	Upregulated	[43]
	Serum	Upregulated	[72]
	Serum	Upregulated	[73]
	Serum	Upregulated	[74]
	Serum	Upregulated	[61]
	Serum	Upregulated	[75]
	Serum	Upregulated	[76]
miR-20a	Serum	Downregulated	[77]
	Serum	Upregulated	[78]
	Serum	Upregulated	[62]
	Serum	Upregulated	[65]
	Serum	Upregulated	[74]
miR-19b	Serum	Upregulated	[43]
miR-92a	Plasma	Upregulated	[79]
	Stool	Upregulated	[54]
	Serum	Upregulated	[75]
	Serum	Upregulated	[80]
	Serum	Upregulated	[63]
	Serum	Upregulated	[81]
	Serum	Upregulated	[82]
	Serum	Upregulated	[83]
	Serum	Upregulated	[84]
	Serum	Downregulated	[84]

Table 1. *Cont.*

miRNA	Biofluid Source	Levels in CRC	Reference
miR-143	Plasma	Downregulated	[85]
	Serum	Downregulated	[74]
	Serum	Downregulated	[86]
	Serum	Upregulated	[87]
	Serum	No Significant Change	[84]
miR-145	Plasma	Downregulated	[85]
	Serum	Downregulated	[88]
	Serum	Downregulated	[74]
	Serum	Downregulated	[86]
	Serum	No Significant Change	[84]
	Serum	Downregulated	[66]
miR-203	Plasma	Upregulated	[89]
	Serum	Downregulated	[84]
	Serum	Upregulated	[90]
	Serum	Upregulated	[91]
	Serum	Downregulated	[92]
	Serum	Upregulated	[93]
	Serum	Upregulated	[94]
	Serum	Upregulated	[95]
	Serum	Upregulated	[96]
miR-200a	None Reported		
miR-200b	Plasma	Upregulated	[89]
miR-200c	Serum	Upregulated	[97]
	Serum	Downregulated	[86]
	Serum	Upregulated	[98]
	Serum	Upregulated	[99]
	Serum	Upregulated	[100]
	Serum	Upregulated	[101]
miR-141	Plasma	Upregulated	[89]
	Serum	Upregulated	[98]
	Serum	Upregulated	[99]
	Serum	Upregulated	[101]
miR-429	Serum	Upregulated	[102]
miR-135a	Stool	Upregulated	[103]
	Stool	Upregulated	[104]
	Serum	Upregulated	[105]
	Serum	Downregulated	[106]
miR-135b	Stool	Upregulated	[107]
	Stool	Upregulated	[108]
	Serum	Upregulated	[109]
miR-96	Plasma	Upregulated	[89]
	Serum	Upregulated	[110]
	Serum	No Significant Change	[69]
miR-183	Serum	Upregulated	[111]
	Serum	Upregulated	[112]
miR-150	Serum	Downregulated	[113]
	Serum	Upregulated	[74]
	Serum	Downregulated	[114]
	Serum	Downregulated	[115]
	Serum	Upregulated	[116]
	Serum	Downregulated	[64]
miR-195	Plasma	Downregulated	[117]
	Serum	Downregulated	[118]

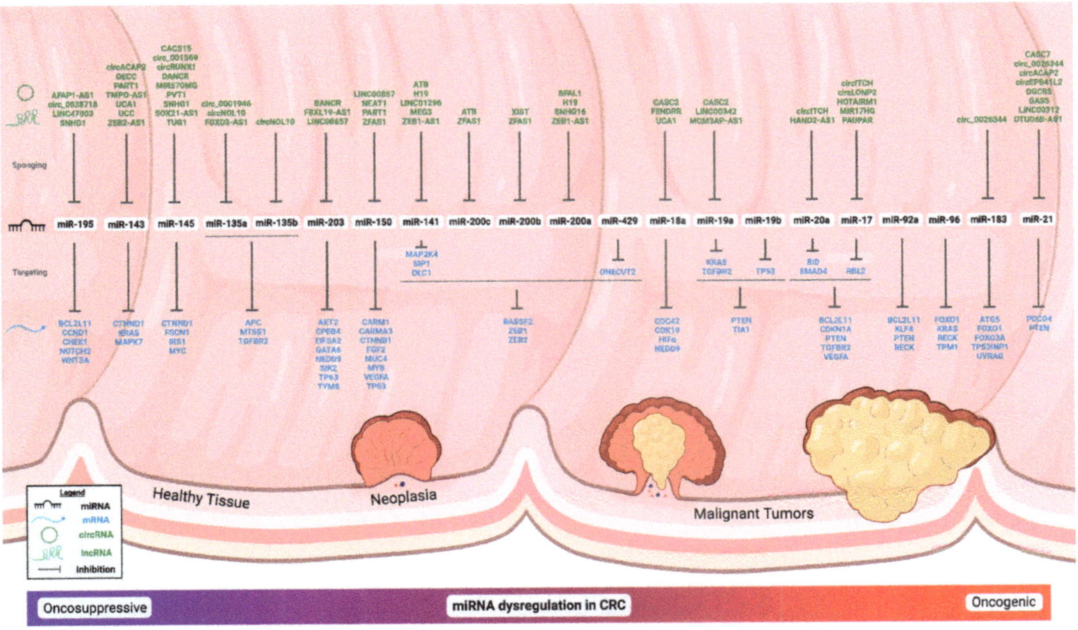

Figure 1. Dynamic relationships between miRNAs and ceRNAs and their connection to CRC. The miRNAs are listed in general/relative order of oncosuppressive to oncogenic in CRC from left to right with miRNA with dual natures being more central. All miRNAs, "sponging" ceRNAs (circRNA and lncRNA), and mRNAs listed have references cited within the main text and/or tables.

2.1. miR-21

The MIR21 gene is located on chromosome 17, and was one of the first miRNAs found whose expression was positively associated with cancers, specifically hepatocellular and breast, via microarray [119–121], and is naturally highly expressed in immune cells (monocytes, macrophages, and dendritic cells) [122,123]. In fact, miR-21 has been found to be related to at least 29 disease conditions, leading to controversy on using miR-21 as a viable biomarker for specific diseases [124]. The global knockout of miR-21 does not appear to cause any phenotypic pathologies outside of elevated levels of some of its target genes in specific cell types [125]. Genomically, MIR21 is found within the VMP1 locus whose protein product is vital in maintaining cell-to-cell connections and the loss of its expression can lead to aggressive colorectal cancer [126,127]. Recently, it was found that miR-21 indirectly represses the expression of VMP1 through the inhibition of miR-21's known tumor suppressing target, PTEN, creating a negative feedback loop on VMP1 with increasing miR-21 levels [128]. The canonical pathway for miR-21-induced oncogenesis is through to be the direct repression of various well-known tumor suppressor genes, including PTEN [120,129] and PDCD4 [130,131], which activate cyclin-dependent kinases, c-MYC, and PI3K/AKT/mTOR pathways, which results in an increased invasion and metastasis [10,125,132–134]. MiR-21 also induces the increased expression of anti-apoptotic proteins such as BCL2 [121,135], and regulates more than twenty-five other known targets [136]. An analysis of resected colonic tissue confirmed the inverse relationship of miR-21 and PDCD4 in colorectal tumors, and their comparative expression levels can predict metastasis [137,138]. In addition to canonical pathways in CRC, miR-21 is known to bind and activate TLR8 protein [35], as well as binding and inhibiting the anti-inflammatory lncRNA GAS5 [36]. Furthermore, miR-21 levels have consistently been shown to be elevated in both the serum [57,58,60,63,84] and fecal [55,56] samples of CRC patients. MiR-21 is implicated in several unique disease states and, thus, the use of its presence alone as an indicator for CRC is likely only a small piece of future miRNA biomarkers for CRC when combined with the expression levels of other ncRNAs and mRNA targets found in CRC patients.

2.2. *miR-17/92 Cluster*

The miR-17/92a cluster is found within the third intron of the C13ORF25 locus on chromosome 13, contains miR-17, miR-18a, miR-19a, miR-20a, miR-19b, and miR-92a, and was initially found as a polycistronic and oncogenic lncRNA (MIR17HG) in lung cancer cells [139,140]. Every member of the miR-17/92 cluster has been found to be associated with either CRC tissue or plasma/serum [68,141–143]. The seed sequences of many of these mature miRNAs are redundant and found in other mature miRNAs. Each miRNA in the miR-17/92 has at least one redundant miRNA from within the miR-17/92a locus or within the miR-106a/363 or miR-106b/25 locus [140]. The deletion of the entire miR-17/92 cluster results in embryos of reduced size, which are fatal immediately after birth and the dual deletion of miR-17/92a and miR-106b/25 causes embryonic lethality, while the singular deletion of either the miR-106b/25 or miR-106a/363 cluster (or their combined ablation) has not produced a similar phenotype [144]. In general, the functional duality of miRNAs being both essential to development and oncogenesis is well-established, as they are both growth-promoting states [145]. In humans, the germline hemizygous deletion of miR-17/92 results in patients with type 2 Feingold syndrome [146]. These results emphasize that the essential nature of the miR-17/92 cluster cannot be rescued with seed sequence redundancy found in other miRNAs at other loci, further implying the importance of noncanonical cellular influence. Similar to many miRNAs, the miRNAs in the miR-17/92 cluster produce their oncogenic effect on several known pathways. In CRC tissue, miR-17 downregulates RBL2, leading to carcinogenic Wnt/ß-catenin induction [147], but has also shown the potential inhibition of colorectal cancer invasiveness when used in isolation [148]. Levels of miR-18a were found to be elevated in the serum of CRC patients and, thus, are a potential biomarker [69], while the application of isolated miR-18a inhibits CRC cell growth through the indirect regulation of the PI3K/AKT pathway [149] and has the potential to be sponged by the tumor-suppressing lncRNA CASC2 [150]. Similar to miR-21, miR-19a directly inhibits the tumor suppressor PTEN [151,152], as well as TIA1 [153], and can be predictive of the effectiveness of chemotherapeutic interventions on CRC [72,152], all while isolated miR-19a seemingly inhibits CRC angiogenesis via KRAS reduction [154]. Both miR-19a and miR-19b share identical seed sequences and miR-19b is also known to canonically inhibit PTEN expression [155,156], yet only miR-19b has been shown to inhibit tumor suppressor TP53 [157], further underscoring the effect of noncanonical miRNA influence. MiR-19b has already been indicated as a putative serum/plasma biomarker for other diseases, including lung cancer [158,159] and diabetic cardiomyopathy [160]. Similar to miR-21 and miR-19a, miR-92a activates the PI3K/AKT cell cycle pathway via PTEN inhibition, as well as activating Wnt/ß-catenin signaling, promoting carcinogenic development [59,161,162] and downregulating the tumor suppressors RECK and KLF4 [163,164]. Both miR-17 and miR-20a share an identical seed sequence and, therefore, share many similar confirmed targets in CRC (BCL2L11, CDKN1A, PTEN, TGFBR2, and VEGFA) [143,165], while also having individualized targets as seen with miR-17 and RBL2 and miR-20a with BID and SMAD4 [78,166], stressing a non-seed sequence-based influence. MiR-92a has already been put forward as a marker of CRC in both serum/plasma [63,80,81] and stool samples [54,167] and whose overexpression is a recognized marker of poor prognosis in CRC tissue [168]. Many studies point to dysregulated miRNAs, including the miR-17/92 cluster, as a predictor of chemotherapy efficacy in CRC patients [152,169,170]. Within the miR-17/92 cluster, there are contradictory results between studies that could be attributable to many issues, including, but not limited to, differences in the cellular context [165], the use of miRNAs in isolation [148,149,154] as opposed to the endogenous polycistronic expression of MIR17HG which increases CRC invasiveness [171], or structural components of pri-miR-19/72 itself that allow for the autoregulation of expression between each individual miRNA within the cluster [172–174].

As with miR-21, the miR-17/92 cluster is intimately tied to CRC at various levels of expression and continues to be an intense and promising area of study for biomarking CRC.

2.3. miR-143 and miR-145

Both miR-143 and miR-145 have different seed sequences and are found on chromosome 5 within the lncRNA CARMN locus [175], with miR-143 being slightly upstream of miR-145 and both under the control of the same SRF/MYOCD/NKX2.5 enhancer region [33]. The complete ablation of miR-143/145 results in the defective development of smooth muscle and aortic tissue [176]. While miR-21 and the miR-17/92 cluster are generally associated to be oncogenic in nature, both miR-143 and miR-145 are considered to be oncosuppressive, specifically within CRC [177–179], by mainly acting on the p53 and RAS/MAPK oncogenic pathways [180]. Both miR-143 and miR-145 are also known to directly downregulate CTNND1, which is a vital piece of Wnt/ß-catenin carcinogenesis [181,182]; thus, showing miR-143/145 are antioncogenic through multiple pathways. Originally thought to be highly expressed in intestinal epithelial cells, it was found that while miR-143/145 are vital to epithelial cell regeneration, their expression was overwhelmingly found in the mesenchyme [183,184]. The experimentally validated targets of miR-143 included reducing MAPK7 [185] and KRAS [177], both members of the RAS/MAPK signaling pathway. MiR-145 is known to canonically inhibit the well-recognized oncogene MYC [6], PI3K activator, IRS1 [186], and metastatic promoter FSCN1 [187]. A reduction in the antioncogenic effect of miR-143/145 can be further exacerbated by their sponging via ncRNAs [188–191]. The sequestration of miR-143/145 via ncRNA sponging adds to the notion that miRNA levels of expression are not absolute in their cellular or histological effects, as there is a need to consider the levels of other ncRNAs that may reduce or eliminate their efficacy. In isolation, reduced tissue levels of miR-143/145 seem to be a marker of CRC, and have been found to be a marker for large tumors [192], while not being diagnostic between clinical stages of malignancy [193]. A few studies, with lower-than-optimal patient numbers, have investigated fecal samples for levels of miR-143/145 and found a lowered expression in CRC patient samples [194,195], but further broader studies are required to confirm these results. In CRC patients, plasma/serum levels of miR-143/145 have been found to be upregulated [98,196], downregulated [74,85,86], or insignificant when compared to control samples [84]. These contradictory findings suggest that a reduced miR-143/145 is likely only a possible indicator of the presence of abnormal colorectal growth. However, the combined analysis of observed miR-143/145 levels, compared to known targets/sponges, as well as other known oncogenic ncRNAs may provide any avenue for more accurate conclusions of the relation between miR-143/145 and CRC.

2.4. miR-200 Family

The miR-200 family can be found across two chromosomes, with miR-200b, miR-200a, and miR-429 on chromosome 1 and miR-200c and miR-141 on chromosome 12 in the intergenic space between PTPN6 and PHB2. MiR-200a and miR-141 share identical seed sequences, while miR-200b, miR-200c, and miR-429 have the same seed sequence, with only a single letter difference between the two groups [197]. Early studies uncovered that canonical targets of the miR-200 family are ZEB1/ZEB2 transcripts [198,199], which are master regulators of the epithelial–mesenchymal transition (EMT) via the inhibition of E-cadherin expression [200,201], and ZEB proteins have a reciprocally negative feedback loop with the miR-200 family [202]. The inter-regulatory relationship of the miR-200 family downregulating ZEB1/ZEB2 that allows for the continued transcription of E-cadherin, which inhibits EMT, is generally accepted as the main carcinogenic pathway downregulated by the miR-200 family, specifically in CRC [203]. However, miR-200c specifically can increase the ability of metastasized CRC cells to proliferate in the liver [204], and the miR-200 family as a whole is oncogenic in some CRC cell lines [205], again emphasizing the temporal and site-specific nature of the consequences of miRNA activity. It is also known that the miR-200 family antagonizes tumor angiogenesis, as shown in several cancer models, through the targeting of the pro-inflammatory CXCL1 [206]. MiR-200 family expression can be seen as tumor-suppressive [198,207], while also promoting metastasis [99,208] and, thus, its relation to CRC is strong but ill-defined. Expanding on the tumor-suppressive nature of the miR-200 family, isolated miR-429 is known to suppress EMT through the downregulation of ONECUT2 [209]. MiR-141 on its own is also known to inhibit the translation of several tumor-suppressor genes [210–212]. As the miR-200 family is located at two distinct chromosomal locations, complete family in vivo knockout is difficult and has only very recently been accomplished in vitro with knockout cells showing increases in senescence and EMT signaling genes with similar expression patterns found in gastric cancer patient RNA-seq samples [213]. The abnormal regulation of the miR-200 family as a whole has been indicated as a marker for CRC in serum/plasma [208,214], and as individual miRNAs: miR-200a [215], miR-200b [89], miR-200c [86,97–101], miR-141 [89,98,99,101,216], and miR-429 [102]. Furthermore, elevated serum levels of miR-200c and miR-141 have been correlated with an increased CRC metastasis [97,99] and poor prognoses in CRC patients [97,99–101,216]. Lowered miR-200a and elevated miR-200b in serum show similar patterns of being correlated with poor outcomes in CRC patients [89,211]. Moreover, the effectiveness of the carcinogenic influence of each member of the miR-200 family can potentially be mitigated by other ncRNAs that competitively inhibit their actions through sponging [217–220]. As in the case of other miRNAs and CRC, the basic presence or absence of miR-200 family transcripts in the biofluids of CRC patients alone is not enough to be diagnostic and must be considered within a cellular and histological framework while also considering expression levels of other contextually relevant and impactful ncRNAs.

2.5. *miR-203*

Within the genome, miR-203 is located intergenically between ASPG and KIF26A on chromosome 14. It first came to research prominence when it was found that its expression was high in colorectal adenocarcinomas compared to noncancerous tissue [221], yet other studies have found lowered miR-203 levels in similar CRC tissue [222,223]. Its expression is a key switch for the differentiation of basal skin epithelial cells by directly downregulating TP63 [224]. In this way, miR-203 acts as a tumor suppressor by directly inhibiting ΔNp63, which, when active, causes nuclear ß-catenin accumulation [225,226]. The whole-body knockout of miR-203 does not cause any overall developmental differences, but does invoke an expansion of proliferating keratinocytes, most notably during embryonic development [227]. In CRC cell lines, miR-203 reduction is required to maintain the stemness quality of cancerous lines [223,228,229], including through the reduction in the overexpressed and known CRC marker, NEDD9 [230,231]. Increases in miR-203 in CRC cells have been shown to canonically inhibit the oncogenic expression of AKT2, SIK2, CPEB4, EIF5A2, and TYMS [92,228,232–235]. Similar to other miRNAs, miR-203's activity can be halted by means of sponging by ncRNAs, such as FBXL19-AS1 [236], BANCR [237], and LINC00657 [238]. Many studies into the serum/plasma levels of miR-203 in CRC patients have discovered that differential levels of miR-203 are most often an indicator of poor prognosis and increased metastasis usually marked by an increase in miR-203 [89–91,93–96], with some showing lowered levels [84,92]. A meta-analyses of 11 papers found a higher miR-203 expression in CRC tissue to be significant, and not in serum, but a combination of both serum and CRC tissue was predictive of a poor outcome [239]. Additionally, such as other miRNAs, miR-203 is known to increase chemosensitivity in CRC cells [232,235]. MiR-203 detection could present as a valuable piece of the puzzle in properly biomarking CRC.

2.6. *miR-135 Family*

The miR-135 family is represented by miR-135a found just upstream of GLYCTK on chromosome 3, and miR-135b, contained within the genomic sequence of both the lncRNA BLACAT1 and the gene LEMD1 on chromosome 1. Both miR-135a and miR-135b share identical mature seed sequences and only one nucleotide difference in total. The miR-135 family was originally tagged as an oncogene by its targeting and downregulating of the tumor suppressor APC (with which miR-135b it has a reciprocally inhibitory relationship [240]) in CRC, which regulates the Wnt/ß-catenin pathway [241,242] as well as priming pancreatic cancer cells for pro-carcinogenic metabolic conditions through PFK1 downregulation [243]. In a CRC-specific context, miR-135a downregulates a metastasis driver MTSS1 [244] and miR-135b downregulates TGFBR2 [245], which is functionally mutated in up to 90% of CRC cases [246]. The effects of miR-135 can be antagonized through lncRNA sponging within a CRC context through circNOL10 [247]. Several studies have concentrated on modified levels of miR-135 family members in stool, tissue, and serum. As expected with the oncogenic nature of miR-135, stool samples show an increase in miR-135 expression [103,107] and even being predictive of later-stage (III-IV) cancer [104]. CRC tissue samples with an increased miR-135 family expression are indicative of poor prognosis and metastatic conditions [240,248–250]. Low serum levels of miR-135 have been correlated with CRC [106], while elevated levels have been able to elucidate differences between polyps and carcinomas [105], and, thus, the use of miR-135 as a serum biomarker has been controversial [111] and, thus, focusing on noninvasive fecal samples has been more productive and a better candidate for the detection of the miR-135 family.

2.7. miR-96 and miR-183

On chromosome 7, miR-96 is found in a cluster with miR-183 and upstream of NRF1. Research interest in miR-96/183-related pathologies and development began with discoveries that both miR-96 and miR-183 are vital to hair cell development and function in the inner ear [251,252], and increases in miR-96 and miR-183 are found within breast and lung cancer tissue while directly downregulating FOXO1 and FOXO3A [253–256], key tumor suppressors in the P13K/AKT pathway of carcinogenesis. The knockout of the miR-96/183 cluster causes increases in target SLC6A6, resulting in dysfunctional photoreception [257]. In vitro CRC experiments have shown that increases in miR-96 and miR-183 are associated with both an increased cell migration [258,259] and invasiveness [260,261]. Alongside this oncogenic character, a high miR-96 and miR-183 expression has also shown to increase resistance to chemotherapy treatments such as oxaliplatin [262], 5-fluoruracil [110], and radiation treatments [263], extending their oncogenic nature. Within studies of CRC tumor tissue samples, miR-96 and miR-183 were found to be elevated in all but one study [248,264–267], and that study found that lowered tumor tissue levels of miR-96 entailed poor patient prognoses [268]. A handful of studies have interrogated miR-96 levels in serum/plasma and have found that high levels [89,110], low levels [269], and no significant association with CRC [69] have been observed despite the fact that high miR-96 serum levels have been tied to both hepatocellular [269] and lung cancers [270]. Serum levels of miR-183 are consistently high in CRC patients [111,112]. The incongruous results from miR-96 could possibly be due to the neutralization of the oncogenic effects of miR-96 through sponging as has been seen in cervical cancer (STXBP5-AS1) [271] and pancreatic ductal adenocarcinoma (TP53TG1) [272]. In this vein, circ_0026344, has recently been shown to abrogate the oncogenic nature of miR-183 in CRC cells [273]. The consideration of miR-96 and miR-183 as biomarkers and targets for CRC diagnosis and treatment should, therefore, always be considered within the context of other mitigating ncRNAs.

2.8. miR-150

Just upstream of RPS11 on chromosome 19 is miR-150. Initial analyses of miR-150 found it to be vital to proper hematopoietic cell differentiation [274], specifically B cell maturation [275], through the downregulation of MYB, which was confirmed in CRC cells [276]. Unsurprisingly, the knockout of miR-150 results in B cell developmental difficulties coupled with obesity-related metabolic dysregulation [277]. In addition to downregulating MYB in CRC, miR-150 also directly lowers the expression of ß-catenin [278], VEGFA [279], and a known marker of poor prognoses in CRC patients, MUC4 [280]. As miR-150 canonically downregulates over several oncogenic pathways, it is generally seen as a tumor suppressor in CRC. In contrast, miR-150 is seen as oncogenic in gastric cancer tissues by downregulating the pro-apoptotic genes P2X$_7$ [281] and EGR2 [282] with a single study showing increases in miR-150 in CRC tissue [283]. Resected tissue samples from CRC patients regularly show a significantly decreased expression of miR-150 when compared to both adenoma and healthy colonic tissue [278,279,284], and these reduced quantities can be indicative of poor prognoses [285]. However, when miR-150 expression was measured within serum samples of CRC patients, there have been observations of both increases [74,116] and decreases [64,113–115], yet again stressing the need for the analysis of miRNA levels within a context of known targets and other ncRNA inhibitors. In this regard, the growth-suppressing consequences of miR-150 expression in CRC can be arrested by means of sponging via the lncRNAs ZFAS1 [286] and PART1 [287]. The previous successes with detecting low levels of miR-150 in CRC patients' serum suggests that the continued study of miR-150 and its relation to CRC, and known sequestering inhibitors, in serum/plasma will continue to be a fruitful area of investigation.

2.9. miR-195

While a member of the miR-15 precursor family based on an identical seed sequence, miR-195 resides away from other family members and is found on chromosome 17 within the genomic location of lncRNA MIR497HG, just upstream of the 3′ UTR of C17ORF49. Modified levels of miR-195 first became an area of research in cardiac studies as increased levels of miR-195 were found to cause cardiac hypertrophy [288], as well as regulating cell cycle checkpoints in cardiomyocytes by targeting CHEK1 [289]. In terms of cancer research, early studies found that miR-195 was downregulated in hepatocellular cancer tissue via microarray [290], and that miR-195 inhibited CCND1 translation in both hepatocellular and colorectal cell lines [291], halting the cell cycling necessary for carcinogenic progression. Recently, miR-195 has also been found to be important in the cellular maintenance of the blood–brain barrier [292]. In addition to inhibiting CHEK1 in CRC cells, miR-195 prevents WNT3A translation [293], which is known to activate the Wnt/ß-catenin oncogenic signaling pathway in CRC cells [294]. MiR-195 also canonically inhibits NOTCH2 [295] and BCL2L11 [296] in CRC cells, providing more pathways of tumor suppression through preventing EMT and promoting apoptosis, respectively. Similar to in vitro studies, miR-195 is consistently downregulated miRNA in CRC tissue samples [117,295–301], and the addition of miR-195 makes CRC cells sensitized to currently used chemotherapy interventions such as 5-fluorouracil [298,302], doxorubicin [303], and radiation therapy [304]. Surprisingly few studies have focused on serum/plasma levels of miR-195 in CRC patients, but results have shown the expected lowered levels of miR-195 [117]. Just as observed with other CRC-associated miRNAs, the tumor-suppressing actions of miR-195 can be annulled when sequestered through sponging by highly expressed lncRNA/circRNAs, with all of them promoting CRC carcinogenesis and progression [305–308]. Being able to couple known miR-195 targets and sponges, with stably reduced miR-195 in CRC tissues insinuates that miR-195 is a ripe area of further research in CRC.

3. Discussion

miRNAs, ceRNAs, and Chemosensitivity in CRC: A War of Attrition

Once the ability for miRNAs to be sponged out of efficacy by other ceRNAs was found to be robust, this provided the incentive to construct synthetic ceRNAs as sponges for oncogenic miRNAs. Methodologies for the building of unique miRNA sponges with multiple seed target sites were quickly introduced [309]. In practice, these sponges have indeed been able to isolate oncogenic miRNAs such as that observed with miR-21 sponging in renal cancer [310], gastric cancer [311], glioblastomas [312], and esophageal carcinomas [313]. Sponging has also been shown to be a direct driver of metastasis in CRC by circHIPK3 isolating miR-7 [314]. Thus, sponging miRNAs, such as the presence of miRNAs themselves, can be both oncogenic and suppressive in CRC. One very interesting operation of miRNAs in CRC is the ability to either sensitize or desensitize tumor cells to established chemotherapies and, therefore, be predictive of patient outcomes [170]. Both increased miR-21 and miR-17/92 expressions are indicative of chemoresistant tumors [110,169], and, thus, utilizing either increased ceRNA or synthetic ncRNA sponges would be ideal in helping sensitize patients to previously resistant chemotherapy or radiation. In contrast, miRNAs such as miR-203, whose increased expression is correlated with chemotolerance in CRC [232,235], represent a ripe opportunity to be used as an adjuvant to make resistant CRC cells susceptible to previously ineffective treatments. These examples epitomize the possible capabilities of targeting miRNA and miRNA-affected pathways in CRC treatments.

Despite observed patterns, one of the most consistent features of biofluid-borne ncRNAs in CRC patients is that their levels are almost never entirely correlative with disease presence, stage of development, metastasis extent, or susceptibility to known treatments. The colon and rectum are incredibly dynamic environments that constantly respond to various environmental cues, nutritional input, hypoxic conditions, and microbial influence, both commensal and pathogenic. This regularly reactive state causes the rapidly modulable nature of RNA to be an ideal molecule for responding to these shifting settings. This can manifest in the inducible power of miRNAs to alter mRNA translation, as well as ceRNAs' ability to negate those same miRNAs' effect on mRNA through competitive sequestration. Early studies on miRNAs and CRC quickly focused on elucidating their relationship to effectors of known carcinogenic pathways, such as mRNAs involved in the Wnt/ß-catenin, PI3K/AKT, p53/c-MYC, and MAPK/ERK pathways. More recent analyses have unveiled the capacity for ceRNAs to mitigate the influence of miRNAs in CRC (Table 2). Neither the isolated investigation of miRNA on mRNA, or ceRNA on miRNA influence would complete the carcinogenic picture due to their attritive relationships.

Table 2. Reported ceRNAs and targeted miRNAs in CRC.

miRNA	ceRNA in CRC	Reference
miR-21	CASC7	[315]
	circ_0026344	[316]
	circACAP2	[317]
	circEPB41L2	[318]
	DGCR5	[319]
	GAS5	[36]
	LINC00312	[320]
	OTUD6B-AS1	[321]
miR-17	circITCH	[12]
	circLONP2	[322]
	HOTAIRM1	[11]
	MIR17HG	[171]
	PAUPAR	[323]
miR-18a	CASC2	[150]
	FENDRR	[324]
	UCA1	[325]
miR-19a	CASC2	[326]
	LINC00342	[327]
	MCM3AP-AS1	[328]
miR-20a	circITCH	[12]
	HAND2-AS1	[329]
miR-19b	None Reported	
miR-92a	None Reported	
miR-143	circACAP2	[330]
	OECC	[331]
	PART1	[190]
	TMPO-AS1	[332]
	UCA1	[189]
	UCC	[188]
	ZEB2-AS1	[333]

Table 2. *Cont.*

miRNA	ceRNA in CRC	Reference
miR-145	CACS15 circ_001569 circRUNX1 DANCR MIR570MG PVT1 SNHG1 SOX2-AS1 TUG1	[334] [191] [335] [336] [337] [338] [339] [340] [341]
miR-203	BANCR FBXL19-AS1 LINC00657	[237] [236] [238]
miR-200a	BFAL1 H19 SNHG16 ZEB1-AS1	[342] [217] [343] [344]
miR-200b	XIST ZFAS1	[345] [218]
miR-200c	ATB ZFAS1	[346] [218]
miR-141	ATB H19 LINC01296 MEG3 ZEB1-AS1	[347] [348] [349] [350] [218]
miR-429	None Reported	
miR-135a	circ_0001946 circNOL10 FOXD3-AS1	[351] [247] [352]
miR-135b	circNOL10	[247]
miR-96	None Reported	
miR-183	circ_0026344	[273]
miR-150	LINC00857 NEAT1 PART1 ZFAS1	[353] [354] [287] [286]
miR-195	AFAP-AS1 circ_0038718 LINC00473 SNHG1	[308] [307] [306] [305]

The molecular CRC environment becomes further complicated as modifications to miRNA and ceRNA are known to alter their effect such as A-to-I editing [355] or m6A methylation [356]. This struggle between mRNA, miRNA, and ceRNAs over the carcinogenic capacity within CRC tumors demands the need for a more comprehensive review of the overall RNA influence in personalized CRC diagnosis and treatment. The main cytological affectations of miRNAs cannot be fully explained except in light of taking downstream targets into account. For example, circACAP2 is capable of sponging both oncogenic miR-21 [317] and suppressive miR-143 [330] and, therefore, circACAP2 in isolation could be considered carcinogenic or suppressive in relation to the sequestration of miR-21 or miR-143 in CRC. Thus, the vital examinations of interactive miRNA/ceRNA networks within a CRC context are becoming more necessary and prevalent [357–359], including studying the epigenetic regulation of miRNA/ceRNA expression [360], instead of focusing on individual cases of ceRNA/miRNA/mRNA interaction in order to provide a more thorough exploration of their collective effect on CRC. These examinations take a wider view of the interactions of multiple RNA species as they operate as a collective axis instead of concentrating on individual RNA species and allow for more personalized medical strategies [361]. Preliminary patterns of ceRNA-miRNA-mRNA networks in CRC samples are beginning to be discovered and provide a better framework for understanding the heterogenous CRC environment [357,362,363], and should be used as templates for the integrative ncRNA analysis in CRC patients. Figure 1 summarizes ceRNA-miRNA-mRNA networks in CRC, in which oncogenic or tumor suppressive miRNAs are dysregulated. The dysregulated miRNAs and ceRNAs can be potentially developed as biomarkers in the diagnosis of CRC as well as therapeutic targets to treat the disease. Continued work into personalized ncRNA-based diagnoses/treatments for the multifaceted and insidious nature of CRC could provide promising new avenues of prevention, diagnosis, and treatment.

Author Contributions: B.G.J. drafted and wrote the original manuscript draft, tables, and figures with intellectual guidance and instruction from S.R. S.R. completed final composition, edits, and design of the figures for the manuscript. All authors have read and agreed to the published version of the manuscript.

Funding: This review received no external funding.

Acknowledgments: Figure 1 adapted from "Benign and Malignant Colon Cancer" template, by BioRender.com (2022). Retrieved from https://app.biorender.com/biorender-templates on 7 January 2022.

Conflicts of Interest: The authors declare no conflict of interest.

References

1. Nigro, J.M.; Cho, K.R.; Fearon, E.R.; Kern, S.E.; Ruppert, J.M.; Oliner, J.D.; Kinzler, K.W.; Vogelstein, B. Scrambled exons. *Cell* **1991**, *64*, 607–613. [CrossRef]
2. Calin, G.A.; Liu, C.; Ferracin, M.; Hyslop, T.; Spizzo, R.; Sevignani, C.; Fabbri, M.; Cimmino, A.; Lee, E.J.; Wojcik, S.E.; et al. Ultraconserved Regions Encoding ncRNAs Are Altered in Human Leukemias and Carcinomas. *Cancer Cell* **2007**, *12*, 215–229. [CrossRef] [PubMed]
3. Calin, G.A.; Dumitru, C.D.; Shimizu, M.; Bichi, R.; Zupo, S.; Noch, E.; Aldler, H.; Rattan, S.; Keating, M.; Rai, K.; et al. Frequent deletions and down-regulation of micro-RNA genes miR15 and miR16 at 13q14 in chronic lymphocytic leukemia. *Proc. Natl. Acad. Sci. USA* **2002**, *99*, 15524–15529. [CrossRef] [PubMed]
4. Kristensen, L.S.; Hansen, T.B.; Venø, M.T.; Kjems, J. Circular RNAs in cancer: Opportunities and challenges in the field. *Oncogene* **2018**, *37*, 555–565. [CrossRef]
5. Takaoka, Y.; Shimizu, Y.; Hasegawa, H.; Ouchi, Y.; Qiao, S.; Nagahara, M.; Ichihara, M.; Lee, J.D.; Adachi, K.; Hamaguchi, M.; et al. Forced expression of miR-143 represses ERK5/c-Myc and p68/p72 signaling in concert with miR-145 in gut tumors of Apcmin mice. *PLoS ONE* **2012**, *7*, e42137. [CrossRef]
6. Sachdeva, M.; Zhu, S.; Wu, F.; Wu, H.; Walia, V.; Kumar, S.; Elble, R.; Watabe, K.; Mo, Y.-Y. p53 represses c-Myc through induction of the tumor suppressor miR-145. *Proc. Natl. Acad. Sci. USA* **2009**, *106*, 3207–3212. [CrossRef]
7. Pan, H.; Li, T.; Jiang, Y.; Pan, C.; Ding, Y.; Huang, Z.; Yu, H.; Kong, D. Overexpression of Circular RNA ciRS-7 Abrogates the Tumor Suppressive Effect of miR-7 on Gastric Cancer via PTEN/PI3K/AKT Signaling Pathway. *J. Cell. Biochem.* **2018**, *119*, 440–446. [CrossRef] [PubMed]
8. Fang, Y.; Xue, J.-L.; Shen, Q.; Chen, J.; Tian, L. MicroRNA-7 inhibits tumor growth and metastasis by targeting the phosphoinositide 3-kinase/Akt pathway in hepatocellular carcinoma. *Hepatology* **2012**, *55*, 1852–1862. [CrossRef]
9. Kino, T.; Hurt, D.E.; Ichijo, T.; Nader, N.; Chrousos, G.P. Noncoding RNA Gas5 Is a Growth Arrest– and Starvation-Associated Repressor of the Glucocorticoid Receptor. *Sci. Signal.* **2010**, *3*, 1–16. [CrossRef]
10. Pickard, M.; Williams, G. Molecular and Cellular Mechanisms of Action of Tumour Suppressor GAS5 LncRNA. *Genes* **2015**, *6*, 484–499. [CrossRef]
11. Ren, T.; Hou, J.; Liu, C.; Shan, F.; Xiong, X.; Qin, A.; Chen, J.; Ren, W. The long non-coding RNA HOTAIRM1 suppresses cell progression via sponging endogenous miR-17-5p/ B-cell translocation gene 3 (BTG3) axis in 5-fluorouracil resistant colorectal cancer cells. *Biomed. Pharmacother.* **2019**, *117*, 109171. [CrossRef]
12. Huang, G.; Zhu, H.; Shi, Y.; Wu, W.; Cai, H.; Chen, X. cir-ITCH Plays an Inhibitory Role in Colorectal Cancer by Regulating the Wnt/β-Catenin Pathway. *PLoS ONE* **2015**, *10*, e0131225. [CrossRef] [PubMed]
13. Huels, D.J.; Sansom, O.J. Stem vs. non-stem cell origin of colorectal cancer. *Br. J. Cancer* **2015**, *113*, 1–5. [CrossRef] [PubMed]

14. Blanpain, C. Tracing the cellular origin of cancer. *Nat. Cell Biol.* **2013**, *15*, 126–134. [CrossRef] [PubMed]
15. Mármol, I.; Sánchez-de-Diego, C.; Dieste, A.P.; Cerrada, E.; Yoldi, M.J.R. Colorectal carcinoma: A general overview and future perspectives in colorectal cancer. *Int. J. Mol. Sci.* **2017**, *18*, 197. [CrossRef] [PubMed]
16. Zauber, A.G.; Winawer, S.J.; O'Brien, M.J.; Lansdorp-Vogelaar, I.; van Ballegooijen, M.; Hankey, B.F.; Shi, W.; Bond, J.H.; Schapiro, M.; Panish, J.F.; et al. Colonoscopic Polypectomy and Long-Term Prevention of Colorectal-Cancer Deaths. *N. Engl. J. Med.* **2012**, *366*, 687–696. [CrossRef]
17. Van Dam, J. Prevention of Colorectal Cancer by Endoscopic Polypectomy. *Ann. Intern. Med.* **1995**, *123*, 949. [CrossRef] [PubMed]
18. Baxter, N.N.; Goldwasser, M.A.; Paszat, L.F.; Saskin, R.; Urbach, D.R.; Rabeneck, L. Association of colonoscopy and death from colorectal cancer. *Ann. Intern. Med.* **2009**, *150*, 1–8. [CrossRef]
19. Baran, B.; Mert Ozupek, N.; Yerli Tetik, N.; Acar, E.; Bekcioglu, O.; Baskin, Y. Difference Between Left-Sided and Right-Sided Colorectal Cancer: A Focused Review of Literature. *Gastroenterol. Res.* **2018**, *11*, 264–273. [CrossRef]
20. Luo, X.; Burwinkel, B.; Tao, S.; Brenner, H. MicroRNA signatures: Novel biomarker for colorectal cancer? *Cancer Epidemiol. Biomarkers Prev.* **2011**, *20*, 1272–1286. [CrossRef]
21. Wang, S.; Xiang, J.; Li, Z.; Lu, S.; Hu, J.; Gao, X.; Yu, L.; Wang, L.; Wang, J.; Wu, Y.; et al. A plasma microRNA panel for early detection of colorectal cancer. *Int. J. Cancer* **2015**, *136*, 152–161. [CrossRef] [PubMed]
22. Ahmed, F.E.; Jeffries, C.D.; Vos, P.W.; Flake, G.; Nuovo, G.J.; Sinar, D.R.; Naziri, W.; Marcuard, S.P. Diagnostic microRNA markers for screening sporadic human colon cancer and active ulcerative colitis in stool and tissue. *Cancer Genom. Proteom.* **2009**, *6*, 281–295.
23. Fan, Y.-H.; Ye, M.-H.; Wu, L.; Wu, M.-J.; Lu, S.-G.; Zhu, X.-G. BRAF-activated lncRNA predicts gastrointestinal cancer patient prognosis: A meta-analysis. *Oncotarget* **2017**, *8*, 6295–6303. [CrossRef]
24. Xu, W.; Zhou, G.; Wang, H.; Liu, Y.; Chen, B.; Chen, W.; Lin, C.; Wu, S.; Gong, A.; Xu, M. Circulating lncRNA SNHG11 as a novel biomarker for early diagnosis and prognosis of colorectal cancer. *Int. J. Cancer* **2020**, *146*, 2901–2912. [CrossRef]
25. Wang, Y.; Li, Z.; Xu, S.; Guo, J. Novel potential tumor biomarkers: Circular RNAs and exosomal circular RNAs in gastrointestinal malignancies. *J. Clin. Lab. Anal.* **2020**, *34*, e23359. [CrossRef]
26. Naeli, P.; Pourhanifeh, M.H.; Karimzadeh, M.R.; Shabaninejad, Z.; Movahedpour, A.; Tarrahimofrad, H.; Mirzaei, H.R.; Bafrani, H.H.; Savardashtaki, A.; Mirzaei, H.; et al. Circular RNAs and gastrointestinal cancers: Epigenetic regulators with a prognostic and therapeutic role. *Crit. Rev. Oncol. Hematol.* **2020**, *145*, 102854. [CrossRef]
27. Liu, J.; Carmell, M.A.; Rivas, F.V.; Marsden, C.G.; Thomson, J.M.; Song, J.-J.; Hammond, S.M.; Joshua-Tor, L.; Hannon, G.J. Argonaute2 Is the Catalytic Engine of Mammalian RNAi. *Science* **2004**, *305*, 1437–1441. [CrossRef]
28. Lewis, B.P.; Shih, I.; Jones-Rhoades, M.W.; Bartel, D.P.; Burge, C.B. Prediction of Mammalian MicroRNA Targets. *Cell* **2003**, *115*, 787–798. [CrossRef]
29. Easow, G.; Teleman, A.A.; Cohen, S.M. Isolation of microRNA targets by miRNP immunopurification. *RNA* **2007**, *13*, 1198–1204. [CrossRef]
30. Baek, D.; Villén, J.; Shin, C.; Camargo, F.D.; Gygi, S.P.; Bartel, D.P. The impact of microRNAs on protein output. *Nature* **2008**, *455*, 64–71. [CrossRef]
31. Bartel, D.P. MicroRNAs: Target Recognition and Regulatory Functions. *Cell* **2009**, *136*, 215–233. [CrossRef] [PubMed]
32. Vasudevan, S.; Steitz, J.A. AU-Rich-Element-Mediated Upregulation of Translation by FXR1 and Argonaute 2. *Cell* **2007**, *128*, 1105–1118. [CrossRef]
33. Cordes, K.R.; Sheehy, N.T.; White, M.P.; Berry, E.C.; Morton, S.U.; Muth, A.N.; Lee, T.H.; Miano, J.M.; Ivey, K.N.; Srivastava, D. MiR-145 and miR-143 regulate smooth muscle cell fate and plasticity. *Nature* **2009**, *460*, 705–710. [CrossRef] [PubMed]
34. Lin, C.-C.; Liu, L.-Z.; Addison, J.B.; Wonderlin, W.F.; Ivanov, A.V.; Ruppert, J.M. A KLF4-miRNA-206 Autoregulatory Feedback Loop Can Promote or Inhibit Protein Translation Depending upon Cell Context. *Mol. Cell. Biol.* **2011**, *31*, 2513–2527. [CrossRef] [PubMed]
35. Fabbri, M.; Paone, A.; Calore, F.; Galli, R.; Gaudio, E.; Santhanam, R.; Lovat, F.; Fadda, P.; Mao, C.; Nuovo, G.J.; et al. MicroRNAs bind to Toll-like receptors to induce prometastatic inflammatory response. *Proc. Natl. Acad. Sci. USA* **2012**, *109*. [CrossRef]
36. Zhang, Z.; Zhu, Z.; Watabe, K.; Zhang, X.; Bai, C.; Xu, M.; Wu, F.; Mo, Y.-Y. Negative regulation of lncRNA GAS5 by miR-21. *Cell Death Differ.* **2013**, *20*, 1558–1568. [CrossRef]
37. Lu, J.; Getz, G.; Miska, E.A.; Alvarez-Saavedra, E.; Lamb, J.; Peck, D.; Sweet-Cordero, A.; Ebert, B.L.; Mak, R.H.; Ferrando, A.A.; et al. MicroRNA expression profiles classify human cancers. *Nature* **2005**, *435*, 834–838. [CrossRef]
38. Volinia, S.; Calin, G.A.; Liu, C.G.; Ambs, S.; Cimmino, A.; Petrocca, F.; Visone, R.; Iorio, M.; Roldo, C.; Ferracin, M.; et al. A microRNA expression signature of human solid tumors defines cancer gene targets. *Proc. Natl. Acad. Sci. USA* **2006**, *103*, 2257–2261. [CrossRef]
39. Bandrés, E.; Cubedo, E.; Agirre, X.; Malumbres, R.; Zárate, R.; Ramirez, N.; Abajo, A.; Navarro, A.; Moreno, I.; Monzó, M.; et al. Identification by Real-time PCR of 13 mature microRNAs differentially expressed in colorectal cancer and non-tumoral tissues. *Mol. Cancer* **2006**, *5*, 1–10. [CrossRef]
40. Ren, A.; Dong, Y.; Tsoi, H.; Yu, J. Detection of miRNA as Non-Invasive Biomarkers of Colorectal Cancer. *Int. J. Mol. Sci.* **2015**, *16*, 2810–2823. [CrossRef]
41. Niu, L.; Yang, W.; Duan, L.; Wang, X.; Li, Y.; Xu, C.; Liu, C.; Zhang, Y.; Zhou, W.; Liu, J.; et al. Biological Implications and Clinical Potential of Metastasis-Related miRNA in Colorectal Cancer. *Mol. Ther. Nucleic Acids* **2021**, *23*, 42–54. [CrossRef] [PubMed]

42. Vautrot, V.; Chanteloup, G.; Elmallah, M.; Cordonnier, M.; Aubin, F.; Garrido, C.; Gobbo, J. Exosomal miRNA: Small Molecules, Big Impact in Colorectal Cancer. *J. Oncol.* **2019**, *2019*, 1–18. [CrossRef] [PubMed]
43. Marcuello, M.; Duran-Sanchon, S.; Moreno, L.; Lozano, J.J.; Bujanda, L.; Castells, A.; Gironella, M. Analysis of A 6-Mirna Signature in Serum from Colorectal Cancer Screening Participants as Non-Invasive Biomarkers for Advanced Adenoma and Colorectal Cancer Detection. *Cancers* **2019**, *11*, 1542. [CrossRef] [PubMed]
44. Slattery, M.L.; Mullany, L.E.; Sakoda, L.; Samowitz, W.S.; Wolff, R.K.; Stevens, J.R.; Herrick, J.S. The NF-κB signalling pathway in colorectal cancer: Associations between dysregulated gene and miRNA expression. *J. Cancer Res. Clin. Oncol.* **2018**, *144*, 269–283. [CrossRef]
45. Tang, X.J.; Wang, W.; Hann, S.S. Interactions among lncRNAs, miRNAs and mRNA in colorectal cancer. *Biochimie* **2019**, *163*, 58–72. [CrossRef] [PubMed]
46. Zhang, H.; Zhu, M.; Shan, X.; Zhou, X.; Wang, T.; Zhang, J.; Tao, J.; Cheng, W.; Chen, G.; Li, J.; et al. A panel of seven-miRNA signature in plasma as potential biomarker for colorectal cancer diagnosis. *Gene* **2019**, *687*, 246–254. [CrossRef]
47. Rapado-González, Ó.; Majem, B.; Álvarez-Castro, A.; Díaz-Peña, R.; Abalo, A.; Suárez-Cabrera, L.; Gil-Moreno, A.; Santamaría, A.; López-López, R.; Muinelo-Romay, L.; et al. A Novel Saliva-Based miRNA Signature for Colorectal Cancer Diagnosis. *J. Clin. Med.* **2019**, *8*, 2029. [CrossRef]
48. Yang, L.; Belaguli, N.; Berger, D.H. MicroRNA and Colorectal Cancer. *World J. Surg.* **2009**, *33*, 638–646. [CrossRef]
49. Huang, X.; Zhu, X.; Yu, Y.; Zhu, W.; Jin, L.; Zhang, X.; Li, S.; Zou, P.; Xie, C.; Cui, R. Dissecting miRNA signature in colorectal cancer progression and metastasis. *Cancer Lett.* **2021**, *501*, 66–82. [CrossRef]
50. Falzone, L.; Scola, L.; Zanghì, A.; Biondi, A.; Di Cataldo, A.; Libra, M.; Candido, S. Integrated analysis of colorectal cancer microRNA datasets: Identification of microRNAs associated with tumor development. *Aging* **2018**, *10*, 1000–1014. [CrossRef]
51. Carter, J.V.; Roberts, H.L.; Pan, J.; Rice, J.D.; Burton, J.F.; Galbraith, N.J.; Eichenberger, M.R.; Jorden, J.; Deveaux, P.; Farmer, R.; et al. A Highly Predictive Model for Diagnosis of Colorectal Neoplasms Using Plasma MicroRNA. *Ann. Surg.* **2016**, *264*, 575–584. [CrossRef] [PubMed]
52. Lai, X.; Friedman, A. Exosomal microRNA concentrations in colorectal cancer: A mathematical model. *J. Theor Biol.* **2017**, *415*, 70–83. [CrossRef] [PubMed]
53. Yang, G.; Zhang, Y.; Yang, J. A Five-microRNA Signature as Prognostic Biomarker in Colorectal Cancer by Bioinformatics Analysis. *Front. Oncol.* **2019**, *9*, 1207. [CrossRef] [PubMed]
54. Wu, C.W.; Ng, S.S.M.; Dong, Y.J.; Ng, S.C.; Leung, W.W.; Lee, C.W.; Wong, Y.N.; Chan, F.K.L.; Yu, J.; Sung, J.J.Y. Detection of miR-92a and miR-21 in stool samples as potential screening biomarkers for colorectal cancer and polyps. *Gut* **2012**, *61*, 739–745. [CrossRef] [PubMed]
55. Yau, T.O.; Tang, C.-M.; Harriss, E.K.; Dickins, B.; Polytarchou, C. Faecal microRNAs as a non-invasive tool in the diagnosis of colonic adenomas and colorectal cancer: A meta-analysis. *Sci. Rep.* **2019**, *9*, 9491. [CrossRef]
56. Link, A.; Balaguer, F.; Shen, Y.; Nagasaka, T.; Lozano, J.J.; Boland, C.R.; Goel, A. Fecal MicroRNAs as Novel Biomarkers for Colon Cancer Screening. *Cancer Epidemiol. Biomarkers Prev.* **2010**, *19*, 1766–1774. [CrossRef]
57. Toiyama, Y.; Takahashi, M.; Hur, K.; Nagasaka, T.; Tanaka, K.; Inoue, Y.; Kusunoki, M.; Boland, C.R.; Goel, A. Serum miR-21 as a diagnostic and prognostic biomarker in colorectal cancer. *J. Natl. Cancer Inst.* **2013**, *105*, 849–859. [CrossRef]
58. Yamada, A.; Horimatsu, T.; Okugawa, Y.; Nishida, N.; Honjo, H.; Ida, H.; Kou, T.; Kusaka, T.; Sasaki, Y.; Yagi, M.; et al. Serum MIR-21, MIR-29a, and MIR-125b are promising biomarkers for the early detection of colorectal neoplasia. *Clin. Cancer Res.* **2015**, *21*, 4234–4242. [CrossRef]
59. Zhang, G.; Zhou, H.; Xiao, H.; Liu, Z.; Tian, H.; Zhou, T. MicroRNA-92a Functions as an Oncogene in Colorectal Cancer by Targeting PTEN. *Dig. Dis. Sci.* **2014**, *59*, 98–107. [CrossRef]
60. Jin, X.-H.; Lu, S.; Wang, A.-F. Expression and clinical significance of miR-4516 and miR-21-5p in serum of patients with colorectal cancer. *BMC Cancer* **2020**, *20*, 241. [CrossRef]
61. Zhu, M.; Huang, Z.; Zhu, D.; Zhou, X.; Shan, X.; Qi, L.W.; Wu, L.; Cheng, W.; Zhu, J.; Zhang, L.; et al. A panel of microRNA signature in serum for colorectal cancer diagnosis. *Oncotarget* **2017**, *8*, 17081–17091. [CrossRef] [PubMed]
62. Gmerek, L.; Martyniak, K.; Horbacka, K.; Krokowicz, P.; Scierski, W.; Golusinski, P.; Golusinski, W.; Schneider, A.; Masternak, M.M. MicroRNA regulation in colorectal cancer tissue and serum. *PLoS ONE* **2019**, *14*, e0222013. [CrossRef] [PubMed]
63. Liu, G.-H.; Zhou, Z.-G.; Chen, R.; Wang, M.-J.; Zhou, B.; Li, Y.; Sun, X.-F. Serum miR-21 and miR-92a as biomarkers in the diagnosis and prognosis of colorectal cancer. *Tumour Biol.* **2013**, *34*, 2175–2181. [CrossRef] [PubMed]
64. Sarlinova, M.; Halasa, M.; Mistuna, D.; Musak, L.; Iliev, R.; Slaby, O.; Mazuchova, J.; Valentova, V.; Plank, L.; Halasova, E. Mir-21, mir-221 and mir-150 are deregulated in peripheral blood of patients with colorectal cancer. *Anticancer Res.* **2016**, *36*, 5449–5454. [CrossRef]
65. Zekri, A.R.N.; Youssef, A.S.E.D.; Lotfy, M.M.; Gabr, R.; Ahmed, O.S.; Nassar, A.; Hussein, N.; Omran, D.; Medhat, E.; Eid, S.; et al. Circulating serum miRNAs as diagnostic markers for colorectal cancer. *PLoS ONE* **2016**, *11*, e0154130. [CrossRef]
66. Li, J.; Liu, Y.; Wang, C.; Deng, T.; Liang, H.; Wang, Y.; Huang, D.; Fan, Q.; Wang, X.; Ning, T.; et al. Serum miRNA expression profile as a prognostic biomarker of stage II/III colorectal adenocarcinoma. *Sci. Rep.* **2015**, *5*, 1–13. [CrossRef]
67. Zhu, J.; Dong, H.; Zhang, Q.; Zhang, S. Combined assays for serum carcinoembryonic antigen and microRNA-17-3p offer improved diagnostic potential for stage I/II colon cancer. *Mol. Clin. Oncol.* **2015**, *3*, 1315–1318. [CrossRef]

68. Zhang, G.-J.; Zhou, T.; Liu, Z.-L.; Tian, H.-P.; Xia, S.-S. Plasma miR-200c and miR-18a as potential biomarkers for the detection of colorectal carcinoma. *Mol. Clin. Oncol.* **2013**, *1*, 379–384. [CrossRef]
69. Vega, A.B.; Pericay, C.; Moya, I.; Ferrer, A.; Dotor, E.; Pisa, A.; Casalots, À.; Serra-Aracil, X.; Oliva, J.-C.; Ruiz, A.; et al. microRNA expression profile in stage III colorectal cancer: Circulating miR-18a and miR-29a as promising biomarkers. *Oncol. Rep.* **2013**, *30*, 320–326. [CrossRef]
70. Farace, C.; Pisano, A.; Griñan-Lison, C.; Solinas, G.; Jiménez, G.; Serra, M.; Carrillo, E.; Scognamillo, F.; Attene, F.; Montella, A.; et al. Deregulation of cancer-stem-cell-associated miRNAs in tissues and sera of colorectal cancer patients. *Oncotarget* **2020**, *11*, 116–130. [CrossRef]
71. Matsumura, T.; Sugimachi, K.; Iinuma, H.; Takahashi, Y.; Kurashige, J.; Sawada, G.; Ueda, M.; Uchi, R.; Ueo, H.; Takano, Y.; et al. Exosomal microRNA in serum is a novel biomarker of recurrence in human colorectal cancer. *Br. J. Cancer* **2015**, *113*, 275–281. [CrossRef]
72. Chen, Q.; Xia, H.W.; Ge, X.J.; Zhang, Y.C.; Tang, Q.L.; Bi, F. Serum miR-19a predicts resistance to FOLFOX chemotherapy in advanced colorectal cancer cases. *Asian Pacific J. Cancer Prev.* **2013**, *14*, 7421–7426. [CrossRef] [PubMed]
73. Yu, F.-B.; Sheng, J.; Yu, J.-M.; Liu, J.-H.; Qin, X.-X.; Mou, B. MiR-19a-3p regulates the Forkhead box F2-mediated Wnt/β-catenin signaling pathway and affects the biological functions of colorectal cancer cells. *World J. Gastroenterol.* **2020**, *26*, 627–644. [CrossRef]
74. Maminezhad, H.; Ghanadian, S.; Pakravan, K.; Razmara, E.; Rouhollah, F.; Mossahebi-Mohammadi, M.; Babashah, S. A panel of six-circulating miRNA signature in serum and its potential diagnostic value in colorectal cancer. *Life Sci.* **2020**, *258*, 118226. [CrossRef] [PubMed]
75. Zheng, G.; Du, L.; Yang, X.; Zhang, X.; Wang, L.; Yang, Y.; Li, J.; Wang, C. Serum microRNA panel as biomarkers for early diagnosis of colorectal adenocarcinoma. *Br. J. Cancer* **2014**, *111*, 1985–1992. [CrossRef] [PubMed]
76. Dokhanchi, M.; Pakravan, K.; Zareian, S.; Hussen, B.M.; Farid, M.; Razmara, E.; Mossahebi-Mohammadi, M.; Cho, W.C.; Babashah, S. Colorectal cancer cell-derived extracellular vesicles transfer miR-221-3p to promote endothelial cell angiogenesis via targeting suppressor of cytokine signaling 3. *Life Sci.* **2021**, *285*, 119937. [CrossRef]
77. Yang, Q.; Wang, S.; Huang, J.; Xia, C.; Jin, H.; Fan, Y. Serum miR-20a and miR-486 are potential biomarkers for discriminating colorectal neoplasia: A pilot study. *J. Cancer Res. Ther.* **2018**, *14*, 1572. [CrossRef]
78. Huang, G.; Chen, X.; Cai, Y.; Wang, X.; Xing, C. miR-20a-directed regulation of BID is associated with the TRAIL sensitivity in colorectal cancer. *Oncol. Rep.* **2017**, *37*, 571–578. [CrossRef]
79. Huang, Z.; Huang, D.; Ni, S.; Peng, Z.; Sheng, W.; Du, X. Plasma microRNAs are promising novel biomarkers for early detection of colorectal cancer. *Int. J. Cancer* **2010**, *127*, 118–126. [CrossRef]
80. Shi, Y.; Liu, Z. Serum miR-92a-1 is a novel diagnostic biomarker for colorectal cancer. *J. Cell. Mol. Med.* **2020**, *24*, 8363–8367. [CrossRef]
81. Elshafei, A.; Shaker, O.; Abd El-motaal, O.; Salman, T. The expression profiling of serum miR-92a, miR-375, and miR-760 in colorectal cancer: An Egyptian study. *Tumor Biol.* **2017**, *39*, 101042831770576. [CrossRef] [PubMed]
82. Shiosaki, J.; Tiirikainen, M.; Peplowska, K.; Shaeffer, D.; Machida, M.; Sakamoto, K.; Takahashi, M.; Kojima, K.; Machi, J.; Bryant-Greenwood, P.; et al. Serum micro-RNA Identifies Early Stage Colorectal Cancer in a Multi-Ethnic Population. *Asian Pacific J. Cancer Prev.* **2020**, *21*, 3019–3026. [CrossRef] [PubMed]
83. Zheng, G.; Wang, H.; Zhang, X.; Yang, Y.; Wang, L.; Du, L.; Li, W.; Li, J.; Qu, A.; Liu, Y.; et al. Identification and validation of reference genes for qPCR detection of serum microRNAs in colorectal adenocarcinoma patients. *PLoS ONE* **2013**, *8*, e83025. [CrossRef] [PubMed]
84. Wang, J.; Huang, S.K.; Zhao, M.; Yang, M.; Zhong, J.L.; Gu, Y.Y.; Peng, H.; Che, Y.Q.; Huang, C.Z. Identification of a circulating microRNA signature for colorectal cancer detection. *PLoS ONE* **2014**, *9*, e87451. [CrossRef]
85. Luo, X.; Stock, C.; Burwinkel, B.; Brenner, H. Identification and Evaluation of Plasma MicroRNAs for Early Detection of Colorectal Cancer. *PLoS ONE* **2013**, *8*, e62880. [CrossRef] [PubMed]
86. Huang, G.; Wei, B.; Chen, Z.; Wang, J.; Zhao, L.; Peng, X.; Liu, K.; Lai, Y.; Ni, L. Identification of a four-microRNA panel in serum as promising biomarker for colorectal carcinoma detection. *Biomark. Med.* **2020**, *14*, 749–760. [CrossRef]
87. Hiyoshi, Y.; Akiyoshi, T.; Inoue, R.; Murofushi, K.; Yamamoto, N.; Fukunaga, Y.; Ueno, M.; Baba, H.; Mori, S.; Yamaguchi, T. Serum miR-143 levels predict the pathological response to neoadjuvant chemoradiotherapy in patients with locally advanced rectal cancer. *Oncotarget* **2017**, *8*, 79201–79211. [CrossRef]
88. Ramzy, I.; Hasaballah, M.; Marzaban, R.; Shaker, O.; Soliman, Z.A. Evaluation of microRNAs-29a, 92a and 145 in colorectal carcinoma as candidate diagnostic markers: An Egyptian pilot study. *Clin. Res. Hepatol. Gastroenterol.* **2015**, *39*, 508–515. [CrossRef]
89. Sun, Y.; Liu, Y.; Cogdell, D.; Calin, G.A.; Sun, B.; Kopetz, S.; Hamilton, S.R.; Zhang, W. Examining plasma microRNA markers for colorectal cancer at different stages. *Oncotarget* **2016**, *7*, 11434–11449. [CrossRef]
90. Kingham, T.P.; Nguyen, H.C.B.; Zheng, J.; Konstantinidis, I.T.; Sadot, E.; Shia, J.; Kuk, D.; Zhang, S.; Saltz, L.; D'Angelica, M.I.; et al. MicroRNA-203 predicts human survival after resection of colorectal liver metastasis. *Oncotarget* **2017**, *8*, 18821–18831. [CrossRef]
91. Takano, Y.; Masuda, T.; Iinuma, H.; Yamaguchi, R.; Sato, K.; Tobo, T.; Hirata, H.; Kuroda, Y.; Nambara, S.; Hayashi, N.; et al. Circulating exosomal microRNA-203 is associated with metastasis possibly via inducing tumor-associated macrophages in colorectal cancer. *Oncotarget* **2017**, *8*, 78598–78613. [CrossRef] [PubMed]

92. Deng, B.; Wang, B.; Fang, J.; Zhu, X.; Cao, Z.; Lin, Q.; Zhou, L.; Sun, X. MiRNA-203 suppresses cell proliferation, migration and invasion in colorectal cancer via targeting of EIF5A2. *Sci. Rep.* **2016**, *6*, 1–11. [CrossRef] [PubMed]
93. Hur, K.; Toiyama, Y.; Okugawa, Y.; Ide, S.; Imaoka, H.; Boland, C.R.; Goel, A. Circulating microRNA-203 predicts prognosis and metastasis in human colorectal cancer. *Gut* **2017**, *66*, 654–664. [CrossRef] [PubMed]
94. Peng, Q.; Shen, Y.; Zhao, P.; Cai, S.; Feng, Z.; Cheng, M.; Wu, Y.; Zhu, Y. Biomarker exploration of microRNA-203 as a promising substrate for predicting poor survival outcome in colorectal cancer. *BMC Cancer* **2020**, *20*, 1–14. [CrossRef] [PubMed]
95. Yuan, Z.; Baker, K.; Redman, M.W.; Wang, L.; Adams, S.V.; Yu, M.; Dickinson, B.; Makar, K.; Ulrich, N.; Böhm, J.; et al. Dynamic plasma microRNAs are biomarkers for prognosis and early detection of recurrence in colorectal cancer. *Br. J. Cancer* **2017**, *117*, 1202–1210. [CrossRef] [PubMed]
96. Nassar, F.J.; Msheik, Z.S.; Itani, M.M.; El Helou, R.; Hadla, R.; Kreidieh, F.; Bejjany, R.; Mukherji, D.; Shamseddine, A.; Nasr, R.R.; et al. Circulating mirna as biomarkers for colorectal cancer diagnosis and liver metastasis. *Diagnostics* **2021**, *11*, 341. [CrossRef] [PubMed]
97. Toiyama, Y.; Hur, K.; Tanaka, K.; Inoue, Y.; Kusunoki, M.; Boland, C.R.; Goel, A. Serum miR-200c Is a Novel Prognostic and Metastasis-Predictive Biomarker in Patients with Colorectal Cancer. *Ann. Surg.* **2014**, *259*, 735–743. [CrossRef]
98. Ardila, H.J.; Sanabria-Salas, M.C.; Meneses, X.; Rios, R.; Huertas-Salgado, A.; Serrano, M.L. Circulating miR-141-3p, miR-143-3p and miR-200c-3p are differentially expressed in colorectal cancer and advanced adenomas. *Mol. Clin. Oncol.* **2019**, *11*, 201–207. [CrossRef]
99. Ding, M.; Zhang, T.; Li, S.; Zhang, Y.; Qiu, Y.; Zhang, B. Correlation analysis between liver metastasis and serum levels of MIR-200 and MIR-141 in patients with colorectal cancer. *Mol. Med. Rep.* **2017**, *16*, 7791–7795. [CrossRef]
100. Tayel, S.I.; Fouda, E.A.M.; Gohar, S.F.; Elshayeb, E.I.; El-sayed, E.H.; El-kousy, S.M. Potential role of MicroRNA 200c gene expression in assessment of colorectal cancer. *Arch. Biochem. Biophys.* **2018**, *647*, 41–46. [CrossRef]
101. Santasusagna, S.; Moreno, I.; Navarro, A.; Martinez Rodenas, F.; Hernández, R.; Castellano, J.J.; Muñoz, C.; Monzo, M. Prognostic impact of miR-200 family members in plasma and exosomes from tumor-draining versus peripheral veins of colon cancer patients. *Oncology* **2018**, *95*, 309–318. [CrossRef] [PubMed]
102. Dong, S.J.; Cai, X.J.; Li, S.J. The clinical significance of MiR-429 as a predictive biomarker in colorectal cancer patients receiving 5-fluorouracil treatment. *Med. Sci. Monit.* **2016**, *22*, 3352–3361. [CrossRef]
103. Koga, Y.; Yasunaga, M.; Takahashi, A.; Kuroda, J.; Moriya, Y.; Akasu, T.; Fujita, S.; Yamamoto, S.; Baba, H.; Matsumura, Y. MicroRNA expression profiling of exfoliated colonocytes isolated from feces for colorectal cancer screening. *Cancer Prev. Res.* **2010**, *3*, 1435–1442. [CrossRef] [PubMed]
104. Li, L.; Wang, A.; Cai, M.; Tong, M.; Chen, F.; Huang, L. Identification of stool miR-135b-5p as a non-invasive diaognostic biomarker in later tumor stage of colorectal cancer. *Life Sci.* **2020**, *260*, 1–7. [CrossRef] [PubMed]
105. Wang, Q.; Zhang, H.; Shen, X.; Ju, S. Serum microRNA-135a-5p as an auxiliary diagnostic biomarker for colorectal cancer. *Ann. Clin. Biochem.* **2017**, *54*, 76–85. [CrossRef] [PubMed]
106. Zhou, X.; Yang, D.; Ding, X.; Xu, P. Clinical value of microRNA-135a and MMP-13 in colon cancer. *Oncol. Lett.* **2021**, *22*, 1–9. [CrossRef]
107. Wu, C.W.; Ng, S.C.; Dong, Y.; Tian, L.; Ng, S.S.M.; Leung, W.W.; Law, W.T.; Yau, T.O.; Chan, F.K.L.; Sung, J.J.Y.; et al. Identification of microrna-135b in stool as a potential noninvasive biomarker for colorectal cancer and adenoma. *Clin. Cancer Res.* **2014**, *20*, 2994–3002. [CrossRef]
108. Phua, L.C.; Chue, X.P.; Koh, P.K.; Cheah, P.Y.; Chan, E.C.Y.; Ho, H.K. Global fecal microRNA profiling in the identification of biomarkers for colorectal cancer screening among Asians. *Oncol. Rep.* **2014**, *32*, 97–104. [CrossRef]
109. Qin, Y.; Li, L.; Wang, F.; Zhou, X.; Liu, Y.; Yin, Y.; Qi, X. Knockdown of miR-135b sensitizes colorectal cancer cells to oxaliplatin-induced apoptosis through increase of FOXO1. *Cell. Physiol. Biochem.* **2018**, *48*, 1627–1637. [CrossRef]
110. Jin, G.; Liu, Y.; Zhang, J.; Bian, Z.; Yao, S.; Fei, B.; Zhou, L.; Yin, Y.; Huang, Z. A panel of serum exosomal microRNAs as predictive markers for chemoresistance in advanced colorectal cancer. *Cancer Chemother. Pharmacol.* **2019**, *84*, 315–325. [CrossRef]
111. Faltejskova, P.; Bocanek, O.; Sachlova, M.; Svoboda, M.; Kiss, I.; Vyzula, R.; Slaby, O. Circulating miR-17-3p, miR-29a, miR-92a and miR-135b in serum: Evidence against their usage as biomarkers in colorectal cancer. *Cancer Biomarkers* **2012**, *12*, 199–204. [CrossRef] [PubMed]
112. Yuan, D.; Li, K.; Zhu, K.; Yan, R.; Dang, C. Plasma miR-183 predicts recurrence and prognosis in patients with colorectal cancer. *Cancer Biol. Ther.* **2015**, *16*, 268–275. [CrossRef] [PubMed]
113. Zhao, Y.J.; Song, X.; Niu, L.; Tang, Y.; Song, X.; Xie, L. Circulating exosomal mir-150-5p and mir-99b-5p as diagnostic biomarkers for colorectal cancer. *Front. Oncol.* **2019**, *9*, 1–10. [CrossRef]
114. Xiao, Y.; Zhong, J.; Zhong, B.; Huang, J.; Jiang, L.; Jiang, Y.; Yuan, J.; Sun, J.; Dai, L.; Yang, C.; et al. Exosomes as potential sources of biomarkers in colorectal cancer. *Cancer Lett.* **2020**, *476*, 13–22. [CrossRef]
115. Aherne, S.T.; Madden, S.F.; Hughes, D.J.; Pardini, B.; Naccarati, A.; Levy, M.; Vodicka, P.; Neary, P.; Dowling, P.; Clynes, M. Circulating miRNAs miR-34a and miR-150 associated with colorectal cancer progression. *BMC Cancer* **2015**, *15*, 1–13. [CrossRef]
116. Ogata-Kawata, H.; Izumiya, M.; Kurioka, D.; Honma, Y.; Yamada, Y.; Furuta, K.; Gunji, T.; Ohta, H.; Okamoto, H.; Sonoda, H.; et al. Circulating exosomal microRNAs as biomarkers of colon cancer. *PLoS ONE* **2014**, *9*, e92921. [CrossRef]

117. Al-Sheikh, Y.A.; Ghneim, H.K.; Softa, K.I.; Al-Jobran, A.A.; Al-Obeed, O.; Mohamed, M.A.V.; Abdulla, M.; Aboul-Soud, M.A.M. Expression profiling of selected microRNA signatures in plasma and tissues of Saudi colorectal cancer patients by qPCR. *Oncol. Lett.* **2016**, *11*, 1406–1412. [CrossRef] [PubMed]
118. Wada, Y.; Shimada, M.; Murano, T.; Takamaru, H.; Morine, Y.; Ikemoto, T.; Saito, Y.; Balaguer, F.; Bujanda, L.; Pellise, M.; et al. A Liquid Biopsy Assay for Noninvasive Identification of Lymph Node Metastases in T1 Colorectal Cancer. *Gastroenterology* **2021**, *161*, 151–162.e1. [CrossRef]
119. Iorio, M.V.; Ferracin, M.; Liu, C.G.; Veronese, A.; Spizzo, R.; Sabbioni, S.; Magri, E.; Pedriali, M.; Fabbri, M.; Campiglio, M.; et al. MicroRNA gene expression deregulation in human breast cancer. *Cancer Res.* **2005**, *65*, 7065–7070. [CrossRef]
120. Meng, F.; Henson, R.; Wehbe–Janek, H.; Ghoshal, K.; Jacob, S.T.; Patel, T. MicroRNA-21 Regulates Expression of the PTEN Tumor Suppressor Gene in Human Hepatocellular Cancer. *Gastroenterology* **2007**, *133*, 647–658. [CrossRef]
121. Si, M.L.; Zhu, S.; Wu, H.; Lu, Z.; Wu, F.; Mo, Y.Y. miR-21-mediated tumor growth. *Oncogene* **2007**, *26*, 2799–2803. [CrossRef] [PubMed]
122. Kasashima, K.; Nakamura, Y.; Kozu, T. Altered expression profiles of microRNAs during TPA-induced differentiation of HL-60 cells. *Biochem. Biophys. Res. Commun.* **2004**, *322*, 403–410. [CrossRef] [PubMed]
123. Cekaite, L.; Clancy, T.; Sioud, M. Increased miR-21 expression during human monocyte differentiation into DCs. *Front. Biosci.* **2010**, *E2*, 143. [CrossRef]
124. Jenike, A.E.; Halushka, M.K. miR-21: A non-specific biomarker of all maladies. *Biomark. Res.* **2021**, *9*, 18. [CrossRef]
125. Ma, X.; Kumar, M.; Choudhury, S.N.; Becker Buscaglia, L.E.; Barker, J.R.; Kanakamedala, K.; Liu, M.-F.; Li, Y. Loss of the miR-21 allele elevates the expression of its target genes and reduces tumorigenesis. *Proc. Natl. Acad. Sci. USA* **2011**, *108*, 10144–10149. [CrossRef]
126. Guo, X.-Z.; Ye, X.-L.; Xiao, W.-Z.; Wei, X.-N.; You, Q.-H.; Che, X.-H.; Cai, Y.-J.; Chen, F.; Yuan, H.; Liu, X.-J.; et al. Downregulation of VMP1 confers aggressive properties to colorectal cancer. *Oncol. Rep.* **2015**, *34*, 2557–2566. [CrossRef]
127. Sauermann, M.; Sahin, Ö.; Sültmann, H.; Hahne, F.; Blaszkiewicz, S.; Majety, M.; Zatloukal, K.; Füzesi, L.; Poustka, A.; Wiemann, S.; et al. Reduced expression of vacuole membrane protein 1 affects the invasion capacity of tumor cells. *Oncogene* **2008**, *27*, 1320–1326. [CrossRef]
128. Wang, C.; Peng, R.; Zeng, M.; Zhang, Z.; Liu, S.; Jiang, D.; Lu, Y.; Zou, F. An autoregulatory feedback loop of miR-21/VMP1 is responsible for the abnormal expression of miR-21 in colorectal cancer cells. *Cell Death Dis.* **2020**, *11*, 1067. [CrossRef]
129. Meng, F.; Henson, R.; Lang, M.; Wehbe, H.; Maheshwari, S.; Mendell, J.T.; Jiang, J.; Schmittgen, T.D.; Patel, T. Involvement of Human Micro-RNA in Growth and Response to Chemotherapy in Human Cholangiocarcinoma Cell Lines. *Gastroenterology* **2006**, *130*, 2113–2129. [CrossRef]
130. Asangani, I.A.; Rasheed, S.A.K.; Nikolova, D.A.; Leupold, J.H.; Colburn, N.H.; Post, S.; Allgayer, H. MicroRNA-21 (miR-21) post-transcriptionally downregulates tumor suppressor Pdcd4 and stimulates invasion, intravasation and metastasis in colorectal cancer. *Oncogene* **2008**, *27*, 2128–2136. [CrossRef]
131. Sheedy, F.J.; Palsson-McDermott, E.; Hennessy, E.J.; Martin, C.; O'Leary, J.J.; Ruan, Q.; Johnson, D.S.; Chen, Y.; O'Neill, L.A.J. Negative regulation of TLR4 via targeting of the proinflammatory tumor suppressor PDCD4 by the microRNA miR-21. *Nat. Immunol.* **2010**, *11*, 141–147. [CrossRef] [PubMed]
132. Wu, Y.; Song, Y.; Xiong, Y.; Wang, X.; Xu, K.; Han, B.; Bai, Y.; Li, L.; Zhang, Y.; Zhou, L. MicroRNA-21 (Mir-21) Promotes Cell Growth and Invasion by Repressing Tumor Suppressor PTEN in Colorectal Cancer. *Cell. Physiol. Biochem.* **2017**, *43*, 945–958. [CrossRef] [PubMed]
133. Long, J.; Yin, Y.; Guo, H.; Li, S.; Sun, Y.; Zeng, C.; Zhu, W. The mechanisms and clinical significance of PDCD4 in colorectal cancer. *Gene* **2019**, *680*, 59–64. [CrossRef] [PubMed]
134. Xiong, B.; Cheng, Y.; Ma, L.; Zhang, C. MiR-21 regulates biological behavior through the PTEN/PI-3 K/Akt signaling pathway in human colorectal cancer cells. *Int. J. Oncol.* **2013**, *42*, 219–228. [CrossRef]
135. Shi, L.; Chen, J.; Yang, J.; Pan, T.; Zhang, S.; Wang, Z. MiR-21 protected human glioblastoma U87MG cells from chemotherapeutic drug temozolomide induced apoptosis by decreasing Bax/Bcl-2 ratio and caspase-3 activity. *Brain Res.* **2010**, *1352*, 255–264. [CrossRef]
136. Buscaglia, L.E.B.; Li, Y. Apoptosis and the target genes of microRNA-21. *Chin. J. Cancer* **2011**, *30*, 371–380. [CrossRef]
137. Ferraro, A.; Kontos, C.K.; Boni, T.; Bantounas, I.; Siakouli, D.; Kosmidou, V.; Vlassi, M.; Spyridakis, Y.; Tsipras, I.; Zografos, G.; et al. Epigenetic regulation of miR-21 in colorectal cancer. *Epigenetics* **2014**, *9*, 129–141. [CrossRef]
138. Chang, K.H.; Miller, N.; Kheirelseid, E.A.H.; Ingoldsby, H.; Hennessy, E.; Curran, C.E.; Curran, S.; Smith, M.J.; Regan, M.; McAnena, O.J.; et al. MicroRNA-21 and PDCD4 expression in colorectal cancer. *Eur. J. Surg. Oncol.* **2011**, *37*, 597–603. [CrossRef]
139. Hayashita, Y.; Osada, H.; Tatematsu, Y.; Yamada, H.; Yanagisawa, K.; Tomida, S.; Yatabe, Y.; Kawahara, K.; Sekido, Y.; Takahashi, T. A Polycistronic MicroRNA Cluster, miR-17-92, Is Overexpressed in Human Lung Cancers and Enhances Cell Proliferation. *Cancer Res.* **2005**, *65*, 9628–9632. [CrossRef]
140. Mendell, J.T. miRiad Roles for the miR-17-92 Cluster in Development and Disease. *Cell* **2008**, *133*, 217–222. [CrossRef]
141. Diosdado, B.; Van De Wiel, M.A.; Terhaar Sive Droste, J.S.; Mongera, S.; Postma, C.; Meijerink, W.J.H.J.; Carvalho, B.; Meijer, G.A. MiR-17-92 cluster is associated with 13q gain and c-myc expression during colorectal adenoma to adenocarcinoma progression. *Br. J. Cancer* **2009**, *101*, 707–714. [CrossRef]

142. Sun, R.; Liang, Y.; Yuan, F.; Nie, X.; Sun, H.; Wang, Y.; Yu, T.; Gao, L.; Zhang, L. Functional polymorphisms in the promoter region of miR-17-92 cluster are associated with a decreased risk of colorectal cancer. *Oncotarget* **2017**, *8*, 82531–82540. [CrossRef]
143. Humphreys, K.J.; Cobiac, L.; Le Leu, R.K.; Van der Hoek, M.B.; Michael, M.Z. Histone deacetylase inhibition in colorectal cancer cells reveals competing roles for members of the oncogenic miR-17-92 cluster. *Mol. Carcinog.* **2013**, *52*, 459–474. [CrossRef] [PubMed]
144. Ventura, A.; Young, A.G.; Winslow, M.M.; Lintault, L.; Meissner, A.; Erkeland, S.J.; Newman, J.; Bronson, R.T.; Crowley, D.; Stone, J.R.; et al. Targeted Deletion Reveals Essential and Overlapping Functions of the miR-17~92 Family of miRNA Clusters. *Cell* **2008**, *132*, 875–886. [CrossRef] [PubMed]
145. Lujambio, A.; Lowe, S.W. The microcosmos of cancer. *Nature* **2012**, *482*, 347–355. [CrossRef] [PubMed]
146. de Pontual, L.; Yao, E.; Callier, P.; Faivre, L.; Drouin, V.; Cariou, S.; Van Haeringen, A.; Geneviève, D.; Goldenberg, A.; Oufadem, M.; et al. Germline deletion of the miR-17∼92 cluster causes skeletal and growth defects in humans. *Nat. Genet.* **2011**, *43*, 1026–1030. [CrossRef] [PubMed]
147. Ma, Y.; Zhang, P.; Wang, F.; Zhang, H.; Yang, Y.; Shi, C.; Xia, Y.; Peng, J.; Liu, W.; Yang, Z.; et al. Elevated oncofoetal miR-17-5p expression regulates colorectal cancer progression by repressing its target gene P130. *Nat. Commun.* **2012**, *3*, 1291. [CrossRef]
148. Ast, V.; Kordaß, T.; Oswald, M.; Kolte, A.; Eisel, D.; Osen, W.; Eichmüller, S.B.; Berndt, A.; König, R. MiR-192, miR-200c and miR-17 are fibroblast-mediated inhibitors of colorectal cancer invasion. *Oncotarget* **2018**, *9*, 35559–35580. [CrossRef]
149. Humphreys, K.J.; McKinnon, R.A.; Michael, M.Z. miR-18a Inhibits CDC42 and Plays a Tumour Suppressor Role in Colorectal Cancer Cells. *PLoS ONE* **2014**, *9*, e112288. [CrossRef]
150. Huang, G.; Wu, X.; Li, S.; Xu, X.; Zhu, H.; Chen, X. The long noncoding RNA CASC2 functions as a competing endogenous RNA by sponging miR-18a in colorectal cancer. *Sci. Rep.* **2016**, *6*, 1–11. [CrossRef]
151. Olive, V.; Bennett, M.J.; Walker, J.C.; Ma, C.; Jiang, I.; Cordon-Cardo, C.; Li, Q.-J.; Lowe, S.W.; Hannon, G.J.; He, L. miR-19 is a key oncogenic component of mir-17-92. *Genes Dev.* **2009**, *23*, 2839–2849. [CrossRef] [PubMed]
152. Zhang, Y.; Liu, X.; Zhang, J.; Xu, Y.; Shao, J.; Hu, Y.; Shu, P.; Cheng, H. Inhibition of miR-19a partially reversed the resistance of colorectal cancer to oxaliplatin via PTEN/PI3K/AKT pathway. *Aging* **2020**, *12*, 5640–5650. [CrossRef] [PubMed]
153. Liu, Y.; Liu, R.; Yang, F.; Cheng, R.; Chen, X.; Cui, S.; Gu, Y.; Sun, W.; You, C.; Liu, Z.; et al. miR-19a promotes colorectal cancer proliferation and migration by targeting TIA1. *Mol. Cancer* **2017**, *16*, 53. [CrossRef]
154. Chen, M.; Lin, M.; Wang, X. Overexpression of miR-19a inhibits colorectal cancer angiogenesis by suppressing KRAS expression. *Oncol. Rep.* **2018**, *39*, 619–626. [CrossRef]
155. Jia, Z.; Wang, K.; Zhang, A.; Wang, G.; Kang, C.; Han, L.; Pu, P. miR-19a and miR-19b Overexpression in Gliomas. *Pathol. Oncol. Res.* **2013**, *19*, 847–853. [CrossRef] [PubMed]
156. Xu, J.; Tang, Y.; Bei, Y.; Ding, S.; Che, L.; Yao, J.; Wang, H.; Lv, D.; Xiao, J. miR-19b attenuates H2O2-induced apoptosis in rat H9C2 cardiomyocytes via targeting PTEN. *Oncotarget* **2016**, *7*, 10870–10878. [CrossRef] [PubMed]
157. Fan, Y.; Yin, S.; Hao, Y.; Yang, J.; Zhang, H.; Sun, C.; Ma, M.; Chang, Q.; Xi, J.J. MiR-19b promotes tumor growth and metastasis via targeting TP53. *RNA* **2014**, *20*, 765–772. [CrossRef]
158. Zaporozhchenko, I.A.; Morozkin, E.S.; Skvortsova, T.E.; Ponomaryova, A.A.; Rykova, E.Y.; Cherdyntseva, N.V.; Polovnikov, E.S.; Pashkovskaya, O.A.; Pokushalov, E.A.; Vlassov, V.V.; et al. Plasma miR-19b and miR-183 as Potential Biomarkers of Lung Cancer. *PLoS ONE* **2016**, *11*, e0165261. [CrossRef]
159. Wu, C.; Cao, Y.; He, Z.; He, J.; Hu, C.; Duan, H.; Jiang, J. Serum Levels of miR-19b and miR-146a as Prognostic Biomarkers for Non-Small Cell Lung Cancer. *Tohoku J. Exp. Med.* **2014**, *232*, 85–95. [CrossRef]
160. Copier, C.U.; León, L.; Fernández, M.; Contador, D.; Calligaris, S.D. Circulating miR-19b and miR-181b are potential biomarkers for diabetic cardiomyopathy. *Sci. Rep.* **2017**, *7*, 13514. [CrossRef]
161. Ke, T.-W.; Wei, P.-L.; Yeh, K.-T.; Chen, W.T.-L.; Cheng, Y.-W. MiR-92a Promotes Cell Metastasis of Colorectal Cancer Through PTEN-Mediated PI3K/AKT Pathway. *Ann. Surg. Oncol.* **2015**, *22*, 2649–2655. [CrossRef] [PubMed]
162. Zhang, G.J.; Li, L.F.; Yang, G.D.; Xia, S.S.; Wang, R.; Leng, Z.W.; Liu, Z.L.; Tian, H.P.; He, Y.; Meng, C.Y.; et al. MiR-92a promotes stem cell-like properties by activating Wnt/β-catenin signaling in colorectal cancer. *Oncotarget* **2017**, *8*, 101760–101770. [CrossRef] [PubMed]
163. Wei, Q.-D.; Zheng, W.-B.; Sun, K.; Xue, Q.; Yang, C.-Z.; Li, G.-X. MiR-92a promotes the invasion and migration of colorectal cancer by targeting RECK. *Int. J. Clin. Exp. Pathol.* **2019**, *12*, 1565–1577. [PubMed]
164. Lv, H.; Zhang, Z.; Wang, Y.; Li, C.; Gong, W.; Wang, X. MicroRNA-92a promotes colorectal cancer cell growth and migration by inhibiting KLF4. *Oncol. Res.* **2016**, *23*, 283–290. [CrossRef]
165. Ma, H.; Pan, J.-S.; Jin, L.-X.; Wu, J.; Ren, Y.-D.; Chen, P.; Xiao, C.; Han, J. MicroRNA-17~92 inhibits colorectal cancer progression by targeting angiogenesis. *Cancer Lett.* **2016**, *376*, 293–302. [CrossRef]
166. Cheng, D.; Zhao, S.; Tang, H.; Zhang, D.; Sun, H.; Yu, F.; Jiang, W.; Yue, B.; Wang, J.; Zhang, M.; et al. MicroRNA-20a-5p promotes colorectal cancer invasion and metastasis by downregulating Smad4. *Oncotarget* **2016**, *7*, 45199–45213. [CrossRef]
167. Sarshar, M.; Scribano, D.; Ambrosi, C.; Palamara, A.T.; Masotti, A. Fecal microRNAs as Innovative Biomarkers of Intestinal Diseases and Effective Players in Host-Microbiome Interactions. *Cancers* **2020**, *12*, 2174. [CrossRef]
168. Zhou, T.; Zhang, G.; Liu, Z.; Xia, S.; Tian, H. Overexpression of miR-92a correlates with tumor metastasis and poor prognosis in patients with colorectal cancer. *Int. J. Colorectal Dis.* **2013**, *28*, 19–24. [CrossRef]

169. Salvi, S.; Molinari, C.; Foca, F.; Teodorani, N.; Saragoni, L.; Puccetti, M.; Passardi, A.; Tamberi, S.; Avanzolini, A.; Lucci, E.; et al. miR-17-92a-1 cluster host gene (MIR17HG) evaluation and response to neoadjuvant chemoradiotherapy in rectal cancer. *Onco. Targets. Ther.* **2016**, *9*, 2735. [CrossRef]
170. Yu, X.; Li, Z.; Yu, J.; Chan, M.T.V.; Wu, W.K.K. MicroRNAs predict and modulate responses to chemotherapy in colorectal cancer. *Cell Prolif.* **2015**, *48*, 503–510. [CrossRef]
171. Xu, J.; Meng, Q.; Li, X.; Yang, H.; Xu, J.; Gao, N.; Sun, H.; Wu, S.; Familiari, G.; Relucenti, M.; et al. Long Noncoding RNA MIR17HG Promotes Colorectal Cancer Progression via miR-17-5p. *Cancer Res.* **2019**, *79*, 4882–4895. [CrossRef] [PubMed]
172. Kooshapur, H.; Choudhury, N.R.; Simon, B.; Mühlbauer, M.; Jussupow, A.; Fernandez, N.; Jones, A.N.; Dallmann, A.; Gabel, F.; Camilloni, C.; et al. Structural basis for terminal loop recognition and stimulation of pri-miRNA-18a processing by hnRNP A1. *Nat. Commun.* **2018**, *9*, 2479. [CrossRef] [PubMed]
173. Chaulk, S.G.; Xu, Z.; Glover, M.J.N.; Fahlman, R.P. MicroRNA miR-92a-1 biogenesis and mRNA targeting is modulated by a tertiary contact within the miR-17~92 microRNA cluster. *Nucleic Acids Res.* **2014**, *42*, 5234–5244. [CrossRef]
174. Chakraborty, S.; Mehtab, S.; Patwardhan, A.; Krishnan, Y. Pri-miR-17-92a transcript folds into a tertiary structure and autoregulates its processing. *RNA* **2012**, *18*, 1014–1028. [CrossRef] [PubMed]
175. Ounzain, S.; Micheletti, R.; Arnan, C.; Plaisance, I.; Cecchi, D.; Schroen, B.; Reverter, F.; Alexanian, M.; Gonzales, C.; Ng, S.Y.; et al. CARMEN, a human super enhancer-associated long noncoding RNA controlling cardiac specification, differentiation and homeostasis. *J. Mol. Cell. Cardiol.* **2015**, *89*, 98–112. [CrossRef] [PubMed]
176. Elia, L.; Quintavalle, M.; Zhang, J.; Contu, R.; Cossu, L.; Latronico, M.V.G.; Peterson, K.L.; Indolfi, C.; Catalucci, D.; Chen, J.; et al. The knockout of miR-143 and -145 alters smooth muscle cell maintenance and vascular homeostasis in mice: Correlates with human disease. *Cell Death Differ.* **2009**, *16*, 1590–1598. [CrossRef]
177. Chen, X.; Guo, X.; Zhang, H.; Xiang, Y.; Chen, J.; Yin, Y.; Cai, X.; Wang, K.; Wang, G.; Ba, Y.; et al. Role of miR-143 targeting KRAS in colorectal tumorigenesis. *Oncogene* **2009**, *28*, 1385–1392. [CrossRef]
178. Su, J.; Liang, H.; Yao, W.; Wang, N.; Zhang, S.; Yan, X.; Feng, H.; Pang, W.; Wang, Y.; Wang, X.; et al. MiR-143 and MiR-145 regulate IGF1R to suppress cell proliferation in colorectal cancer. *PLoS ONE* **2014**, *9*, e0114420. [CrossRef]
179. Michael, M.Z.; O' Connor, S.M.; van Holst Pellekaan, N.G.; Young, G.P.; James, R.J. Reduced accumulation of specific microRNAs in colorectal neoplasia. *Mol. Cancer Res.* **2003**, *1*, 882–891.
180. Pagliuca, A.; Valvo, C.; Fabrizi, E.; di Martino, S.; Biffoni, M.; Runci, D.; Forte, S.; De Maria, R.; Ricci-Vitiani, L. Analysis of the combined action of miR-143 and miR-145 on oncogenic pathways in colorectal cancer cells reveals a coordinate program of gene repression. *Oncogene* **2013**, *32*, 4806–4813. [CrossRef]
181. Xing, A.Y.; Wang, Y.W.; Su, Z.X.; Shi, D.B.; Wang, B.; Gao, P. Catenin-δ 1, negatively regulated by miR-145, promotes tumour aggressiveness in gastric cancer. *J. Pathol.* **2015**, *236*, 53–64. [CrossRef] [PubMed]
182. Ding, X.; Du, J.; Mao, K.; Wang, X.; Ding, Y.; Wang, F. MicroRNA-143-3p suppresses tumorigenesis by targeting catenin-δ1 in colorectal cancer. *Onco. Targets. Ther.* **2019**, *12*, 3255–3265. [CrossRef] [PubMed]
183. Chivukula, R.R.; Shi, G.; Acharya, A.; Mills, E.W.; Zeitels, L.R.; Anandam, J.L.; Abdelnaby, A.A.; Balch, G.C.; Mansour, J.C.; Yopp, A.C.; et al. An Essential Mesenchymal Function for miR-143/145 in Intestinal Epithelial Regeneration. *Cell* **2014**, *157*, 1104–1116. [CrossRef] [PubMed]
184. Kent, O.A.; McCall, M.N.; Cornish, T.C.; Halushka, M.K. Lessons from miR-143/145: The importance of cell-type localization of miRNAs. *Nucleic Acids Res.* **2014**, *42*, 7528–7538. [CrossRef]
185. Akao, Y.; Nakagawa, Y.; Naoe, T. MicroRNAs 143 and 145 are possible common onco-microRNAs in human cancers. *Oncol. Rep.* **2006**, *16*, 845–850. [CrossRef]
186. Yin, Y.; Yan, Z.-P.; Lu, N.-N.; Xu, Q.; He, J.; Qian, X.; Yu, J.; Guan, X.; Jiang, B.-H.; Liu, L.-Z. Downregulation of miR-145 associated with cancer progression and VEGF transcriptional activation by targeting N-RAS and IRS1. *Biochim. Biophys. Acta* **2013**, *1829*, 239–247. [CrossRef]
187. Feng, Y.; Zhu, J.; Ou, C.; Deng, Z.; Chen, M.; Huang, W.; Li, L. MicroRNA-145 inhibits tumour growth and metastasis in colorectal cancer by targeting fascin-1. *Br. J. Cancer* **2014**, *110*, 2300–2309. [CrossRef]
188. Huang, F.T.; Chen, W.Y.; Gu, Z.Q.; Zhuang, Y.Y.; Li, C.Q.; Wang, L.Y.; Peng, J.F.; Zhu, Z.; Luo, X.; Li, Y.H.; et al. The novel long intergenic noncoding RNA UCC promotes colorectal cancer progression by sponging MIR-143. *Cell Death Dis.* **2017**, *8*, e2778. [CrossRef]
189. Luan, Y.; Li, X.; Luan, Y.; Zhao, R.; Li, Y.; Liu, L.; Hao, Y.; Oleg Vladimir, B.; Jia, L. Circulating lncRNA UCA1 Promotes Malignancy of Colorectal Cancer via the miR-143/MYO6 Axis. *Mol. Ther. Nucleic Acids* **2020**, *19*, 790–803. [CrossRef]
190. Hu, Y.; Ma, Z.; He, Y.; Liu, W.; Su, Y.; Tang, Z. PART-1 functions as a competitive endogenous RNA for promoting tumor progression by sponging miR-143 in colorectal cancer. *Biochem. Biophys. Res. Commun.* **2017**, *490*, 317–323. [CrossRef]
191. Xie, H.; Ren, X.; Xin, S.; Lan, X.; Lu, G.; Lin, Y.; Yang, S.; Zeng, Z.; Liao, W.; Ding, Y.-Q.; et al. Emerging roles of circRNA_001569 targeting miR-145 in the proliferation and invasion of colorectal cancer. *Oncotarget* **2016**, *7*, 26680–26691. [CrossRef] [PubMed]
192. Slaby, O.; Svoboda, M.; Fabian, P.; Smerdova, T.; Knoflickova, D.; Bednarikova, M.; Nenutil, R.; Vyzula, R. Altered Expression of miR-21, miR-31, miR-143 and miR-145 Is Related to Clinicopathologic Features of Colorectal Cancer. *Oncology* **2007**, *72*, 397–402. [CrossRef] [PubMed]
193. Akao, Y.; Nakagawa, Y.; Hirata, I.; Iio, A.; Itoh, T.; Kojima, K.; Nakashima, R.; Kitade, Y.; Naoe, T. Role of anti-oncomirs miR-143 and -145 in human colorectal tumors. *Cancer Gene Ther.* **2010**, *17*, 398–408. [CrossRef] [PubMed]

194. Ahmed, F.E.; Ahmed, N.C.; Vos, P.W.; Bonnerup, C.; Atkins, J.N.; Casey, M.; Nuovo, G.J.; Naziri, W.; Wiley, J.E.; Mota, H.; et al. Diagnostic microRNA markers to screen for sporadic human colon cancer in stool: I. Proof of principle. *Cancer Genomics Proteomics* **2013**, *10*, 93–113. [PubMed]
195. Li, J.M.; Zhao, R.H.; Li, S.T.; Xie, C.X.; Jiang, H.H.; Ding, W.J.; Du, P.; Chen, W.; Yang, M.; Cui, L. Down-regulation of fecal miR-143 and miR-145 as potential markers for colorectal cancer. *Saudi Med. J.* **2012**, *33*, 24–29. [PubMed]
196. Arndt, G.M.; Dossey, L.; Cullen, L.M.; Lai, A.; Druker, R.; Eisbacher, M.; Zhang, C.; Tran, N.; Fan, H.; Retzlaff, K.; et al. Characterization of global microRNA expression reveals oncogenic potential of miR-145 in metastatic colorectal cancer. *BMC Cancer* **2009**, *9*, 374. [CrossRef] [PubMed]
197. Humphries, B.; Yang, C. The microRNA-200 family: Small molecules with novel roles in cancer development, progression and therapy. *Oncotarget* **2015**, *6*, 6472–6498. [CrossRef] [PubMed]
198. Korpal, M.; Lee, E.S.; Hu, G.; Kang, Y. The miR-200 Family Inhibits Epithelial-Mesenchymal Transition and Cancer Cell Migration by Direct Targeting of E-cadherin Transcriptional Repressors ZEB1 and ZEB2. *J. Biol. Chem.* **2008**, *283*, 14910–14914. [CrossRef]
199. Gregory, P.A.; Bert, A.G.; Paterson, E.L.; Barry, S.C.; Tsykin, A.; Farshid, G.; Vadas, M.A.; Khew-Goodall, Y.; Goodall, G.J. The miR-200 family and miR-205 regulate epithelial to mesenchymal transition by targeting ZEB1 and SIP1. *Nat. Cell Biol.* **2008**, *10*, 593–601. [CrossRef]
200. Browne, G.; Sayan, A.E.; Tulchinsky, E. ZEB proteins link cell motility with cell cycle control and cell survival in cancer. *Cell Cycle* **2010**, *9*, 886–891. [CrossRef]
201. Drápela, S.; Bouchal, J.; Jolly, M.K.; Culig, Z.; Souček, K. ZEB1: A Critical Regulator of Cell Plasticity, DNA Damage Response, and Therapy Resistance. *Front. Mol. Biosci.* **2020**, *7*, 1–10. [CrossRef] [PubMed]
202. Bracken, C.P.; Gregory, P.A.; Kolesnikoff, N.; Bert, A.G.; Wang, J.; Shannon, M.F.; Goodall, G.J. A double-negative feedback loop between ZEB1-SIP1 and the microRNA-200 family regulates epithelial-mesenchymal transition. *Cancer Res.* **2008**, *68*, 7846–7854. [CrossRef] [PubMed]
203. O'Brien, S.J.; Carter, J.V.; Burton, J.F.; Oxford, B.G.; Schmidt, M.N.; Hallion, J.C.; Galandiuk, S. The role of the miR-200 family in epithelial-mesenchymal transition in colorectal cancer: A systematic review. *Int. J. Cancer* **2018**, *142*, 2501–2511. [CrossRef] [PubMed]
204. Hur, K.; Toiyama, Y.; Takahashi, M.; Balaguer, F.; Nagasaka, T.; Koike, J.; Hemmi, H.; Koi, M.; Boland, C.R.; Goel, A. MicroRNA-200c modulates epithelial-tomesenchymal transition (EMT) in human colorectal cancer metastasis. *Gut* **2013**, *62*, 1315–1326. [CrossRef]
205. Carter, J.V.; O'Brien, S.J.; Burton, J.F.; Oxford, B.G.; Stephen, V.; Hallion, J.; Bishop, C.; Galbraith, N.J.; Eichenberger, M.R.; Sarojini, H.; et al. The microRNA-200 family acts as an oncogene in colorectal cancer by inhibiting the tumor suppressor RASSF2. *Oncol. Lett.* **2019**, *18*, 3994–4007. [CrossRef]
206. Pecot, C.V.; Rupaimoole, R.; Yang, D.; Akbani, R.; Ivan, C.; Lu, C.; Wu, S.; Han, H.D.; Shah, M.Y.; Rodriguez-Aguayo, C.; et al. Tumour angiogenesis regulation by the miR-200 family. *Nat. Commun.* **2013**, *4*, 1–14. [CrossRef]
207. Schliekelman, M.J.; Gibbons, D.L.; Faca, V.M.; Creighton, C.J.; Rizvi, Z.H.; Zhang, Q.; Wong, C.H.; Wang, H.; Ungewiss, C.; Ahn, Y.H.; et al. Targets of the tumor suppressor miR-200 in regulation of the epithelial-mesenchymal transition in cancer. *Cancer Res.* **2011**, *71*, 7670–7682. [CrossRef]
208. Yu, C.; Wan, H.; Shan, R.; Wen, W.; Li, J.; Luo, D.; Wan, R. The prognostic value of the MiR-200 family in colorectal cancer: A meta-analysis with 1882 patients. *J. Cancer* **2019**, *10*, 4009–4016. [CrossRef]
209. Sun, Y.; Shen, S.; Liu, X.; Tang, H.; Wang, Z.; Yu, Z.; Li, X.; Wu, M. MiR-429 inhibits cells growth and invasion and regulates EMT-related marker genes by targeting Onecut2 in colorectal carcinoma. *Mol. Cell. Biochem.* **2014**, *390*, 19–30. [CrossRef]
210. Ding, L.; Yu, L.L.; Han, N.; Zhang, B.T. miR-141 promotes colon cancer cell proliferation by inhibiting MAP2K4. *Oncol. Lett.* **2017**, *13*, 1665–1671. [CrossRef]
211. Hu, M.; Xia, M.; Chen, X.; Lin, Z.; Xu, Y.; Ma, Y.; Su, L. MicroRNA-141 regulates smad interacting protein 1 (SIP1) and inhibits migration and invasion of colorectal cancer cells. *Dig. Dis. Sci.* **2010**, *55*, 2365–2372. [CrossRef] [PubMed]
212. Wu, P.P.; Zhu, H.Y.; Sun, X.F.; Chen, L.X.; Zhou, Q.; Chen, J. MicroRNA-141 regulates the tumour suppressor DLC1 in colorectal cancer. *Neoplasma* **2013**, *62*, 705–712. [CrossRef] [PubMed]
213. Yu, L.; Cao, C.; Li, X.; Zhang, M.; Gu, Q.; Gao, H.; Balic, J.J.; Xu, D.; Zhang, L.; Ying, L.; et al. Complete loss of miR-200 family induces EMT associated cellular senescence in gastric cancer. *Oncogene* **2021**, *41*, 26–36. [CrossRef] [PubMed]
214. Maierthaler, M.; Benner, A.; Hoffmeister, M.; Surowy, H.; Jansen, L.; Knebel, P.; Chang-Claude, J.; Brenner, H.; Burwinkel, B. Plasma miR-122 and miR-200 family are prognostic markers in colorectal cancer. *Int. J. Cancer* **2017**, *140*, 176–187. [CrossRef] [PubMed]
215. Pichler, M.; Ress, A.L.; Winter, E.; Stiegelbauer, V.; Karbiener, M.; Schwarzenbacher, D.; Scheideler, M.; Ivan, C.; Jahn, S.W.; Kiesslich, T.; et al. MiR-200a regulates epithelial to mesenchymal transition-related gene expression and determines prognosis in colorectal cancer patients. *Br. J. Cancer* **2014**, *110*, 1614–1621. [CrossRef]
216. Cheng, H.; Zhang, L.; Cogdell, D.E.; Zheng, H.; Schetter, A.J.; Nykter, M.; Harris, C.C.; Chen, K.; Hamilton, S.R.; Zhang, W. Circulating plasma MiR-141 is a novel biomarker for metastatic colon cancer and predicts poor prognosis. *PLoS ONE* **2011**, *6*, e17745. [CrossRef]

217. Liang, W.C.; Fu, W.M.; Wong, C.W.; Wang, Y.; Wang, W.M.; Hu, G.X.; Zhang, L.; Xiao, L.J.; Wan, D.C.C.; Zhang, J.F.; et al. The LncRNA H19 promotes epithelial to mesenchymal transition by functioning as MiRNA sponges in colorectal cancer. *Oncotarget* **2015**, *6*, 22513–22525. [CrossRef]
218. O'Brien, S.J.; Fiechter, C.; Burton, J.; Hallion, J.; Paas, M.; Patel, A.; Patel, A.; Rochet, A.; Scheurlen, K.; Gardner, S.; et al. Long non-coding RNA ZFAS1 is a major regulator of epithelial-mesenchymal transition through miR-200/ZEB1/E-cadherin, vimentin signaling in colon adenocarcinoma. *Cell Death Discov.* **2021**, *7*, 1–14. [CrossRef]
219. Wu, G.; Xue, M.; Zhao, Y.; Han, Y.; Li, C.; Zhang, S.; Zhang, J.; Xu, J. Long noncoding rna zeb1-as1 acts as a sponge of mir-141-3p to inhibit cell proliferation in colorectal cancer. *Int. J. Med. Sci.* **2020**, *17*, 1589–1597. [CrossRef]
220. Cheng, Z.; Li, Z.; Ma, K.; Li, X.; Tian, N.; Duan, J.; Xiao, X.; Wang, Y. Long non-coding RNA XIST promotes glioma tumorigenicity and angiogenesis by acting as a molecular sponge of miR-429. *J. Cancer* **2017**, *8*, 4106–4116. [CrossRef]
221. Schetter, A.J.; Leung, S.Y.; Sohn, J.J.; Zanetti, K.A.; Bowman, E.D.; Yanaihara, N.; Yuen, S.T.; Chan, T.L.; Kwong, D.L.W.; Au, G.K.H.; et al. MicroRNA Expression Profiles Associated With Prognosis and Therapeutic Outcome in Colon Adenocarcinoma. *J. Am. Med. Assoc.* **2008**, *299*, 425–436. [CrossRef] [PubMed]
222. Fu, Q.; Zhang, J.; Xu, X.; Qian, F.; Feng, K.; Ma, J. miR-203 is a predictive biomarker for colorectal cancer and its expression is associated with BIRC5. *Tumor Biol.* **2016**, *37*, 15989–15995. [CrossRef] [PubMed]
223. Chiang, Y.; Song, Y.; Wang, Z.; Chen, Y.; Yue, Z.; Xu, H.; Xing, C.; Liu, Z. Aberrant Expression of miR-203 and Its Clinical Significance in Gastric and Colorectal Cancers. *J. Gastrointest. Surg.* **2011**, *15*, 63–70. [CrossRef] [PubMed]
224. Yi, R.; Poy, M.N.; Stoffel, M.; Fuchs, E. A skin microRNA promotes differentiation by repressing "stemness". *Nature* **2008**, *452*, 225–229. [CrossRef]
225. DeCastro, A.J.; Dunphy, K.A.; Hutchinson, J.; Balboni, A.L.; Cherukuri, P.; Jerry, D.J.; DiRenzo, J. MiR203 mediates subversion of stem cell properties during mammary epithelial differentiation via repression of δnP63α and promotes mesenchymal-to-epithelial transition. *Cell Death Dis.* **2013**, *4*, 1–10. [CrossRef]
226. Patturajan, M.; Nomoto, S.; Sommer, M.; Fomenkov, A.; Hibi, K.; Zangen, R.; Poliak, N.; Califano, J.; Trink, B.; Ratovitski, E.; et al. ΔNp63 induces β-catenin nuclear accumulation and signaling. *Cancer Cell* **2002**, *1*, 369–379. [CrossRef]
227. Riemondy, K.; Wang, X.; Torchia, E.C.; Roop, D.R.; Yi, R. MicroRNA-203 represses selection and expansion of oncogenic Hras transformed tumor initiating cells. *Elife* **2015**, *4*, e07004. [CrossRef]
228. Lai, H.T.; Tseng, W.K.; Huang, S.W.; Chao, T.C.; Su, Y. MicroRNA-203 diminishes the stemness of human colon cancer cells by suppressing GATA6 expression. *J. Cell. Physiol.* **2020**, *235*, 2866–2880. [CrossRef]
229. Ju, S.Y.; Chiou, S.H.; Su, Y. Maintenance of the stemness in CD44+ HCT-15 and HCT-116 human colon cancer cells requires miR-203 suppression. *Stem Cell Res.* **2014**, *12*, 86–100. [CrossRef]
230. Li, P.; Zhou, H.; Zhu, X.; Ma, G.; Liu, C.; Lin, B.; Mao, W. High expression of NEDD9 predicts adverse outcomes of colorectal cancer patients. *Int. J. Clin. Exp. Pathol.* **2014**, *7*, 2565–2570.
231. Han, R.; Sun, Q.; Wu, J.; Zheng, P.; Zhao, G. Sodium Butyrate Upregulates miR-203 Expression to Exert Anti-Proliferation Effect on Colorectal Cancer Cells. *Cell. Physiol. Biochem.* **2016**, *39*, 1919–1929. [CrossRef]
232. Li, J.; Chen, Y.; Zhao, J.; Kong, F.; Zhang, Y. MiR-203 reverses chemoresistance in p53-mutated colon cancer cells through downregulation of Akt2 expression. *Cancer Lett.* **2011**, *304*, 52–59. [CrossRef] [PubMed]
233. Liu, Y.; Gao, S.; Chen, X.; Liu, M.; Mao, C.; Fang, X. Overexpression of miR-203 sensitizes paclitaxel (Taxol)-resistant colorectal cancer cells through targeting the salt-inducible kinase 2 (SIK2). *Tumor Biol.* **2016**, *37*, 12231–12239. [CrossRef] [PubMed]
234. Zhong, X.; Xiao, Y.; Chen, C.; Wei, X.; Hu, C.; Ling, X.; Liu, X. MicroRNA-203-mediated posttranscriptional deregulation of CPEB4 contributes to colorectal cancer progression. *Biochem. Biophys. Res. Commun.* **2015**, *466*, 206–213. [CrossRef]
235. Li, T.; Gao, F.; Zhang, X.P. MiR-203 enhances chemosensitivity to 5-fluorouracil by targeting thymidylate synthase in colorectal cancer. *Oncol. Rep.* **2015**, *33*, 607–614. [CrossRef]
236. Shen, B.; Yuan, Y.; Zhang, Y.; Yu, S.; Peng, W.; Huang, X.; Feng, J. Long non-coding RNA FBXL19-AS1 plays oncogenic role in colorectal cancer by sponging miR-203. *Biochem. Biophys. Res. Commun.* **2017**, *488*, 67–73. [CrossRef] [PubMed]
237. Ma, S.; Yang, D.; Liu, Y.; Wang, Y.; Lin, T.; Li, Y.; Yang, S.; Zhang, W.; Zhang, R. LncRNA BANCR promotes tumorigenesis and enhances adriamycin resistance in colorectal cancer. *Aging* **2018**, *10*, 2062–2078. [CrossRef]
238. Zhao, L.; Liu, C.; Yan, S.; Hu, G.; Xiang, K.; Xiang, H.; Yu, H. LINC00657 promotes colorectal cancer stem-like cell invasion by functioning as a miR-203a sponge. *Biochem. Biophys. Res. Commun.* **2020**, *529*, 500–506. [CrossRef]
239. Ye, H.; Hao, H.; Wang, J.; Chen, R.; Huang, Z. miR-203 as a novel biomarker for the diagnosis and prognosis of colorectal cancer: A systematic review and meta-analysis. *Onco. Targets. Ther.* **2017**, *10*, 3685–3696. [CrossRef]
240. Valeri, N.; Braconi, C.; Gasparini, P.; Murgia, C.; Lampis, A.; Paulus-Hock, V.; Hart, J.R.; Ueno, L.; Grivennikov, S.I.; Lovat, F.; et al. MicroRNA-135b promotes cancer progression by acting as a downstream effector of oncogenic pathways in colon cancer. *Cancer Cell* **2014**, *25*, 469–483. [CrossRef]
241. Nagel, R.; Le Sage, C.; Diosdado, B.; Van Der Waal, M.; Oude Vrielink, J.A.F.; Bolijn, A.; Meijer, G.A.; Agami, R. Regulation of the adenomatous polyposis coli gene by the miR-135 family in colorectal cancer. *Cancer Res.* **2008**, *68*, 5795–5802. [CrossRef] [PubMed]
242. Yang, X.; Wang, X.; Nie, F.; Liu, T.; Yu, X.; Wang, H.; Li, Q.; Peng, R.; Mao, Z.; Zhou, Q.; et al. MiR-135 family members mediate podocyte injury through the activation of Wnt/β-catenin signaling. *Int. J. Mol. Med.* **2015**, *36*, 669–677. [CrossRef] [PubMed]

243. Yang, Y.; Ishak Gabra, M.B.; Hanse, E.A.; Lowman, X.H.; Tran, T.Q.; Li, H.; Milman, N.; Liu, J.; Reid, M.A.; Locasale, J.W.; et al. MiR-135 suppresses glycolysis and promotes pancreatic cancer cell adaptation to metabolic stress by targeting phosphofructokinase-1. *Nat. Commun.* **2019**, *10*, 1–15. [CrossRef]
244. Zhou, W.; Li, X.; Liu, F.; Xiao, Z.; He, M.; Shen, S.; Liu, S. MiR-135a promotes growth and invasion of colorectal cancer via metastasis suppressor 1 in vitro. *Acta Biochim. Biophys. Sin.* **2012**, *44*, 838–846. [CrossRef] [PubMed]
245. Li, J.; Liang, H.; Bai, M.; Ning, T.; Wang, C.; Fan, Q.; Wang, Y.; Fu, Z.; Wang, N.; Liu, R.; et al. miR-135b promotes cancer progression by targeting transforming growth factor beta receptor II (TGFBR2) in colorectal cancer. *PLoS ONE* **2015**, *10*, e130194. [CrossRef]
246. Parsons, R.; Myeroff, L.L.; Liu, B.; Willson, J.K.; Markowitz, S.D.; Kinzler, K.W.; Vogelstein, B. Microsatellite instability and mutations of the transforming growth factor beta type II receptor gene in colorectal cancer. *Cancer Res.* **1995**, *55*, 5548–5550. [PubMed]
247. Zhang, Y.; Zhang, Z.; Yi, Y.; Wang, Y.; Fu, J. CircNOL10 acts as a sponge of miR-135a/b-5p in suppressing colorectal cancer progression via regulating KLF9. *Onco. Targets. Ther.* **2020**, *13*, 5165–5176. [CrossRef]
248. Xu, X.M.; Qian, J.C.; Deng, Z.L.; Cai, Z.; Tang, T.; Wang, P.; Zhang, K.H.; Cai, J.P. Expression of miR-21, miR-31, miR-96 and miR-135b is correlated with the clinical parameters of colorectal cancer. *Oncol. Lett.* **2012**, *4*, 339–345. [CrossRef]
249. Vickers, M.M.; Bar, J.; Gorn-Hondermann, I.; Yarom, N.; Daneshmand, M.; Hanson, J.E.L.; Addison, C.L.; Asmis, T.R.; Jonker, D.J.; Maroun, J.; et al. Stage-dependent differential expression of microRNAs in colorectal cancer: Potential role as markers of metastatic disease. *Clin. Exp. Metastasis* **2012**, *29*, 123–132. [CrossRef]
250. Jia, L.; Luo, S.; Ren, X.; Li, Y.; Hu, J.; Liu, B.; Zhao, L.; Shan, Y.; Zhou, H. miR-182 and miR-135b Mediate the Tumorigenesis and Invasiveness of Colorectal Cancer Cells via Targeting ST6GALNAC2 and PI3K/AKT Pathway. *Dig. Dis. Sci.* **2017**, *62*, 3447–3459. [CrossRef]
251. Mencía, A.; Modamio-Høybjør, S.; Redshaw, N.; Morín, M.; Mayo-Merino, F.; Olavarrieta, L.; Aguirre, L.A.; Del Castillo, I.; Steel, K.P.; Dalmay, T.; et al. Mutations in the seed region of human miR-96 are responsible for nonsyndromic progressive hearing loss. *Nat. Genet.* **2009**, *41*, 609–613. [CrossRef]
252. Pierce, M.L.; Weston, M.D.; Fritzsch, B.; Gabel, H.W.; Ruvkun, G.; Soukup, G.A. MicroRNA-183 family conservation and ciliated neurosensory organ expression. *Evol. Dev.* **2008**, *10*, 106–113. [CrossRef]
253. Guttilla, I.K.; White, B.A. Coordinate regulation of FOXO1 by miR-27a, miR-96, and miR-182 in breast cancer cells. *J. Biol. Chem.* **2009**, *284*, 23204–23216. [CrossRef]
254. Gao, F.; Wang, W. MicroRNA-96 promotes the proliferation of colorectal cancer cells and targets tumor protein p53 inducible nuclear protein 1, forkhead box protein O1 (FOXO1) and FOXO3a. *Mol. Med. Rep.* **2015**, *11*, 1200–1206. [CrossRef]
255. Suzuki, R.; Amatya, V.J.; Kushitani, K.; Kai, Y.; Kambara, T.; Takeshima, Y. miR-182 and miR-183 promote cell proliferation and invasion by targeting FOXO1 in mesothelioma. *Front. Oncol.* **2018**, *8*, 1–9. [CrossRef]
256. Huangfu, L.; Liang, H.; Wang, G.; Su, X.; Li, L.; Du, Z.; Hu, M.; Dong, Y.; Bai, X.; Liu, T.; et al. miR-183 regulates autophagy and apoptosis in colorectal cancer through targeting of UVRAG. *Oncotarget* **2016**, *7*, 4735–4745. [CrossRef]
257. Xiang, L.; Chen, X.J.; Wu, K.C.; Zhang, C.J.; Zhou, G.H.; Lv, J.N.; Sun, L.F.; Cheng, F.F.; Cai, X.B.; Jin, Z.B. MiR-183/96 plays a pivotal regulatory role in mouse photoreceptor maturation and maintenance. *Proc. Natl. Acad. Sci. USA* **2017**, *114*, 6376–6381. [CrossRef]
258. He, P.Y.; Yip, W.K.; Jabar, M.F.; Mohtarrudin, N.; Dusa, N.M.; Seow, H.F. Effect of the miR-96-5p inhibitor and mimic on the migration and invasion of the SW480-7 colorectal cancer cell line. *Oncol. Lett.* **2019**, *18*, 1949–1960. [CrossRef]
259. Li, P.; Sheng, C.; Huang, L.; Zhang, H.; Huang, L.; Cheng, Z.; Zhu, Q. MiR-183/-96/-182 cluster is up-regulated in most breast cancers and increases cell proliferation and migration. *Breast Cancer Res.* **2014**, *16*, 1–17. [CrossRef]
260. Iseki, Y.; Shibutani, M.; Maeda, K.; Nagahara, H.; Fukuoka, T.; Matsutani, S.; Hirakawa, K.; Ohira, M. MicroRNA-96 promotes tumor invasion in colorectal cancer via RECK. *Anticancer Res.* **2018**, *38*, 2031–2035. [CrossRef]
261. Bi, D.P.; Yin, C.H.; Zhang, X.Y.; Yang, N.N.; Xu, J.Y. MIR-183 functions as an oncogene by targeting ABCA1 in colon cancer. *Oncol. Rep.* **2016**, *35*, 2873–2879. [CrossRef]
262. Ge, T.; Xiang, P.; Mao, H.; Tang, S.; Zhou, J.; Zhang, Y. Inhibition of miR-96 enhances the sensitivity of colorectal cancer cells to oxaliplatin by targeting TPM1. *Exp. Ther. Med.* **2020**, *20*, 2134–2140. [CrossRef]
263. Zheng, S.; Zhong, Y.F.; Tan, D.M.; Xu, Y.; Chen, H.X.; Wang, D. miR-183-5p enhances the radioresistance of colorectal cancer by directly targeting ATG5. *J. Biosci.* **2019**, *44*, 1–11. [CrossRef]
264. Rapti, S.M.; Kontos, C.K.; Papadopoulos, I.N.; Scorilas, A. High miR-96 levels in colorectal adenocarcinoma predict poor prognosis, particularly in patients without distant metastasis at the time of initial diagnosis. *Tumor Biol.* **2016**, *37*, 11815–11824. [CrossRef]
265. Yue, C.; Chen, J.; Li, Z.; Li, L.; Chen, J.; Guo, Y. microRNA-96 promotes occurrence and progression of colorectal cancer via regulation of the AMPKα2-FTO-m6A/MYC axis. *J. Exp. Clin. Cancer Res.* **2020**, *39*, 1–15. [CrossRef]
266. Zhang, Q.; Ren, W.; Huang, B.; Yi, L.; Zhu, H. MicroRNA-183/182/96 cooperatively regulates the proliferation of colon cancer cells. *Mol. Med. Rep.* **2015**, *12*, 668–674. [CrossRef]
267. Zhou, T.; Zhang, G.J.; Zhou, H.; Xiao, H.X.; Li, Y. Overexpression of microRNA-183 in human colorectal cancer and its clinical significance. *Eur. J. Gastroenterol. Hepatol.* **2014**, *26*, 229–233. [CrossRef]

268. Ress, A.L.; Stiegelbauer, V.; Winter, E.; Schwarzenbacher, D.; Kiesslich, T.; Lax, S.; Jahn, S.; Deutsch, A.; Bauernhofer, T.; Ling, H.; et al. MiR-96-5p influences cellular growth and is associated with poor survival in colorectal cancer patients. *Mol. Carcinog.* **2015**, *54*, 1442–1450. [CrossRef]
269. Ning, S.; Liu, H.; Gao, B.; Wei, W.; Yang, A.; Li, J.; Zhang, L. MiR-155, miR-96 and miR-99a as potential diagnostic and prognostic tools for the clinical management of hepatocellular carcinoma. *Oncol. Lett.* **2019**, *18*, 3381–3387. [CrossRef]
270. Wu, H.; Zhou, J.; Mei, S.; Wu, D.; Mu, Z.; Chen, B.; Xie, Y.; Ye, Y.; Liu, J. Circulating exosomal microRNA-96 promotes cell proliferation, migration and drug resistance by targeting LMO7. *J. Cell. Mol. Med.* **2017**, *21*, 1228–1236. [CrossRef]
271. Shao, S.; Wang, C.; Wang, S.; Zhang, H.; Zhang, Y. LncRNA STXBP5-AS1 suppressed cervical cancer progression via targeting miR-96-5p/PTEN axis. *Biomed. Pharmacother.* **2019**, *117*, 109082. [CrossRef] [PubMed]
272. Zhang, Y.; Yang, H.; Du, Y.; Liu, P.; Zhang, J.; Li, Y.; Shen, H.; Xing, L.; Xue, X.; Chen, J.; et al. Long noncoding RNA TP53TG1 promotes pancreatic ductal adenocarcinoma development by acting as a molecular sponge of microRNA-96. *Cancer Sci.* **2019**, *110*, 2760–2772. [CrossRef] [PubMed]
273. Shen, T.; Cheng, X.; Liu, X.; Xia, C.; Zhang, H.; Pan, D.; Zhang, X.; Li, Y. Circ_0026344 restrains metastasis of human colorectal cancer cells via miR-183. *Artif. Cells Nanomed. Biotechnol.* **2019**, *47*, 4038–4045. [CrossRef] [PubMed]
274. Monticelli, S.; Ansel, K.M.; Xiao, C.; Socci, N.D.; Krichevsky, A.M.; Thai, T.H.; Rajewsky, N.; Marks, D.S.; Sander, C.; Rajewsky, K.; et al. MicroRNA profiling of the murine hematopoietic system. *Genome Biol.* **2005**, *6*, 1–15. [CrossRef]
275. Xiao, C.; Calado, D.P.; Galler, G.; Thai, T.H.; Patterson, H.C.; Wang, J.; Rajewsky, N.; Bender, T.P.; Rajewsky, K. MiR-150 Controls B Cell Differentiation by Targeting the Transcription Factor c-Myb. *Cell* **2007**, *131*, 146–159. [CrossRef]
276. Feng, J.; Yang, Y.; Zhang, P.; Wang, F.; Ma, Y.; Qin, H.; Wang, Y. miR-150 functions as a tumour suppressor in human colorectal cancer by targeting c-Myb. *J. Cell. Mol. Med.* **2014**, *18*, 2125–2134. [CrossRef]
277. Ying, W.; Tseng, A.; Chang, R.C.A.; Wang, H.; Lin, Y.L.; Kanameni, S.; Brehm, T.; Morin, A.; Jones, B.; Splawn, T.; et al. MiR-150 regulates obesity-Associated insulin resistance by controlling B cell functions. *Sci. Rep.* **2016**, *6*, 1–12. [CrossRef]
278. He, Z.; Dang, J.; Song, A.; Cui, X.; Ma, Z.; Zhang, Y. The involvement of miR-150/β-catenin axis in colorectal cancer progression. *Biomed. Pharmacother.* **2020**, *121*, 109495. [CrossRef]
279. Chen, X.; Xu, X.; Pan, B.; Zeng, K.; Xu, M.; Liu, X.; He, B.; Pan, Y.; Sun, H.; Wang, S. miR-150-5p suppresses tumor progression by targeting VEGFA in colorectal cancer. *Aging* **2018**, *10*, 3421–3437. [CrossRef]
280. Wang, W.H.; Chen, J.; Zhao, F.; Zhang, B.R.; Yu, H.S.; Jin, H.Y.; Dai, J.H. MiR-150-5p suppresses colorectal cancer cell migration and invasion through targeting MUC4. *Asian Pacific J. Cancer Prev.* **2014**, *15*, 6269–6273. [CrossRef]
281. Zhou, L.; Qi, X.; Potashkin, J.A.; Abdul-Karim, F.W.; Gorodeski, G.I. MicroRNAs miR-186 and miR-150 down-regulate expression of the pro-apoptotic purinergic P2X7 receptor by activation of instability sites at the 3′-untranslated region of the gene that decrease steady-state levels of the transcript. *J. Biol. Chem.* **2008**, *283*, 28274–28286. [CrossRef] [PubMed]
282. Wu, Q.; Jin, H.; Yang, Z.; Luo, G.; Lu, Y.; Li, K.; Ren, G.; Su, T.; Pan, Y.; Feng, B.; et al. MiR-150 promotes gastric cancer proliferation by negatively regulating the pro-apoptotic gene EGR2. *Biochem. Biophys. Res. Commun.* **2010**, *392*, 340–345. [CrossRef] [PubMed]
283. Liu, F.; Di Wang, X. miR-150-5p represses TP53 tumor suppressor gene to promote proliferation of colon adenocarcinoma. *Sci. Rep.* **2019**, *9*, 1–7. [CrossRef] [PubMed]
284. Slattery, M.L.; Herrick, J.S.; Mullany, L.E.; Samowitz, W.S.; Sevens, J.R.; Sakoda, L.; Wolff, R.K. The co-regulatory networks of tumor suppressor genes, oncogenes, and miRNAs in colorectal cancer. *Genes Chromosom. Cancer* **2017**, *56*, 769–787. [CrossRef] [PubMed]
285. Yanlei, M.; Zhang, P.; Wang, F.; Zhang, H.; Yang, J.; Peng, J.; Liu, W.; Qin, H. miR-150 as a potential biomarker associated with prognosis and therapeutic outcome in colorectal cancer. *Gut* **2012**, *61*, 1447–1453. [CrossRef]
286. Chen, X.; Zeng, K.; Xu, M.; Hu, X.; Liu, X.; Xu, T.; He, B.; Pan, Y.; Sun, H.; Wang, S. SP1-induced lncRNA-ZFAS1 contributes to colorectal cancer progression via the miR-150-5p/VEGFA axis. *Cell Death Dis.* **2018**, *9*, 1–18. [CrossRef]
287. Zhou, T.; Wu, L.; Ma, N.; Tang, F.; Zong, Z.; Chen, S. LncRNA PART1 regulates colorectal cancer via targeting miR-150-5p/miR-520h/CTNNB1 and activating Wnt/β-catenin pathway. *Int. J. Biochem. Cell Biol.* **2020**, *118*, 105637. [CrossRef]
288. Van Rooij, E.; Sutherland, L.B.; Liu, N.; Williams, A.H.; McAnally, J.; Gerard, R.D.; Richardson, J.A.; Olson, E.N. A signature pattern of stress-responsive microRNAs that can evoke cardiac hypertrophy and heart failure. *Proc. Natl. Acad. Sci. USA* **2006**, *103*, 18255–18260. [CrossRef]
289. Porrello, E.R.; Johnson, B.A.; Aurora, A.B.; Simpson, E.; Nam, Y.J.; Matkovich, S.J.; Dorn, G.W.; Van Rooij, E.; Olson, E.N. MiR-15 family regulates postnatal mitotic arrest of cardiomyocytes. *Circ. Res.* **2011**, *109*, 670–679. [CrossRef]
290. Murakami, Y.; Yasuda, T.; Saigo, K.; Urashima, T.; Toyoda, H.; Okanoue, T.; Shimotohno, K. Comprehensive analysis of microRNA expression patterns in hepatocellular carcinoma and non-tumorous tissues. *Oncogene* **2006**, *25*, 2537–2545. [CrossRef]
291. Xu, T.; Zhu, Y.; Xiong, Y.; Ge, Y.Y.; Yun, J.P.; Zhuang, S.M. MicroRNA-195 suppresses tumorigenicity and regulates G1/S transition of human hepatocellular carcinoma cells. *Hepatology* **2009**, *50*, 113–121. [CrossRef] [PubMed]
292. Chen, C.Y.; Chao, Y.M.; Lin, H.F.; Chen, C.J.; Chen, C.S.; Yang, J.L.; Chan, J.Y.H.; Juo, S.H.H. miR-195 reduces age-related blood–brain barrier leakage caused by thrombospondin-1-mediated selective autophagy. *Aging Cell* **2020**, *19*, 1–17. [CrossRef] [PubMed]
293. Li, B.; Wang, S.; Wang, S. MiR-195 suppresses colon cancer proliferation and metastasis by targeting WNT3A. *Mol. Genet. Genom.* **2018**, *293*, 1245–1253. [CrossRef] [PubMed]

294. Kang, D.W.; Min, D.S. Positive feedback regulation between phospholipase D and Wnt signaling promotes Wnt-Driven anchorage-independent growth of colorectal cancer cells. *PLoS ONE* **2010**, *5*, e12109. [CrossRef]
295. Lin, X.; Wang, S.; Sun, M.; Zhang, C.; Wei, C.; Yang, C.; Dou, R.; Liu, Q.; Xiong, B. MiR-195-5p/NOTCH2-mediated EMT modulates IL-4 secretion in colorectal cancer to affect M2-like TAM polarization. *J. Hematol. Oncol.* **2019**, *12*, 1–14. [CrossRef]
296. Liu, L.; Chen, L.; Xu, Y.; Li, R.; Du, X. MicroRNA-195 promotes apoptosis and suppresses tumorigenicity of human colorectal cancer cells. *Biochem. Biophys. Res. Commun.* **2010**, *400*, 236–240. [CrossRef]
297. Wang, X.; Wang, J.; Ma, H.; Zhang, J.; Zhou, X. Downregulation of miR-195 correlates with lymph node metastasis and poor prognosis in colorectal cancer. *Med. Oncol.* **2012**, *29*, 919–927. [CrossRef]
298. Jin, Y.; Wang, M.; Hu, H.; Huang, Q.; Chen, Y.; Wang, G. Overcoming stemness and chemoresistance in colorectal cancer through miR-195-5p-modulated inhibition of notch signaling. *Int. J. Biol. Macromol.* **2018**, *117*, 445–453. [CrossRef]
299. Zhang, X.; Xu, J.; Jiang, T.; Liu, G.; Wang, D.; Lu, Y. MicroRNA-195 suppresses colorectal cancer cells proliferation via targeting FGF2 and regulating Wnt/β-catenin pathway. *Am. J. Cancer Res.* **2016**, *6*, 2631–2640.
300. Wang, L.; Qian, L.; Li, X.; Yan, J. MicroRNA-195 inhibits colorectal cancer cell proliferation, colony-formation and invasion through targeting CARMA3. *Mol. Med. Rep.* **2014**, *10*, 473–478. [CrossRef]
301. Luo, Q.; Zhang, Z.; Dai, Z.; Basnet, S.; Li, S.; Xu, B.; Ge, H. Tumor-suppressive microRNA-195-5p regulates cell growth and inhibits cell cycle by targeting cyclin dependent kinase 8 in colon cancer. *Am. J. Transl. Res.* **2016**, *8*, 2088–2096. [PubMed]
302. Feng, C.; Zhang, L.; Sun, Y.; Li, X.; Zhan, L.; Lou, Y.; Wang, Y.; Liu, L.; Zhang, Y. GDPD5, a target of miR-195-5p, is associated with metastasis and chemoresistance in colorectal cancer. *Biomed. Pharmacother.* **2018**, *101*, 945–952. [CrossRef] [PubMed]
303. Qu, J.; Zhao, L.; Zhang, P.; Wang, J.; Xu, N.; Mi, W.; Jiang, X.; Zhang, C.; Qu, J. MicroRNA-195 chemosensitizes colon cancer cells to the chemotherapeutic drug doxorubicin by targeting the first binding site of BCL2L2 mRNA. *J. Cell. Physiol.* **2015**, *230*, 535–545. [CrossRef] [PubMed]
304. Zheng, L.; Chen, J.; Zhou, Z.; He, Z. miR-195 enhances the radiosensitivity of colorectal cancer cells by suppressing CARM1. *Onco. Targets. Ther.* **2017**, *10*, 1027–1038. [CrossRef] [PubMed]
305. Bai, J.; Xu, J.; Zhao, J.; Zhang, R. lncRNA SNHG1 cooperated with miR-497/miR-195-5p to modify epithelial–mesenchymal transition underlying colorectal cancer exacerbation. *J. Cell. Physiol.* **2020**, *235*, 1453–1468. [CrossRef] [PubMed]
306. Li, S.; Lv, C.; Li, J.; Xie, T.; Liu, X.; Zheng, Z.; Qin, Z.; Hui, X.; Yu, Y. LncRNA LINC00473 promoted colorectal cancer cell proliferation and invasion by targeting miR-195 expression. *Am. J. Transl. Res.* **2021**, *13*, 6066–6075.
307. Gu, H.; Xu, Z.; Zhang, J.; Wei, Y.; Cheng, L.; Wang, J. circ_0038718 promotes colon cancer cell malignant progression via the miR-195-5p/Axin2 signaling axis and also effect Wnt/β-catenin signal pathway. *BMC Genom.* **2021**, *22*, 1–17. [CrossRef]
308. Li, Y.; Zhu, Z.; Hou, X.; Sun, Y. LncRNA AFAP1-AS1 Promotes the Progression of Colorectal Cancer through miR-195-5p and WISP1. *J. Oncol.* **2021**, *2021*, 6242798. [CrossRef]
309. Ebert, M.S.; Neilson, J.R.; Sharp, P.A. MicroRNA sponges: Competitive inhibitors of small RNAs in mammalian cells. *Nat. Methods* **2007**, *4*, 721–726. [CrossRef]
310. Dey, N.; Das, F.; Ghosh-Choudhury, N.; Mandal, C.C.; Parekh, D.J.; Block, K.; Kasinath, B.S.; Abboud, H.E.; Choudhury, G.G. microRNA-21 governs TORC1 activation in renal cancer cell proliferation and invasion. *PLoS ONE* **2012**, *7*, e37366. [CrossRef]
311. Liu, X.; Abraham, J.M.; Cheng, Y.; Wang, Z.; Wang, Z.; Zhang, G.; Ashktorab, H.; Smoot, D.T.; Cole, R.N.; Boronina, T.N.; et al. Synthetic Circular RNA Functions as a miR-21 Sponge to Suppress Gastric Carcinoma Cell Proliferation. *Mol. Ther. Nucleic Acids* **2018**, *13*, 312–321. [CrossRef] [PubMed]
312. Monfared, H.; Jahangard, Y.; Nikkhah, M.; Mirnajafi-Zadeh, J.; Mowla, S.J. Potential therapeutic effects of exosomes packed with a miR-21-sponge construct in a rat model of glioblastoma. *Front. Oncol.* **2019**, *9*, 1–11. [CrossRef]
313. Wang, Z.; Ma, K.; Cheng, Y.; Abraham, J.M.; Liu, X.; Ke, X.; Wang, Z.; Meltzer, S.J. Synthetic circular multi-miR sponge simultaneously inhibits miR-21 and miR-93 in esophageal carcinoma. *Lab. Investig.* **2019**, *99*, 1442–1453. [CrossRef] [PubMed]
314. Zeng, K.; Chen, X.; Xu, M.; Liu, X.; Hu, X.; Xu, T.; Sun, H.; Pan, Y.; He, B.; Wang, S. CircHIPK3 promotes colorectal cancer growth and metastasis by sponging miR-7 article. *Cell Death Dis.* **2018**, *9*. [CrossRef] [PubMed]
315. Zhang, Z.; Fu, C.; Xu, Q.; Wei, X. Long non-coding RNA CASC7 inhibits the proliferation and migration of colon cancer cells via inhibiting microRNA-21. *Biomed. Pharmacother.* **2017**, *95*, 1644–1653. [CrossRef] [PubMed]
316. Yuan, Y.; Liu, W.; Zhang, Y.; Zhang, Y.; Sun, S. CircRNA circ_0026344 as a prognostic biomarker suppresses colorectal cancer progression via microRNA-21 and microRNA-31. *Biochem. Biophys. Res. Commun.* **2018**, *503*, 870–875. [CrossRef]
317. He, J.H.; Li, Y.G.; Han, Z.P.; Zhou, J.B.; Chen, W.M.; Lv, Y.B.; He, M.L.; Zuo, J.D.; Zheng, L. The CircRNA-ACAP2/Hsa-miR-21-5p/Tiam1 Regulatory Feedback Circuit Affects the Proliferation, Migration, and Invasion of Colon Cancer SW480 Cells. *Cell. Physiol. Biochem.* **2018**, *49*, 1539–1550. [CrossRef]
318. Jiang, Z.; Hou, Z.; Li, L.; Liu, W.; Yu, Z.; Chen, S. Exosomal circEPB41L2 serves as a sponge for miR-21-5p and miR-942-5p to suppress colorectal cancer progression by regulating the PTEN/AKT signalling pathway. *Eur. J. Clin. Investig.* **2021**, *51*, e13581. [CrossRef]
319. Huang, H.; Yang, X.; Chen, J.; Fu, J.; Chen, C.; Wen, J.; Mo, Q. lncRNA DGCR5 inhibits the proliferation of colorectal cancer cells by downregulating miR-21. *Oncol. Lett.* **2019**, *18*, 3331–3336. [CrossRef]
320. Li, G.; Wang, C.; Wang, Y.; Xu, B.; Zhang, W. LINC00312 represses proliferation and metastasis of colorectal cancer cells by regulation of miR-21. *J. Cell. Mol. Med.* **2018**, *22*, 5565–5572. [CrossRef]

321. Cai, Y.; Li, Y.; Shi, C.; Zhang, Z.; Xu, J.; Sun, B. LncRNA OTUD6B-AS1 inhibits many cellular processes in colorectal cancer by sponging miR-21-5p and regulating PNRC2. *Hum. Exp. Toxicol.* **2021**, *40*, 1463–1473. [CrossRef] [PubMed]
322. Han, K.; Wang, F.W.; Cao, C.H.; Ling, H.; Chen, J.W.; Chen, R.X.; Feng, Z.H.; Luo, J.; Jin, X.H.; Duan, J.L.; et al. CircLONP2 enhances colorectal carcinoma invasion and metastasis through modulating the maturation and exosomal dissemination of microRNA-17. *Mol. Cancer* **2020**, *19*, 1–18. [CrossRef] [PubMed]
323. Wen, R.; Chen, C.; Zhong, X.; Hu, C. PAX6 upstream antisense RNA (PAUPAR) inhibits colorectal cancer progression through modulation of the microRNA (miR)-17-5p / zinc finger protein 750 (ZNF750) axis. *Bioengineered* **2021**, *12*, 3886–3899. [CrossRef] [PubMed]
324. Yin, S.L.; Xiao, F.; Liu, Y.F.; Chen, H.; Guo, G.C. Long non-coding RNA FENDRR restrains the aggressiveness of CRC via regulating miR-18a-5p/ING4 axis. *J. Cell. Biochem.* **2020**, *121*, 3973–3985. [CrossRef]
325. Horita, K.; Kurosaki, H.; Nakatake, M.; Ito, M.; Kono, H.; Nakamura, T. Long noncoding RNA UCA1 enhances sensitivity to oncolytic vaccinia virus by sponging miR-18a/miR-182 and modulating the Cdc42/filopodia axis in colorectal cancer. *Biochem. Biophys. Res. Commun.* **2019**, *516*, 831–838. [CrossRef]
326. Li, Y.; Lv, S.; Ning, H.; Li, K.; Zhou, X.; Xv, H.; Wen, H. Down-regulation of CASC2 contributes to cisplatin resistance in gastric cancer by sponging miR-19a. *Biomed. Pharmacother.* **2018**, *108*, 1775–1782. [CrossRef]
327. Shen, P.; Qu, L.; Wang, J.; Ding, Q.; Zhou, C.; Xie, R.; Wang, H.; Ji, G. LncRNA LINC00342 contributes to the growth and metastasis of colorectal cancer via targeting miR-19a-3p/NPEPL1 axis. *Cancer Cell Int.* **2021**, *21*, 1–11. [CrossRef]
328. Dai, W.; Zeng, W.; Lee, D. lncRNA MCM3AP-AS1 inhibits the progression of colorectal cancer via the miR-19a-3p/FOXF2 axis. *J. Gene Med.* **2021**, *23*, 1–13. [CrossRef]
329. Jiang, Z.; Li, L.; Hou, Z.; Liu, W.; Wang, H.; Zhou, T.; Li, Y.; Chen, S. LncRNA HAND2-AS1 inhibits 5-fluorouracil resistance by modulating miR-20a/PDCD4 axis in colorectal cancer. *Cell. Signal.* **2020**, *66*, 109483. [CrossRef]
330. Zhang, G.; Liu, Z.; Zhong, J.; Lin, L. Circ-ACAP2 facilitates the progression of colorectal cancer through mediating miR-143-3p/FZD4 axis. *Eur. J. Clin. Investig.* **2021**, *51*, e13607. [CrossRef]
331. Huang, F.; Wen, C.; Zhuansun, Y.; Huang, L.; Chen, W.; Yang, X.; Liu, H. A novel long noncoding RNA OECC promotes colorectal cancer development and is negatively regulated by miR-143-3p. *Biochem. Biophys. Res. Commun.* **2018**, *503*, 2949–2955. [CrossRef] [PubMed]
332. Zhao, L.; Li, Y.; Song, A. Inhibition of lncRNA TMPO-AS1 suppresses proliferation, migration and invasion of colorectal cancer cells by targeting miR-143-3p. *Mol. Med. Rep.* **2020**, *22*, 3245–3254. [CrossRef] [PubMed]
333. Liu, A.; Liu, L. Long non-coding RNA ZEB2-AS1 promotes proliferation and inhibits apoptosis of colon cancer cells via miR-143/bcl-2 axis. *Am. J. Transl. Res.* **2019**, *11*, 5240–5248. [PubMed]
334. Gao, R.; Fang, C.; Xu, J.; Tan, H.; Li, P.; Ma, L. LncRNA CACS15 contributes to oxaliplatin resistance in colorectal cancer by positively regulating ABCC1 through sponging miR-145. *Arch. Biochem. Biophys.* **2019**, *663*, 183–191. [CrossRef] [PubMed]
335. Chen, Z.L.; Li, X.N.; Ye, C.X.; Chen, H.Y.; Wang, Z.J. Elevated levels of circrunx1 in colorectal cancer promote cell growth and metastasis via MiR-145-5p/IGF1 signalling. *Onco. Targets. Ther.* **2020**, *13*, 4035–4048. [CrossRef] [PubMed]
336. Bahreini, F.; Saidijam, M.; Mousivand, Z.; Najafi, R.; Afshar, S. Assessment of lncRNA DANCR, miR-145-5p and NRAS axis as biomarkers for the diagnosis of colorectal cancer. *Mol. Biol. Rep.* **2021**, *48*, 3541–3547. [CrossRef] [PubMed]
337. Wei, F.; Wang, M.; Li, Z.; Wang, Y.; Zhou, Y. Long Non-coding RNA MIR570MG Causes Regorafenib Resistance in Colon Cancer by Repressing miR-145/SMAD3 Signaling. *Front. Oncol.* **2020**, *10*, 1–13. [CrossRef]
338. Wang, Z.; Su, M.; Xiang, B.; Zhao, K.; Qin, B. Circular RNA PVT1 promotes metastasis via miR-145 sponging in CRC. *Biochem. Biophys. Res. Commun.* **2019**, *512*, 716–722. [CrossRef]
339. Tian, T.; Qiu, R.; Qiu, X. SNHG1 promotes cell proliferation by acting as a sponge of miR- 145 in colorectal cancer. *Oncotarget* **2018**, *9*, 2128–2139. [CrossRef]
340. Wei, A.W.; Li, L.F. Long non-coding RNA SOX21-AS1 sponges miR-145 to promote the tumorigenesis of colorectal cancer by targeting MYO6. *Biomed. Pharmacother.* **2017**, *96*, 953–959. [CrossRef]
341. Wang, X.; Bai, X.; Yan, Z.; Guo, X.; Zhang, Y. The lncRNA TUG1 promotes cell growth and migration in colorectal cancer via the TUG1–miR-145-5p–TRPC6 pathway. *Biochem. Cell Biol.* **2021**, *99*, 249–260. [CrossRef] [PubMed]
342. Bao, Y.; Tang, J.; Qian, Y.; Sun, T.; Chen, H.; Chen, Z.; Sun, D.; Zhong, M.; Chen, H.; Hong, J.; et al. Long noncoding RNA BFAL1 mediates enterotoxigenic Bacteroides fragilis-related carcinogenesis in colorectal cancer via the RHEB/mTOR pathway. *Cell Death Dis.* **2019**, *10*, 1–14. [CrossRef] [PubMed]
343. Li, Y.; Lu, Y.; Chen, Y. Long non-coding RNA SNHG16 affects cell proliferation and predicts a poor prognosis in patients with colorectal cancer via sponging MIR-200a-3p. *Biosci. Rep.* **2019**, *39*, 1–10. [CrossRef] [PubMed]
344. Jiao, M.; Ning, S.; Chen, J.; Chen, L.; Jiao, M.; Cui, Z.; Guo, L.; Mu, W.; Yang, H. Long non-coding RNA ZEB1-AS1 predicts a poor prognosis and promotes cancer progression through the miR-200a/ZEB1 signaling pathway in intrahepatic cholangiocarcinoma. *Int. J. Oncol.* **2020**, *56*, 1455–1467. [CrossRef] [PubMed]
345. Chen, D.L.; Chen, L.Z.; Lu, Y.X.; Zhang, D.S.; Zeng, Z.L.; Pan, Z.Z.; Huang, P.; Wang, F.H.; Li, Y.H.; Ju, H.Q.; et al. Long noncoding rna xist expedites metastasis and modulates epithelial–mesenchymal transition in colorectal cancer. *Cell Death Dis.* **2017**, *8*, e3011. [CrossRef]
346. Gao, Z.; Zhou, H.; Wang, Y.; Chen, J.; Ou, Y. Regulatory effects of lncRNA ATB targeting miR-200c on proliferation and apoptosis of colorectal cancer cells. *J. Cell. Biochem.* **2020**, *121*, 332–343. [CrossRef]

347. Liu, X.; Wang, C. Long non-coding RNA ATB is associated with metastases and promotes cell invasion in colorectal cancer via sponging miR-141-3p. *Exp. Ther. Med.* **2020**, *20*, 1. [CrossRef]
348. Ren, J.; Ding, L.; Zhang, D.; Shi, G.; Xu, Q.; Shen, S.; Wang, Y.; Wang, T.; Hou, Y. Carcinoma-associated fibroblasts promote the stemness and chemoresistance of colorectal cancer by transferring exosomal lncRNA H19. *Theranostics* **2018**, *8*, 3932–3948. [CrossRef]
349. Sun, Z.; Shao, B.; Liu, Z.; Dang, Q.; Guo, Y.; Chen, C.; Guo, Y.; Chen, Z.; Liu, J.; Hu, S.; et al. LINC01296/miR-141-3p/ZEB1-ZEB2 axis promotes tumor metastasis via enhancing epithelial-mesenchymal transition process. *J. Cancer* **2021**, *12*, 2723–2734. [CrossRef]
350. Wang, H.; Li, H.; Zhang, L.; Yang, D. Overexpression of MEG3 sensitizes colorectal cancer cells to oxaliplatin through regulation of miR-141/PDCD4 axis. *Biomed. Pharmacother.* **2018**, *106*, 1607–1615. [CrossRef]
351. Deng, Z.; Li, X.; Wang, H.; Geng, Y.; Cai, Y.; Tang, Y.; Wang, Y.; Yu, X.; Li, L.; Li, R. Dysregulation of CircRNA_0001946 Contributes to the Proliferation and Metastasis of Colorectal Cancer Cells by Targeting MicroRNA-135a-5p. *Front. Genet.* **2020**, *11*, 1–12. [CrossRef] [PubMed]
352. Wu, Q.; Shi, M.; Meng, W.; Wang, Y.; Hui, P.; Ma, J. Long noncoding RNA FOXD3-AS1 promotes colon adenocarcinoma progression and functions as a competing endogenous RNA to regulate SIRT1 by sponging miR-135a-5p. *J. Cell. Physiol.* **2019**, *234*, 21889–21902. [CrossRef]
353. Zhou, D.; He, S.; Zhang, D.; Lv, Z.; Yu, J.; Li, Q.; Li, M.; Guo, W.; Qi, F. LINC00857 promotes colorectal cancer progression by sponging miR-150-5p and upregulating HMGB3 (high mobility group box 3) expression. *Bioengineered* **2021**, *12*, 12107–12122. [CrossRef] [PubMed]
354. Wang, X.; Jiang, G.; Ren, W.; Wang, B.; Yang, C.; Li, M. LncRNA NEAT1 regulates 5-fu sensitivity, apoptosis and invasion in colorectal cancer through the MiR-150-5p/CPSF4 axis. *Onco. Targets. Ther.* **2020**, *13*, 6373–6383. [CrossRef] [PubMed]
355. Shoshan, E.; Mobley, A.K.; Braeuer, R.R.; Kamiya, T.; Huang, L.; Vasquez, M.E.; Salameh, A.; Lee, H.J.; Kim, S.J.; Ivan, C.; et al. Hypo adenosine-to-inosine miR-455-5p editing promotes melanoma growth and metastasis. *Nat. Cell Biol.* **2015**, *17*, 311–321. [CrossRef]
356. Yang, X.; Zhang, S.; He, C.; Xue, P.; Zhang, L.; He, Z.; Zang, L.; Feng, B.; Sun, J.; Zheng, M. METTL14 suppresses proliferation and metastasis of colorectal cancer by down-regulating oncogenic long non-coding RNA XIST. *Mol. Cancer* **2020**, *19*, 1–16. [CrossRef]
357. Huang, Q.R.; Pan, X. Bin Prognostic lncRNAs, miRNAs, and mRNAs Form a Competing Endogenous RNA Network in Colon Cancer. *Front. Oncol.* **2019**, *9*, 712. [CrossRef]
358. Shuwen, H.; Qing, Z.; Yan, Z.; Xi, Y. Competitive endogenous RNA in colorectal cancer: A systematic review. *Gene* **2018**, *645*, 157–162. [CrossRef]
359. Wang, L.; Cho, K.B.; Li, Y.; Tao, G.; Xie, Z.; Guo, B. Long noncoding RNA (LncRNA)-mediated competing endogenous RNA networks provide novel potential biomarkers and therapeutic targets for colorectal cancer. *Int. J. Mol. Sci.* **2019**, *20*, 5758. [CrossRef]
360. Jung, G.; Hernández-Illán, E.; Moreira, L.; Balaguer, F.; Goel, A. Epigenetics of colorectal cancer: Biomarker and therapeutic potential. *Nat. Rev. Gastroenterol. Hepatol.* **2020**, *17*, 111–130. [CrossRef]
361. Qi, X.; Lin, Y.; Liu, X.; Chen, J.; Shen, B. Biomarker Discovery for the Carcinogenic Heterogeneity Between Colon and Rectal Cancers Based on lncRNA-Associated ceRNA Network Analysis. *Front. Oncol.* **2020**, *10*, 1–15. [CrossRef] [PubMed]
362. Liu, J.; Li, H.; Zheng, B.; Sun, L.; Yuan, Y.; Xing, C. Competitive Endogenous RNA (ceRNA) Regulation Network of lncRNA–miRNA–mRNA in Colorectal Carcinogenesis. *Dig. Dis. Sci.* **2019**, *64*, 1868–1877. [CrossRef] [PubMed]
363. Yuan, W.; Li, X.; Liu, L.; Wei, C.; Sun, D.; Peng, S.; Jiang, L. Comprehensive analysis of lncRNA-associated ceRNA network in colorectal cancer. *Biochem. Biophys. Res. Commun.* **2019**, *508*, 374–379. [CrossRef] [PubMed]

MDPI
St. Alban-Anlage 66
4052 Basel
Switzerland
Tel. +41 61 683 77 34
Fax +41 61 302 89 18
www.mdpi.com

International Journal of Molecular Sciences Editorial Office
E-mail: ijms@mdpi.com
www.mdpi.com/journal/ijms